The Complete Book of Locks and Locksmithing

Bill Phillips

Sixth Edition

McGraw-Hill
New York Chicago San Francisco Lisbon London Madrid
Mexico City Milan New Delhi San Juan Seoul
Singapore Sydney Toronto

The McGraw-Hill Companies

Cataloging-in-Publication Data is on file with the Library of Congress

10 11 12 13 14 15 DOC/DOC 1 5 4 3

ISBN 0-07-144829-2

The sponsoring editor for this book was Cary Sullivan, the editing supervisor was Caroline Levine, and the production supervisor was Pamela Pelton. The art director for the cover was Anthony Landi. It was set in Century Schoolbook by Kim Sheran of McGraw-Hill Professional's Hightstown, N.J., composition unit.

Printed and bound by RR Donnelley.

McGraw-Hill books are available at special quantity discounts to use as premiums and sales promotions, or for use in corporate training programs. For more information, please write to the Director of Special Sales, McGraw-Hill Professional, Two Penn Plaza, New York, NY 10121-2298. Or contact your local bookstore.

Contents

To Patricia Bruce not only because she is a good friend and a fine writer but also because she listened . . . and listened . . . and listened.

Foreword

The knowledge, skill, and expertise of any true professional often can be gauged by the library of literature that he or she keeps and has read. In the often mysterious, secretive, and intriguing world of locks and locksmithing, no true professional, aficionado, or curiosity seeker would be without *The Complete Book of Locks and Locksmithing,* 6th edition. Just as the title states, this is by far the most complete, informative, and all-encompassing piece of literature ever assembled between two covers on this subject.

As a 21-year veteran of the locksmith trade, I can tell you that the scope of this book (covering one of the oldest trades known to humans) is much broader than one would think. The locksmith profession is also often one of the most challenging and rewarding. There are few things in life that are dearer to anyone than the security of their family and worldly possessions. Having the knowledge and ability to instill and incorporate security, eliminating the helpless sense of vulnerability, is what locksmithing is all about. However, to accomplish this task, knowledge of all the available security devices and an intimate knowledge of their operation and applications are necessary. Enveloping yourself within the pages of this book certainly will set you on the right path.

The test of any good source of information is that it is one that you refer to again and again. This is sure to be the case with *The Complete Book of Locks and Locksmithing,* 6th edition. This is so much more than a "book" on locks and locksmithing. It is an encyclopedia of locks and locksmithing. It is a reference manual from which anyone at any level of skill or experience can glean a wealth of information. It is an informative, well-studied, impeccably presented piece of work that required an exhaustive amount of research, cataloging, organizing, and writing to assemble and publish. This is a daunting task, demanding the talents of a skilled writer and locksmith, and Bill Phillips certainly possesses the necessary requirements to accomplish it.

Bill Phillips is a knowledgeable locksmith and talented writer. He is one of those individuals who possess the rare ability to make a wide variety of complicated topics enjoyable to read and easy to understand. This is a sure sign of a talented, versatile writer.

When Bill Phillips asked me to write this Foreword, I was honored and flattered. I am privileged to have Bill Phillips' name included on my masthead of contributing writers. As editor of *The National Locksmith* magazine, I cherish his contributions on security-related topics. *The National Locksmith* is the oldest and most respected locksmith trade journal in the industry, serving locksmiths and security professionals since 1929. This monthly magazine has a group of the most experienced and well-respected writers in the field of locksmithing. In 2004, Bill Phillips became one of the distinguished elite to join *The National Locksmith's* outstanding team of contributing writers, and I couldn't be more excited to have him.

Bill Phillips and *The National Locksmith* share one common goal—to provide informative, timely, useful, and cutting-edge information to beginning and experienced locksmiths. Bill Phillips accomplishes this through his writings and personal presentations. *The National Locksmith* does it through its monthly publication, as well as books, software, *The Institutional Locksmith* magazine, the www.TheNationalLocksmith.com Web site, and membership organizations such as the National Safeman's Organization (NSO) and the National Locksmith Automobile Organization (NLAA).

Few people have as much knowledge or experience writing about locks or locksmithing as Bill Phillips does. He has written for most of the locksmith trade journals, many general consumer magazines, and the *World Book Encyclopedia* (the "Lock" article), and he has been published by one of the most respected and distinguished publishing houses—McGraw Hill. Those are claims and accomplishments few others can make.

If you want to become a locksmith or are just fascinated with locks and want an insider's view, you won't find a better book than *The Complete Book of Locks and Locksmithing,* 6th edition. It's a tremendous treasure of information that you will refer to for years to come.

Greg Mango

Greg Mango is a 21-year veteran of the locksmith industry. He is currently editor of The National Locksmith *magazine and director of the National Locksmith Automobile Organization (NLAA). He has authored numerous security-related articles for multiple publications over the last 15 years and his monthly "Mango's Message" editorial column is read by thousands.*

Preface

I've read a lot of great reviews for earlier editions of *The Complete Book of Locks and Locksmithing*. For the last edition, the most memorable one came from Alan T. Peto, who works for the Las Vegas Metropolitan Police Department. He wrote

> This is a great "catch-all" book about locksmithing. Although it looks like it is designed for beginners, it still has invaluable information on a variety of locks, etc., so it will remain in your shop for years to come. . . . The book is also invaluable to homeowners . . . and anyone who wants to know more about locks, safes, and more!

Mr. Peto's book review summed up what *The Complete Book of Locks and Locksmithing* has been about since the first edition in 1976. Today, it's the world's best-selling locksmithing book, and it is used often as a textbook in locksmithing schools and programs.

I know what beginning locksmiths and seasoned professionals want from a book. My first exposure to locksmithing was through reading books. In those days, the books that were available didn't provide much practical information, so I took a couple of locksmithing correspondence courses—which were virtually worthless. I ended up moving to New York City for about a year to study at the National School of Locksmithing and Alarms. That was the best investment I ever made, because I learned a lot.

A certified master locksmith, who was an avid writer of locksmithing articles, hired me right out of school. He taught me some things, but we never did get along. We were both opinionated, and we often had toe-to-toe shouting matches about how things should be done. Therefore, we parted ways. I worked for another shop for a while, and then I started my own shop. I've been working for myself ever since.

Over the years, I went on to write 13 security books and hundreds of articles. My work appeared in *Consumers Digest* (reviewed over 150 door locks, safes, and car alarms), *Keynotes* (the trade journal of the Associated Locksmiths of America), the *Los Angeles Times*, *Locksmith Ledger International* (was a contributing editor), *The National Locksmith* (am a contributing writer), *Safe and Vault Technology* (was a contributing editor), and many other periodicals. I also wrote the "Lock" article for the *World Book Encyclopedia*—the world's most read print encyclopedia.

I love to teach, and I love to learn. I'm constantly learning new things about locksmithing, which is why I felt the need to revise this book. This sixth edition continues the tradition of being heavily illustrated and providing lots of useful information about locks and locking devices. I updated every page and added four new chapters—including one on automotive lock servicing and one on lock picking and impressioning. I added information about selling safes and finding a job as a locksmith.

This edition includes a new chapter on frequently asked questions (Chap. 24) with candid answers to lots of questions such as: Do I need to be licensed or certified to work as a locksmith? What can I do if a customer refuses to pay me? How can I get started writing about locksmithing? What are the best locksmithing resources on the Internet? Are pick guns worth the money? Which are the best trade journals? What keys are illegal to duplicate?

I've also included a new Registered Professional Locksmith (RPL) test that you can complete to earn a certificate. The certificate can help you to find work and will allow you to enter restricted areas on the Internet. Also, four appendixes are included. The resources all have been updated with current addresses, Web sites, telephone numbers, and so on.

If you have questions or comments about the book, contact me at Box 2044, Erie, PA 16512-2044, or e-mail me at *LocksmithWriter@aol.com.*

Acknowledgments

This sixth edition of *The Complete Book of Locks and Locksmithing* is the result of help from a lot of people, too many to mention each by name. I appreciate everyone's help. Special thanks go to Greg Mango, editor of *The National Locksmith;* Earl Halls, president of Adesco Safe Company in Paramount, California; Gordon Little, president of Gordon Safe and Lock, in Houston, Texas; Joe Esposito, of Liberty Lock & Safe in Las Vegas, Nevada; Peter Burns, inventor of "The Bolt" High Security Lock; Billy B. Edwards, CML ("TheLockMan"); Thomas F. Hennessy, curator of the Lock Museum of America; Alex Krai; Bill Morrison; Alan T. Peto; Alice Petty; and Donald Streeter. Thanks also go to all the readers of prior editions who shared their thoughts about the book with me. Their criticisms and suggestions were invaluable in making this edition.

The following companies were especially helpful in providing technical information: A-l Security Manufacturing Corp., Abus Lock Co., Adams Rite Manufacturing Co., Alarm Lock Systems, Inc., Arrow Lock, Belwith International, Ltd., Black & Decker U.S. Power Tools Group, Dominion Lock Co., Folger Adam Co., Framon Manufacturing Co., Inc., Gardall Safe Corp., Ilco Unican Corp., International Association of Home Safety and Security Professionals, Keedex Manufacturing, Kryptonite Corporation, Kustom Key, Inc., Kwikset Corp., Lock Corporation of America, M. A. G. Engines & Manufacturing, Inc., Makita U.S.A., Inc., Master Lock Co., M. K. Morse Co., Medeco Security Locks, Milwaukee Electric Tool Corp., Monarch Tool and Manufacturing Co., the National School of Locksmithing and Alarms, Porter Cable Corp., Presto-Matic Lock Co., R&D Tool Co., Rofu International Corp., Schlage Lock Co., Securitech Group, Securitron, Security Engineering, Inc., Sentry Door Lock Guards, Sentry Group, Simplex Access Controls Co., Taylor Lock Co., Trine Consumer Products Division, Square "D" Co., and Vaughan & Bushnell Manufacturing Co.

I also appreciate the support I received from family and friends, including Janet L. Griffin, Merlynn Smith-Coles, and Jonathan Gavin for his help with photographs.

Special thanks also go to Cary Sullivan and Carol Levine, my editors at McGraw-Hill.

Bill Phillips

Chapter

1

A Short History of the Lock

This chapter traces the development of the lock from earliest times to the present, focusing on the most important models. Every locksmith should be familiar with them because they form the building blocks for all other locks. Many of the lock types and construction principles mentioned here are looked at in more detail in later chapters. This is a quick overview to help you better understand and appreciate the world of locks.

Who Invented the Lock?

The earliest locks may no longer be around, and there may be no written records of them. How likely it is for old locks to be found depends on the materials they were made from and on the climate and various geological conditions they have been subjected to over the years. There is evidence to suggest that different civilizations probably developed the lock independently of each other. The Egyptians, Romans, and Greeks are credited with inventing the oldest known types of locks.

Egypt

The oldest known lock was found in 1842 in the ruins of Emperor Sargon II's palace in Khorsabad, Persia. The ancient Egyptian lock was dated to be about 4000 years old. It relied on the same pin tumbler principle that is used by many of today's most popular locks.

The Egyptian lock consisted of three basic parts: a wood crossbeam, a vertical beam with tumblers, and a large wood key. The crossbeam ran horizontally across the inside of the door and was held in place by two vertically mounted wooden staples. Part of the length of the crossbeam was hollowed out, and the vertical beam intersected it along that hollowed out side. The vertical beam contained metal tumblers that locked the two pieces of wood together. Near the tumbler edge of the door there was a hole accessible from outside the door that was large enough for someone to insert the key and an arm. The spoon-shaped key was about 14 inches to 2 feet long with pegs sticking out of one end. After the key was inserted in the keyhole (or "armhole"), it was pushed into the hollowed out part of the crossbeam until its pegs were aligned with their corresponding tumblers. The right key allowed all the tumblers to be lifted into a position between the crossbeam and vertical beam so that the pins no longer obstructed movement of the crossbeam. Then the crossbeam (bolt) could be pulled into the open position. To see how the lock looked and operated see Fig. 1.1.

Greece

Most early Greek doors pivoted at the center and were secured with rope tied in intricate knots. The cleverly tied knots, along with beliefs about being cursed for tampering with them, provided some security. When more security was needed, doors were secured by bolts from the inside. In the few cases where locks were used, they were primitive and easy to defeat. The Greek locks used a notched boltwork and were operated by inserting the blade of an iron sickle-shaped key, about a foot long, in a key slot and twisting it 180° to work the bolt (Fig. 1.2). They could be defeated just by trying a few different-size keys.

In about 800 B.C.E. the Greek poet Homer described that Greek lock in his poem *The Odysseus:*

> She went upstairs and got the store room key, which was made of bronze and had a handle of ivory; she then went with her maidens into the store room at the end of the house, where her husband's treasures of gold, bronze, and wrought iron were kept....She loosed the strap from the handle of the door, put in the key, and drove it straight home to shoot back the bolts that held the doors.

Rome

Like the Greeks, the Romans used notched boltwork. But the Romans improved on the lock design in many ways, such as by putting the boltwork in an iron case and using keys of iron or bronze. Because iron rusts and corrodes, few early Roman locks are in existence. But a lot of the keys are around. Often the keys were ornately designed to be worn as jewelry, either as finger rings or as necklaces using string (because togas didn't have pockets). Figure 1.3 shows some early Roman finger rings.

Two of the most important innovations of the Roman locks were the spring-loaded bolt and the use of wards on the case. The extensive commerce during the time of Julius Caesar led to a great demand for locks among the many wealthy merchants and politicians. The type of lock used by the Romans, the warded bit-key lock, is still being used today in many older homes. Because the lock provides so little security, typically it's found on interior doors, such as closets and sometimes bedrooms.

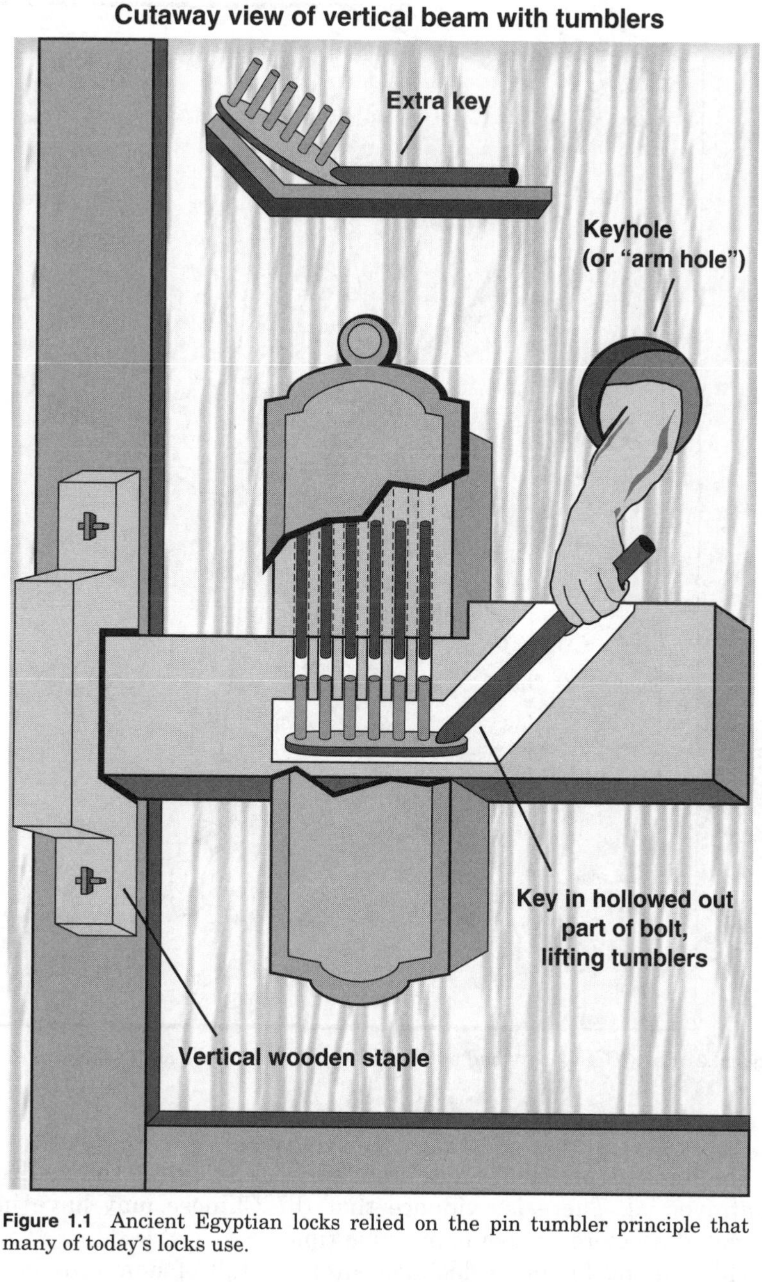

Figure 1.1 Ancient Egyptian locks relied on the pin tumbler principle that many of today's locks use.

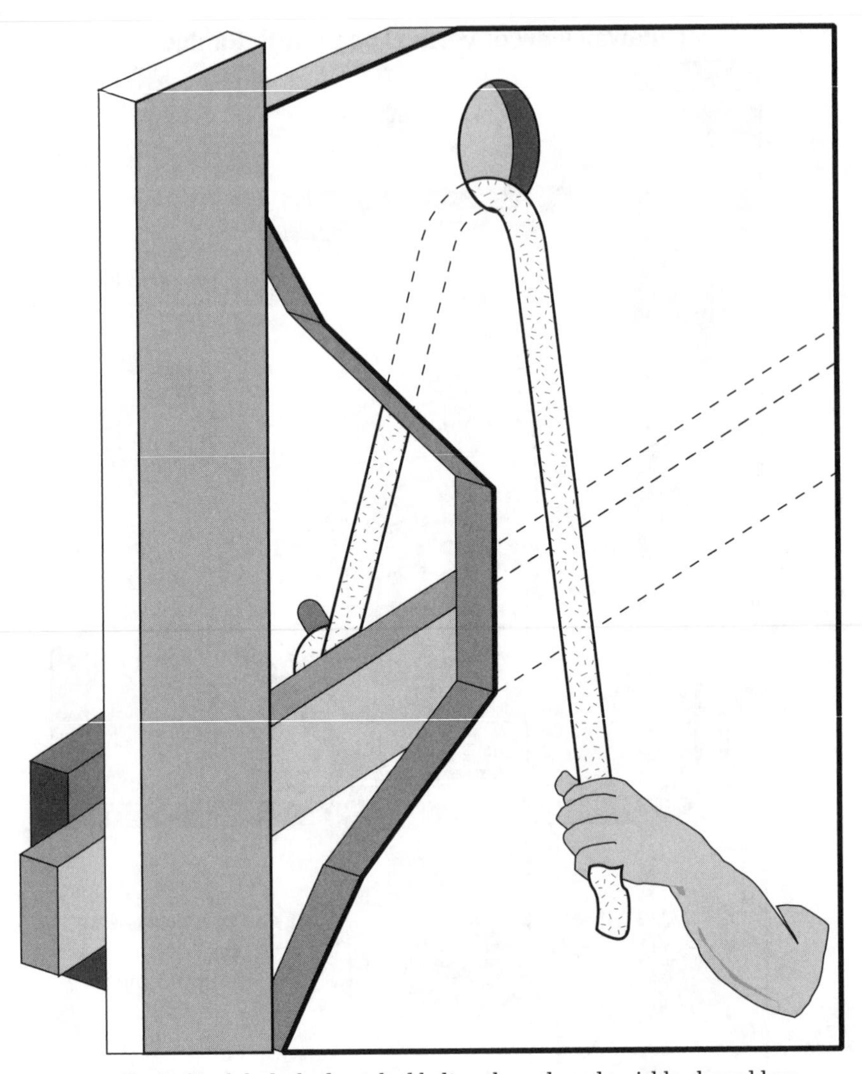

Figure 1.2 Early Greek locks had notched boltwork, and used a sickle-shaped key.

The Romans are sometimes credited with inventing the padlock, but that's controversial. There is evidence that the Chinese may have independently invented it before or about the same time.

The demand for locks declined after the fall of Rome in the fifth century because people had little property to protect. The few locks used during the period were specially ordered for nobility and the handful of wealthy merchants.

Figure 1.3 Many early Roman keys were made to be worn as rings, because clothing of Romans didn't have pockets. (*Courtesy of TheLockMan.com/The First Internet Lock Museum.*)

Europe

During the Middle Ages, metal workers in England, Germany, and France continued to make warded locks, with no significant security changes. They focused on making elaborate ornately designed cases and keys. Locks became works of art.

Keys were made that could move about a post and shift the position of a movable bar (the locking bolt). The first obstacles to unauthorized use of the lock were internal wards. Medieval and renaissance craftsmen improved on the warded lock by using many interlocking wards and more complicated keys. But many of the wards could easily be bypassed.

In France in 1767, the treatise *The Art of the Locksmith* was published; it described examples of the lever tumbler lock. The inventor of the lock is unknown. As locksmithing advanced, locks were designed with multiple levers, each of which had to be lifted and properly aligned before the bolt could move to the unlocked position.

In the fourteenth century the locksmith's guilds came into prominence. They required journeymen locksmiths to create and submit a working lock and key

Figure 1.4 A masterpiece lock was made to be displayed in the guild hall. (*Courtesy of TheLockMan.com/The First Internet Lock Museum.*)

to the guild before being accepted as a master locksmith. The locks and keys weren't made to be installed, but to be displayed in the guild hall. The guild's work resulted in some beautiful locks and keys (Fig. 1.4). The problem with the locksmith guilds is that they gained too much control over locksmiths, including the regulating of techniques and prices. The guilds became corrupt and didn't encourage technological advances. Few significant security innovations were made because of the locksmiths' guilds. The innovations included things like false and hidden keyholes. A fish-shaped lock, for instance, might have the keyhole hidden behind a fin.

England

Little progress was made in lock security until the eighteenth century. Incentive was given in the form of cash awards and honors to those who could pick open newer and more complex locks. That resulted in more secure lock designs. In the forefront of lock designing were three Englishmen: Robert Barron, Joseph Bramah, and Jeremiah Chubb.

The first major improvement over warded locks was patented in England in 1778 by Robert Barron. He added the tumbler principle to wards for increased security. His double-acting lever tumbler lock was more secure than other locks during that time and remains today the basic design for lever tumbler locks. Like other lever tumbler locks, Barron's used wards. But he also used a series of lever tumblers, each of which was acted upon by a separate step of the key. If any tumbler wasn't raised to the right height by the key, its contact with a bolt stump would obstruct bolt movement (Fig. 1.5). Barron's lock corrected the shortcomings of earlier lever tumbler locks, which could easily be circumvented

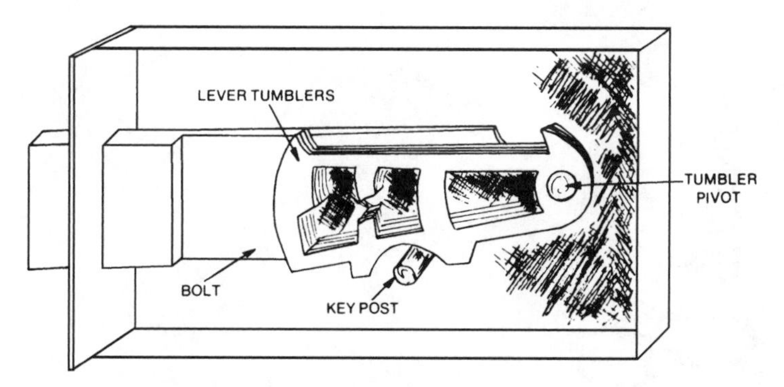

Figure 1.5 An eighteenth century lever tumbler lock. The width of the tumblers corresponded to top slots in the key.

by any key or instrument thin enough to bypass the wards. Barron added up to six of these double-lever actions to his lock and thought it was virtually impossible to open it except by the proper key. He soon found out differently.

Another Englishman, Joseph Bramah, wrote *A Dissertation on the Construction of Locks,* which exposed the many weaknesses of existing so-called thiefproof locks and pointed out that many of them could be picked by a good specialist or criminal with some training in locks and keys. Bramah admitted that Barron's lock had many good points but also revealed its major fault: The levers, when in the locked position, gave away the lock's secret. The levers had uneven edges at the bottom; thus, a key coated with wax could be inserted into the lock and a new key could be made by filing where the wax had been pressed down or scraped away. Several tries could create a key that matched the lock. Bramah pointed out that the bottom edges of the levers showed exactly the depths to which the new key should be cut to clear the bolt. Bramah suggested that the lever bottoms should be cut unevenly. Then only a master locksmith should be able to open it.

Using those guidelines, Bramah patented a barrel-shaped lock in 1798 that employed multiple sliders around the lock that were to be aligned with corresponding notches around the barrel of its key. The notches on the key were of varying heights. When the right key was pushed into the lock, all the notches lined up with the sliders, allowing the barrel to rotate to the unlocked position. It was the first to use the rotating element in the lock itself (Fig. 1.6).

During this period, burglary was a major problem. After the Portsmouth, England, dockyard was burglarized in 1817, the British Crown offered a reward to anyone who could make an unpickable lock. A year later, Jeremiah Chubb patented his lock and won the prize money.

Jeremiah Chubb's detector lock was a four-lever tumbler rim lock that used a barrel key. It had many improvements over Barron's. One of the improvements was a metal "curtain" that fell across the keyhole when the mechanism began to turn, making the lock hard to pick. Chubb's lock also added a detec-

Figure 1.6 A Bramah radial lever lock (circa 1790).

tor lever that indicated whether the lock had been tampered with. A pick or an improperly cut key would raise one of the levers too high for the bolt gate. That movement engaged a pin that locked the detector lever. The lever could be cleared by turning the correct key backward and then forward.

Chubb's lock got much attention. It was recorded that a convict who had been a lockmaker was on board one of the prison ships at Portsmouth Dockyard and said he had easily picked open some of the best locks and that he could easily pick open Chubb's detector lock. He was given one of the locks and all the tools that he asked for, including key blanks fitted to the drill pin of the lock. As incentive to pick open the lock, Mr. Chubb offered the convict a reward of £100, and the government offered a free pardon if he succeeded. After trying for several months to pick the lock, he gave up. He said that Chubb's lock was the most secure lock he had ever met with and that it was impossible for anyone to pick or open it with false instruments. The lock was improved on by Jeremiah's brother, Charles Chubb, and Charles' son John Chubb in several ways, including the addition of two levers and false notches on the levers.

The lock was considered unpickable until it was picked open in 1851 at the International Industrial Exhibition in London by an American locksmith named Alfred C. Hobbs. At that event, Hobbs picked open both the Bramah and the Chubb locks in less than half an hour.

America

During America's early years, England had a policy against its skilled artisans leaving the country. That was to keep them from running off and starting competing foreign companies. Locks made by early American locksmiths didn't sell well. In the mid-1700s few colonists used door locks, and most that were used were copies of European models. More often, Americans used lock bolts mounted on the inside of the door that could be opened from the outside by a latchstring, hence the phrase, "the latchstring's always out." At night, the string would be pulled inside, "locking" the door. Of course, someone had to be inside to release the bolt. An empty house was left unlocked. As the country settled, industry progressed and theft increased, increasing the demand for more and better locks. American locksmiths soon greatly improved on the English locks and were making some of the most innovative locks in the world. Before 1920, American lockmakers patented about 3000 different locking devices.

In 1805, an American physician, Abraham O. Stansbury, was granted an English patent for a pin tumbler lock that was based on the principles of both the Egyptian and Bramah locks. Two years later, the design was granted the first lock patent by the U.S. Patent and Trademark office. Stansbury's lock used segmented pins that automatically relocked when any tumbler was pushed too far. The double-acting pin tumbler lock was never manufactured for sale.

In 1836, a New Jersey locksmith, Solomon Andrews, developed a lock that had adjustable tumblers and keys, which allowed the owner to rekey the lock anytime. Because the key could also be modified, there was no need to use a new key to operate a rekeyed lock. But few homeowners used the lock because rekeying it required dexterity, practice, and skill. The lock was of more interest to banks and businesses.

In the 1850s, two inventors, Andrews and Newell, were granted patents on an important new feature—removable tumblers that could be disassembled and scrambled. The keys had interchangeable bits that matched the various tumbler arrangements. After locking up for the night, a prudent owner would scramble the key bits. Even if a thief got possession of the key, it would take hours to stumble onto the right combination. In addition to removable tumblers, this lock featured a double set of internal levers.

Newell was so proud of this lock that he offered a reward of $500 to anyone who could open it. A master mechanic took him up on the offer and collected the money. This experience convinced Newell that the only secure lock would have its internals sealed off from view. Ultimately, the sealed locks appeared on bank safes in the form of combination locks.

Until the time of A. C. Hobbs, who picked the famed English locks with ease, locks were opened by making a series of false keys. If the series was complete, one of the false keys would match the original. Of course, this procedure took time. Thousands of hours might pass before the right combination was found. Hobbs depended upon manual dexterity. He applied pressure on the bolt while manipulating one lever at a time with a small pick inserted through the keyhole. As each lever tumbler unlatched, the bolt moved a hundredth of an inch or so.

Hobbs patented what he called "Protector" locks. They weren't invincible either. In 1854, one of Chubb's locksmiths used special tools to pick open one of Hobb's locks.

Until the early nineteenth century, locks were made by hand. Each locksmith had his own ideas about the type of mechanism—the number of lever tumblers, wards, and internal cams to put into a given lock. Keys contained the same individuality. A lock could have 20 levers and weigh as much as 5 pounds.

In 1844, Linus Yale, Sr., of Middletown, Connecticut, patented his "Quadruplex" bank lock, which incorporated a combination of ancient Egyptian design features and mechanical principles of the Bramah and Stansbury locks. The Quadruplex had a cylinder subassembly that denied access to the lock bolt. In 1848 Yale patented another pin tumbler design based on the Egyptian and the Bramah locks. His early models had the tumblers built into the case of the lock and had a round fluted key. His son, Linus Yale, Jr., improved on the lock design and is credited with inventing the modern pin tumbler lock. Figure 1.7 illustrates the working principle of the lock cylinder.

Arguably, the most important modern lock development is the Yale Mortise Cylinder Lock, U.S. patent 48,475, issued on June 27, 1865, to Linus Yale, Jr. It turned the lock-making industry upside down and established a new standard. Yale, Jr.'s, lock not only could easily be rekeyed, but it also provided a high level of security, could easily be mass produced, and could be used on doors of various thicknesses. His lock design meant that keys no longer had to pass through the thickness of the door to reach the tumblers or bolt mechanism, which allowed the keys to be made thinner and smaller. (Linus Yale,

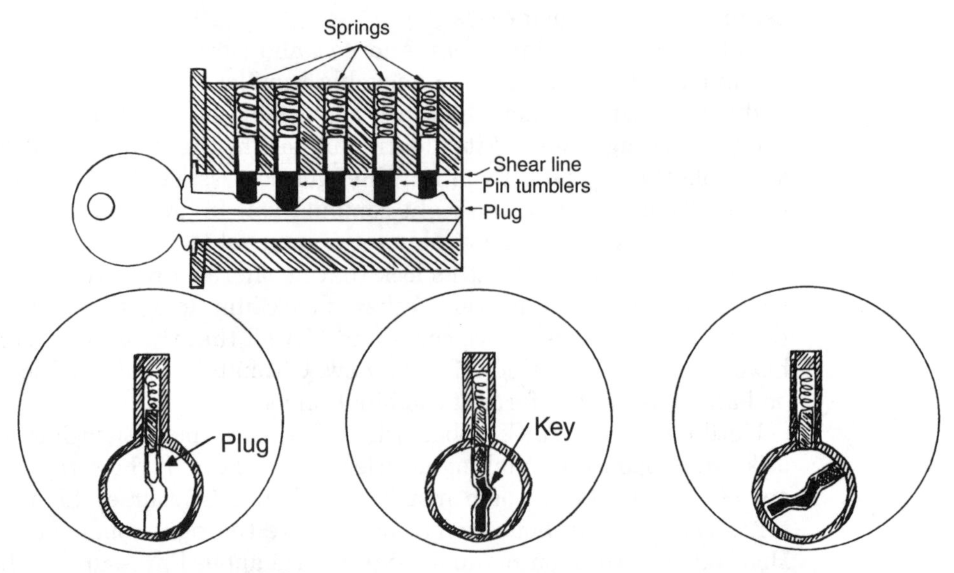

Figure 1.7 The operation of the Yale pin tumbler cylinder. The key raises the tumblers to the shear line, which frees the plug to be rotated to the locked or unlocked position.

Jr.'s, first pin tumbler locks used a flat steel key rather than the paracentric cylinder type often used today.)

Since 1865, there have been few major changes to the basic design of mechanical lock cylinders. Most cylinder refinements since that time have been limited to using unique keyways (along with corresponding shaped keys), adding tumblers, varying tumbler positions, varying tumbler sizes and shapes, and combining two or more basic types of internal construction—such as the use of both pin tumblers and wards. Most major changes in lock design have centered around the shape and installation methods of the lock.

In 1916 Samuel Segal, a former New York City police officer, invented the jimmyproof rim lock (or "interlocking deadbolt"). The surface-mounted lock has vertical bolts that interlock with "eye-loops" of its strike, locking the two parts together in such a way that you would have to break the lock to pry them apart.

In 1920 Frank E. Best received his first patent for an interchangeable core lock. It allows you to rekey a lock just by using a control key and switching cores. The core was made to fit into padlocks, mortise cylinders, deadbolts, key-in-knobs, and other types of locks.

In 1833, three brothers, the Blake brothers, were granted a patent for a unique door latch that had two connecting doorknobs. It was installed by boring two connecting holes. The larger hole, which was drilled through the door face, was for the knob mechanism. The smaller hole, which was drilled through the door edge, was for the latch. The big difference between their latch and others of their time was that all the door locks were installed by being surface mounted to the inside surface of a door. In 1834, the brothers formed the Blake Brothers Lock Company to produce and sell their unusual latch. At that time, the brothers probably never imagined that nearly 100 years later their development would be used to revolutionize lock designs.

In 1928 Walter Schlage patented a cylindrical lock that incorporates a locking mechanism between the two knobs. Schlage's was the first knob-type lock to have mass appeal. Today, key-in-knob locks are commonplace.

In 1933 Chicago Lock Company introduced its tubular key lock, called the Chicago Ace Lock. The lock was based on the pin tumbler principle but used a circular keyway. The odd keyway made it hard to pick open without using special tools. For a long time many locksmiths referred to all tubular key locks as "Ace Locks," not realizing that was only a brand name. Today the lock is made by many manufacturers and is used on vending machines and in padlocks and bicycle locks.

A recent innovation in high-security mechanical locks came in 1967, with the introduction of the Medeco high-security cylinder. The cylinder, made by Roy C. Spain and his team, used chisel-pointed rotating pins and restricted angularly bitted keys that made picking and impressing harder. To open the lock, a key had to not only simultaneously lift each pin to the proper height, but also rotate each one to the proper position to allow a sidebar to retract. The name "Medeco" was based on the first two letters of each word of the name Mechanical Development Company. The Medeco Security Lock was the largest and most talked about high-security lock. In the early 1970s the company offered a reward for anyone who could pick open one, two, or three of its cylinders within a set amount of time. In 1972 Bob McDermott, a New York City police detective, picked one open in time and collected the reward. That feat didn't slow the demand for Medeco locks. Much of the general public never heard about the contest and still considered Medeco locks to be invincible. In 1986 Medeco won a patent infringement lawsuit against a locksmith who was making copies of Medeco keys. That ruling stopped most other locksmiths from making the keys without signing up with Medeco. The patent for the original Medeco key blank ran out, and now anyone can make keys for those cylinders. In 1988 the company received a new patent for its "biaxial" key blanks. The big difference is that the biaxial brand gave Medeco a new patent (which can be helpful for preventing unauthorized key duplication). The company's newest lock is the Medeco^3.

Early American Lock Companies

In 1832, the English lockmaker Stephen G. Bucknall became the first trunk and cabinet lock manufacturer in America. He made about 100 cabinet lock models but didn't sell very many. After his company folded, Bucknall went to work for Lewis, McKee & Company. William E. McKee was a major investor in the company. In 1835 Bucknall left and received financial assistance from McKee to form the first trunk lock company in America, called the Bucknall, McKee Company. A couple years later, Bucknall sold the business and went to work for North & Stanley Company.

One of the greatest successes and failures of the American lock industry was the Eagle Lock Company, formed in 1854. It was the result of a merger between the James Terry Company and Lewis Lock Company. Eagle Lock had money to burn and was quick to buy out its competitors, such as American Lock Company, Gaylord, and Eccentric Lock Company. In 1922 the company had 1800 employees and several large warehouses. In 1961 the company introduced a popular line of locks and cylinders called "Supr-Security." They were highly pick-resistant, and their keys couldn't be duplicated on standard key machines. The company was sold to people who didn't know a lot about locks, and many bad decisions were made that sent the company on a downward path. Profits were being siphoned off without considering the long-term needs of the business. Top management quit. By the early 1970s, the company was barely holding on. In 1973 a businessperson bought it from Penn-Akron Corporation, which had gone bankrupt. In 1974 Eagle Lock lost a bid to the Lori Corporation for a large order for cylinders for the U.S. Postal Service. A few months later the company folded. The Lori Corporation bought the Eagle Lock equipment at a public auction, and the Eagle Lock plants were burned, ending 122 years of lockmaking.

For historical information about current lockmakers, see Appendix A.

A Brief History of Automotive Locks in the U.S.

While the car has been with us since the beginning of the 20th century, the automobile lock was adopted slowly. By the late 1920s, however, nearly every vehicle had an ignition lock, and closed cars had door locks as well. Current models can be secured with half a dozen locks. This section gives an overview of the history of automobile locks in the U.S. The information is especially useful when you're working on older vehicles. [For in-depth information on servicing and opening all types of automobiles, see chapters 15 and 16.]

1935 General Motors began using sidebar locks. There was only one keyway, with 6 cuts and 4 depths.

1959 Chrysler began using sidebar trunk locks.

1966 Chrysler stopped using sidebar trunk locks, and began using pin tumbler locks that were the same size as the door locks.

1967 General Motors added a 5th depth to their codes and introduced two new key blanks—P1098A and S1098B.

1968 General Motors introduced two new key blanks, P1098C and S1098D.

1969 General Motors began using steering column mounted ignition locks and introduced two new key blanks—P1098E and S1098H.

1970 American Motors, Chrysler and Ford began using column mounted ignition locks. General Motors stopped putting codes on door locks and introduced two new blanks—P1098J and S1098K.

1972 Chrysler began using General Motors' Saginaw tilt steering columns with sidebar ignition locks.

1973 Chrysler began using trunk locks retained by large nuts.

1974 General Motors began keying locks so that the primary key fit the ignition lock only and all other locks on the vehicle used the secondary key.

1977 Ford stopped putting codes on door locks.

1978 General Motors stopped putting codes on glove compartment locks.

1979 General Motors changed ignition locks from being spring tab retained to screw retained.

1979 Ford began using fixed pawls on door locks.

1980 General Motors began using fixed pawls on door locks.

1981 Ford began keying locks so that the primary key fit the ignition and the secondary key fit all other locks. The company stopped putting codes on glove compartment locks.

1983 American Motors began making the Alliance, using X116 ignition key and X92 door key, both previously known as foreign auto keys.

1985	Ford began using sidebar ignition locks and wafer tumbler door locks, which worked with Ford 10-cut keys. Primary keys operate doors and ignition; secondary keys operate glove compartment and trunk.
1986	VATS or "Vehicle Anti-theft System" were introduced by General Motors on the 1986 Corvette.
1989	Chrysler introduced its double-sided wafer tumbler locks.
1992	General Motors introduced the MATS or "Mechanical Anti-theft System" in its full sized rear wheel drive cars—Oldsmobile Custom Cruiser Station Wagon, Buick Roadmaster and Chevrolet Caprice.
1996	The PATS or "Passive Anti-theft System." A radio frequency identification system that was introduced on the 1996 Ford Taurus and Mercury Sable.

Chapter

2

Tools of the Trade

To make money in locksmithing, you have to work quickly and professionally, which requires using proper tools and supplies. In addition to many of the hand and power tools commonly used by carpenters and electricians, locksmiths use a variety of special tools. Which special items you'll need depends on which jobs you will perform.

The more services you want to offer, the more special tools you'll need. For tools, quality is often more important than quantity. A tool that breaks, bends, or stops working at the wrong time can cause you to lose a lot of money. It can also cost you money by damaging your customer's property. On the other hand, a tool that's designed to last a lifetime will pay for itself many times. Even if you have little money to invest in tools, always get the highest-quality products that you can afford.

This chapter lists all the basic tools and supplies you'll need to perform any locksmithing task. It also tells what to look for when buying tools and supplies.

Electric Drills

It's important to choose the right drill because when doing locksmithing work, you'll use your drill more often than any other type of portable power tool (Fig. 2.1). Drills range in price from under $50 to more than $200. Although price is often an indication of the quality of a drill, the only way to be sure you're getting the most for your money is to know the important ways drills differ from one another.

Three common drill sizes are available: ¼, ⅜, and ½ inch. Drill size is determined by the largest diameter drill bit shank the drill's chuck can hold without an adapter. For example, a drill whose chuck can hold a drill bit shank of up to a ½ inch is called a *½-inch drill*.

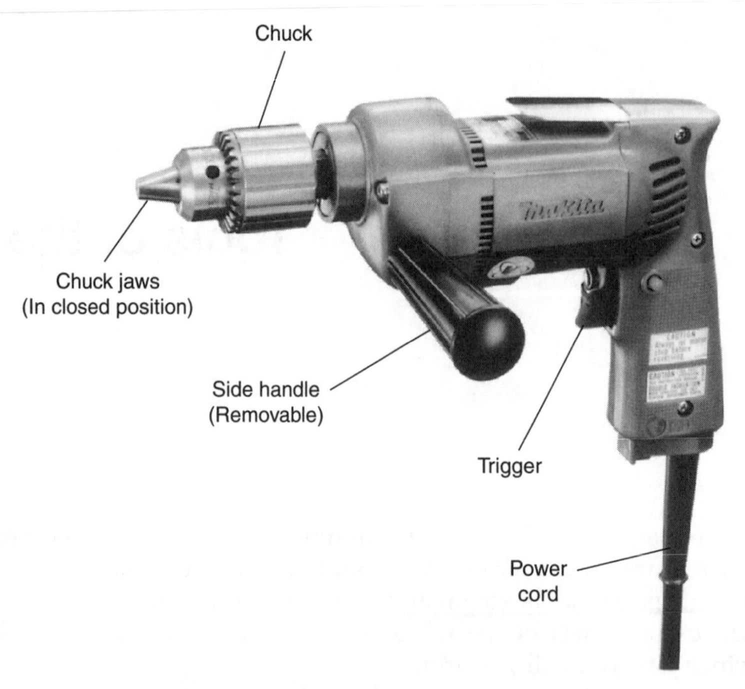

Figure 2.1 A ½-inch electric drill is one of the most useful tools a locksmith can have. (*Makita U.S.A., Inc.*)

A drill's power is a combination of chuck speed and torque. Speed is measured in rpm (revolutions per minute when the chuck is spinning freely in the air). *Torque* refers to the twisting force at the chuck when the drill is being used to drill a hole. Any drill will slow down while drilling a hole, but the more torque it has the easier it will bore through material.

Chuck speed and torque are largely determined by the type of *reduction* gears a drill has. Reduction gears in a drill work somewhat like car gears. One gear, for example, lets the car move quickly on a flat road. Another gear gives the car more power when climbing a steep hill. This analogy isn't perfect, however, because a drill comes with a fixed type of reduction gear set. You can't shift the gears of a drill.

A *single-stage* reduction gear set lets a drill's chuck spin extremely fast in the air but supplies little torque. A *two-stage* reduction gear set provides fewer rpms, and more torque. A *three-stage* reduction gear set further reduces rpms and increases torque. Greater torque is especially useful when drilling hardwood, steel, or other hard materials. It's also useful when drilling large holes, such as those needed to install deadbolt locks.

Most ¼-inch drills have a single-stage reduction gear set. Their chucks commonly spin at 2500 rpms or more. Such drills are lightweight and most often used for drilling plastic, softwood, and sheet metal. Using a ¼-inch drill to drill hardwood or steel would be time-consuming and could damage the drill.

Most ⅜-inch drills have either a single- or two-stage set of reduction gears. The shanks of many popular hole saws and drilling accessories won't fit into a ⅜-inch chuck without using an adapter.

The ½-inch drill is by far the most popular size among locksmiths. A ½-inch drill usually has a two- or three-stage set of reduction gears, and its chuck typically spins at up to 600 rpm. The ½-inch can drill any material that can be drilled by a smaller drill.

Not all ½-inch drills are alike. They can differ greatly in quality and price. Many manufacturers call their drills "heavy-duty," "professional," "commercial," and the like. Such labels are primarily for promotional purposes; they have no industry-standard meanings. It's best to look beyond such terms for specific features. The most important features are two- or three-stage reduction gears, variable speed reversing, double insulation, antifriction bearings, and at least 4 amps.

A *variable speed* feature provides maximum control over drilling speed, allowing you to drill different materials at different speeds. Without variable speed control, you might be limited to drilling at one or two speeds only. Many drills also have a switch that lets you reverse the direction that the chuck turns. That is useful for backing out screws or a stuck drill bit. A drill that has a *variable speed feature* and is reversible is called variable speed reversible (VSR). VSR drills are well worth the few extra dollars they cost.

Double insulation means the drill is housed in nonconductive material (such as plastic) and the motor is isolated from other parts of the drill by a nonconductive material. Double insulation protects you from getting shocked. Most high-quality drills are double insulated. Don't mistake plastic housing as a sign of low-quality. All-metal housing is a sign of low-quality.

Antifriction (*ball* or *needle*) *bearings* help a drill run smoothly and make it long-lived. All high-quality drills use some antifriction bearings. Low-quality models use all plain sleeve bearings.

An *amp* (short for *amperes*) is a unit of electric current. In general, the more amps used, the more powerful the drill. High-quality drills usually use at least 4 amps. (Drills that list horsepower rather than amps are usually low-quality.)

Any drill that has all of these features is a high-quality model that should last a lifetime if it's properly used and maintained. Some locksmiths have additional considerations such as a drill's color, size, weight, feel, brand, and country of origin. Those additional considerations are personal and have little to do with a drill's quality.

Cordless Drills

A cordless drill can conveniently insert screws and drill small holes (Fig. 2.2). It's typically lighter than an electric drill and can be used in places with no electrical outlet. As a rule, cordless drills are less powerful than their electric

Figure 2.2 A cordless drill can be useful when no electricity is available. (*Milwaukee Tool Co.*)

counterparts. Another problem with cordless drills is that their batteries have to be recharged and replaced. Although a cordless drill can be great as a backup or extra drill, it isn't a good choice if you can afford only one drill.

Key Cutting Machines

Because cutting keys is often a major source of income for a locksmith, key cutting (or *duplicating*) *machines* are among the most important tools a locksmith buys. Expensive machines often can cut a wide variety of keys for standard and high-security locks. Two popular types of key cutting machines are shown in Fig. 2.3.

If you have little money to invest in key machines, it's probably best to start with a low-cost model that cuts cylinder keys—the most common type of keys for home and car locks. Later, you might want to add a low-cost machine that cuts flat keys (such as those used for safe deposit boxes) or tubular keys (such as those used for vending machines and laundromat equipment). Another way to save money is to buy only portable dual-voltage models that can be used either in the shop or in a service vehicle.

Workbench

Whether you practice locksmithing in a shop or at home, you'll need a workbench. You can use a table or desk on a temporary basis, but a workbench is

Figure 2.3 Different types of key machines are needed to cut cylinder keys (*top*) and tubular key lock keys (*bottom*).

more practical and comfortable. If you have basic woodworking tools and skills, you should have no trouble making your own workbench.

The workbench should be:

- Long enough to ensure adequate work space
- Strong enough to support a key machine at one end, out of the way
- Solid enough to keep the key machine in alignment
- High enough to allow you to work without stooping
- Wide enough to store parts and supplies; 30 inches is the comfortable maximum
- Lit from overhead and behind or from the sides (never have the light too close to the bench)

Workbench location

Consider where you want to locate your workbench. If possible, place it in a well-ventilated area, away from general traffic, near frequently used equipment and supplies, with an easy access from several directions. Many locksmiths place storage bins near their workbenches to hold tools and supplies (Fig. 2.4).

Examples of workbench designs are shown in Figs. 2.5 and 2.6. No one type of workbench is ideal for everybody; individual tastes and needs differ.

Install a vise at one end, out of the way. Mount the vise on a swivel base, sturdily built, with jaws at least 3 inches wide. The jaws should open to at least 5 inches.

List 1: Common hand and power tools all locksmiths should have:

- Allen wrench set
- Bench grinder with wire wheel
- Bolt cutters, 16-inch
- C-clamps (Fig 2.7)
- Center punch set
- Code books, general set
- Combination square
- Coping saw and blades
- Dent puller
- Disc grinder

Figure 2.4 Bins can be helpful for sorting lock parts and supplies.

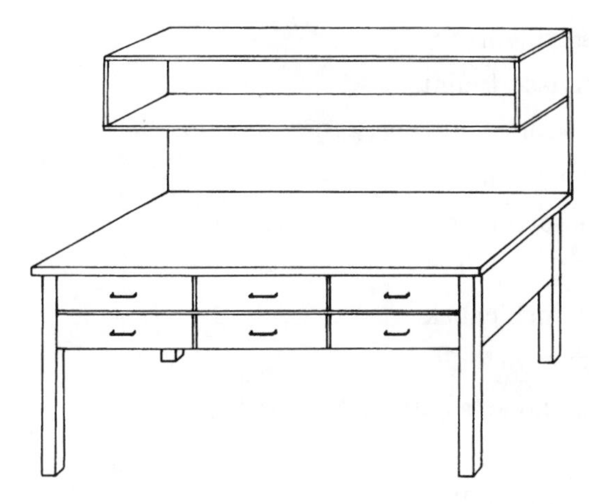

Figure 2.5 A simple bench is adequate for a student locksmith.

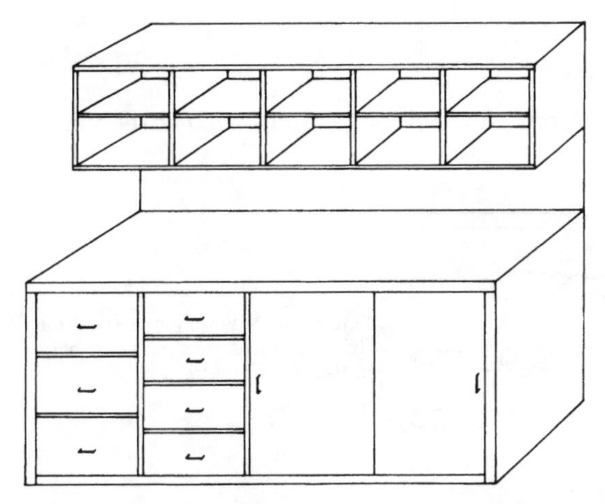

Figure 2.6 Benches with a full complement of drawers and partitioned overhead bins are ideal for professional locksmiths.

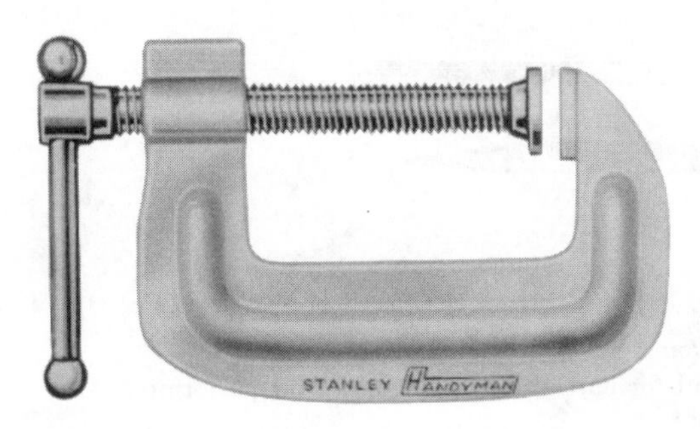

Figure 2.7 C-clamps usually range in size from 1 to 8 inches.

- Drill bits, auger assortment
- Drill bits, expansion assortment
- Drill bits, masonry assortment (Fig. 2.8)
- Drill bits, spade assortment
- Drill bits, straight assortment
- Drill, cordless
- Drill, electric with ½-inch chuck
- Extension cord, 50-foot
- Files, assorted types and sizes (Fig. 2.9)
- Flashlight
- Hacksaw and blades (Fig. 2.10)
- Hammers: claw, ball-peen, and soft-face (Fig. 2.11)
- Hand cleaner
- Hollow mill rivet set
- Lever, carpenter's, from 18 to 96 inches long with three vials

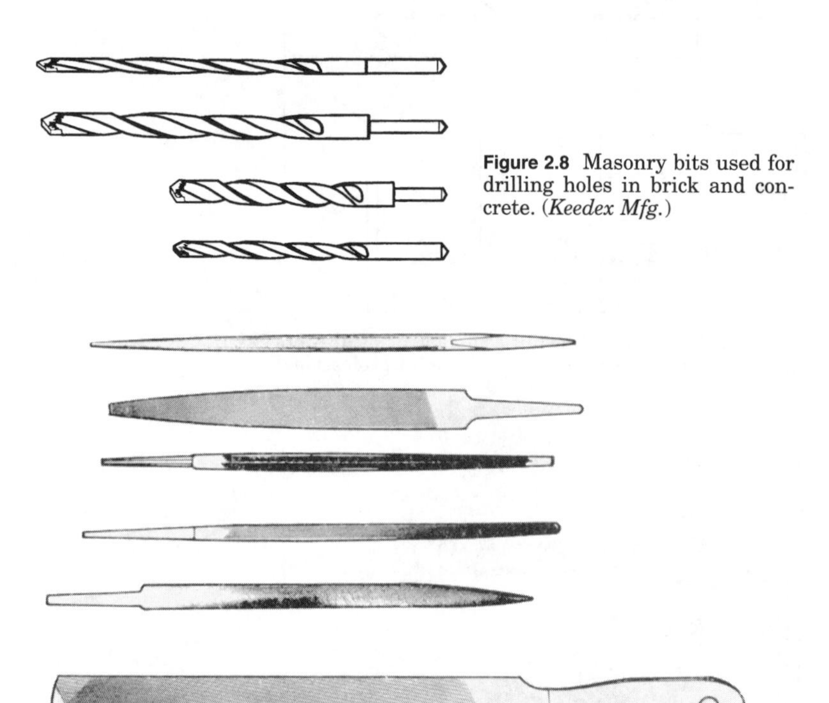

Figure 2.8 Masonry bits used for drilling holes in brick and concrete. (*Keedex Mfg.*)

Figure 2.9 Files can be useful for cutting keys by hand and performing many other locksmithing tasks.

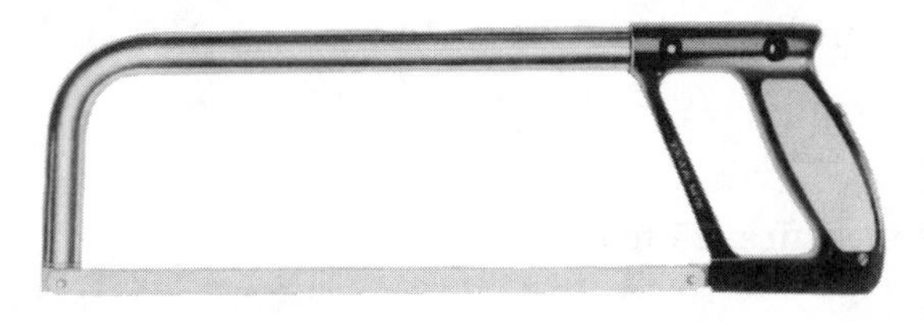

Figure 2.10 A hacksaw for cutting metals and other nonwood materials.

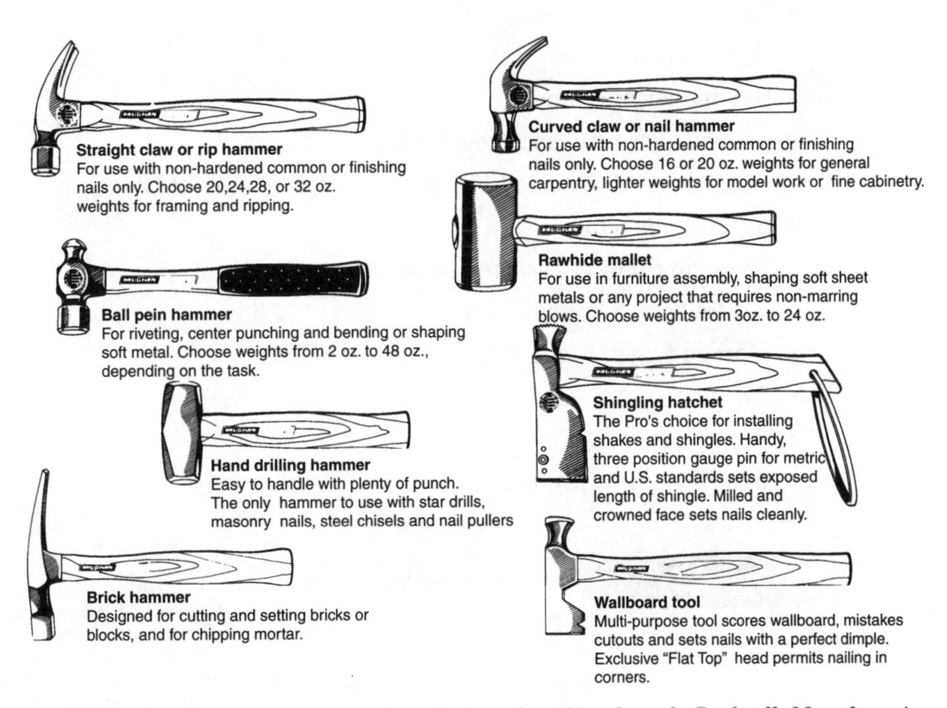

Straight claw or rip hammer
For use with non-hardened common or finishing nails only. Choose 20,24,28, or 32 oz. weights for framing and ripping.
Curved claw or nail hammer
For use with non-hardened common or finishing nails only. Choose 16 or 20 oz. weights for general carpentry, lighter weights for model work or fine cabinetry.
Rawhide mallet
For use in furniture assembly, shaping soft sheet metals or any project that requires non-marring blows. Choose weights from 3oz. to 24 oz.
Ball pein hammer
For riveting, center punching and bending or shaping soft metal. Choose weights from 2 oz. to 48 oz., depending on the task.
Shingling hatchet
The Pro's choice for installing shakes and shingles. Handy, three position gauge pin for metric and U.S. standards sets exposed length of shingle. Milled and crowned face sets nails cleanly.
Hand drilling hammer
Easy to handle with plenty of punch. The only hammer to use with star drills, masonry nails, steel chisels and nail pullers
Brick hammer
Designed for cutting and setting bricks or blocks, and for chipping mortar.
Wallboard tool
Multi-purpose tool scores wallboard, mistakes cutouts and sets nails with a perfect dimple. Exclusive "Flat Top" head permits nailing in corners.

Figure 2.11 Striking tools are useful to locksmiths. (*Vaughan & Bushnell Manufacturing Company*)

- Lever, torpedo, up to 9 inches long
- Lubricant (such as WD-40)
- Mallet, rubber
- Masking tape
- Nails and screws, assorted
- One-way screw removal tool
- Paint scraper
- Pencils
- Pliers, adjustable
- Pliers, cutting

- Pliers, long nose, 7-inch
- Pliers, locking
- Pliers, slip-joint
- Retractable tape measure, 25-foot
- Rivet assortment
- Safety glasses
- Scratch awl
- Sandpaper and emery cloth
- Screwdriver bits, assorted Phillips and slotted
- Screwdrivers, assorted Phillips and slotted (Fig. 2.12)
- Scissors for paper
- Snap ring pliers, assorted sizes
- Socket sets, ½- and ¼-inch
- Storage trays (Fig. 2.13)
- Tap set
- Toolboxes
- Wrenches, adjustable and pipe

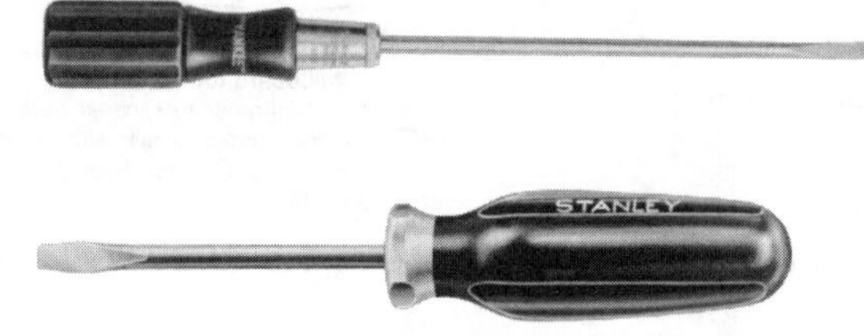

Figure 2.12 Electrician's screwdrivers (*top*) are designed for precision work; standard screwdrivers are used on heavier jobs (*bottom*). (*Stanley Works*)

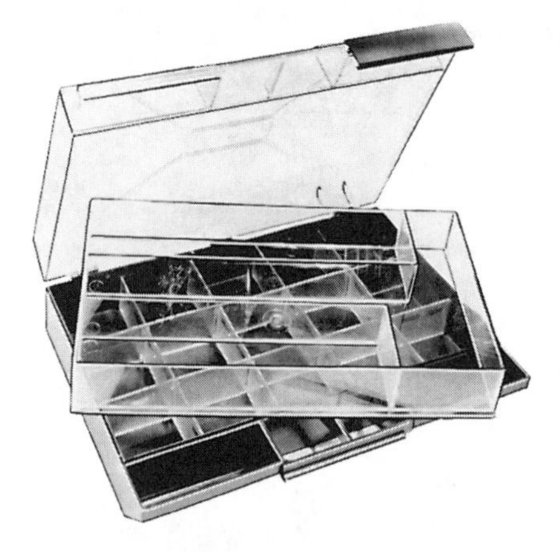

Figure 2.13 Parts and storage boxes are needed for work in and out of the shop.

List 2: Tools for shop/bench work (needed in addition to all of the tools in List 1)

- Code books
- Cylinder cap remover
- Dial caliper (Fig. 2.14)
- Flat steel spring stock
- Interchangeable core capping machine (Fig. 2.15)
- Interchangeable core lock service kit (Fig. 2.16)
- Key blanks, assorted
- Key cutting machine, electric
- Key duplicating machine, electric
- Key marking tools
- Lock parts, assorted
- Lock pick gun
- Lock pick set (Fig. 2.17)
- Lock reading tool
- Mortise cylinder clamp (Fig. 2.18)
- Pin kit (Fig. 2.19)
- Pin tray
- Pin tumbler tweezers (Fig. 2.20)
- Plug followers, assorted sizes (Fig. 2.21)
- Plug holders (Fig. 2.22)
- Plug spinner (Fig. 2.23)
- Retainer ring assortment
- Round spring steel, assorted sizes
- Shim stock (Fig. 2.24)
- Spindle assortment
- Spring assortment
- Tension wrenches (Fig. 2.25)
- Tubular-key decoder (Fig. 2.26)
- Tubular-key lock picks (Fig. 2.27)
- Tubular-key lock saw (Fig. 2.28)
- Vise
- Whisk broom
- Workbench

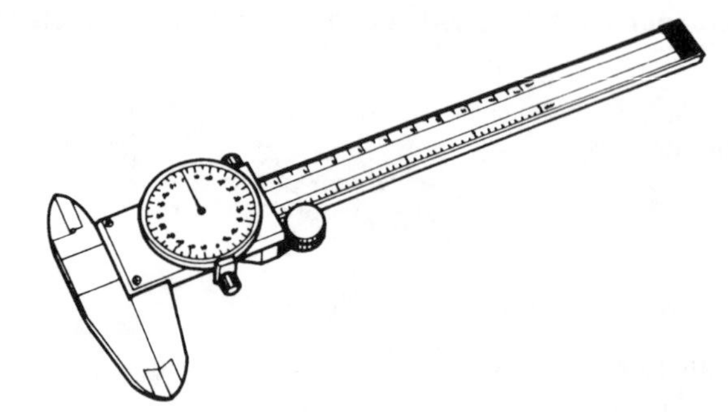

Figure 2.14 A dial caliper allows locksmiths to accurately measure pin tumblers, key blanks, key cut depths, and plugs. (*Ilco Unican Corp.*)

Figure 2.15 An interchangeable core capping machine is needed for capping IC cores. (*Arrow Mfg. Company*)

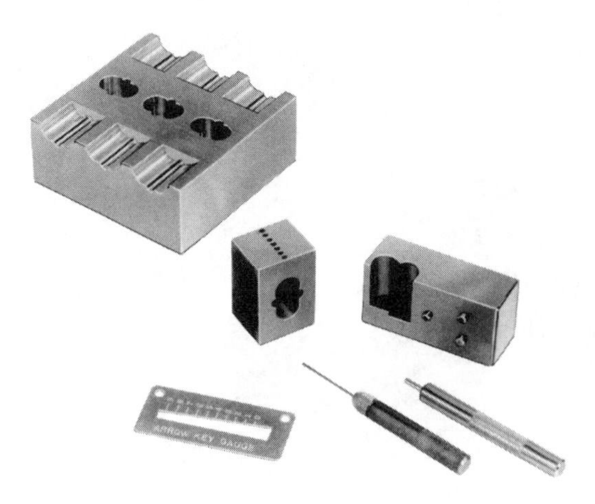

Figure 2.16 An interchangeable core cylinder and key stamping fixture and supplies needed for servicing IC cores. (*Arrow Mfg. Company*)

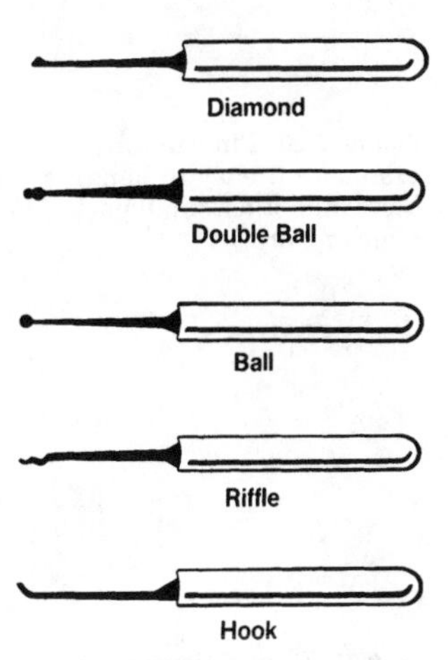

Figure 2.17 A variety of lock picks allow the locksmith to choose the best one for a given lock. (*A-1 Security Manufacturing Corp.*)

Figure 2.18 A cylinder removal tool makes it easy to remove stubborn mortise cylinders without stripping their threads. (*A-1 Security Manufacturing Corp.*)

Figure 2.19 A pin kit holds tumblers, springs, and other supplies needed to rekey lock cylinders. (*Arrow Mfg. Company*)

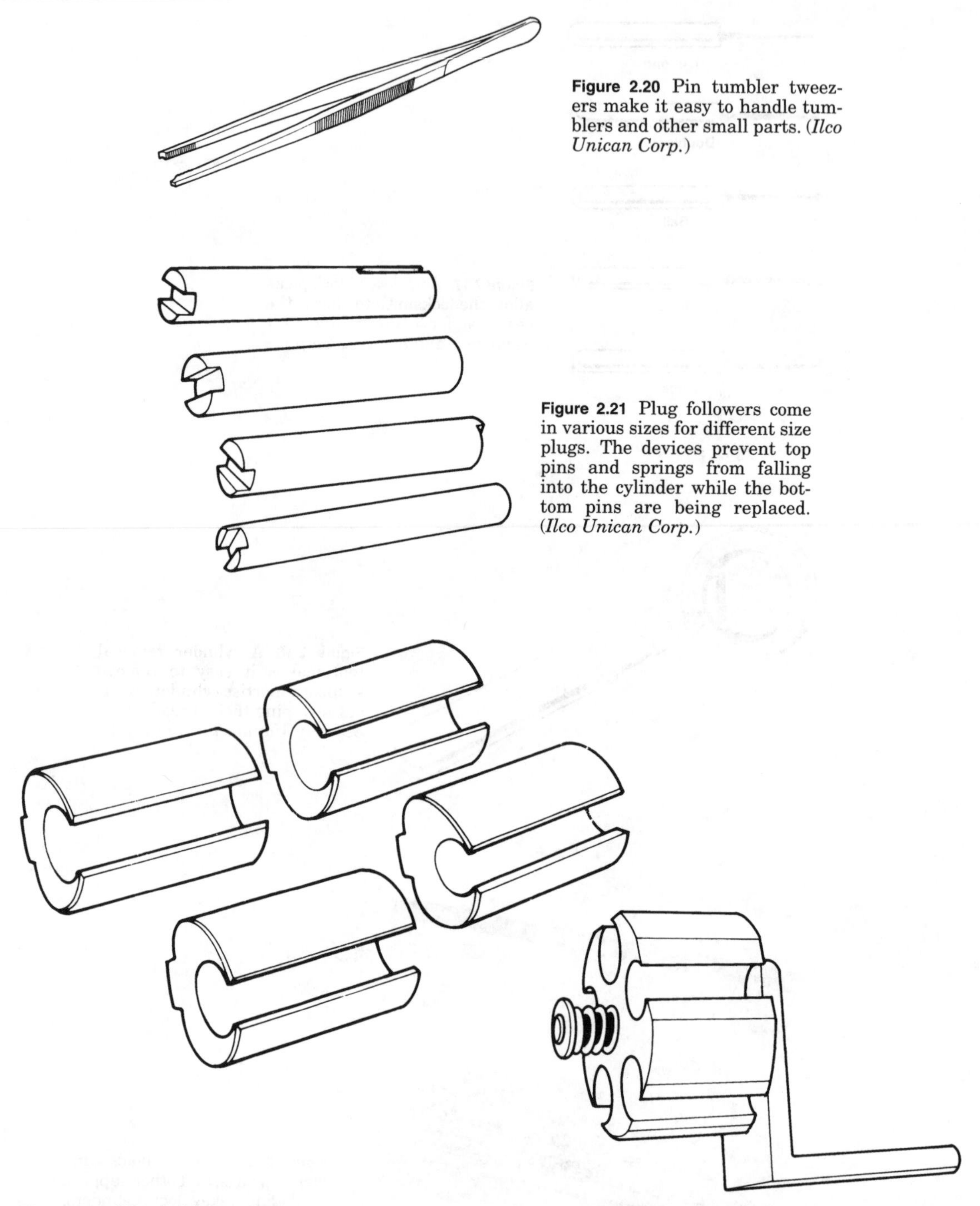

Figure 2.20 Pin tumbler tweezers make it easy to handle tumblers and other small parts. (*Ilco Unican Corp.*)

Figure 2.21 Plug followers come in various sizes for different size plugs. The devices prevent top pins and springs from falling into the cylinder while the bottom pins are being replaced. (*Ilco Unican Corp.*)

Figure 2.22 Plug holders come in various sizes; some models can hold several plugs of different sizes. In addition to holding a plug that's being serviced, plug holders allow a locksmith to make sure a rekeyed plug will rotate in the cylinder properly before inserting it back into the cylinder. (*A-1 Security Manufacturing Corp.*)

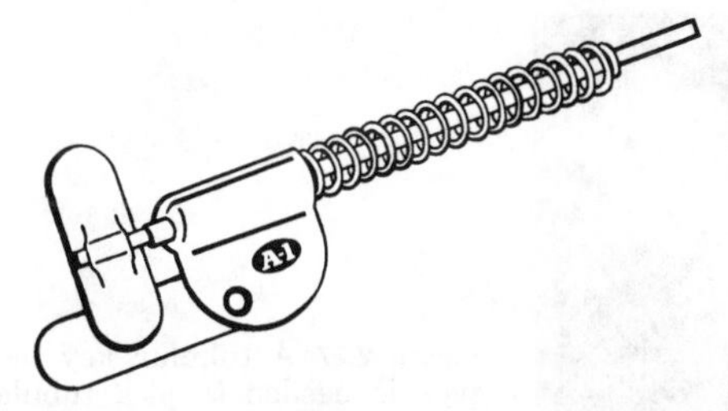

Figure 2.23 A plug spinner is designed to be inserted into a keyway and to use spring pressure to quickly rotate a plug clockwise or counterclockwise. The device is used when a lock has been picked in the wrong direction. (*A-1 Security Manufacturing Corp.*)

Figure 2.24 Shims are used when disassembling pin tumbler cylinders. (*Ilco*)

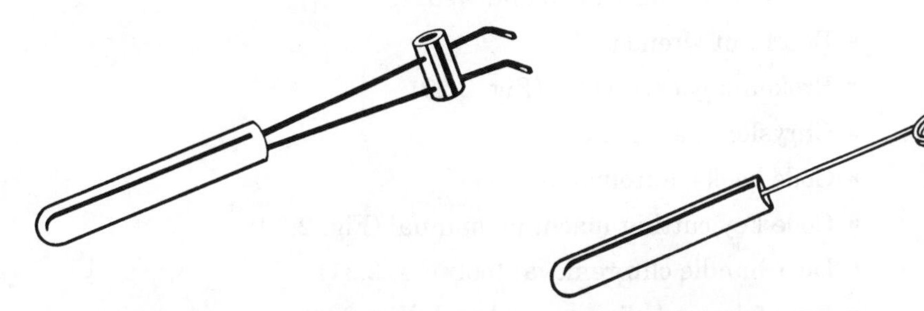

Figure 2.25 Tension wrenches come in various shapes and sizes. They're designed to partially enter a keyway along with a lock pick so that when all the tumblers are at the shear line, the tension wrench can rotate the cylinder as the proper key would. (*A-1 Security Manufacturing Corp.*)

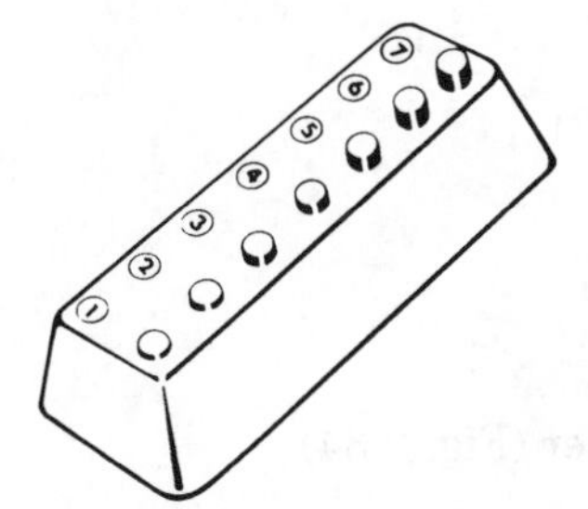

Figure 2.26 A tubular key decoder makes it easy for a locksmith to determine the bitting of tubular keys. (*A-1 Security Manufacturing Corp.*)

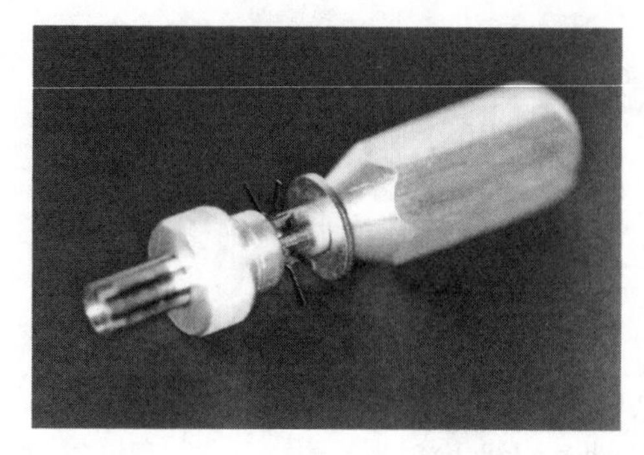

Figure 2.27 A tubular key lock pick is needed to pick tubular key locks.

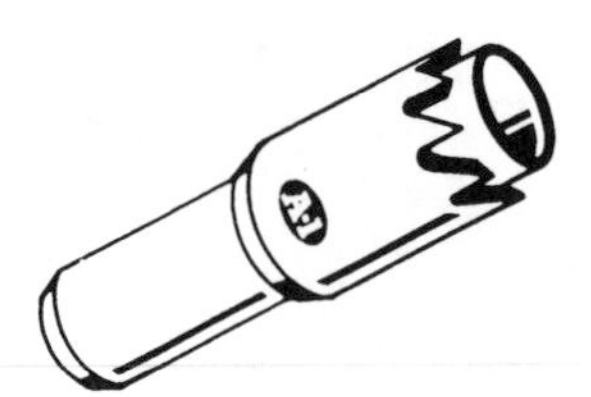

Figure 2.28 A tubular key lock saw is used to drill through the tumblers of tubular key locks. *(A-1 Security Manufacturing Corp.)*

List 3: Tools for automotive work (needed in addition to all of the tools in Lists 1 and 2)

- Automobile entry tools and wedges
- Bezel nut wrench
- Broken-key extractors (Fig. 2.29)
- Chrysler shaft puller
- Code books, automotive
- Code key cutting machine, manual (Fig. 2.30)
- Door handle clip removal tool (Fig. 2.31)
- Door trim pad clip removal tool (Fig. 2.32)
- Face caps
- Face cap pliers
- Flexible light
- General Motors lock decoder (Fig. 2.33)
- Lock plate compressor
- Steering column lock plate compressor
- Steering wheel pullers
- VATS/PASSKey decoder or key analyzer (Fig. 2.34)

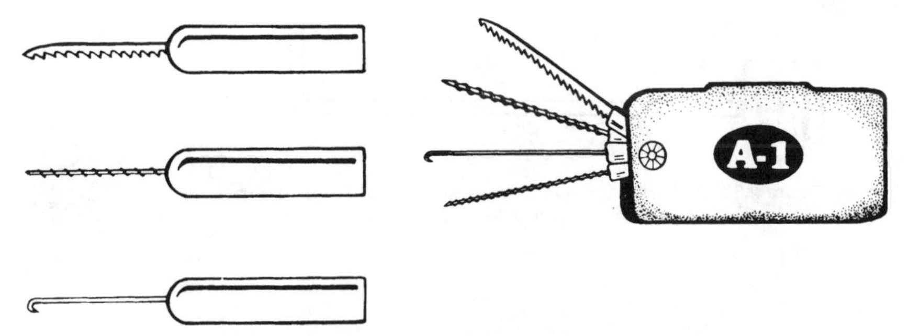

Figure 2.29 Broken-key extractors are useful for removing broken key parts from locks. (*A-1 Security Manufacturing Corp.*)

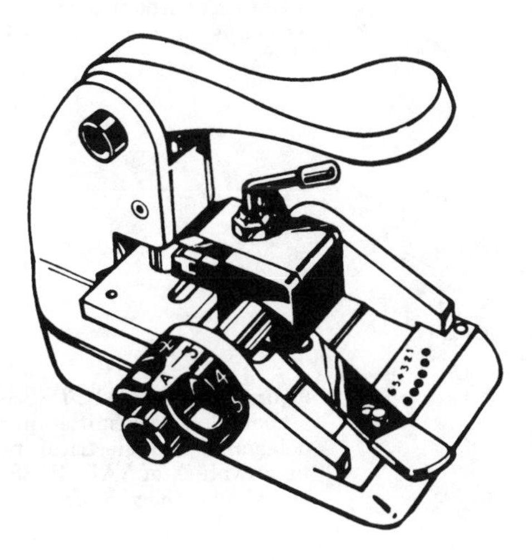

Figure 2.30 A mechanical code key cutter lets locksmiths make accurate keys without having a key to duplicate. (*Ilco*)

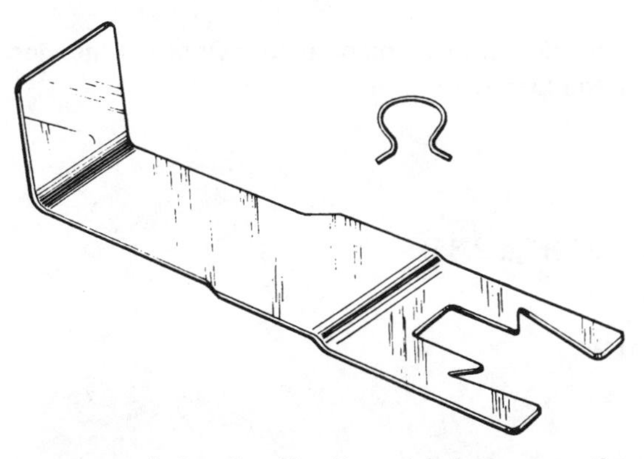

Figure 2.31 A door handle clip tool helps remove the retainer clip that secures an automobile's door handle to the door. (*A-1 Security Manufacturing Corp.*)

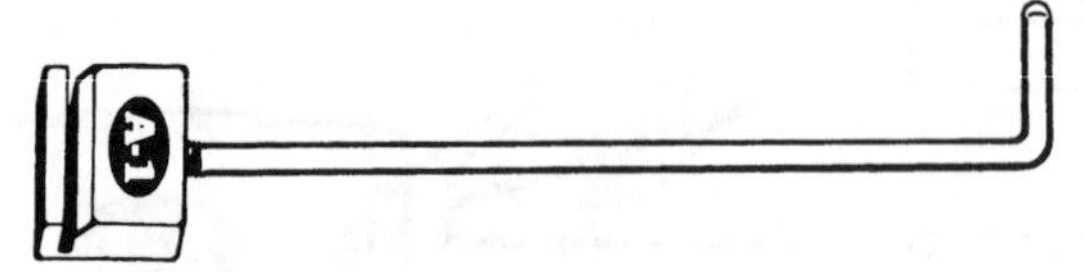

Figure 2.32 A door trim pad clip removal tool is used for automotive work. (*A-1 Security Manufacturing Corp.*)

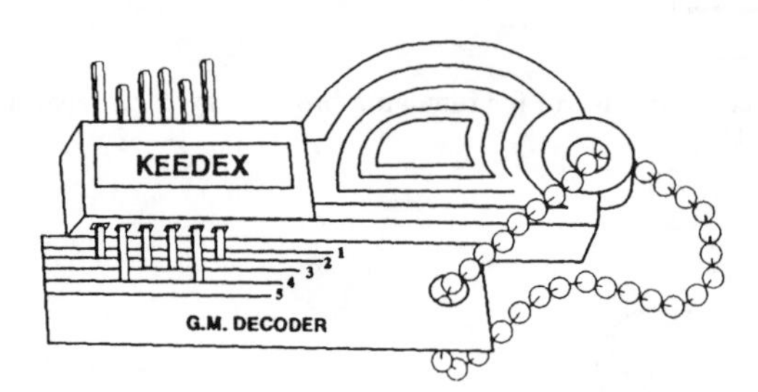

Figure 2.33 A General Motors Decoder decodes the tumblers in a GM lock without complete disassembly of the lock. (*Keedex Mfg.*)

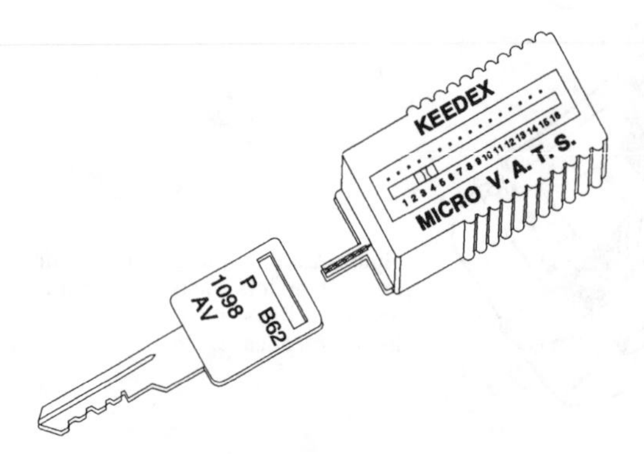

Figure 2.34 A VATS/PASSKey decoder lets locksmiths quickly determine the electrical resistance values of VATS/PASSKey key blanks. (*Keedex Mfg.*)

List 4: Tools for servicing safes, vaults, and safe deposit boxes (needed in addition to all of the tools in List 1 and List 2)

- Borescope
- Carbide drill bits
- Change keys, assorted (Fig. 2.35)
- Door puller
- Drill rig (Fig. 2.36)
- Hammer drill
- Nose puller
- Safe-moving equipment
- Sledgehammer

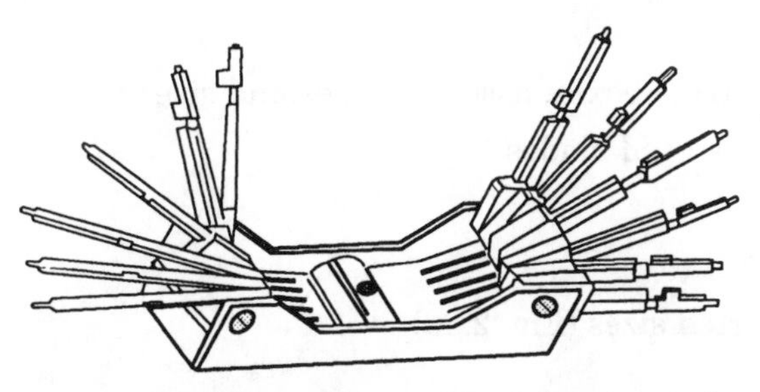

Figure 2.35 Safe change keys are needed to change the combinations of certain types of safes. (*Keedex Mfg.*)

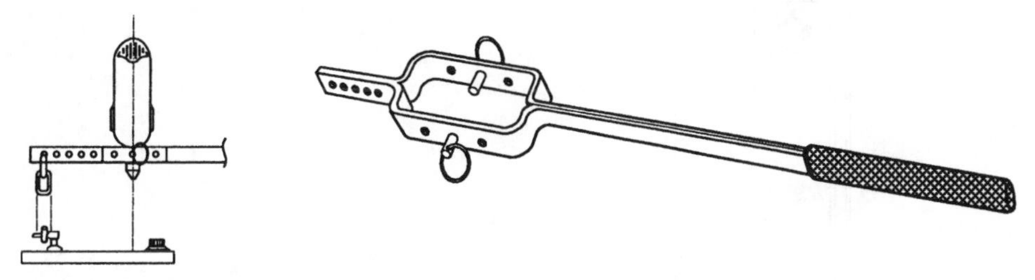

Figure 2.36 A drill rig holds a drill in place while a safe is being drilled. (*Keedex Mfg.*)

List 5: Tools/supplies for installing door locks and other door hardware (needed in addition to all of the tools in Lists 1 and 2)

- Boring jigs (Fig. 2.37)
- Broom and dustpan
- Compass (or keyhole) saw
- Door reinforcers, assorted sizes and finishes
- Drop cloth
- Drywall (or wallboard) saw
- Filler plates
- Hole saws, assorted sizes (Fig. 2.38)
- Kwikset cylinder removal tool
- Lever
- Nails, assorted types and sizes
- Pry bar
- Reciprocating saw
- Screw gun
- Screws, assorted sizes
- Screws, one-way

- Shovel
- Strike plates and strike boxes, assorted types and sizes
- Utility knife and assorted blades
- Vacuum cleaner
- Weiser shim pick
- Wood chisels, assorted sizes (Fig. 2.39)
- Wood glue

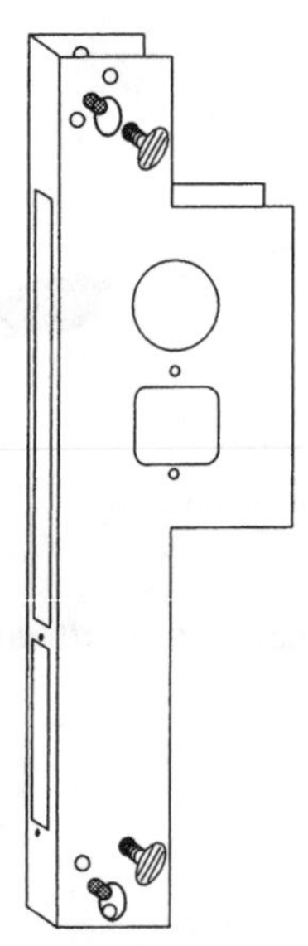

Figure 2.37 A boring jig is a template that acts as a guide for drilling precise installation holes and cutouts for locks and door hardware. (*Keedex Mfg.*)

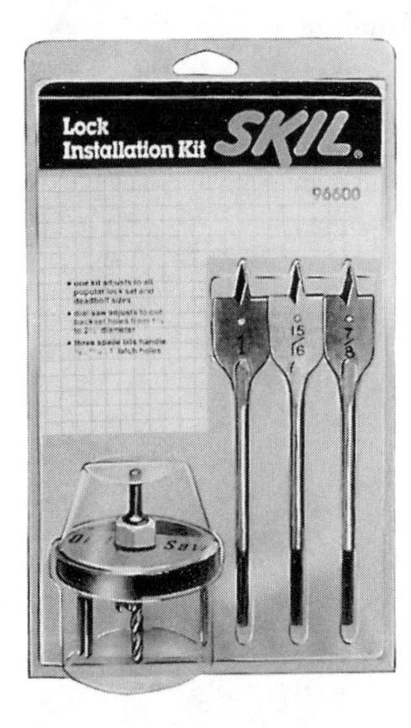

Figure 2.38 Hole saws and spade bits are used to install locks on doors. (*Skil Corporation*)

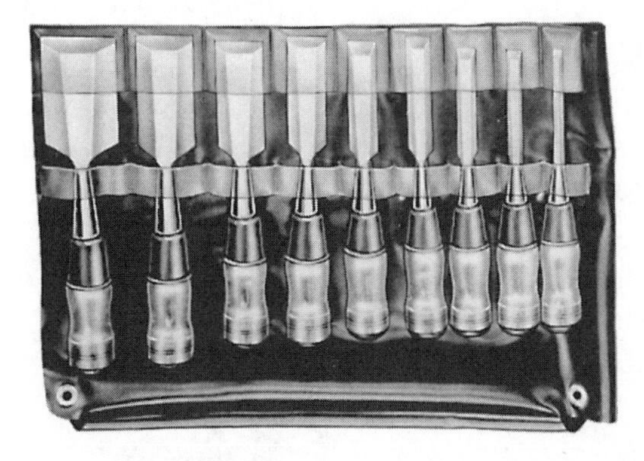

Figure 2.39 Wood chisels of various sizes are needed when installing locks and other hardware on wood doors. (*Stanley Works.*)

List 6: Tools for installing alarms and electronic security devices (needed in addition to all of the tools in Lists 1 and 2)

- Electrical tape
- Fish tape
- Flexible drill bits and extensions
- Multimeter
- Staple guns (or wiring tackers) and staples, assorted sizes for wire and coaxial cable
- Twist connectors
- Under-carpet tape
- Voltage tester
- Wire stripper
- Wire cutters

It isn't necessary to have all of the listed tools to practice the lessons in this book. As you study each chapter, you'll learn which tools you need and which you can do without. With all the tools listed in this chapter, you'll have enough equipment and supplies to start a locksmithing business.

Chapter

3

Types of Locks and Keys

Most people have a general understanding of what a lock is, but that understanding is often limited to the few models they've owned or seen. Few nonlocksmiths know much about the wide variety of locks that exist throughout the world or how locking technology has changed over the years. That's why many general consumer books and articles about locks (which are almost always written by nonlocksmiths) use definitions of the term *lock* that are too narrow, broad, or outdated. Although laypersons might not need to be concerned about the misleading definitions, it's important for a locksmith to know what a lock really is and isn't.

What Is a Lock?

Coming up with clear and precise definitions of *lock* isn't as easy as you might think. To see what I mean, write down your best definitions before reading further. After you read this introduction, look at them again.

Stephen Tchudi, a University of Nevada English professor, had a hard time figuring out what a lock is. In his book on locks, he suggested that a stone rolled in front of a cave may have been the first "lock," and that rolling the rock away was "picking" the lock. He also called guard crocodiles in moats "living locks," and wrote that drugging them is picking the locks. "A lock, after all, is simply a barrier or closure, a way of sealing up an entryway, of keeping what you want in, in, what you want out, out," Tchudi wrote. "A stick or doorstop that you wedge under a door...is also a lock, though, again, not a reliable or unpickable one." Although it's creative, such a broad all-encompassing definition has no practical value for the locksmith or anyone who works with, uses, or needs to buy a lock. Also, it shows that that person doesn't know what *lock picking* means because a rock, crocodile, and stick are unpickable—as is anything without a keyway.

Major dictionaries are more precise in their definitions. *The Random House Dictionary of the English Language,* 2d ed. unabridged, says a lock is "A device for securing a door gate, lid, drawer, or the like in position when closed, consisting of a bolt or system of bolts, propelled and withdrawn by a mechanism operated by a key, dial, etc." *Webster's Third New International Dictionary,* unabridged, says a lock is "A fastening (as for a door, box, trunk lid, drawer) in which a bolt is secured by any of various mechanisms and can be released by inserting and turning a key or by operating a special device (as a combination, time lock, automatic release button, magnetic solenoid)."

Those dictionary definitions are much better than the "locks are all things that hold, hide, fasten, or bite" kinds of definitions. High-quality dictionaries don't just make up definitions; they try to keep up with standard usage, based on a wide range of sources. As the editors explained in the preface of the *Merriam Webster's Collegiate Dictionary,* 10th ed.: "The ever-expanding vocabulary of our language exerts inexorable pressure on the contents of any dictionary. Words and senses are born at a far greater rate than that at which they die out." The locksmithing field progresses faster than the dictionaries update the related terms, which is why dictionary definitions tend to be a bit dated and exclude some current types of locks and locking technology.

A more precise definition of *lock* is given by the International Association of Home Safety and Security Professionals: "A device that incorporates a bolt, cam, shackle or switch to secure an object—such as a door, drawer or machine—to a closed, opened, locked, off, or on position, and that provides a restricted means of releasing the object from that position." One important difference of that from many of the other definitions is the matter of "restricted means." If anyone can just turn the doorknob and walk in, there's no restriction. That's why a set of doorknobs isn't a lock but a key-in-knob is.

If you tried to make up your own definition of *lock,* ask yourself three things about it: (1) Does your definition include padlocks? (2) Does it include car ignition locks? (3) Does it exclude a chair wedged under a doorknob? If you can honestly answer Yes to all three, you could be an English professor.

Lock Names

Most locks have several names, which are usually based on a lock's common uses, appearances, major security features, installation method, internal construction, technology, or manufacturer. Many of the names have overlapping meanings. Some names based on common usage, like *trailer lock* and *bicycle lock,* are specific enough to be used by locksmiths as well as laypersons (Figs. 3.1 and 3.2) because there are few significant variations among such locks. That is to say, one trailer lock isn't much different from another trailer lock.

Other names based on common usage, however, like *house lock* and *car lock,* refer to too many different types of locks. A locksmith would be confused if, say, a homeowner simply asked for a new "house lock." The homeowner would need to be more specific. He or she would need to ask for a key-in-knob or lever lock (names based on the style of handle a lock uses) (Figs. 3.3 and 3.4). Other common "house locks" might include an *interlocking deadlock* or *deadbolt lock,* whose names are based on important security features (Figs. 3.5 and 3.6).

While talking with one another about servicing locks, locksmiths often use names that describe a lock's installation method or internal construction, such as rim, mortise, and bored. A *rim lock* is any lock designed to be mounted on the surface (or rim) of a door or object. The interlocking deadbolt shown in Fig. 3.6 is one type of rim lock. A *mortise lock* is installed in a hollowed out (or mortised) cavity (Fig. 3.7). A *bored* (or *bored-in*) *lock* is installed by cross-boring two holes—one for the cylinder and one for the bolt mechanism. Warded, lever tumbler, disc tumbler, and pin tumbler are names that describe a lock's internal construction.

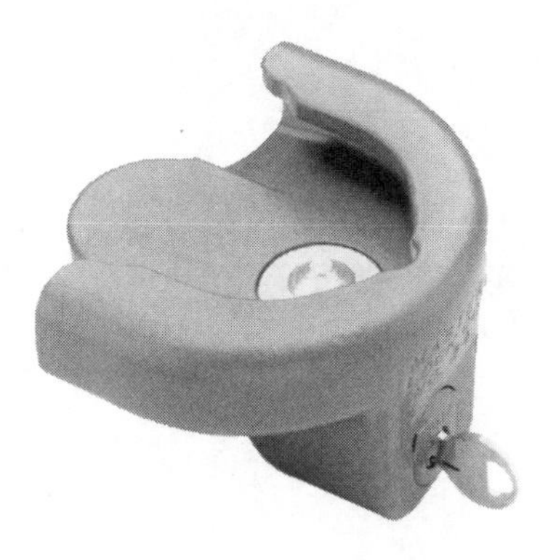

Figure 3.1 A trailer lock guards against tow-away theft of unattended boat, camping, snowmobile, horse, mobile home, and utility trailers. (*Master Lock Company*)

Figure 3.2 The U-shape design is common for bicycle locks.

Figure 3.3 A key-in-knob lock is one of the most common locks used on homes. (*Master Lock Company*)

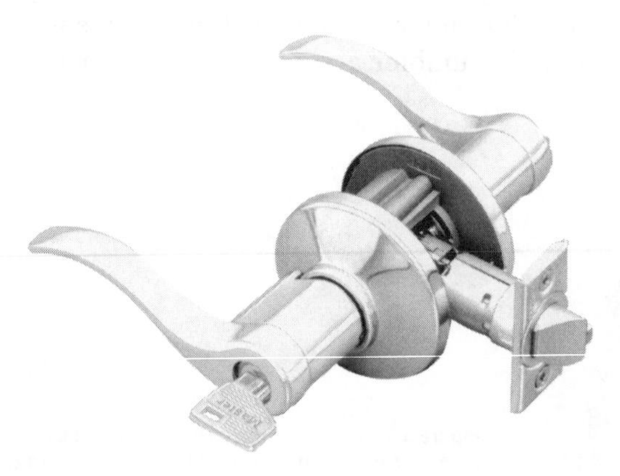

Figure 3.4 A lever lock is basically a key-in-knob lock with levers rather than knobs. (*Master Lock Company*)

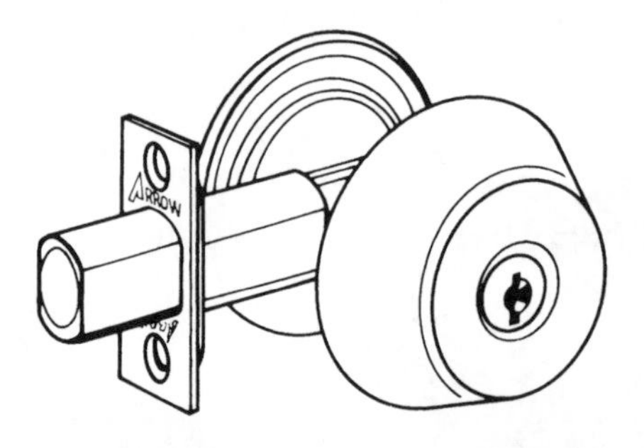

Figure 3.5 A deadbolt lock is one of the most secure types of locks commonly found on homes. (*Arrow Lock Company*)

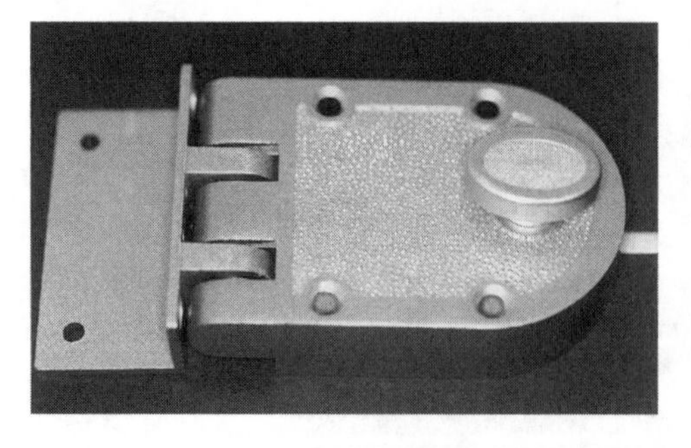

Figure 3.6 An interlocking deadbolt, or "jimmyproof deadbolt," is often used on front doors of homes.

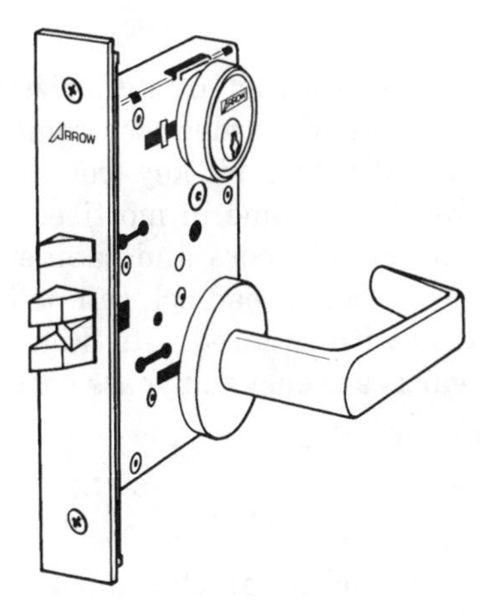

Figure 3.7 Most of a mortise lock is installed in a cavity or cutout. (*Arrow Lock Company*)

Figure 3.8 Warded bit key locks come in rim and mortised styles. (*Taylor Lock Company*)

Figure 3.9 Warded padlocks are for low-security applications. (*Master Lock Company*)

Warded locks

A *ward* is a fixed projection designed to obstruct unauthorized keys from entering or operating the lock. One old type of warded lock comes in a metal case, has a large keyhole, and is operated with a bit key (commonly called a *skeleton key*). Such locks, called *bit-key locks,* come in mortised and surface-mounted styles, and are often used on closet doors and cabinets (Fig. 3.8). Some low-cost padlocks are also warded. Such a padlock can be identified by its wide, sawtoothlike keyway and keys with squared cuts (Fig. 3.9). Warded locks provide little security because wards are easy to bypass with a stiff piece of wire or thin strip of metal.

Tumbler locks

Tumblers are small objects, usually made of metal, that move within a lock cylinder in ways that obstruct a lock's operation until an authorized key or combination moves them into alignment. There are several types of tumblers; they come in a variety of shapes and sizes and move in different ways. Because tumblers generally provide more security than wards, most locks today use some type of tumbler arrangement either instead of or in addition to wards.

A typical key-operated cylinder consists of a cylinder case (or housing), a plug (the part with a keyway), springs, and tumblers. The springs are positioned in a way that makes them apply pressure to the tumblers. The tumblers are positioned so that when no key is inserted, or when the wrong key is inserted, the spring pressure forces one or more of the tumblers into a position that blocks the plug from being rotated. When the proper key is inserted into the keyway, however, the key moves the tumblers to a position that frees the plug to turn.

A lock can have more than one cylinder. A key-operated, single-cylinder lock has a cylinder on one side of the door only (usually the exterior side) so that no key is needed to operate it from the other side. Typically, it can be operated from the noncylinder side by pushing a button or by turning a knob, handle, or turn piece. Key-operated, double-cylinder locks require a key on both sides of a door. Many local building and fire codes restrict the use of double-cylinder locks on doors leading to the outside because the locks can make it hard for people to exit quickly during a fire or other emergency.

Types of tumbler locks. There are three basic types of tumblers: lever, disc, and pin. Most *lever tumbler locks,* such as those used on luggage, brief cases, private mailboxes, and lockers, offer a low level of security. The lever tumbler locks commonly found on bank safe-deposit boxes are specially designed to offer a high level of security. *Disc tumbler locks* offer a medium level of security. They're often used on desks, file cabinets, and automobile doors and glove compartments. *Pin tumbler locks* can provide medium to high security, but in general they offer more security than do other types of tumbler locks. Many prison locks and virtually all house locks and high-security padlocks use pin tumbler cylinders. Some automobile door and ignition locks also have pin tumblers.

A special type of pin tumbler lock, called a *tubular key lock* (or tubular lock), has its tumblers arranged in a circular keyway (Fig. 3.10). It uses a tubular key to push the tumblers into proper alignment. Because of its odd appearance, a tubular key lock is harder for most people to pick open than are standard pin tumbler locks. Sometimes erroneously called "Ace Locks" (which is Chicago Lock Company's trade name for some of its tubular key locks), tubular key locks are often found on vending machines, laundromat equipment, bicycle locks, and high-security padlocks.

Another type of pin tumbler system is found in *interchangeable core* (IC) locks. Although they come in the form of deadbolts, key-in-knobs, rim locks, mortise locks, padlocks, and desk and cabinet locks, all locks in an IC system can either use the same key or be masterkeyed. Some examples of IC locks are shown in Fig. 3.11. The common feature of IC locks is a figure 8-shaped core that houses the tumblers and springs. The cores can be easily removed and replaced (Fig. 3.12). Any IC lock can be rekeyed simply by inserting new cores in the locks.

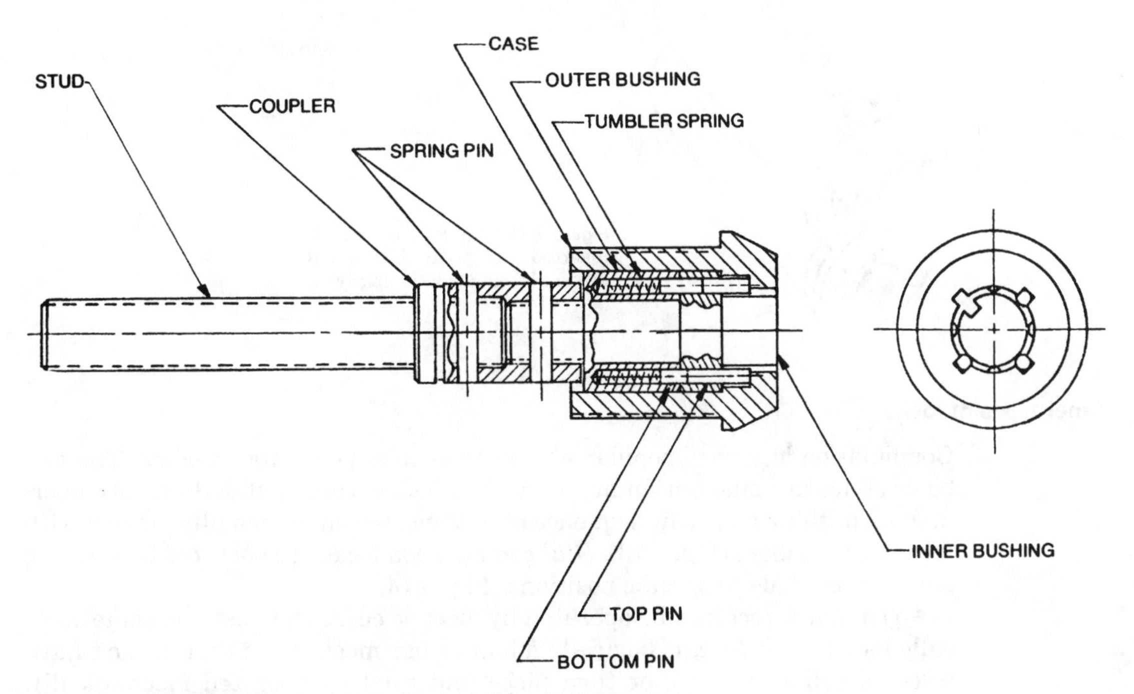

Figure 3.10 A cross-section of a tubular key lock. (*Desert Publications*)

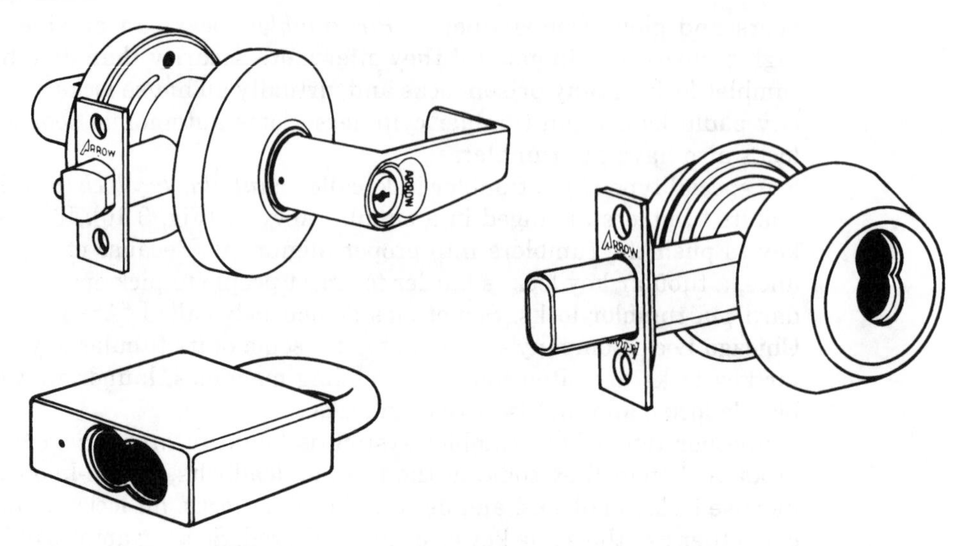

Figure 3.11 Types of IC locks. (*Arrow Lock Company*)

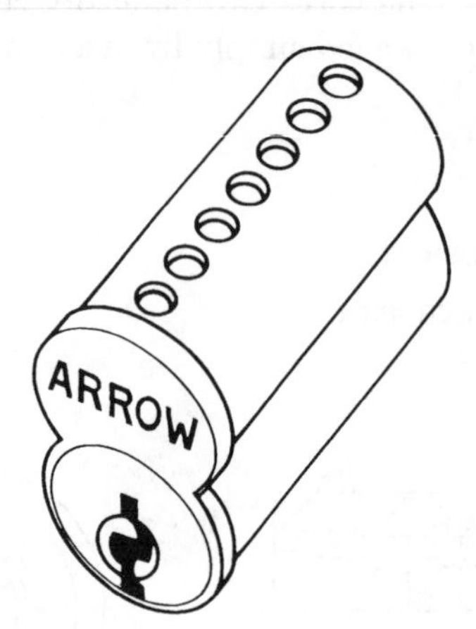

Figure 3.12 An IC core can be inserted into many different IC locks. (*Arrow Lock Company*)

Other types of locks

Combination locks are popular alternatives to key-operated models. The two basic styles are pushbutton and dial. *Pushbutton* combination locks are operated by pushing a specific sequence of buttons, which are usually labeled with letters or numbers (Fig. 3.13). *Dial* combination locks are operated by rotating one or more dials to specific positions (Fig. 3.14).

An *electrical lock* can be operated by electric current. One type, sometimes called an *electric lock,* is basically a bolt or bar mechanism that doesn't have a keyed cylinder, knob, or turn piece and can't be operated mechanically.

Another type, called an *electrified lock,* is a modified mechanical lock that can be operated either mechanically or with electricity. *Electric switch locks* complete and break an electric current when an authorized key is inserted and turned. An automobile ignition lock is an example of such a lock; after the key is turned, electricity flows from the battery to the car's starter. Similar locks are also used on alarm system control boxes to arm and disarm the system.

Time locks are designed to be opened only at certain times on certain days. They're commonly installed inside bank vaults and safes. *Biometric locks* unlock only after a computer has verified a physical feature, such as a fingerprint, signature, voiceprint, hand geometry, or the pattern of the retina of the eye.

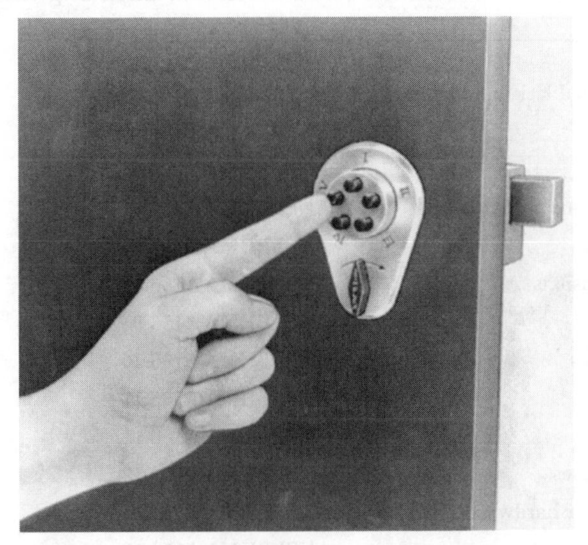

Figure 3.13 Some combination locks use pushbuttons. (*Simplex Access Controls Corp.*)

Figure 3.14 Many padlocks have dial combinations. (*Master Lock Company*)

Lock Grades

The American National Standards Institute, Inc. (ANSI) determines manufacturing standards for a wide variety of building hardware and other products used in the United States. Many manufacturers make sure their products meet or exceed ANSI standards because architects, home builders, locksmiths, and other professionals specify products based on the standards.

As a rule, ANSI doesn't create standards. Interested industry associations usually create and propose standards and ANSI reviews them for possible adoption. The current standards for door hardware, listed under ANSI section 156, were proposed by the Builders Hardware Manufacturers Association, Inc. (BHMA).

ANSI 156 is of special importance to locksmiths. That section includes standards for many products locksmiths sell and install, including the following:

Butts and hinges	ANSI/BHMA A156.1
Bored and preassembled locks and latches	ANSI/BHMA A156.2
Exit devices	ANSI/BHMA A156.3
Door closers	ANSI/BHMA A156.4
Auxiliary locks and associated products	ANSI/BHMA A156.5
Architectural door trim	ANSI/BHMA A156.6
Template hinge dimensions	ANSI/BHMA A156.8
Cabinet hardware	ANSI/BHMA A156.9
Power pedestrian doors	ANSI/BHMA A156.10
Cabinet locks	ANSI/BHMA A156.11
Interconnected locks	ANSI/BHMA A156.12
Mortise locks and latches	ANSI/BHMA A156.13
Sliding and folding door hardware	ANSI/BHMA A156.14
Closer holder release devices	ANSI/BHMA A156.15
Auxiliary hardware	ANSI/BHMA A156.16
Hinges and pivots	ANSI/BHMA A156.19
Strap and tee hinges and hasps	ANSI/BHMA A156.20
Thresholds	ANSI/BHMA A156.21
Electromagnetic locks	ANSI/BHMA A156.23
Delayed egress locks	ANSI/BHMA A156.24

ANSI 156.2 includes provisions for locks to be Grade Certified. For a model to receive a grade, which ranges from 1 to 3, a sample lock must pass many rigorous tests that examine the lock's performance, strength, and durability. Grade 1 locks are the strongest and are often specified for industrial applications. Grade 2 locks are for light commercial and residential uses. Grade 3 locks are the lightest and are primarily for residential applications. Some of the

things the grading tests measure include bolt strength, turning torque needed to retract the latch, how well the finish holds up against salt sprays and other corrosives, and how many times the lock can be operated before failure.

Operational test

The operational test examines the amount of torque needed to retract the latchbolt with and without a key. It's performed first by depressing the deadlatch plunger (if necessary), then slowly applying torque force to the outside knob of an unlocked lockset until the latch is fully retracted. For a key-in-knob to receive a grade, torque used may not exceed 9 lbf-in. For a lever lock to be graded, torque cannot exceed 28 lbf-in. (Key-in-knobs are then tested in the opposite direction.) The test is repeated for the inside knob or lever.

The lock is then put in the locked position and the deadlatch plunger is again depressed (if applicable). Then the key is inserted into the keyway and is slowly rotated until the latch is fully retracted. Torque may not exceed 9 lbf-in. The test is repeated in the opposite direction if the lockset is designed to allow such movement. If the inside knob or lever is key-operated, the test is also applied to the inside knob or lever.

Strength test

The strength test examines how much forcible turning force a lock in the locked position can withstand. To be graded, a lock must stay locked after the minimum force has been applied. For a Grade 1, a key-in-knob must hold up to 300 lbf-in.; a lever lock must withstand 450 lbf-in. For a Grade 2, a key-in-knob must hold up to 150 lbf-in.; a lever lock must hold up to 225 lbf-in. A Grade 3 requires a key-in-knob to withstand at least 120 lbf-in. and a lever lock to withstand at least 180 lbf-in.

Cycle test

The cycle test examines how many times the lock can be operated before failure. For a Grade 1, a lock must complete 800,000 cycles. For a Grade 2, a lock must complete 400,000 cycles. Grade 3 locks complete at least 200,000 cycles.

Key Types

There are six basic types of keys: bit, barrel, flat, corrugated, tubular lock, and cylinder. Although different types of keys have different parts, virtually all keys have a bow (rhymes with toe). The bow (or handle) is the part of the key that a person holds while inserting the key into a lock. Bows come in a variety of shapes and often have identifying information or advertising imprinted on them. Figure 3.15 shows some common types of keys.

A *bit* (or skeleton) key is usually made of iron, brass, steel, or aluminum. Major parts of the key include the bow, shank, shoulder, post, and bit. Many

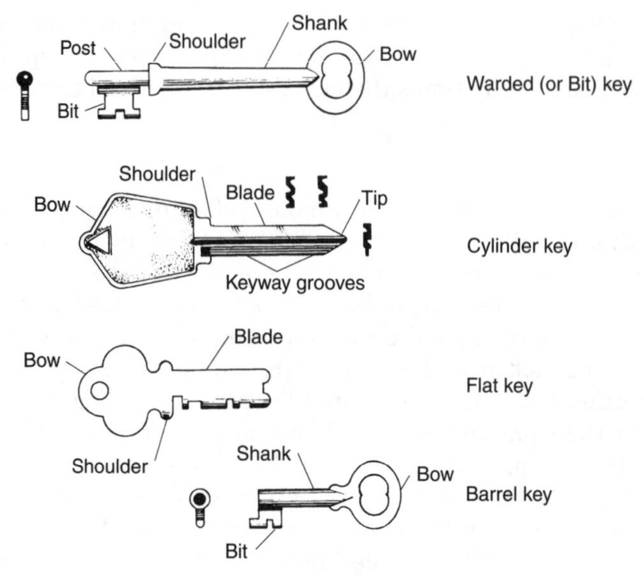

Figure 3.15 Different types of keys have different parts.

barrel keys look similar to bit keys, but barrel keys have a hollow shank and often don't have a shoulder.

A *flat* key is flat on both sides. Most are made of steel or nickel silver. Such keys are used for operating lever tumbler locks. *Corrugated* keys look similar to flat keys. Both types usually have the same parts. However, corrugated keys have corrugations (ripples) along the length of their blades. Corrugated keys are most often used to operate warded padlocks.

A *tubular* key has a short tubular blade with cuts (depressions) milled in a circle around the end of the blade. The key is used to operate tubular key locks. Parts of a tubular key include the bow, blade, tumbler cuts, and nib. The *nib* is a small protrusion at the tip. It shows the position the key must be at to enter and operate the lock.

The most common type of key is the *cylinder* key. It's used for operating most pin tumbler and disc tumbler locks. Major parts of a cylinder key include the bow, shoulder (one or two), blade, keyway (or milling) grooves, and tip. The shoulders of a key are generally used as stops when cutting keys. When a cylinder key doesn't have a shoulder, its tip is used as the stop. The keyway grooves are millings along the length of the blade that allow the key to fit into a lock's keyway.

Key Blank Identification

It's as important for locksmiths to be able to identify keys and key blanks as it is for them to identify locks. A *key blank* is basically an uncut key. Before duplicating a key, a locksmith must find a matching blank.

Important factors to consider when choosing a blank to duplicate a bit key or barrel key are thickness of the bit and diameter of the shank and post. Because there's usually a lot of tolerance in locks that use these types of keys, the blank might not have to match perfectly.

Important factors when choosing a blank for flat keys and corrugated keys are thickness, length, width, and shape of blade. The key and blank should match closely in those areas. A blank for a corrugated key should have the same corrugated configuration that the key has.

Choosing a blank for a tubular key is simple because there are few significant differences among tubular keys. The important areas of such keys are the size of the nib and the inside and outside diameters of the shank. If you find a blank that closely matches the key in those respects, you can duplicate the key.

Cylinder key blank identification

The trickiest type of key to find a blank for is the cylinder key, the type that locksmiths are most often asked to duplicate. Three parts that are generally common to cylinder keys and their corresponding key blanks can be helpful when searching for a key blank. These are the bow, blade length, and keyway grooves.

The *bow* is the head of the key or blank. It's the part that you grip when using a key to operate a lock. Most cylinder lock manufacturers use distinctive key bows and many aftermarket key blanks copy their basic shapes. Based on the bow alone, a locksmith can often either quickly find a matching key blank or at least narrow the choices to a few key blanks.

The *blade* is the part of the key that enters a lock's keyway. Generally, the key and key blank should have blades of the same length. To compare lengths, you can either hold the key and blank together or use the illustrations in a key blank catalog. When using a catalog, lay the key over the illustration and align the shoulders. You can then observe the length at a glance because the key catalog drawings are the size of the original keys.

Keyway grooves (or millings) are critical. Only when the key and key blank have the same, or very similar, grooves can both fit into the same keyway. One way to compare millings is to try to insert the key blank into the lock's keyway. If it can be inserted, the millings are the same or similar. You can also compare millings by holding the key and key blank side by side and looking at the blade tips. Finally, use the catalog to compare blade millings. To use the catalog, stand the key directly over the cross section that appears under the blank's illustration. The grooves must match exactly to be considered the same.

Key manufacturers publish catalogs that identify their key blanks and sizes and show a cross section of each different blank. In addition, they may have a cross-reference section that refers to other manufacturers' keys that are comparable to each of the keys in that particular catalog. These cross-reference sections are valuable because they obviate reference to a wide variety of catalogs in order to determine the manufacturer of a given key blank. Merely refer to the key type and cross-reference it to the one that you have.

Some manufacturers have literally hundreds of key blanks and the cross-reference section becomes unwieldy at times. It is best to use a catalog that carries the most popular and commonly used blanks. About 95 percent of the keys you will duplicate will be in such catalogs. Other catalogs can be used for general reference but are not absolutely necessary to have for your shop.

Key Blank Examples

On the following pages are illustrations of various types of cylinder key blanks. Figure 3.16 shows key blanks from the Taylor Lock Company. These blanks will fit a variety of American locks. Notice that in Fig. 3.17 the blank

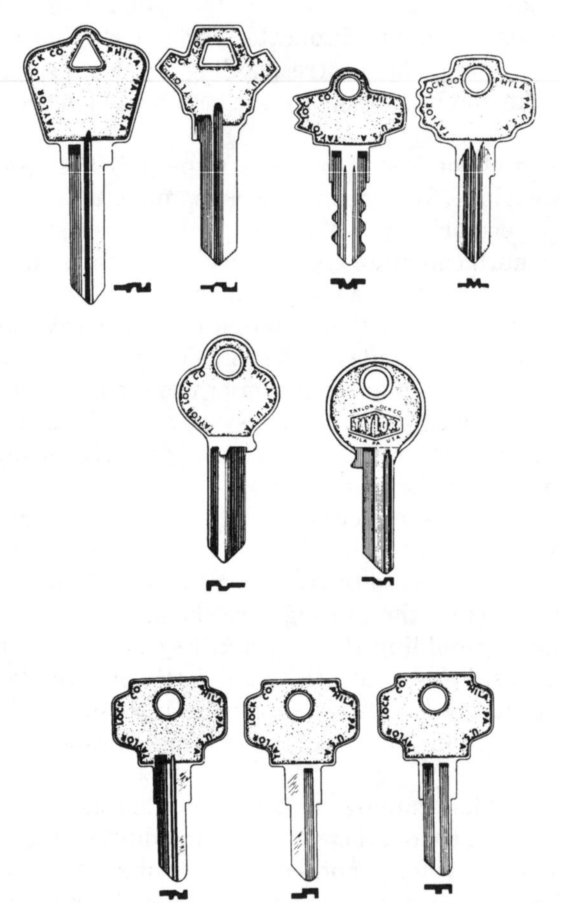

Figure 3.16 Representative sampling of standard key blanks. (*Taylor Lock Company*)

is further laid out to incorporate a master key system and that, from the cross section, you can identify the various keyways that will fall into this particular master key system.

Figure 3.18 shows another keyway system with the applicable blanks. Notice that within this key series there are three different key lengths, which allow for wider use of the master key system. The length also allows more pins

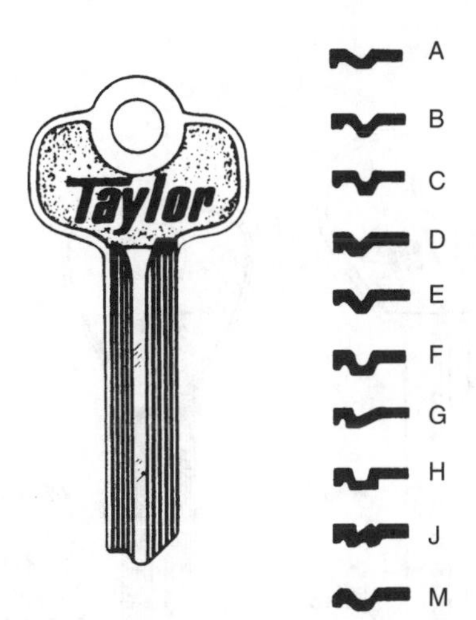

Figure 3.17 Six-pin cylinder key and various associated keyways. Keyway M is used only for the master key. (*Taylor Lock Company*)

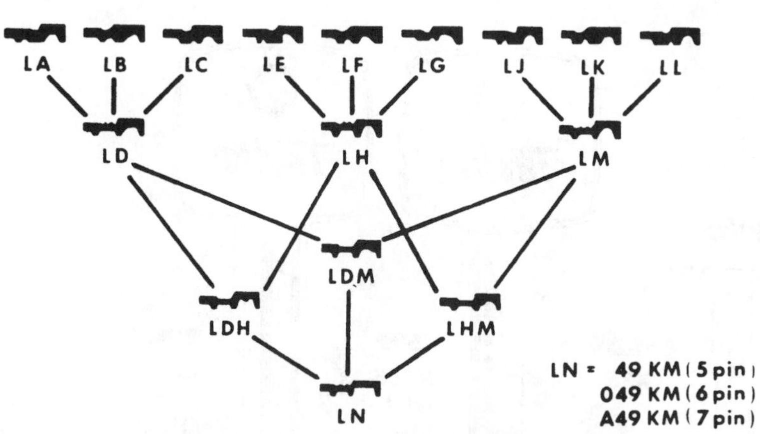

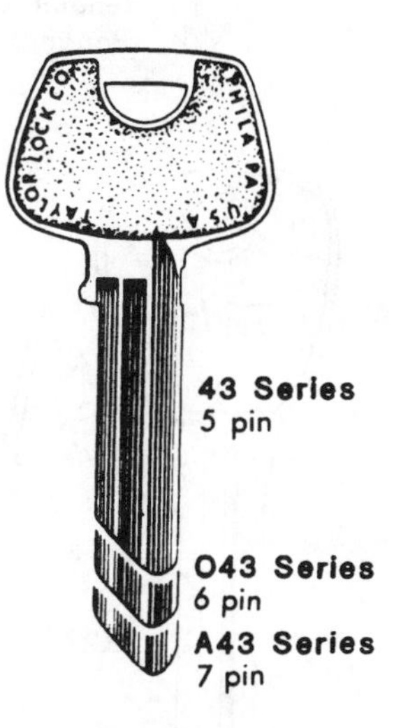

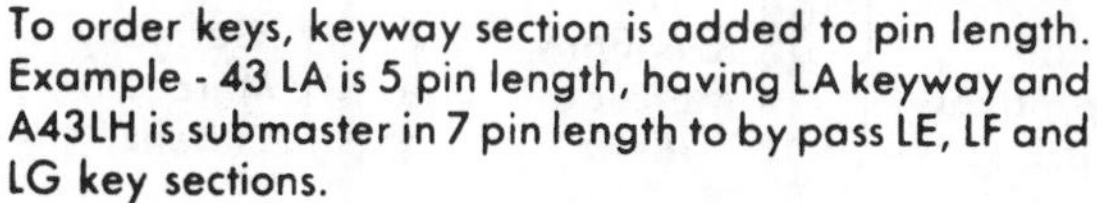
To order keys, keyway section is added to pin length.
Example - 43 LA is 5 pin length, having LA keyway and A43LH is submaster in 7 pin length to by pass LE, LF and LG key sections.

Figure 3.18 A large master key system provides a great variety of key uses and control features through submasters and individual use of keys. (*Taylor Lock Company*)

to be used in the cylinders. This means a greater variety in the number of possible key combinations available to the system.

Foreign automobiles (and, naturally, their keys) are increasing in popularity, so it is necessary to have a well-stocked variety of foreign lock keys in the locksmith shop. Figure 3.19 provides illustrative details on a variety of the more common key blanks in use. You should have these types and a slightly wider supply of other foreign auto key blanks available to service the needs of your customers.

Figure 3.19 Key blanks for foreign automobiles. (*Taylor Lock Company*)

Automotive Key Blanks

Duplicating automotive keys is more prevalent in recreational areas and very large cities. Lost or stolen keys mean duplication and/or rekeying on a daily basis for many locksmiths. It is imperative that, as a locksmith, you have in your shop an ample selection of automotive key blanks on hand. Keep a selection of the most common and popular automotive blanks in your van plus a few others that might be required. (*Note:* Every locksmith shop should have a standard selection of automotive key blanks, but in large cities and in recreational areas such as beaches, parks, etc., the need is greater than usual. Thus, the number and variety of blanks needed will be greater in these areas than in others. Always consult with your distributor or manufacturer's representative to obtain an accurate view of the types and quantities of blanks that you're likely to need; don't buy a larger selection than required just because you think you may someday need it.)

Figure 3.20 is a list of foreign automotive blanks identified by the number and the vehicle they accompany. These are, by far, the most common vehicles for which you will be cutting keys. (This is a representational selection of popular keys that require frequent duplication; it is not all-inclusive.) Refer to your key catalog for illustrations of all the automotive keys and develop a working memory of them. After a period of time you should be able to look at a key that is already cut (or even a blank) and determine with relative accuracy its make of automobile and where the key blanks are located on your key blank board. Figure 3.21 shows some domestic key blanks. Figure 3.22 lists the most common domestic keys by number of automotive type and Figure 3.23 shows a variety of foreign key blanks.

AD1	Audi 100 and 100LS—1971 on
CP1	Capri Ign. 1971 on—Ign./Trunk 1975 on
CP2	Capri Door/Trunk 1971-74
CP3	Capri—Supplements CP2
CP4	Capri—Fiesta
5DA1	Datsun
5DA2	Datsun 1970-72, Mazda 1971 on, Eng. 1969 on
DA3	Datsun—1970 on
DA4	Datsun & Subaru—1970 on
LDC1	Long Head Dodge Colt, Arrow, Chall., Sapporo, Jap. Opel—1970 on
5FT1	Fiat Ignition—1967 on
FT2	Fiat Trunk 1967 on—Pañtera-Ferrari
FT3	Fiat—Supplements FT2
HN1	Honda Ignition 1970 on, MG Ignition 1972
HN2-3	Honda MASTER (Codes #1001-1700, 2001-2700) —1976
HN4	Honda Trunk 1973-76
HN5	Honda Cycles—1977 on
LU1	GM LUV Ign./Door & Trunk/Glove—1973 on
MZ1	Mazda Door/Trunk—1971 on
MZ2	Mazda—Supplements MZ1
MZ3	Mazda, Courier Trucks 1971-73
MZ4	Mazda/Courier R100/1200-Door/Trunk 1970-72
MZ5	Mazda—Supplements MZ4
RP1	Renault—Peugeot
5RP2	Renault-Peugeot Ignition 1971 on
TO1	Toyota
TO2	Toyota Ignition—1969 on
UN2	Union (Hillman—Vauxhall—English Ford)
UN3	English 1960 on, Volvo Door/Trunk 1970
UN4	Union—MG—Triumph
6VL1	Volvo 1960 to 1965-6 Pin
VW1	Volkswagen
VW2	Volkswagen 1965-70
VW3	Volkswagen Beetle—1971 on
VW4	Volkswagen VW411-412-DA-RA-SC, Fox, Audi 4000/ 5000, Porsche, Jetta
VW5	Volkswagen Bus—1971 on

Figure 3.20 Foreign automobile key blanks. (*Star Key and Lock Co.*)

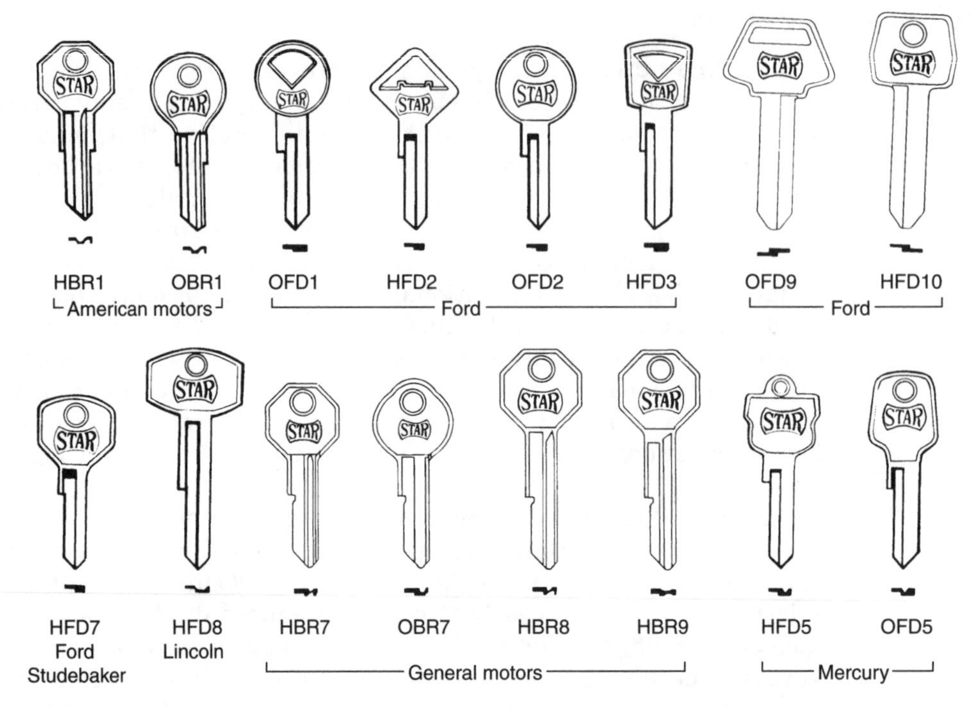

Figure 3.21 Sampling of common domestic automobile key blanks. (*Star Key and Lock Co.*)

Look-Alike Keys

You will find key blanks that are *look-alike* keys—blanks made by a manufacturer other than the original with bows that look like the bows of the original key blanks. These provide you with a great advantage; as look-alikes, they are quickly identified with a specific lock manufacturer, enabling you to quickly select the proper key for duplication.

Because not all keys are look-alikes you will need comparison or cross-reference key charts—important reference tools. Figure 3.24 is a manufacturers' key comparison chart for automotive and house keys.

Suppose you have a customer who brings in two keys for duplication—one for her house and one for her car. The house key is National/Curtis number IN1 and the car key is 1127DP. As an up-and-coming locksmith, you carry a wide assortment of the most commonly used keys and you have a cross-reference key chart listing. You can quickly determine whether or not you have the particular blanks for these two keys.

Go to the cross-reference list and find that the IN1 goes to the Star blank 5IL1 (Ilco house lock). The 1127DP is an Ilco key for which you have the Star blank, HFD4—a Ford auto ignition key. Rapidly looking at the key blanks, you quickly select the two proper keys. Within minutes you have another satisfied customer, thanks to the cross-reference listing.

Blank	Application
HBR1	American Motors Ignition/Door to 1959
OBR1	American Motors Trunk/Glove to 1959, GAS locks
OBR1DB	Briggs & Stratton—GAS & utility locks
HBR2	GM Ignition/Door—all years to 1966
OBR2	GM Trunk/Glove—all years to 1966
HBR3	American Motors Ignition/Door—1960 on
OBR3	American Motors Trunk/Glove—1960 on
OBR4	GM Long Head Ignition/Door—all years to 1966
HBR5	GM Ignition/Door "A"—1967
HBR5M	GM Ignition/Door Master "A/C"—1967/68
OBR5	GM Trunk/Glove "B"—1967
HBR6	GM Ign./Door & Trunk/Glove Master—hexagon head—1967
OBR6	GM Ign./Door & Trunk/Glove Master—oval head—1967
HBR7	GM Ignition/Door "C"—1968
OBR7	GM Trunk/Glove "D"—1968
HBR8	GM Long Head Ign./Door Master "A/C"—1967/68/71/72 (Ign. only 1975/76/79/80)
HBR9	GM Long Head Ign./Door & Trunk/Glove Master "E/H"—1969/73/77/81
OBR9	GM Oval Head Ign./Door & Trunk/Glove Master "E/H"—1969/73/77/81
HBR9E	GM Ign./Door "E"—1969/73 (Ign. only—1977/81)
OBR9H	GM Trunk/Glove "H"—1969/73 (Door also—1977/81)
HBR10J	GM Ign./Door "J"—1970 (Ign. only—1974/78/82)
OBR10K	GM Trunk/Glove "K"—1970 (Door also—1974/78/82)
HBR11	American Motors Ign./Door—1970 square head—1960 on
OBR11	American Motors Trunk/Glove—1970 ellipse head—1960 on
HBR12A	GM Ign./Door "A"—1967/71 (Ign. only—1975/79)
OBR12B	GM Trunk/Glove "B"—1967/71 (Door also—1975/79)
OBR13	GM Ign./Door & Trunk/Glove Master "A/B"—1967/71/75/79
OBR13S	GM Ign./Trunk—fits most years—1967 on—FOR LOCKSMITHS ONLY—NOT FOR DUPLICATION
HBR14C	GM Ign./Door "C"—1968/72 (Ign. only—1976/80)
OBR14D	GM Trunk/Glove "D"—1968/72 (Door also—1976/80)

Blank	Application
HFD1	Ford Ignition/Door—1952 to 1954
OFD1	Ford Trunk/Glove—1952 to 1954
HFD2	Ford Ignition/Door—to 1951
OFD2	Ford Trunk/Glove—to 1951
HFD3	Ford Ignition/Door—1955 to 1958
OFD3	Ford Trunk/Glove—1955 to 1958
HFD4	Ford Ignition/Door—1952 to 1964
OFD4	Ford Trunk/Glove—1952 to 1964
HFD5	Mercury Ignition/Door—1952 to 1964
OFD5	Mercury Trunk/Glove—1952 to 1964
HFD6	Ford, Linc., Merc. Ign./Trunk Square Head Master—1952 to 1964
OFD6	Ford, Linc., Merc. Ign./Trunk Round Head Master—1952 to 1964
HFD7	Ford Ign./Door to 1951—Studebaker 1947 on
HFD8	Lincoln Ignition/Door—1952 to 1964
HFD9	Ford Ignition/Door—1965/66 only (double side)
OFD9	Ford Trunk/Glove—1965/66 only (double side)
HFD10	Ford Ignition/Door—1965 on (double side)
OFD10	Ford Trunk/Glove—1965 on (double side)
HPL1	Chrysler, Plymouth, Dodge Ign./Door 1949-55
OPL1	Chrysler, Plymouth, Dodge Trunk/Glove 1949-58
HPL2	Chrysler, Plymouth, Dodge Ign./Door to 1948
OPL2	Chrysler, Plymouth, Dodge Trunk/Glove to 1948
HPL3	Chrysler, Plymouth, Dodge Ign./Door 1956-59
OPL4	Chrysler, Dodge Trunk/Glove 1959-65
OPL5	Plymouth Trunk/Glove 1959-65
HPL6	Chrysler, Plymouth, Dodge Ign./Door 1960-67
OPL7	(Master) Long GM to 1966 & Chrysler Trunk 1959-65
OPL8	Chrysler Trunk/Glove—1966/67
HPL68	Chrysler Ign./Door Master 1956 on—hexagon head
OPL68	Chrysler Trunk/Glove Master—1966 on
OPL70	Chrysler Ign./Door & Trunk/Glove Master—1966 on
HPL73	Chrysler Ign./Door Master 1956 on—diamond head
HYA4	Kaiser Frazer Ign./Door—1949/50
HYA5	Studebaker Ign./Door—1940 on

Figure 3.22 Domestic automobile key blanks. (*Star Key and Lock Co.*)

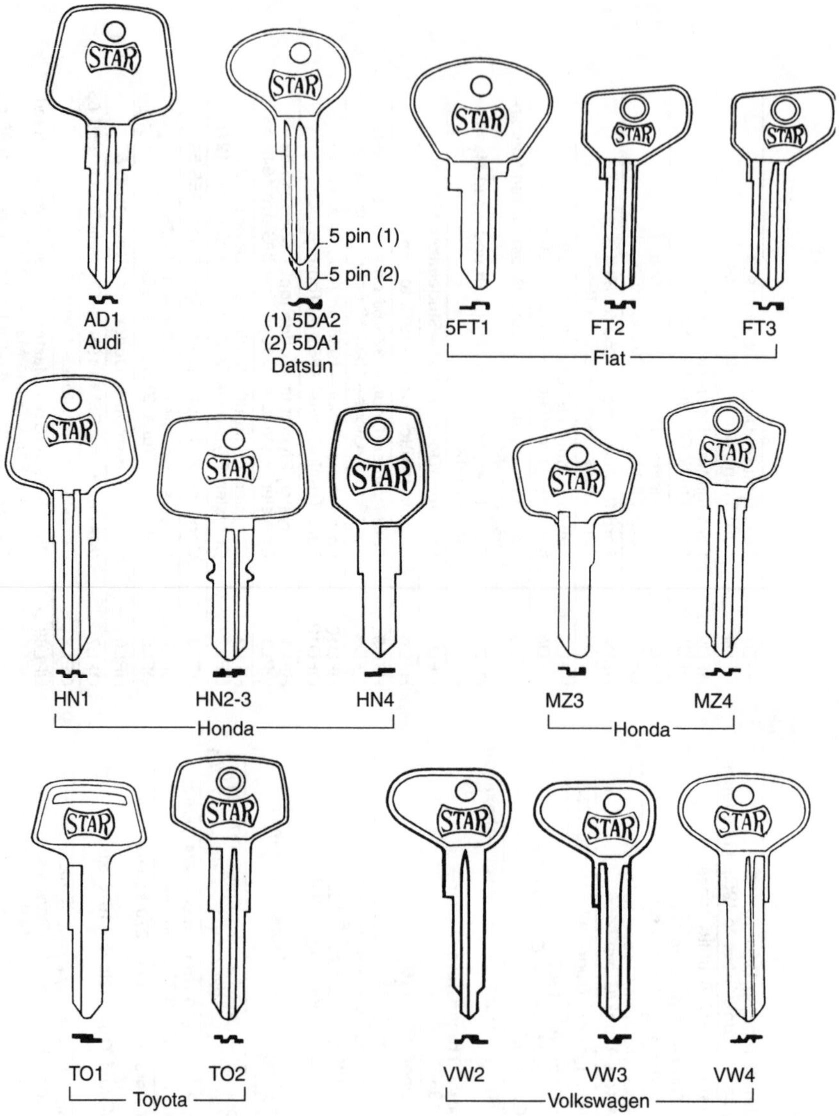

5 pin (1)
5 pin (2)
AD1
Audi
(1) 5DA2
(2) 5DA1
Datsun
5FT1
FT2
FT3
Fiat
HN1
HN2-3
HN4
Honda
MZ3
MZ4
Honda
TO1
TO2
Toyota
VW2
VW3
VW4
Volkswagen

Figure 3.23 Sampling of common foreign automobile key blanks. (*Star Key and Lock Co.*)

The cross-reference listing is available from your key blank representative, or you can spend between $20 and $50 to get a complete cross-reference book. This is both an advantage and a disadvantage. With the full book, you may have to look in a number of sections whereas with the smaller, individual cross-reference breakout from your key distributor/manufacturer, you are more likely to find the required key quickly.

AUTOMOTIVE

STAR	Ilco	Natl/Curtis	Make
HBR2	H1098LA	B10	GM Ignition to 1966
OBR2	01098LA	B11	GM Trunk to 1966
HBR3	H1098NR	B24,RA2	Amer. Mtrs. Ign. 1960 on
OBR3	1098NR	B4,RA1	Amer. Mtrs. Trunk 1960 on
HBR5	H1098A	B40	GM Ignition "A"-1967
HBR9E	P1098E	B44	GM Ignition "E" - 1969 & 1973 (Ign. only-1977/81)
OBR9H	S1098H	B45	GM Trunk "H" - 1969 & 1973 (Door also-1977/81)
HBR10J	P1098J	B46	GM Ignition "J" - 1970 (Ignition only-1974/78/82)
OBR10K	S1098K	B47	GM Trunk "K" - 1970 (Door also-1974/78/82)
HBR11	1970AM	RA4	Amer. Mtrs. Ignition - 1970 square head - 1960 on
OBR11	S1970AM	RA3	Amer. Mtrs. Trunk - 1970 ellipse head - 1960 on
HBR12A	P1098A	B48	GM Ignition "A" - 1971 & 1967 (Ign. only-1975/79)
OBR12B	S1098B	B49	GM Trunk "B"-1971 & 1967 (Door also-1975/79)
HBR14C	P1098C	B50	GM Ignition "C" - 1968/72 (Ign. only-1976/80)
OBR14D	S1098D	B51	GM Trunk "D" - 1968/72 (Door also-1976/80)
HFD4	1127DP	H27	Ford Ignition 1952-64
OFD4	1127ES	H26	Ford Trunk 1952-64
HFD10	1167FD	H33,H51	Ford Ign.Double Side 1965 on
OFD10	S1167FD	H32,H50	Ford Trk.Double Side 1965 on
OPL4	1759L/1764S	Y141,B29-31	Chrysler Trunk 1959-65
HPL68	1768/69/70CH	Y152/150/148/146	Chrysler Ign.Master 1956 on
OPL68	S1768/69/70CH	Y151/149/138	Chrysler Trk.Master 1966 on

HOUSE

STAR	Ilco	Natl/Curtis	Make
5CO1	1001EN	CO7	Corbin X1-67-5
5DE3	D1054K	DE6	Dexter 67
5EA1	1014F,X1014F	EA27,EA50	Eagle 11929 - Harloc
5HO1	1170B	HO3/HO1	Hollymade - Challenger K1
5IL1	1054F	IN1	Ilco 1054F - Keil 159AA
5IL2	1054K	IN3	Ilco - Dexter - Weslock
5IL7	X1054F	IN21,IN18	Ilco X1054F
5KE1	1079B	K2	Keil 2KK
5KW1	1176	KW1,PT1	Kwikset 1-1063 - Donner - Petco
5LO1	1004	L1	Lockwood B308 - 5 Pin
MA1	1092	M1	Master 1K,77K
5RO4	R1064D	NA5, NA6	Rockford 411-31A
5RU2	1011P	RU4	Russwin 981B
5SA1	01010	S4	Sargent 265U
5SE1	1022	SE1	Segal K9 - Norwalk - Star
5SH1	1145	SC1	Schlage 35-100C(923C)
SH2	1307A	SC6	Schlage 35-180(920A)
SH6	1307W	SC22	Schlage 35-200(927W)
5TA4	1141GE	T4,T7	Taylor 111GE
5WK1	1175, 1175N	WK1/WK2	Weslock 4425
5WR2	1054WB	WR2,WR3	Weiser 1556 - Falcon
4YA1	999B	Y145,Y220	Yale 9½
5YA1	999	Y1	Yale 8 - Acrolock

Figure 3.24 Key comparison chart for automotive and residential key blanks. (*Star Key and Lock Co.*)

A copy of a cross-reference listing is a prime requisite for every shop. Remember, your priority is the listing obtained from your key manufacturer; you can get the book later if you find you need it.

Neuter Bows

Because most nonlocksmiths can only identify key blanks by their bows, locksmiths often use *neuter* (or security) bow key blanks to make keys harder to duplicate. Such bows have a generic shape and style and provide no

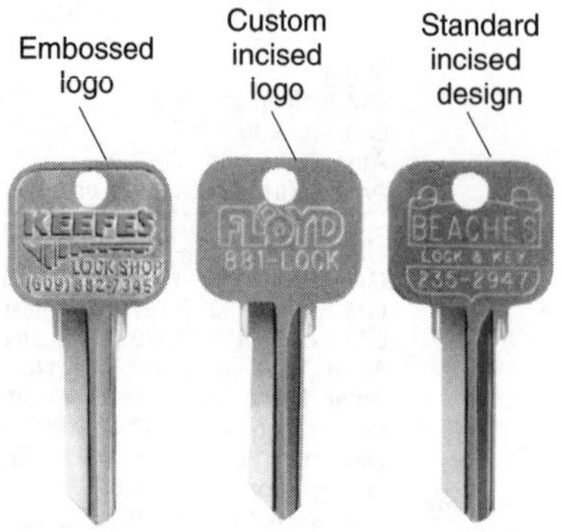

Figure 3.25 Neuter bow key blanks can be used for advertising purposes. (*Kustom Key, Inc.*)

information that identifies a lockmaker. That prevents unauthorized people who may have a key for a short time from running to any hardware or department store and getting the key copied. In addition to increasing security, such bows provide space to imprint advertising, increasing the likelihood that the customer will return to the same locksmith to get duplicate keys. Figure 3.25 shows some neuter bow key blanks.

Chapter

4

Warded Locks

The warded lock is the oldest lock still commonly in use and is found in all corners of the world. It employs a single or multiple warding system. Because of its simple design, its straightforward internal structure, and its easily duplicated key, this lock is an excellent training aid for locksmiths. This same simplicity means that warded locks give very little security. Use these locks only in low-risk applications such as storage sheds and rooms where high security is not essential.

At one time warded locks were used on most doors. They are still found in abundance in buildings still standing in older metropolitan neighborhoods such as Center City Philadelphia, Market Street in San Francisco, the Old Town section of Chicago, and the East Side of New York City.

The oldest of these buildings have cast-iron locks on the doors, some of which date back to the last century. Later locks were made of medium-gauge sheet metal. The casing consists of two stampings: the cover plate and the back plate. The latter mounts the internal mechanism and forms the sides.

The warded lock derives its name from the word *ward*, meaning to *guard*. The interior of the lock case has protruding ridges or wards that help protect against the use of an unauthorized or improperly cut key. Normally there are two interior wards positioned directly across from each other on the inside of the cover and backing plates.

This lock is sometimes mistakenly referred to as a *skeleton-key* lock. The proper and full name is the *warded-bit* key lock.

Types

Two types of warded locks are currently in use: the *surface-mounted* (or *rim*) lock and the *mortised* lock (Fig. 4.1). While both types are similar in structure and size, they give a varying degree of security. The internal mechanisms of both operate on the same principle, but the mortise lock may have several additional parts. Differences between these locks are as follows:

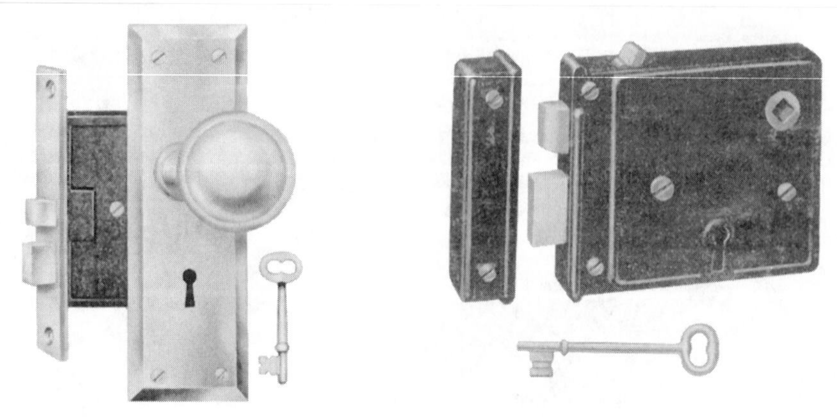

Figure 4.1 Bit key locks are available in mortised (left) and surface-mounted styles. (*Taylor Lock Co.*)

Surface-mounted (rim) lock	Mortised ward lock
Mounted on door surface	Mounted inside of door
Secured by screws in the door face	Secured by screws in the side of the door at the lock face plate
Door can be any thickness	Door must be thick enough to accommodate
Thin case	Fairly thick case
Short latchbolt throw	Up to 1-inch latchbolt throw
Lock from either side	Locked from either side
Strike can be removed with door closed	Strike cannot be removed with door closed
Very restricted range of key	Restricted range of key changes
Very weak security	Weak security

Construction

The basic interior mechanism is drawn in Fig. 4.2. Since the relative security of any lock lies in the type of key used, the number of key variations possible, and the amount of access to the locking mechanism afforded by the keyhole, the warded lock is the least secure.

The keyhole is an access route to the interior mechanism of the lock. The larger the keyhole, the easier it is to insert a pick or other tool and release the bolt.

If a lock was designed to have no more than 10 different key patterns (changes) and 1000 locks were made, 10 different keys would open all 1000 locks. By the same token, one key would open the lock it was sold with and about 99 others. Furthermore, it is often possible to cut away parts of a key to pass (negotiate) the wards of all 1000 locks. You can see that lock security is related to the kind and number of key changes built into the system when it is initially designed.

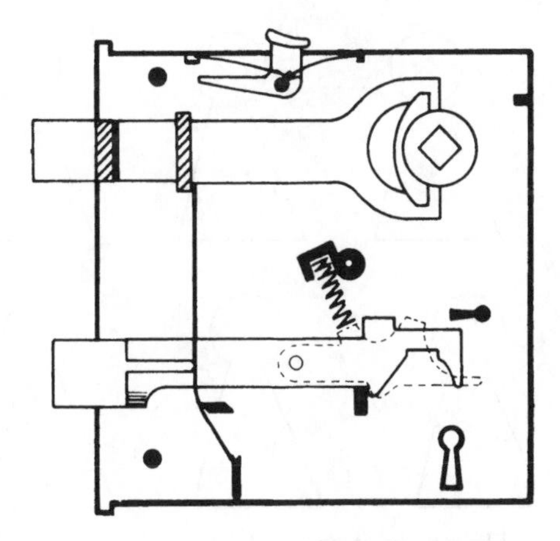

Figure 4.2 The internal parts of a typical bit key lock.

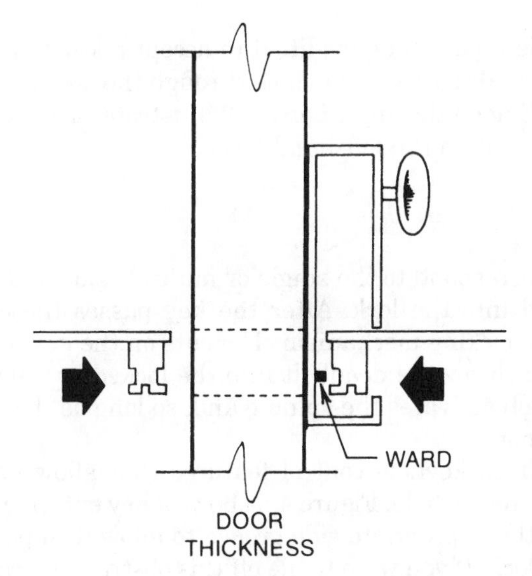

Figure 4.3 The key on the right is for a lock activated from one side of a door; the key on the left can pass the wards from either side.

In theory, each warded lock can be designed to accept 50, or even 100, slightly different keys. In practice, these locks tend to become more selective as they age and wear. The lock might respond to the original key or to one very much like it, but keys that would have worked when the mechanism was new no longer fit. While this might seem fine and well for the lock owner, excessive wear increases the potential of both key breakage within the lock and jams in the open, partly open, or closed positions. It can also mean that a new lock will have to be installed.

Most surface-mounted and mortised locks are intended to be operated from both sides of the door. Keyholes and doorknob spindle holes extend through both sides of the lock body. Occasionally you will encounter a surface-mounted lock with a doorknob spindle and keyhole only on one side. The other side is

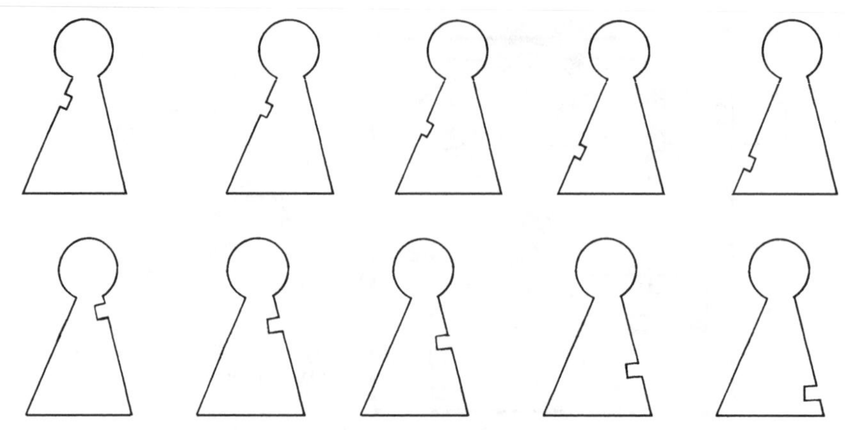

Figure 4.4 Case wards can be at many different positions.

blanked off. A lock of this type can be modified to accept a key from the other side. This modification entails cutting a keyhole through the door and lock body and may require some filing on the key. Figure 4.3 illustrates the differences in keys. Note the additional cut on the left-hand key.

Operation

The key must be cut to correspond to the single or multiple side and end wards that have been designed into the lock. After the key passes these wards, it comes in contact with the locking mechanism. The cuts on the key lift the lever to the correct height and throw the deadbolt into the locked or unlocked position. Turning the doorknob activates the spindle and, so long as the deadbolt is retracted, releases the door.

Figure 4.4 depicts various keyhole control features that allow only certain types of cut keys to enter the keyhole. Figure 4.5 shows a key entering a keyhole. Notice that the key has the appropriate side groove to allow it to pass through the keyhole and into the lock. If you were to file off this obstruction (called a *case ward*), any key thin enough to pass could enter. (Some ward bit keys are quite thick.) By the same token, a very thin key can pass whether or not the side ward is present (Fig. 4.6). The common skeleton key is a prime example of this; it is thin enough to pass most case wards but it will not necessarily open the lock.

Figure 4.7 shows a key engaging the bolt. While there is only one set of wards in this particular lock, the positioning of this ward gives more security than a lock with no ward or one with a ward that has been worn down to almost nothing.

Repair

Because it is usually cheaper to replace them, warded locks are not repaired to any great extent. You should, however, have a supply of spare parts for these locks.

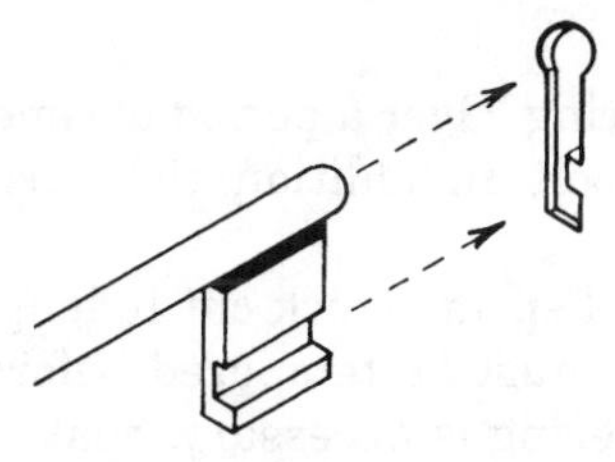

Figure 4.5 A slot milled on the edge of the key allows the key to pass the case ward.

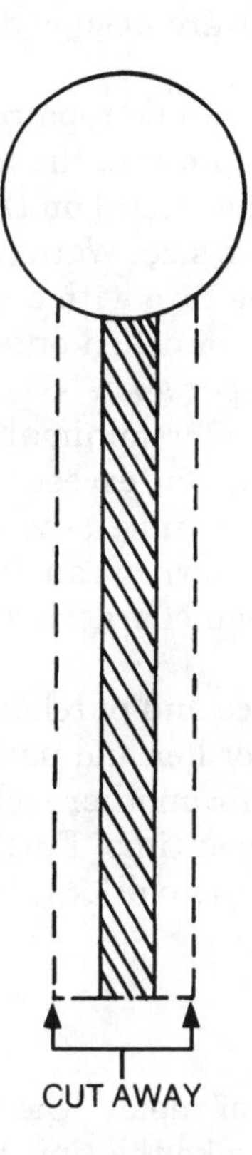

Figure 4.6 A skeleton key is a key that has been ground down to bypass the case ward.

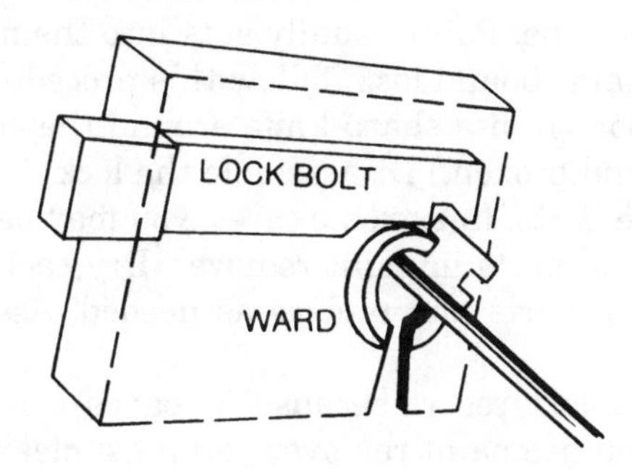

Figure 4.7 Even a single ward limits the number of keys that can operate a bit key lock.

Broken locks

The most frequent failure is a broken spring. Over a period of time, the spring may crystallize where it mates with the bolt. In addition, the wards can break or wear down into uselessness.

Replace a broken spring with a piece of spring stock cut to length and bent to the correct angle. Some spring stock must be tempered before use; other springs come already tempered. If tempering is necessary, heat the spring to cherry red, then quench it in oil. You can save time by purchasing standard springs already bent into a variety of shapes that are designed to fit almost all locks.

Worn or broken wards on locks with cast cases can be repaired by drilling a small hole in the case and forcing a short brass pin into the lock case. The best technique for brittle cases is to braze a piece of metal on the case at the appropriate spot and file it down to the appropriate size. Wards on locks with sheet metal cases can be renewed by indenting the case with a punch ground to a fairly sharp point. If the factory has already punched out the wards, it would be best to braze a piece of metal at the proper spot.

Since most of these locks are inexpensive and offer minimal security, you should remind customers that the cost of repair may far exceed the cost of the lock. Purchasing a new and more secure lock has definite advantages. If you are already at his or her home on a call, the homeowner can save money by asking you to install a new lock immediately instead of ordering one and having you make a second trip to install it.

Should the homeowner decide to take your advice and purchase a new lock, ask to keep the broken one; it is of no use to him or her and parts are always nice to have. Sooner or later you will have to repair another lock of the same type; having the correct parts at hand will save you time. Furthermore, you have made a sale and, by being allowed to keep the old lock, have obtained parts at no cost.

Removing paint

Removing paint from warded locks is also a form of repair. The home painter is a major cause of lock failure. He often does not take the time to remove the lock or to cover it with masking tape. Paint usually gets into the mechanism, freezing its works. To clean a paint-bound lock, follow this procedure:

Remove the lock from the door. (Run a sharp knife around the edges so the new paint will not be cracked and broken.) Disassemble the lock. Using a wire brush, scrape the paint from the parts. In extreme cases, you may have to use a small knife or soak the individual parts in paint remover. Dry each part thoroughly. Check for rust and worn parts and replace as needed. Assemble and mount the lock.

Use paint remover only as a last resort because it leaves a residue that attracts dust and lint. When you use paint remover you must clean each part before assembly.

Lubricating locks

Locks that are difficult to operate usually have not been lubricated in a long time, if ever. Never use oil to lubricate a lock. The professional approach is to use a flake or powdered graphite. Apply the lubricant sparingly. Remember, a little bit goes a long way. This is especially true of graphite. Should you overuse it, you may have to explain to the homeowner why there is a dark patch on the carpet that cannot be cleaned. Graphite stains are almost impossible to remove.

Warded Keys

Warded keys are made of iron, steel, brass, and aluminum. Iron and aluminum keys have a tendency to break or bend within a relatively short time; steel and brass keys can outlast the lock. The warded key has seven parts, as shown in Fig. 4.8. The configuration of the bow, length of the shank, and the relative thickness of the shoulder are not critical to the selection or the cutting of the key.

Types of warded keys

Warded keys come in various types, including the simple warded key, the standard warded key, the multicut key, and the antique key. *Simple* warded keys are often factory-made precut keys that fit several different keyholes. *Multicut* keys, on the other hand, are designed for specific locks. The *standard* warded key is usually mass-produced, but it has more precut ward and end cuts than the simple warded key. Standard warded keys can be easily converted into master keys by cutting. The *antique* key may have several kinds of cuts: ward cuts, end cuts, and even side (or bullet) groove cuts extending the length of the bit. Antique keys usually go to older locks, but these keys are still manufactured.

Selection of key blanks

A key blank is a key that has not been cut or shaped to fit a specific locking mechanism. When selecting a blank for a duplicate warded key, the following should be considered:

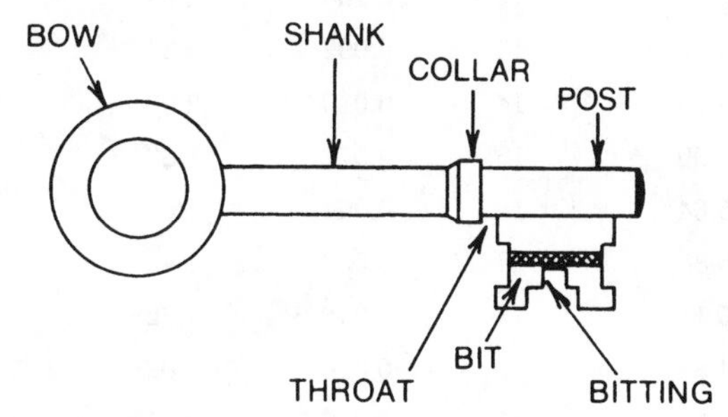

Figure 4.8 Parts of a bit key.

Post size. The post of both keys (original and duplicate) must be the same diameter. If your eyesight cannot correctly determine whether or not they are the same size, use either calipers or a paper clip. Use the calipers to compare the diameter of the original key with the diameter of the duplicate. You can use a paper clip that has been wrapped around the original key to check the diameter of the duplicate.

Length. From the collar to the end of the pin, both keys should be approximately the same length. This is important because this portion of the key enters the keyhole and operates the lock.

Height. The height of the bitting (cuts in the bit) must be the same on both keys. If the bitting in the duplicate is higher, it should be filed down; if it is lower, another blank should be selected.

Bow. The bows need not be identical but generally should be closely matched.

Width. The width of the bittings should be approximately the same. If the bitting on the duplicate is too wide, the extra thickness may prevent the duplicate from entering the lock.

Thickness. The thickness of the bits should be the same. If the original key bit is tapered, the bit of the duplicate should also be tapered. You may

TABLE 4.1 Standard Wire Gauges (inches and millimeters). 1 mm = 0.03937 inches; 1 inch = 25.4 mm

Standard wire gauge number	Inches	Millimeters	Standard wire gauge number	Inches	Millimeters
4/0	0.400	10.16	7	0.176	4.47
3/0	0.372	9.45	7½	0.168	4.27
2/0	0.348	8.84	8	0.160	4.06
0	0.324	8.23	9	0.144	3.66
1	0.300	7.60	10	0.128	3.25
2	0.276	7.01	11	0.116	2.95
2½	0.264	6.71	12	0.104	2.64
3	0.252	6.40	13	0.092	2.34
3½	0.242	6.15	14	0.080	2.03
4	0.232	5.89	15	0.072	1.83
4½	0.222	5.64	16	0.064	1.63
5	0.212	5.38	17	0.056	1.42
5½	0.202	5.13	18	0.048	1.22
6	0.192	4.88	19	0.040	1.02
6½	0.184	4.67	20	0.036	0.91

have to select a blank with a thick bit that you can file down to the correct taper.

If you don't have a micrometer, Table 4.1, which shows standard wire diameters, can help you determine the approximate diameter of warded key blanks. You can take a standard piece of wire with a known diameter and compare it with any key blank. Also, use a drill to determine the approximate diameter of warded key blanks. Insert the pin of a blank into the hole that matches the blank's diameter.

Duplicating a warded key by hand

1. Select the proper key blank.
2. Wrap a strip of aluminum approximately 1½ inches × 2¼ inches or 2½ inches around the pin and bit of the original key with one edge against the collar (Fig. 4.9).
3. Clamp the original key (wrapped in the aluminum) into a vise, bitting edge up. Ensure that the aluminum fits snugly around the key bit and pin.
4. Cut off excess aluminum around the bit of the key. Remove the aluminum strip and smoke the key with a candle.
5. Place the strip back on the bit and reclamp.
6. Using a warding file, cut down the aluminum strip—not the original key—until it is in the shape of the original key. Since the aluminum bends easily, use the file only in one direction—away from you. The stroke should be firm and steady at first. As you file closer to the cuts of the original key, the strokes should be shorter and lighter. When the file just barely touches the original key and starts to remove the candle black, stop and go no further. If the candle black is removed and the shiny surface of the original key is revealed, you have filed too deeply.
7. Fit the aluminum strip onto the key blank.
8. File down the exposed areas on the bit until it matches the outline of the aluminum strip. Be careful not to cut into the strip. Again, use shorter and lighter strokes as you get closer to finishing each cut.

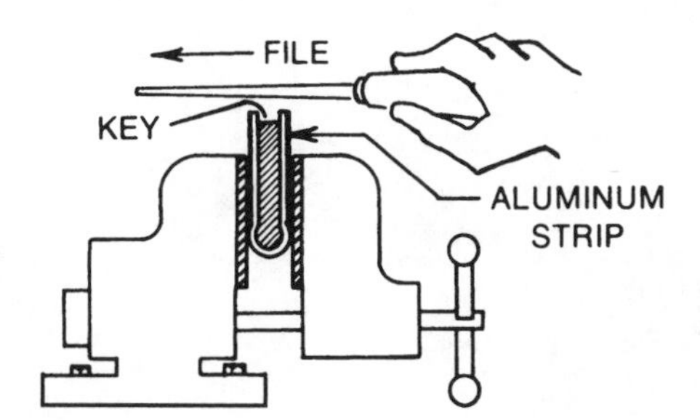

Figure 4.9 File into the aluminum strip in the direction shown.

9. If the original key has a side groove that matches a keyhole ward, this too must be cut. Use another strip of aluminum. If the original key has two grooves, you must wrap the strip around the bit so that both grooves are covered.
10. Using a scriber, scratch the metal strip to indicate the top and bottom of the groove(s). Fit the strip onto the key blank and mark the positions of the groove(s) on both ends of the bit. By connecting the marks with lines, you know exactly where to file.
11. To determine the depth of a groove cut, put one edge of a metal strip into the groove of the original key and scribe the depth of the groove on it. This mark will be your depth guide when filing the groove on the duplicate key.

Pass keys

Skeleton-type pass keys (master keys) are sold in variety stores. These keys will fit many old locks and more than a few new ones. As such, they are convenient. However, few locksmiths stock them. Why? Many locksmiths believe the ethics of the profession forbid it. You certainly do not want to supply someone with a key that could open her neighbor's lock. Nor should you duplicate a pass key without authorization from the owner. Be leery of a customer who wants a key duplicated but with additional cuts. Locksmiths have lost their licenses for less. Don't let it happen to you.

Chapter

5

Lever Tumbler Locks

Lever tumbler locks (or lever locks) have many uses in light security roles. Available in a variety of sizes and shapes, these locks are found on desks, mailboxes, lockers, bank deposit boxes, and other devices.

Figure 5.1 illustrates a popular example of a lever tumbler lock. The circular "window" on the back of the case aids the locksmith by revealing the heights of the levers, so you aren't forced to dismantle the lock. Thanks to the window, it is relatively easy to make a key "in the blind."

Parts

A lever tumbler lock has six basic parts:

- Cover boss
- Cover
- Trunnion
- Lever tumblers (top, middle, and bottom)
- Bolt (bolt stop, notch, and post)
- Base

These parts are shown in Fig. 5.2.

Operation

This lock requires a standard flat key. When the key is turned, the various key cuts raise corresponding lever tumblers to the correct height. As the levers are raised, the gates (Fig. 5.3) of the levers align and release the bolt. The bolt stop

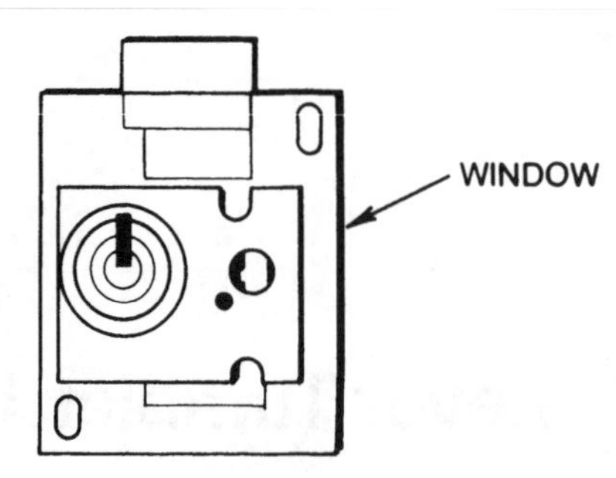

Figure 5.1 Notice the "window" on the cover of this lever tumbler lock.

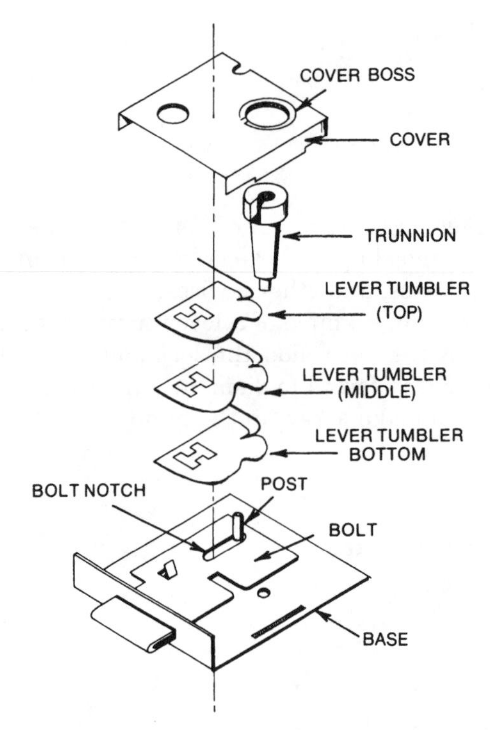

Figure 5.2 A lever tumbler lock in exploded view. This model has three tumblers; others might have a dozen or more.

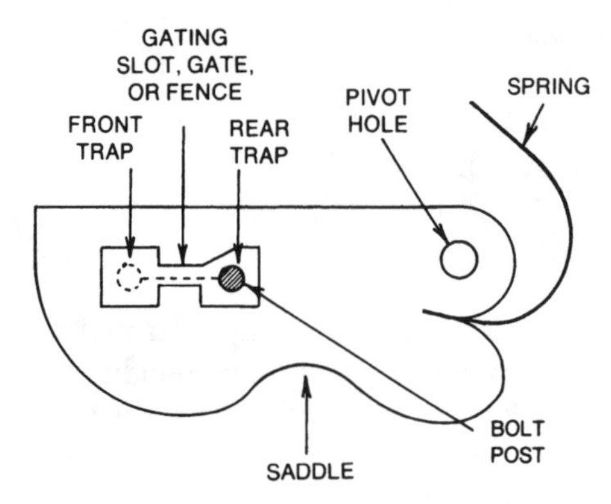

Figure 5.3 Lever tumbler nomenclature. The key bears against the saddle.

(or post) must pass through the gating from the rear to the front trap or vice versa, either unlocking or locking the lock.

Lever tumblers

The number of lever tumblers varies. Most locks have no more than five, although deposit box locks have more. The lever tumbler consists of six parts:

- Saddle (or belly)
- Pivot hole
- Spring
- Gating slot
- Front trap
- Rear trap

Over the years, manufacturers have developed a variety of lever types (Fig. 5.4). The operating principle is the same for all of them.

Each *lever* is a flat plate that is held in place by a pivot pin and a flat spring. Each lever has a gate cut into it. The gates are located at various heights with the saddle either aligned for all levers or staggered. The latter approach is antiquated. When the levers are raised to the proper position, the gates are open and the bolt post can shift from one trap to another, thus locking or unlocking the lock.

Because the bolt post meets no resistance at the gating, the lock works properly. On some designs the edge of the lever has serrated notches. The bolt post has corresponding notches. The notches on the lever and the bolt jam together if an improperly cut key is used. This effectively stops the bolt from passing through the gating and keeps the lock secure. Only a perfectly and properly cut key will open this type of lock. This feature adds immensely to the security of the lock.

The width of the gate is a critical factor in the operation of the lock. Some gates are just wide enough for the pin to pass through. A duplicate key, even slightly off on a single cut, will not work on the lock.

The saddle of the lever is also important. Recall from Chap. 1 that staggered saddles make it possible to cut a key by observation. In the case of modern

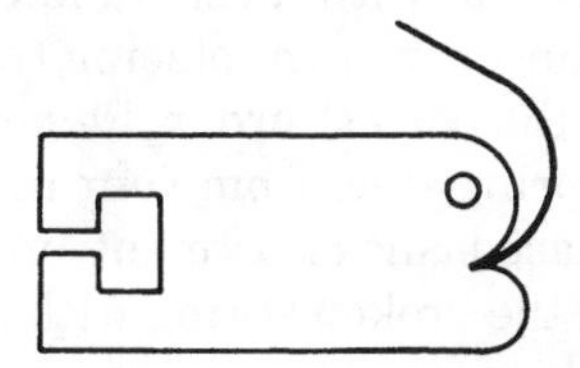

Figure 5.4 Some lever tumblers have an open gating.

lever tumblers, the gate traps have different heights, which leaves the saddles in perfect alignment.

There are two methods for making gating changes. The most common is to substitute a lever tumbler with a different gate dimension. Locksmith supply houses stock a variety of levers, so all you are required to do is change the original for one with a higher or lower gate. The second method is to alter the tumbler by filing the saddle. This approach is used with levers whose movement is restricted at the gate. Unless the tumbler gating varies greatly, the curvature of the saddle must vary with the shape of the key.

When working on the lever tumbler lock, as on other types, it is best to have two available: one for disassembly to determine internal working parts and one for actual problem solving.

The typical lever tumbler lock contains two, three, five, or six levers. Bank deposit-box locks can have as many as 14 levers. Lever tumbler locks can be keyed individually, alike (two or more with the same key), or master keyed, depending on the wishes of the buyer.

Repairs

General-purpose lever tumbler locks come in three styles. In order of popularity, these are the following:

- Solid case, usually spot-welded or riveted
- Pressed form with the back and sides of the case one piece—small tabs from the sides bend to hold the cover in place
- Cover plate secured by a screw

Riveted or spot-welded locks should be discarded when they fail. It is cheaper to purchase a new lock; the time required to drill out the rivets or chisel through the spot welds costs the customer more than the lock is worth.

Locks secured by tabs can be disassembled and reassembled quite easily. To disassemble, insert a thin tool or small screwdriver under the flanges and pry upward. Remove the cover. Look for a small object jammed in the keyway; if you find something, remove it. At times like this you may wonder if being a locksmith is really worth it, but a locksmith must have patience with the small jobs as well as the big ones.

Broken springs are another common problem with lever tumbler locks. If it is not the top lever, carefully remove each lever in turn, placing them in a logical order so you can reassemble them in the correct order. Remove the lever with the broken spring. Select a piece of spring steel from your inventory. Cut and bend it to shape. If you have purchased an assortment of ready-made springs, select the proper one, then replace the broken spring with the new one. Replace the levers and reassemble the lock.

Varieties of Lever Tumbler Locks

Lever tumbler locks are used in a variety of light security applications. Some of them include safe-deposit boxes, suitcases, lockers, and cabinets.

Safe-deposit box locks

Lever tumbler safe-deposit box locks normally have at least 6 and as many as 12 or even 14 levers. These locks require two keys. One key is assigned to the individual who rents the box; the second (or guard) key is held by the bank. Both keys are needed to open the lock. If one key could be turned to the open position by itself, it would mean that the key was faulty, the lock mechanism was in some way evaded, or the bolt post was bent or broken.

Disassembly is simple. Remove the screws and lift off the plate. Notice that the lock is constructed differently than previously discussed lever tumbler locks. There are two sets of lever tumblers; two bolt pins must pass through the lever gates at the same time. Notice the unique lever shape (Fig. 5.5). The levers have a V-shaped ridge that matches a similar V cut in the bolt post. Another key, even one that is 0.001 inch off on any cut, will mesh the notches in the post and lever. This is another built-in security feature of these locks.

Many safe-deposit locks have a compression spring bearing against the upper lever that forces the lever stack down, allowing no play between them. This spring is an integral part of the lock's security mechanism. Without the spring, the levers would be able to move a fraction of an inch when a key or pick "irritates" them. This movement is enough to provide clues for the lock pick artist.

Broken springs are the main difficulty. The levers may also be at fault. The saddles may wear enough to affect the gate. In cases like this, replace the lever with a new one. Do not file the gate cut wider to compensate. Sometimes a pivot post will work loose. Repair it by rapping the post with a light hammer or by

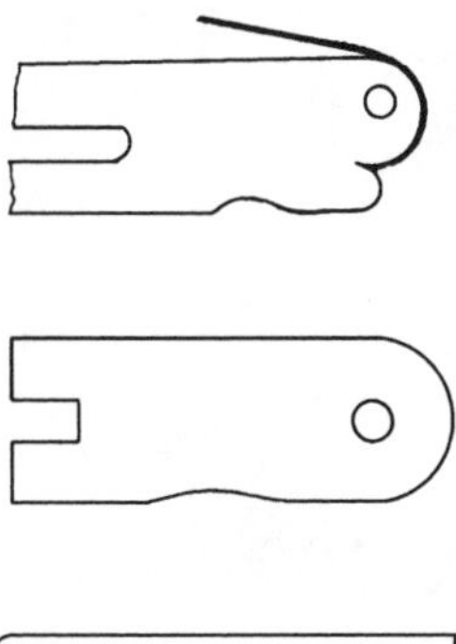

Figure 5.5 Notched tumblers are a security feature in some locks.

brazing. Never use solder. You can straighten slightly bent parts, but it is best to replace the entire damaged part.

Safe-deposit locks are serviced by locksmiths who enjoy high standings within the community. Unless an emergency arises, a beginning locksmith or a newcomer to the area would not get the job. This reluctance reflects the conservatism of bankers. It does not reflect on the skills of students and those working locksmiths who have had only a few months' experience. Many of these men and women could repair a safe-deposit box lock without difficulty.

Suitcase locks

Suitcase locks are essentially simple but have been built to accept a staggering variety of keys. Ninety-nine locks out of a hundred are warded with a primitive bolt mechanism to keep the case closed. Only a few, such as the Yale luggage lock, use a lever-type mechanism. These locks offer better security than warded locks.

The lock size, the depth the key is inserted, and the number of cuts are important clues to the lock type. A key for a lever tumbler lock will go ¼ inch or deeper into the lock before turning. Warded locks are shallower.

Opening suitcase locks is relatively simple. Many times one key will open several different suitcase and luggage locks. Many of these locks on the market are not designed for security. By cutting down almost any suitcase key, it is possible to make a skeleton key for emergencies.

Locker and cabinet locks

The Lock Corporation of America. The Lock Corporation of America offers a locker and cabinet with a locking spring design accompanied by a free-turning keyway that combine to make the locks pick resistant (Fig. 5.6).

The locks are made of a one-piece, heavy-duty steel case, and the hard-wrought steel keys and the lock case are electroplated. Another advantage to

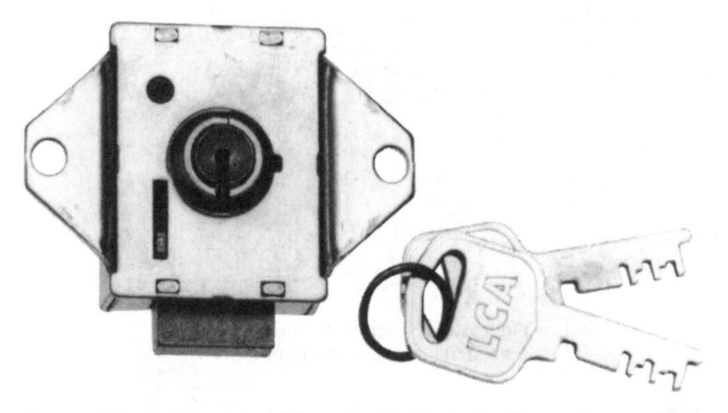

Figure 5.6 A lever tumbler deadbolt lock for lockers. (*Lock Corporation*)

these locks is that they are designed to fit into current master key systems. They will fit left- or right-hand steel locker or cabinet doors that have the standard piercings.

The keys can be removed from either the deadbolt or springbolt locks when the locker door is opened or closed. The deadbolt locks are available in models with the key removable in the locked or unlocked mode or with the key removable in the locked position only. Also, with these locks, you have a self-locking springbolt with the key removable in the locked position only.

Series 4000 and 5000. The series 4000 and 5000 flat key built-in lever tumbler (Fig. 5.7) specifications are as follows:

- Mounts in a standard three-hole piercing on steel lockers, cabinets, and other shop equipment
- Solid-locking bolt moves horizontally with a $\frac{5}{16}$-inch throw
- Masterkeyed, keyed different, keyed alike, or group keyed
- Key removable when locked or when locked or unlocked (must be specified when ordered)
- $\frac{5}{16}$-inch-gauge, heavy-duty, one-piece steel case
- Free-turning keyway
- Phosphor bronze locking springs
- Five hard-wrought steel levers

The dimensions for the lock and the standard locker door punchings are shown in Fig. 5.8.

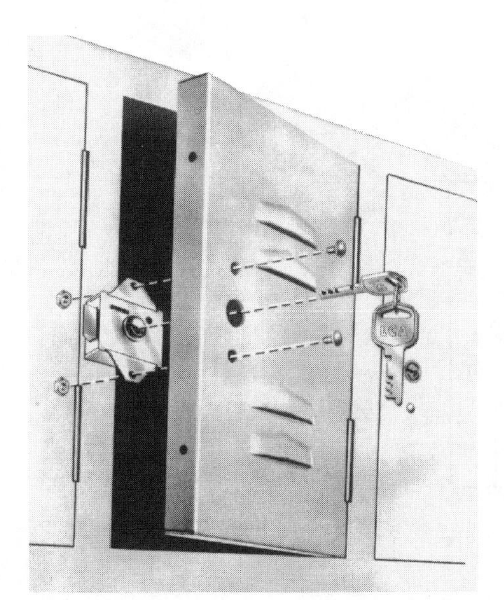

Figure 5.7 Notice how a lever tumbler lock can be installed on a locker. (*Lock Corporation*)

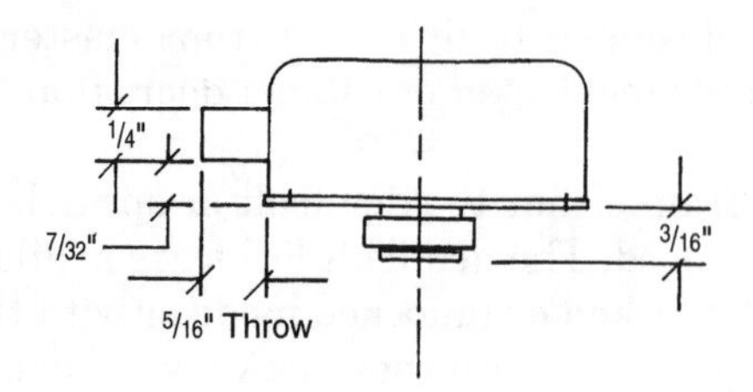

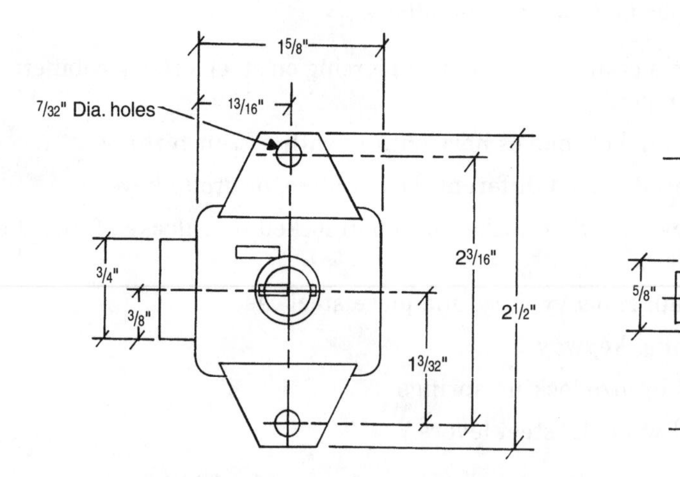

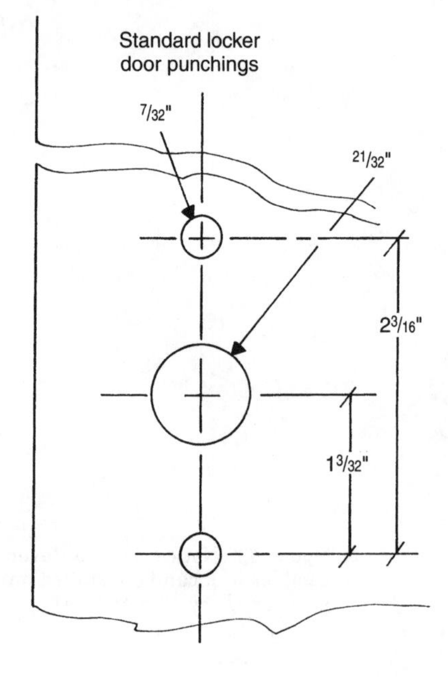

Figure 5.8 Specifications for the LCA series 4000 lever tumbler lock. (*Lock Corporation*)

The model 5001 MK locker lock is a built-in, flat-key lock with a deadbolt locking mechanism; the key is removable in either the locked or unlocked position. A free-turning plug (with a minimum of five lever tumblers) provides a minimum of 1400 different key changes. The locks (within a key change range specified) can be masterkeyed if desired. The lock will not be operated by keys of another lock within the specified key range, except a master key. The casing is of wrought steel and is 1⅜ × 1⅝ inches with top and bottom attaching ears to fit the standard steel locker piercings. All parts are of phosphor bronze, zinc, or rustproofed steel. The backset is 1 inch or less, and the solid locking bolt is ¾ × ¼ inch with a 5⁄16-inch throw.

Specifications are the same for the 5000 MK except for a key removable in the locked position only; the 4000 MK differs only in having a solid springbolt locking mechanism. The internal parts' relationship and positioning are illustrated in Fig. 5.9.

Torsion lever locks. The LCA torsion lever flat-key lock contains a patented, multimovable lever, flat-key operated mechanism that has the latest improvements to the world's oldest and still most popular means of security—the warded lock.

Seven levers (including one master lever, if required) engage and restrain the locking bolt. The novel uni-spring construction is arranged in two banks of seven staggered, interlocking arms constantly exerting a force on all the levers. When you insert and turn the precision-cut key in the rotatable key guide, the latched levers move from securing the deadbolt, enabling the lock to open.

Note: The model 4000 (basically the same) has a bolt split into two parts, including a separate spring that allows the bolt to retract without using a key to close or slam shut.

The grooved key, built-in torsion tumbler cylinder lock (Fig. 5.10) has the following specifications:

- Standard three-hole piercings for mounting
- Beveled locking bolt, moving horizontally

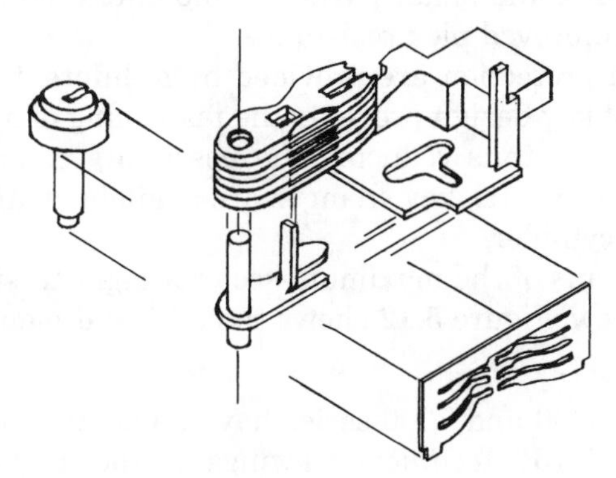

Figure 5.9 Internal parts for the 4000 and 5000 series lever tumbler lock. (*Lock Corporation*)

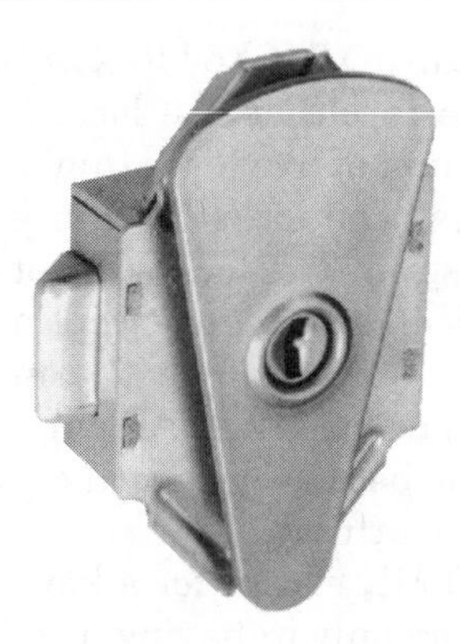

Figure 5.10 Built-in torsion tumbler cylinder lock. (*Lock Corporation*)

- Locking bolt is ¾ × ⅜ inch with a 5⁄16-inch throw
- Over 1 million computerized key changes
- Eight nickel silver torsion tumblers (to provide maximum security)
- Regular, submaster, masterkeyed, keyed different, keyed alike, or group keyed
- The master key design prevents the use of a regular key as a master key
- The keys are embossed, nickel plated, and grooved
- Cylinder plug of solid brass
- ⅝-inch-gauge, heavy-duty, one-piece steel case construction

Each of the eight nickel silver torsion tumblers is manufactured with its own integral spring; it is made strong and durable by an exclusive forging process. This tumbler-spring design protects against unfavorable climatic conditions such as temperature and humidity changes because the torsion spring is heavy and sturdy.

Variation in the eight tumblers (arranged four to each side of the cylinder) and in the double-bitted key provide more than 1 million key change possibilities. The open construction of a one-piece tumbler and spring eliminates clogging due to accumulated dirt and small particles, and offers more reliability, smoother operation, and improved pick resistance.

Master key control and protection are provided by a different master key configuration. This special key design prevents the fashioning of a master key from regular keys or blanks, thus reducing the possibility of unauthorized entry. Practically unlimited master key changes are available without diminishing the security of the cylinder.

Figure 5.11 provides details of the maximum-security eight-torsion tumbler cylinder in an exploded view. Figure 5.12 shows the critical dimensions of the cylinder.

Series 6000 and 7000. The 6000 and 7000 series have a patented snap-on face plate for installation (Fig. 5.13). Technical drawings for the snap-on installa-

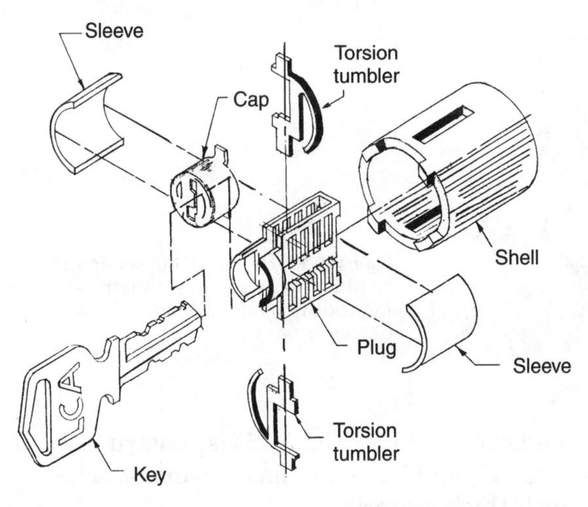

Figure 5.11 Exploded view of the maximum-security, eight-torsion tumbler cylinder. (*Lock Corporation*)

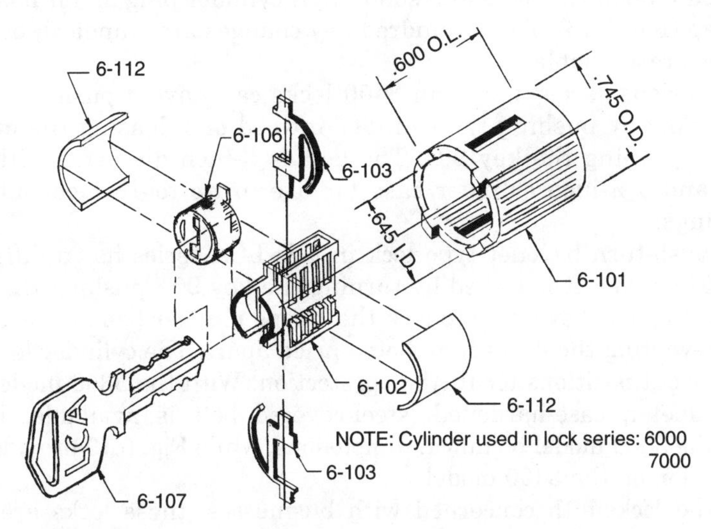

Figure 5.12 Critical cylinder dimensions of the eight-torsion tumbler cylinder lock. (*Lock Corporation*)

tion are in Fig. 5.14, along with the locker door hole punchings. Note that the series 6000 has a springbolt, while the 7000 operates with a deadbolt.

The 7001 locker lock has a built-in high-security type of ruggedness with a deadbolt locking mechanism and the cover plate; it is masterkeyed with the key removable in either the locked or unlocked position. Over 100,000 possible different key changes are available. The bolt is ¾ × ⅜ inch with a 5⁄16-inch throw.

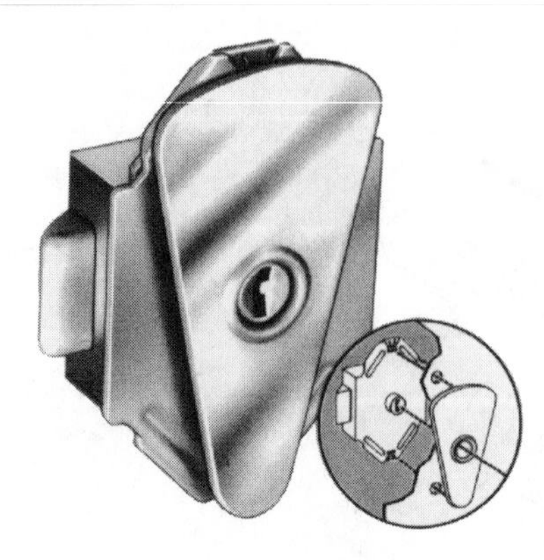

Figure 5.13 Series 6000 torsion tumbler lock with inset illustrating the snap-on face plate. (*Lock Corporation*)

The LCA drawer and cabinet lock (Fig. 5.15) is constructed of high-pressure solid Zamak unicast casing and base and has a ⅞-inch-diameter cylinder case for use in up to 1⅜-inch-thick material.

Push locks. The LCA push lock for sliding doors and showcases uses the same patented torsion tumblers and solid brass cylinder plug and it has a stainless steel cap (Fig. 5.16). Three hundred key changes are standard, but unlimited changes are available.

The various series 3300 and 3400 locks can have a push bolt, automatically locked by pushing the cylinder inward or releasing the cylinder and catch by turning the key 90°. The bolt is ⅜-inch diameter with a 15⁄32-inch throw and a ⅞-inch cylinder case for use in up to 1⅛-inch-thick material mountings.

The push-turn bayonet type lock in this LCA series has a different strike plate (Fig. 5.17). It is locked by turning the key 90°, pushing the cylinder in and rotating the key again to lock the T-type plunger behind the strike plate, thus preventing the doors from being pried apart. The cylinder locks in either the in or out positions for further protection. With the 3350 model lock, a 7⁄16-inch-diameter, case-hardened, steel-covered bolt is available. Figure 5.18 shows the 3300 model technical dimensions, while Fig. 5.19 provides the same information on the 3400 model.

For the locksmith concerned with businesses, these locks are a stock in trade, constantly proving themselves over and over.

Lever Tumbler Lock Keys

Unlike the warded or cylinder key, the lever tumbler lock key used by the average person almost never has a keyway groove running along its side; it is a flat key. A number of different flat keys exist. Figures 5.20 and 5.21 show some of them. Figure 5.22 identifies the various parts of a typically cut lever tumbler lock key.

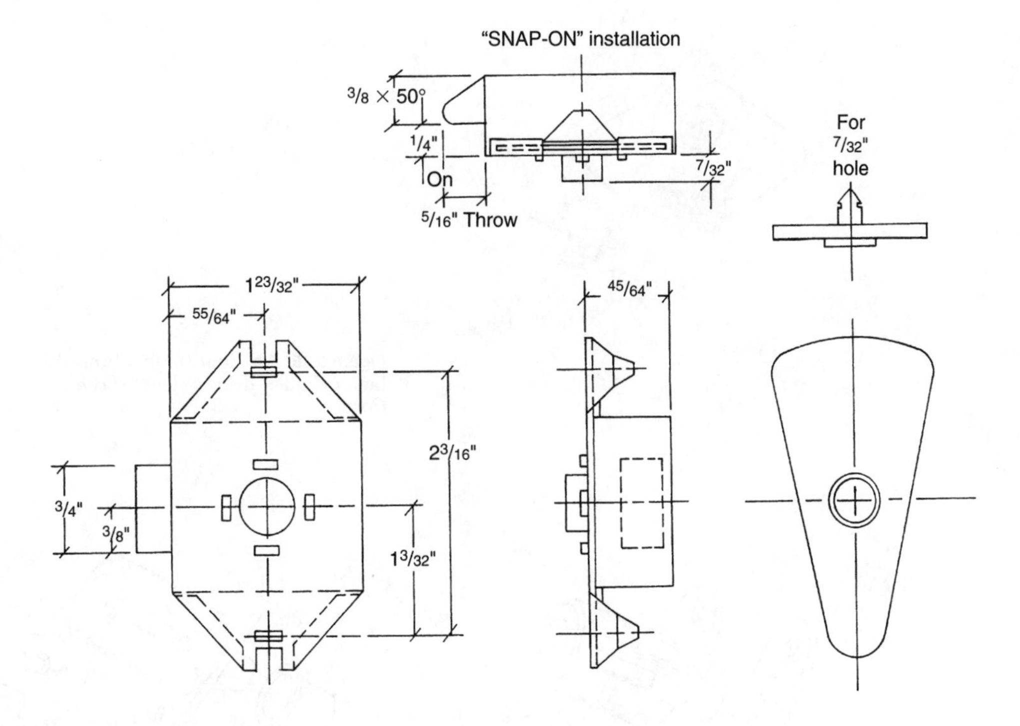

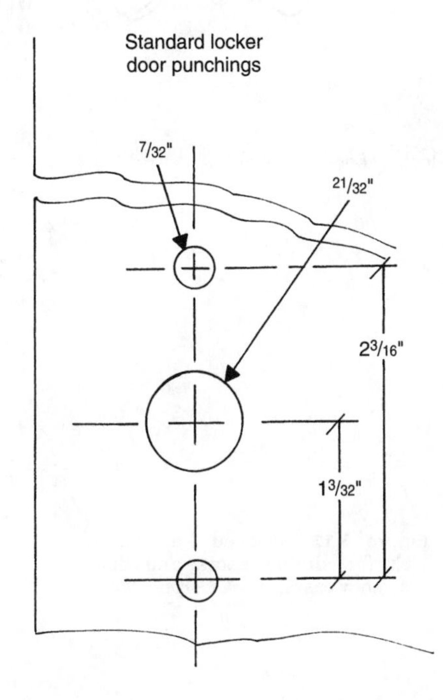

Figure 5.14 Specifications for series 6000/7000 "snap-on" installation lock. (*Lock Corporation*)

Figure 5.15 An eight-torsion tumbler cylinder drawer lock. (*Lock Corporation*)

Figure 5.16 Sliding door and showcase locks. (*Lock Corporation*)

Figure 5.17 Grooved key push lock for sliding doors and display/showcases. (*Lock Corporation*)

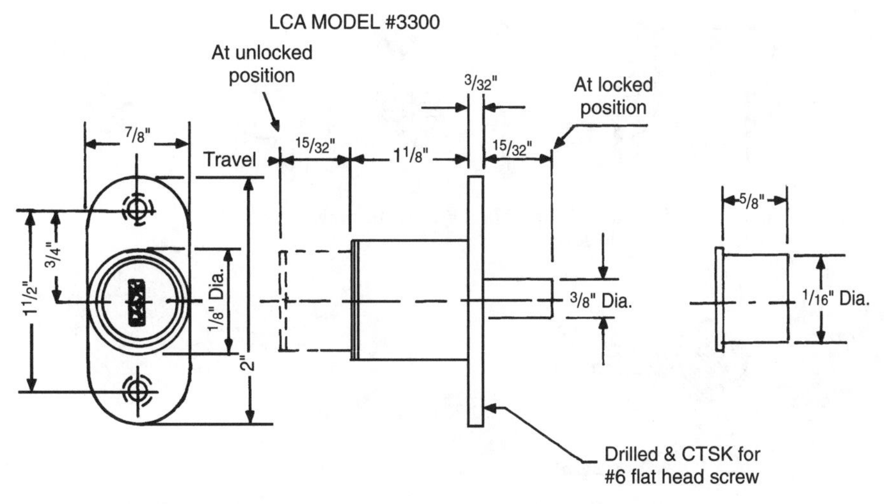

Figure 5.18 Technical dimensions for the sliding door pushbutton lock. (*Lock Corporation*)

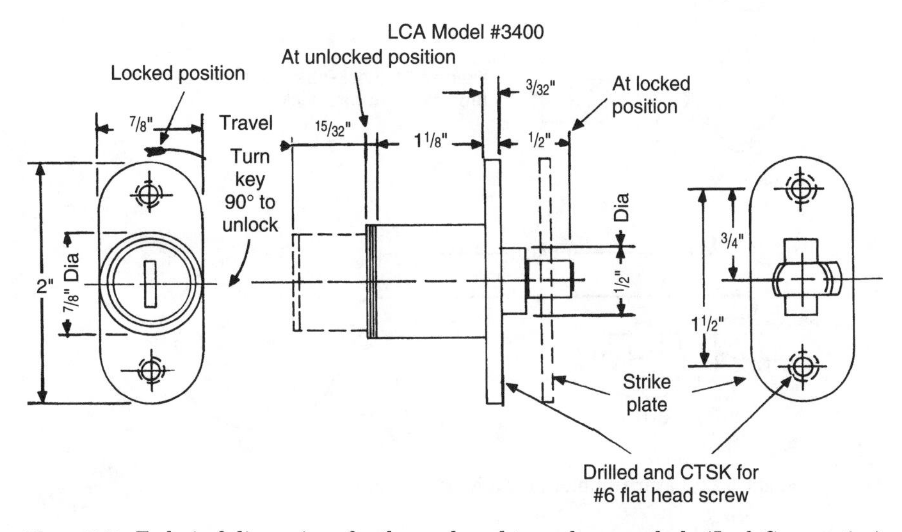

Figure 5.19 Technical dimensions for the push and turn showcase lock. (*Lock Corporation*)

Cutting keys

The lever tumbler lock key can be cut by hand or machine. To make the key, you must have the proper key blank. The three critical dimensions of the lever tumbler lock key are the thickness, length, and height. If the key blank is slightly higher or wider than the original, the blank should be filed down to the proper size. If it is thicker, select another blank. Filing down the thickness of the blank weakens it structurally.

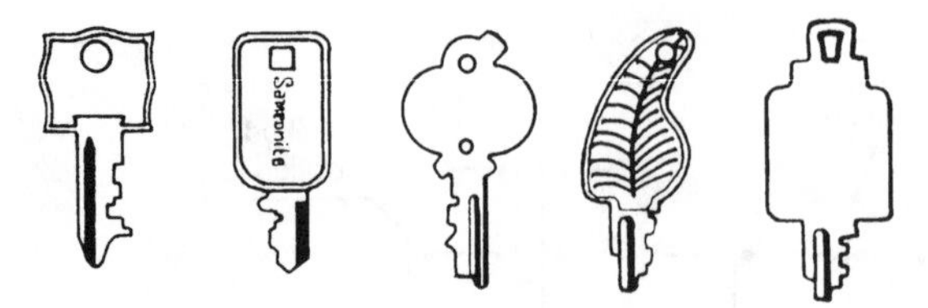

Figure 5.20 Typical flat key blanks used to make keys for lever tumbler locks.

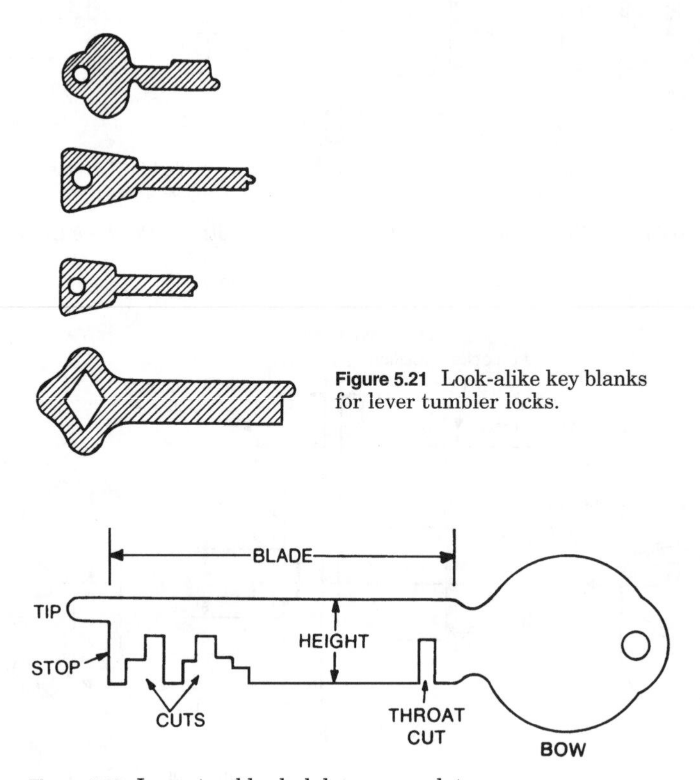

Figure 5.21 Look-alike key blanks for lever tumbler locks.

Figure 5.22 Lever tumbler lock key nomenclature.

Lay the keys (original and blank) side by side, then run your finger lightly across them. If the blank is thicker than the original, you will feel your finger catch as it passes from one to the other. When you insert the key into the keyway, it should not fit tightly or bind.

The first cut is called the *throat cut*, which enables the key to turn within the keyway. To make a throat cut, insert the blank in the keyway and scribe each side of the blank where it comes in contact with the cover boss. Determine the point where the trunnion or pin of the lock turns. Draw a vertical line there to indicate the depth of the throat cut. Remove the key from the lock, place it in a vise, and use a 4-inch warding file to cut alongside the vertical line to the proper depth. Cut on the side of the line toward the tip of the key.

There is a small round window in the back of most lever tumbler lock cases. The window is positioned so you can see where the bolt pin meets the lever gates. By observing the lever action through the window, you can get a general idea of the proper cuts to make on a key blank.

After the throat cut is filed, make the other cuts.

1. Smoke the blank, insert it in the keyway, and turn. Remove it. The lever locations are marked on the blank.
2. File the marks slightly, starting with the marks closest to the throat cut.
3. Insert the key and turn it. Notice (through the window) the height to which each lever comes up and the position of the pin in relation to the lever gates. The distance from each gate to the pin indicates the cut depth for each lever. File each key cut a little at a time, periodically inserting and turning the key to check the gate/pin relationship. Be sure to file thin cuts. Continue until you can insert the key and have the gate and pin line up exactly.

If the key in the keyway binds, observe the levers. One or more may not be correctly aligned. If the pin is too high, you have cut too deeply; if the pin is too low, you have not cut deeply enough. It may be necessary to resmoke the blank and reinsert it. The point where the key binds hardest will be indicated by the shiniest spot on the key. Just a touch with the file will usually alleviate the problem. Ensure that each filed cut is directly under its own lever.

At this point, you should correct any variations between the original key and the blank. The dimensions must be identical: the key height, thickness, and length.

You need a vise for holding the two keys, a small C-clamp, a warding file, a candle, and pliers. Follow this procedure:

1. Hold the original key over a candle flame and smoke it thoroughly.
2. Allow a few minutes for the key to cool; clamp the original and the blank at the bows. Most locksmiths use a C-clamp for this initial alignment.
3. Once aligned, place the key and blank in a vise. If you wish, you can leave the C-clamp in position.
4. Use the warding file to make the tip cut first. File in even and steady strokes, bearing down in the forward or cutting stroke. Keep a careful eye on the original. Stop when the file just disturbs the blackening.
5. Once the tip cut is completed, move to the next cut.
6. Remove the keys from the vise and inspect the cuts. Each should be rectangular and flat.

7. Use emery paper to lightly sand away the burrs on the edges of the cuts. Wipe off the candle black from the original key.
8. Test the duplicate in the lock. Should it stick, blacken the duplicate and try it again in the lock. Breaks in the blackening will show where the key is binding. A light stroke with the file should correct this.

Reading lever tumbler locks

After you learn how to read lever tumbler locks, you'll be able to make working keys without disassembling the locks. As you know, a lever tumbler lock keyway is narrow, and the view of the tumblers is further restricted by the trunnion. While this is a handicap, it can, in part, be overcome with the help of an appropriate reading tool.

The positions of the lever tumbler saddles are one clue to reading the lock. The saddle of the lock can tell you quite a bit. The wider the saddle, the deeper the cut; the narrower the saddle, the shallower the cut. Using the reading tool, you will feel the different saddle widths to determine the cuts and develop some idea of the cuts and of the key shape.

To do this properly, you must have an appreciation of the internal workings of the lock. Depth key sets can be useful when you are ready to cut the keys. Lever cuts are usually in 0.015- to 0.025-inch increments.

Practice is required. Begin with a lock that has a window. If possible, obtain extra levers so you can change the keying of the lock at will.

Raise each lever with the reading tool and determine the proper position for each one. Remember the general rule: A wide saddle requires more movement than a narrow saddle. Once you are satisfied that you have read the lock, disassemble it and examine the lever tumblers. You may have misread the tumblers but do not be disheartened. The only way to achieve competence in this skill is to practice.

Other tumblers are designed with uniform saddle widths. Keying is determined by the positions of the gates in the tumblers. These locks can be quite difficult to read.

Working with the reading tool, raise one of the levers as high as it will go. While this movement is not a direct indication of the depth of the key cut, it is important. The amount of upward movement establishes the minimum key cut; a shallower cut will jam the levers against the top of the lock case. In addition, the individual gates and traps are usually in some rough alignment. You will usually find that two are in almost perfect alignment.

A cut for a lever that has the post on the upper half of the trap will be shallow; a cut for a lever that has its post on the lower half will be deep. A lever that has its gate in the intermediate position will require a key cut between these extremes.

Once you have established the general topography of the cuts, refer to your set of depth keys for the exact dimensions. Established locksmiths have reference manuals that may simplify the work.

Chapter

6

Disc Tumbler Locks

The disc tumbler (or cam) lock gets its name from the shape of the tumblers. These are about as secure as lever locks and as such are superior to warded and other simple locks. However, disc tumbler locks are far less secure than pin tumbler locks.

Disc tumbler locks are found in automobiles, desks, and a variety of coin-operated machines. Some padlocks are built on this principle. Because the cost of manufacture is very low, replacement is cheaper than repair.

While similar to the pin tumbler lock in outside appearance and in the broad principle of operation, the internal design is unique.

The disc tumblers are steel stampings arranged in slots in the cylinder core. Figure 6.1 shows a typical disc tumbler. The rectangular *hole* (or *cutout*) in the center of the disc matches a notch on the key bit. The protrusion on the side, known as the *hook*, locates the spring. The disc stack is arranged with alternating hooks, one on the right of the cylinder and one on the left. The lock is illustrated in Fig. 6.2.

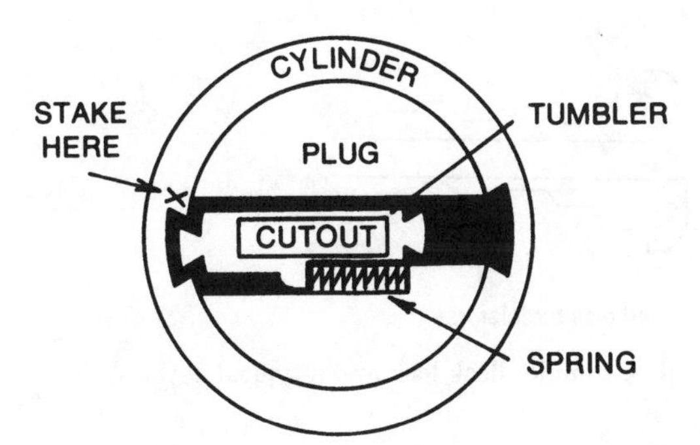

Figure 6.1 Disc tumbler lock in cross section. The position of the cutout determines the depth of the key cut.

Operation

The disc lock employs a rotating core, as does the more familiar pin tumbler design. The disc tumbler core is cast so the tumblers protrude through the core and into slots on the inner diameter of the cylinder. So long as the tumblers are in place, the core is locked to the cylinder.

The key has cuts that align with the cutouts in each tumbler. The key should raise the tumblers high enough to clear the lower cylinder slot but not so high as to enter the upper cylinder slot. In other words, the correct key will arrange the tumblers along the upper and lower shear lines (Fig. 6.2). The plug is free to rotate and, in the process, throw the bolt.

The key resembles a cylinder pin-tumbler key except that it is generally smaller and has five cuts. A cylinder pin tumbler key might have six or seven cuts.

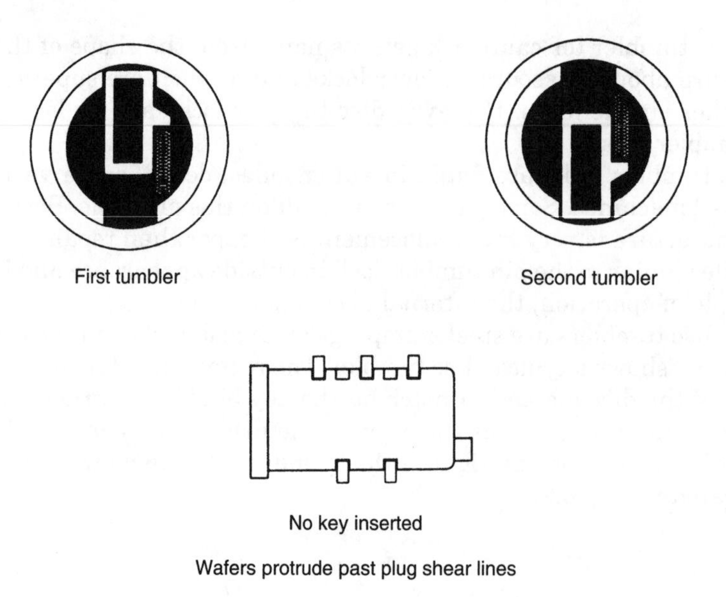

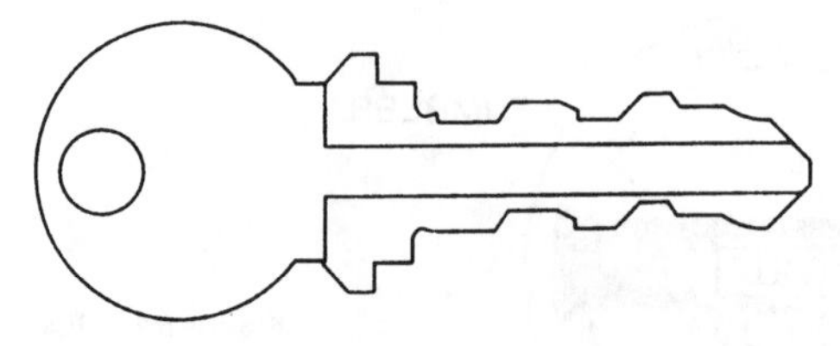

Figure 6.2 Parts of a common disk tumbler desk lock and a typical key for such a lock.

Disassembly

Good quality disc locks have a small hole on the face of the plug, usually just to the right of the keyhole. To disassemble the lock, insert a length of piano wire into the hole and press the retainer clip. Turn the plug slightly to release it. The key gives enough purchase to withdraw the plug. If a key is not available, you can extract the plug with the help of a second length of piano wire inserted into the keyhole. Bend the end of the wire into a small hook. Other locks attach the plug to the cylinder with the same screw that secures the bolt-actuating cam. Others (fortunately, a minority) have the plug and cylinder brazed together. File off the brass.

Keying

Manufacturers have agreed on five possible positions for the cutouts relative to the tumblers. Keying is a matter of arranging the tumblers in a sequence that matches the key cuts. Once the sequence has been identified, install the tumbler springs over their respective hooks and mount the tumblers in the plug. The tumblers are spring-loaded and, until the plug is installed in the cylinder, are free to pop out. Lightly stake them in place with a punch or the corner of a small screwdriver blade. One or two pips are enough because you will break the tumblers free once the assembly is inside the cylinder. Inserting the key is enough to release the tumblers.

Security

These locks usually have no more than five tumblers, and each tumbler cutout has five possible positions. These variations allow 3125 or 5^5 key changes. In practice, the manufacturer will discard some combinations as inappropriate and may further simplify matters by limiting the key changes to 500 or less. Obviously, disc tumbler locks are not high-security devices.

Cam Locks

Cam locks are used for a variety of general and special purposes (Fig. 6.3). More than likely, you will see approximately 90 percent of these units in offices. From the face (or front) these locks look pretty much the same after they have been installed. Inside, however, they may differ in various regards. The illustrations on the next several pages are exploded views of different models and the parts associated with each. These models are all five-disc tumbler locks, each with a possibility of 200 different key combinations. Because of the key variations possible, these locks are often masterkeyed prior to purchase by the customer. Only by disassembling the lock or viewing the tumblers through the keyway will you know for sure whether or not the lock is masterkeyed (Figs. 6.4 through 6.9).

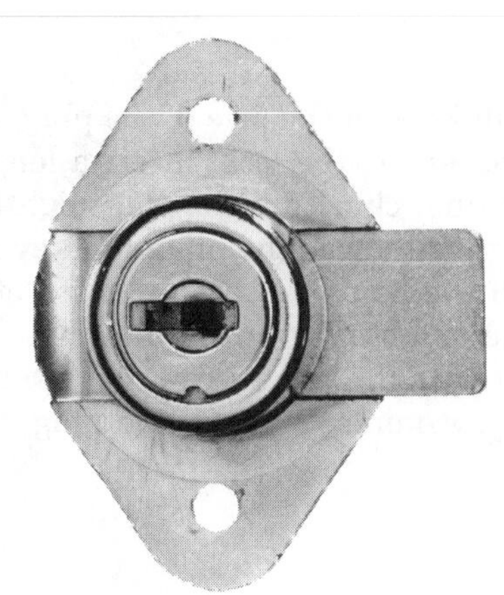

Figure 6.3 Disc tumbler cam lock. (*Dominion Lock Co.*)

Two other core-type cam locks are shown in Figs. 6.10 and 6.11. These cam locks with removable cores are of the seven-pin variety that uses the Ace-type circular key.

A variety of cams is used with these locks. Figure 6.12 is a chart showing the various types of cams available. The length and offset specifications are also included. Figure 6.13 shows the hook, bent, double-ended, and other miscellaneous cams that you may come across in the course of your work.

The standard thickness of a cam is $\frac{3}{32}$ inch (2.667 mm). Cams are made of steel; many are cadmium-plated for durability and longevity.

Reading Disc Tumbler Locks

It's not unusual for a locksmith to be asked to make a key for a lock whose original key was lost or misplaced. If the lock does not have a code number on its face or if the owner neglected to write down the number on the key, the locksmith has three choices: Pick the lock, impression a key, or "read" the lock.

Like impressioning, reading a lock is a skill that must be developed through patient practice. You cannot expect to master this skill quickly, nor can you expect to remain proficient in it without constant practice. At first, practice daily, then weekly or twice weekly to maintain your skill.

When called on to fit a key, either by impressioning or reading of the lock, the locksmith invariably looks into the lock keyway. A quick glance determines whether it is a lever, disc, or pin tumbler lock.

Disc locks

The view down the keyway of a disc tumbler lock shows a row of discs with their centers cut away and staggered. The cutaways are at different heights; the discs themselves are the same diameter. The cutaway looks like a small staircase with a surrounding wall (formed by the vertical edges) around the steps. Since each disc is the same size, only the height of the "steps" varies, and this

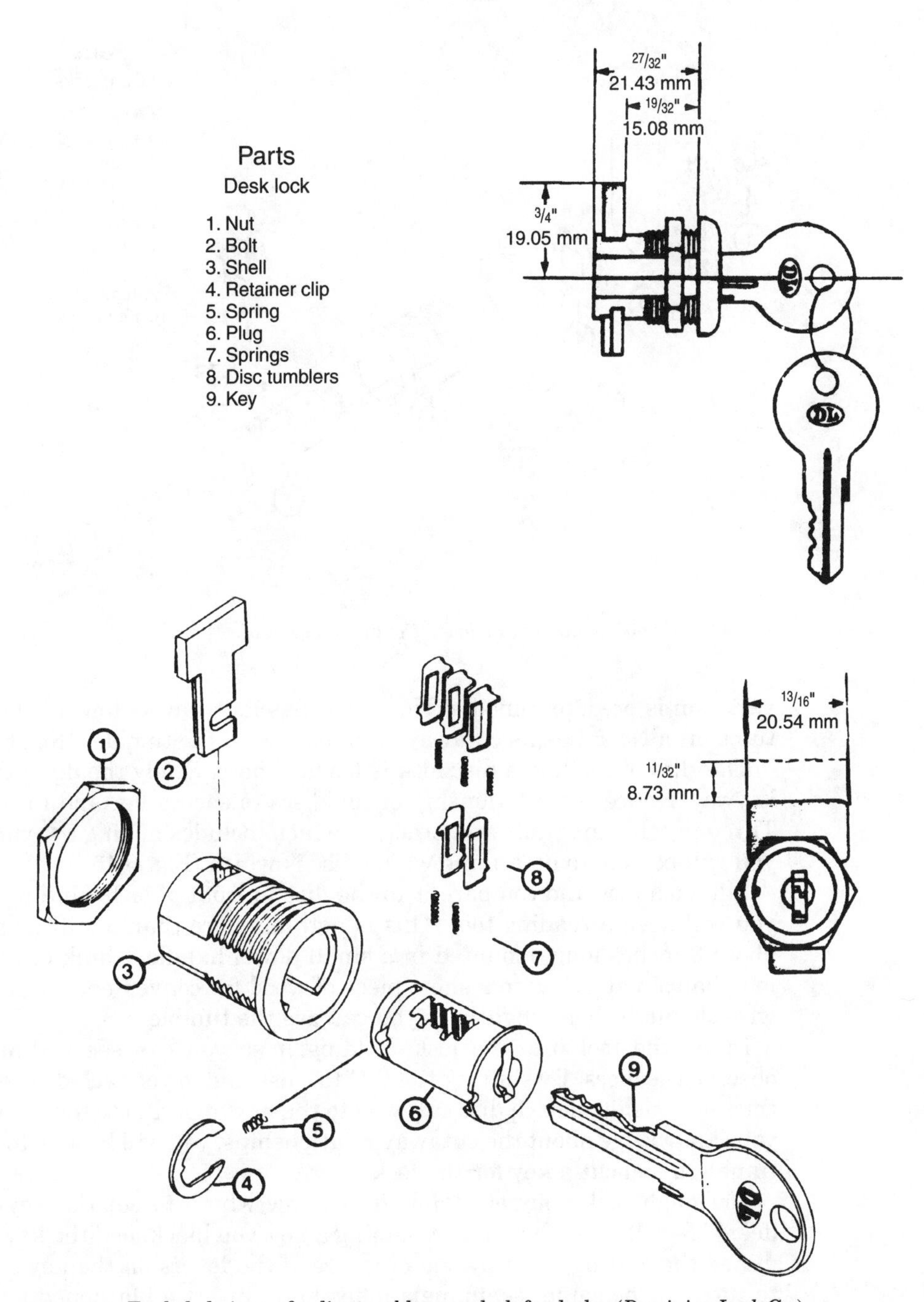

Figure 6.4 Exploded views of a disc tumbler cam lock for desks. (*Dominion Lock Co.*)

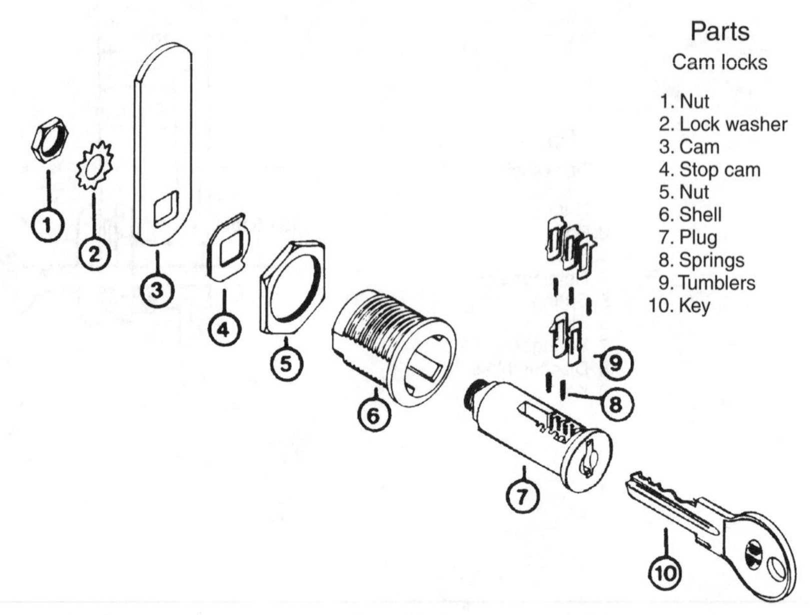

Figure 6.5 Exploded view of cam lock. (*Dominion Lock Co.*)

variation is predetermined. A No. 1 disc has its cutaway toward the top of the tumbler; a No. 5 has its cutaway situated low on the tumbler (Fig. 6.14).

The skill of reading a disc lock is learned here. Study the discs through the keyway. Notice the relationship of the discs to each other and to the keyhole. Through constant study and practice, which includes mixing the various discs, you will be able to determine which disc is which (Fig. 6.15).

Lift each disc and compare it to the disc in front of or behind it. To do this, you will need a reading tool. This is nothing more than a stiff length of wire about 3 inches long, mounted in a small dowel handle. Think of the tool as a long hairpin attached to a short piece of wood for convenience in holding. The wire should be bent slightly so you can see the tumblers.

Insert the tool into the lock, holding it so you can see the interior and observe the discs. By shifting the tool to raise and lower each disc, you can see the relationship of each disc cutaway to the next disc and to the keyway. Using your knowledge about the cutaway relationships, you will be able to decode the tumblers and cut a key for the lock.

You might ask yourself, "How do I know where to cut the key—and how deep?" Recall that when impressioning a key, you blackened the key and determined the cutting point by the pressure of the levers on the key as you tried to turn it. The same technique applies here. Insert a blackened key and give it a slight turn. This brings the key in contact with the sides of each tumbler cutout. The cutout leaves a mark on the blank, indicating the portion of each cut.

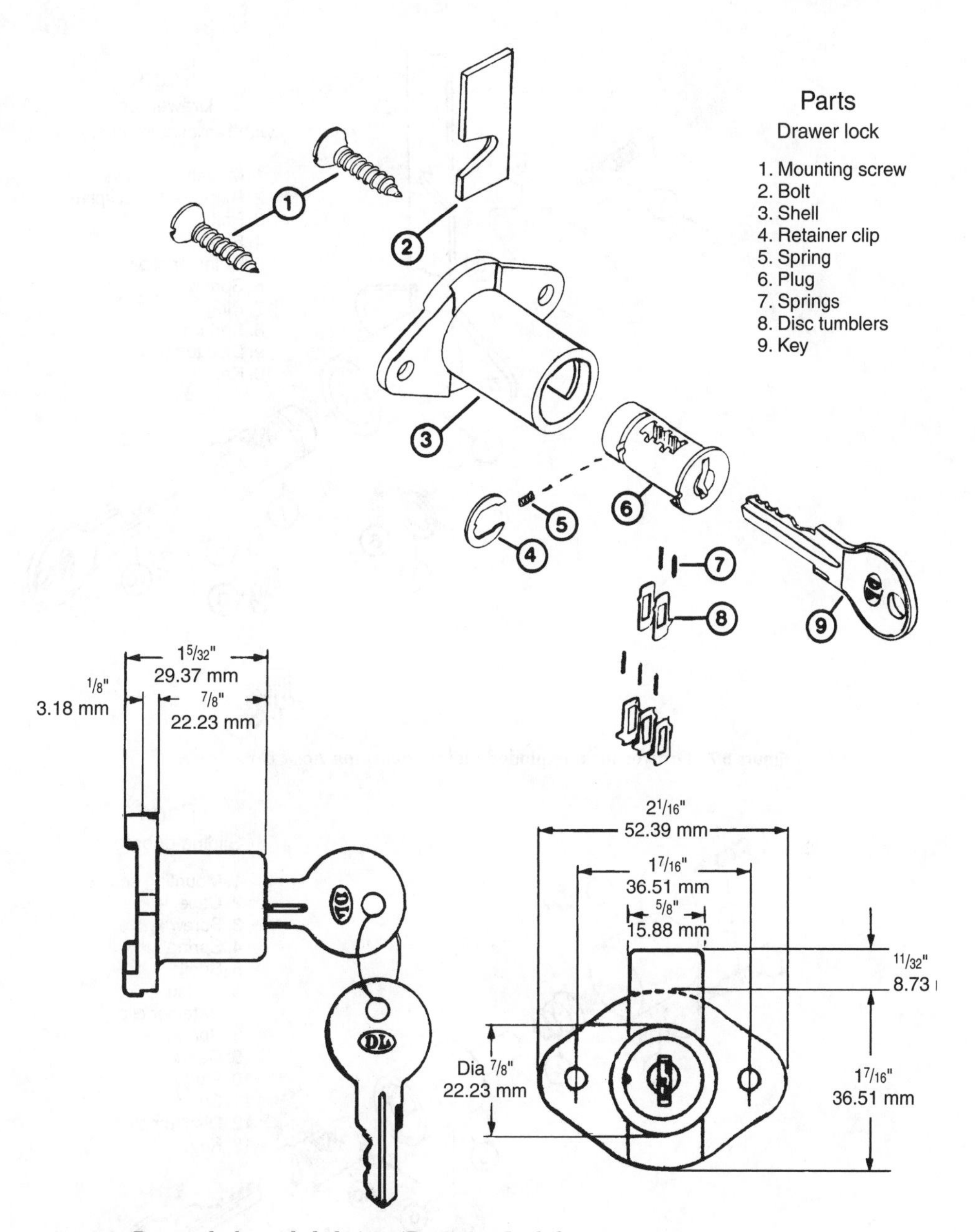

Figure 6.6 Drawer lock, exploded view. (*Dominion Lock Co.*)

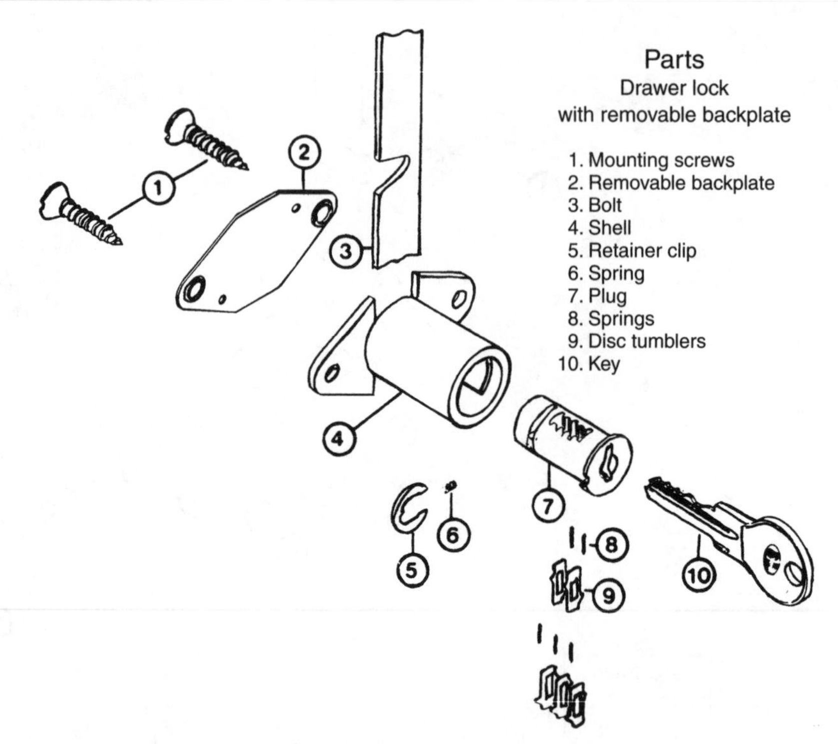

Figure 6.7 Drawer lock, exploded view. (*Dominion Lock Co.*)

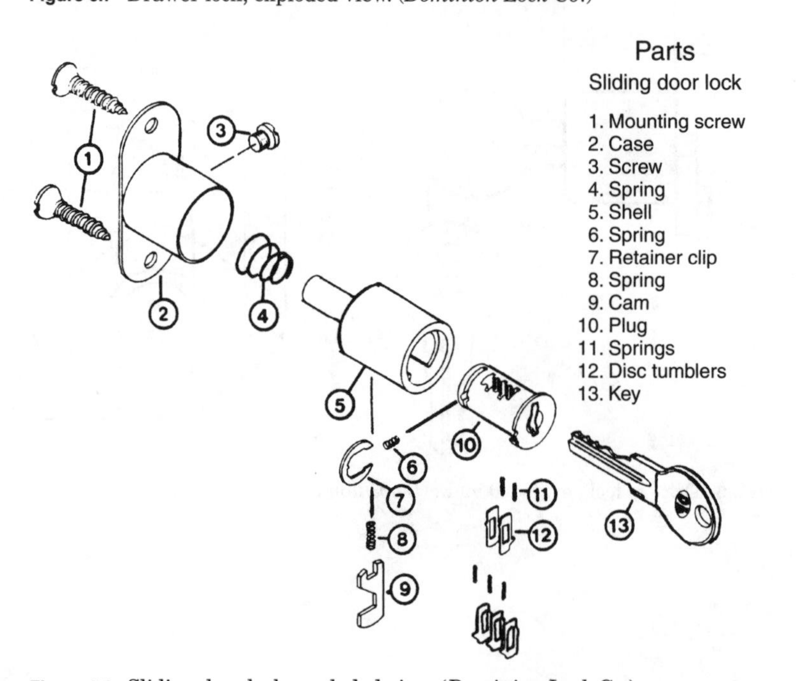

Figure 6.8 Sliding door lock, exploded view. (*Dominion Lock Co.*)

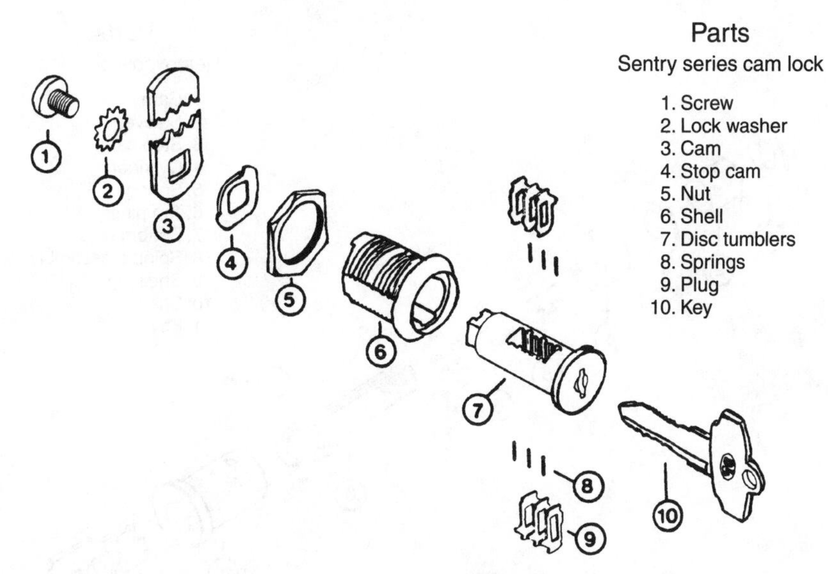

Figure 6.9 Sentry series cam lock, exploded view. (*Dominion Lock Co.*)

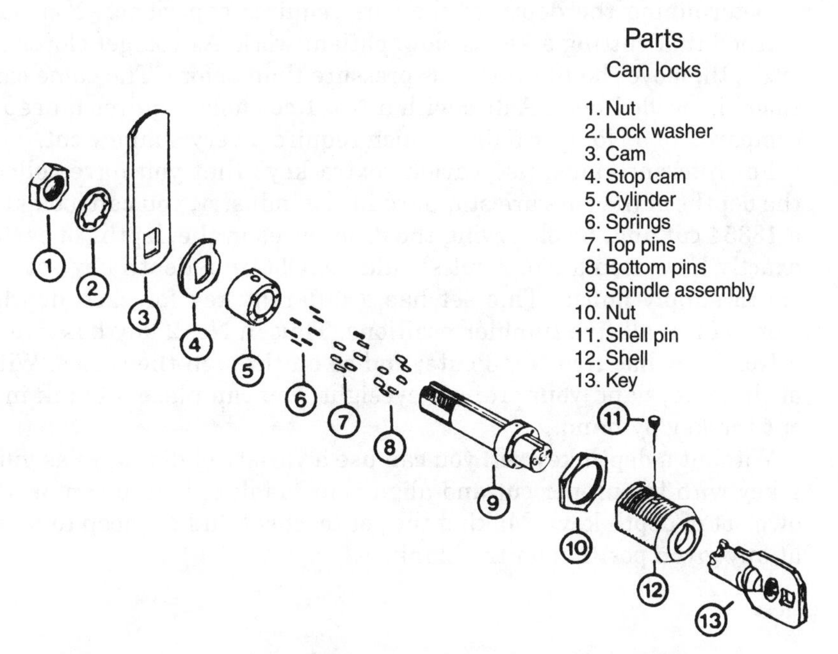

Figure 6.10 Tubular key cam lock, exploded view. (*Dominion Lock Co.*)

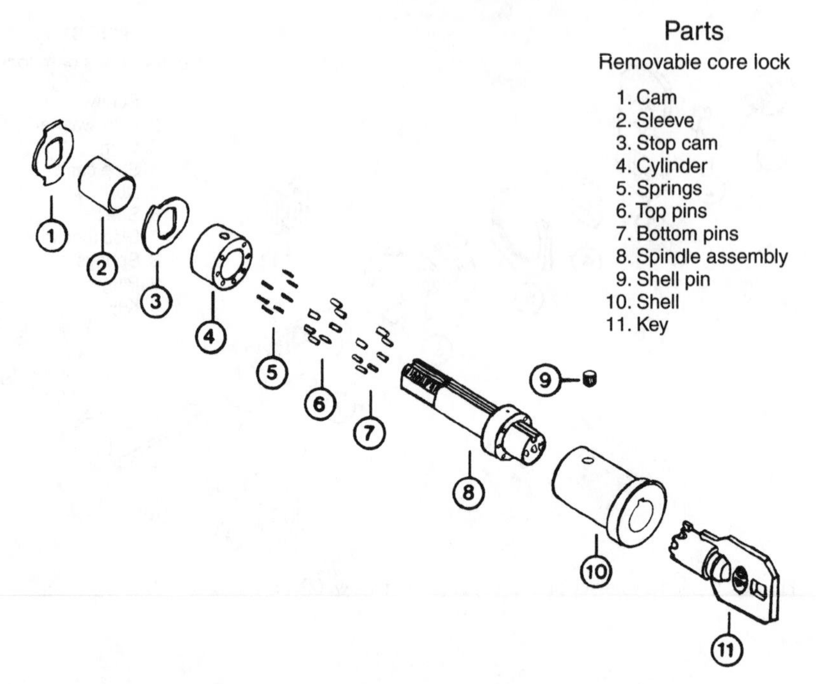

Figure 6.11 Removable core lock, exploded view. (*Dominion Lock Co.*)

Determining the depth of the cuts requires experience. You have already learned that cutting a key is slow, patient work. As you get closer to the proper depth, move the file with less pressure than before. The same care and precision is needed here. A disc with a No. 1 cut hole requires a deep key cut as compared to a No. 4 or 5 disc, which require a very shallow cut.

For reference aids, use various extra keys that you have collected. Since the depths of the cuts are standard in the industry, you can look at a key with a 13354 cut and, by observing the differences in the depths of each cut, know exactly how deep a No. 2 cut should be. Obtain a depth key set from a locksmith supply house. This set has a different key for each depth, with the same cut in all five tumbler positions. Thus, a No. 2 key has five No. 2 cuts; a No. 3 key has five No. 3 cuts; and so on through the series. With the help of these keys and your trained eyesight, you can place a blank in a vise and cut the key by hand.

Without a depth key set, you can use a variety of disc keys as guides. Select a key with the proper cut and align your blank to it. You can also make your own set of depth keys. Making the set teaches you how deep to make each cut at any given position on the blank.

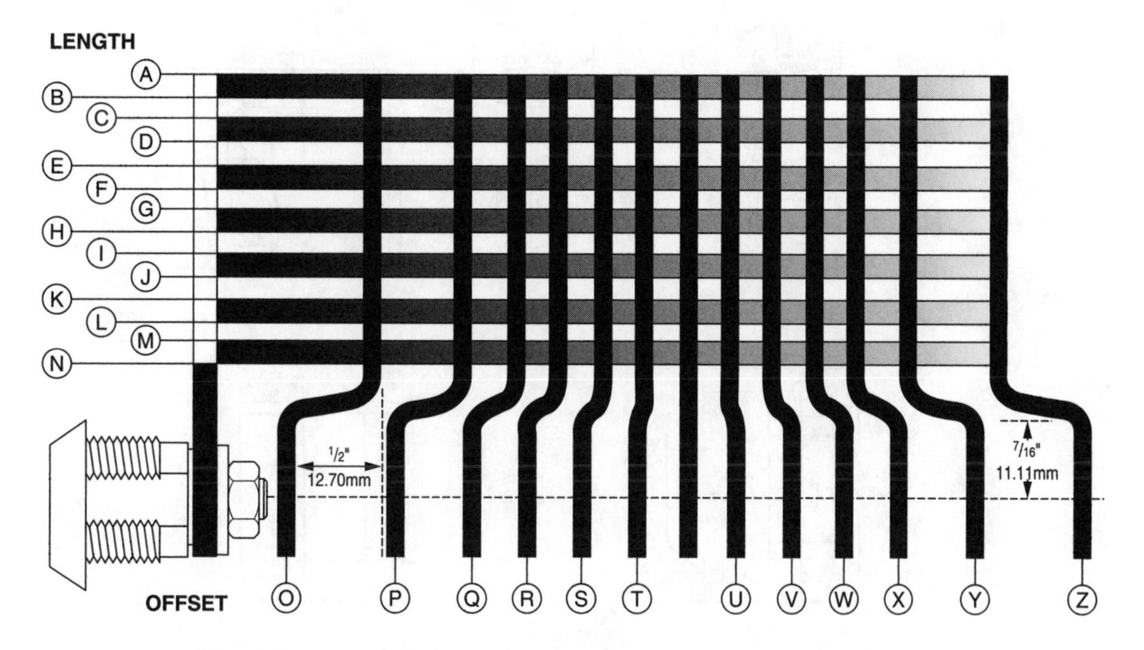

LENGTH			
	A	2 3/8"	60.325 mm
	B	2 1/4"	57.150 mm
	C	2 1/8"	53.975 mm
	D	2"	50.800 mm
	E	1 7/8"	47.625 mm
	F	1 3/4"	44.450 mm
	G	1/5/8"	41.275 mm
	H	1 1/2"	38.100 mm
	I	1 3/8"	34.925 mm
	J	1 1/4"	31.750 mm
	K	1 1/8"	28.575 mm
	L	1"	25.400 mm
	M	7/8"	22.225 mm
	N	3/4"	19.050 mm
OFFSET			
OUTSIDE	O	1/2"	12.700 mm
	P	3/8"	9.525 mm
	Q	1/4"	6.350 mm
	R	3/16"	4.775 mm
	S	1/8"	3.175 mm
	T	1/16"	1.600 mm
INSIDE	U	1/16"	1.600 mm
	V	1/8"	3.175 mm
	W	3/16"	4.775 mm
	X	1/4"	6.350 mm
	Y	3/8"	9.525 mm
	Z	1/2"	12.700 mm

Figure 6.12 Cam lengths and offsets. (*Dominion Lock Co.*)

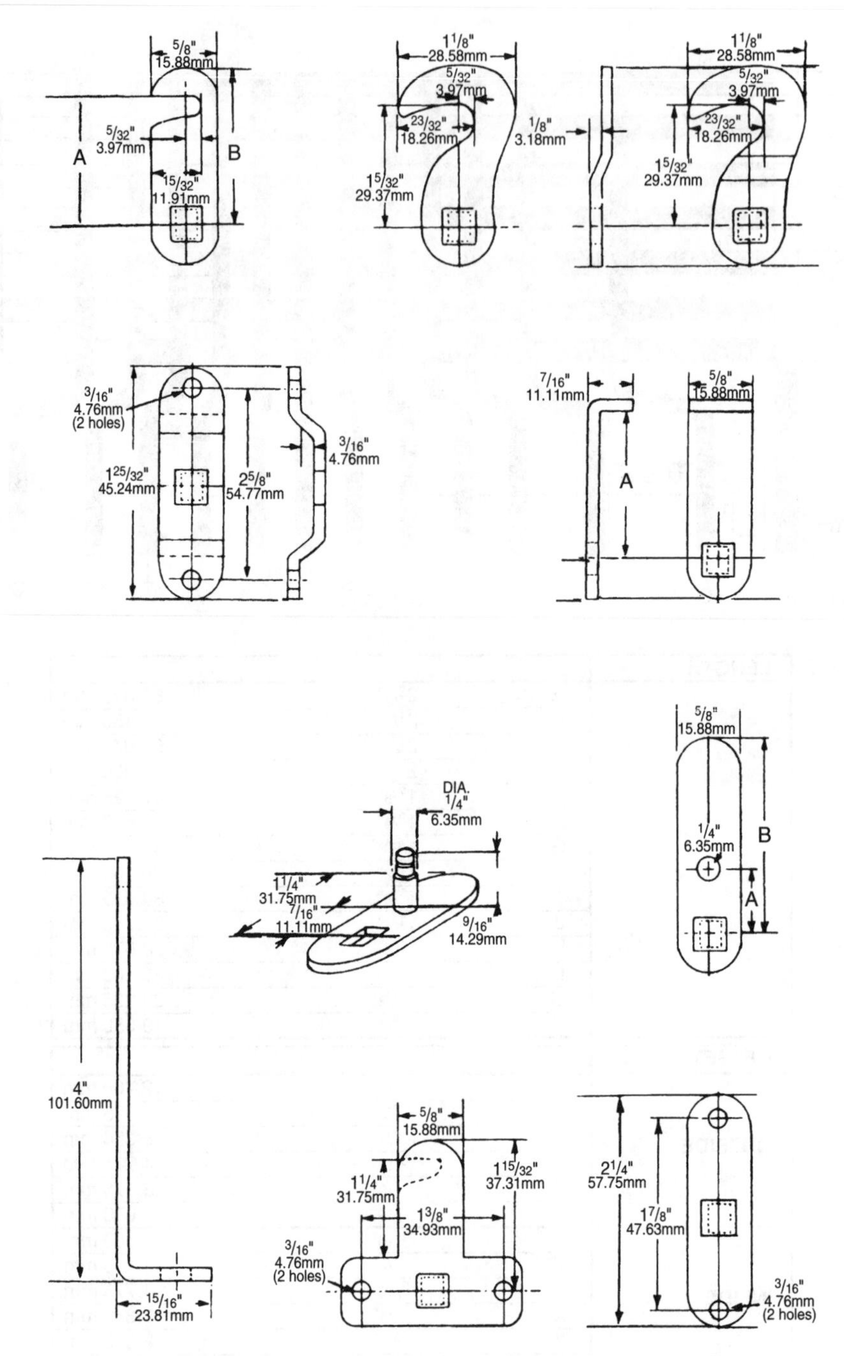

Figure 6.13 Hook, offset, double-ended, and miscellaneous-type cams. (*Dominion Lock Co.*)

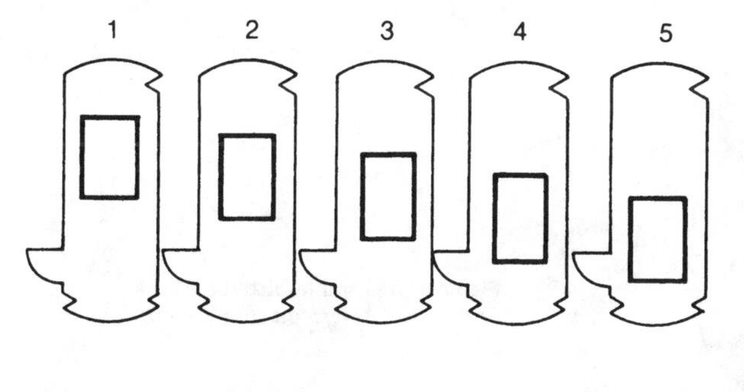

Figure 6.14 Disc tumblers have five variations.

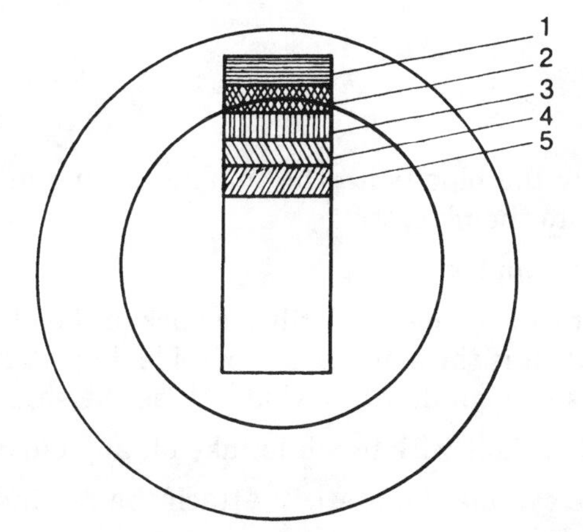

Figure 6.15 The position of the tumblers gives a general idea of the key profile.

Double-Bitted Disc Tumbler Locks

The Junkunc Brothers American double-bitted disc tumbler lock is often found in padlocks and office and utility locks (Fig. 6.16).

Operation

When the double-bitted key is inserted, it passes through the center of the tumblers (as in a disc lock) and aligns them to the shear line, allowing the plug to rotate. But, the key and the tumbler arrangement is different from that of the regular disc lock.

The key cuts are wavy in appearance; thus the tumblers have to align in a wavelike configuration for the lock to open. Further, there are no definitive tumbler cuts on the key. This is because the key holds 10 or more tumblers compressed together and held in a locked position by means of a Z-shaped wire within the tumblers.

All the tumblers are *uncoded*, meaning they are all of a standard cut. For them to turn within the cylinder, the tumblers have to be cut down. A special keying tool is used for this purpose.

Figure 6.16 Double-bitted padlock and key. (*American Lock Company*)

Cutting down the tumblers

1. Insert the tumblers into the plug (with the tumbler spring in place), then insert the precut key into the plug.
2. Mount the plug firmly in your vise.
3. Attach the keying tool to a ¼-inch drill. Drill the back of the plug. Since the inside diameter of the drill is the same as the outside diameter of the plug, the individual tumblers will cut down to what will be the shear line.
4. Trim the tumblers with a light wire brush to take off any burrs.
5. Insert the plug into the cylinder and test it. Attach the retainer screw and withdraw the key.
6. Since they are uncoded, the tumblers can be used within any plug. The tumbler spring, because of the shape, holds the various tumblers in the locked position. Only with the insertion of a key, which forces the tumblers into another position, can the lock be opened.

Keys and keyways

The double-bitted lock takes four basic key sections (Fig. 6.17). These sections, of course, match the shapes of the keyways. Keyway one is referred to as a K4 and the center point is at the center of the tumbler. Keyway two is referred to as a K4L; the center point is just left of center. Keyway three is called a K4R; the center point is right of center. Keyway four, called a K4W, is shaped like a W. The keyway shape does not reflect the tumbler types that are within any given plug.

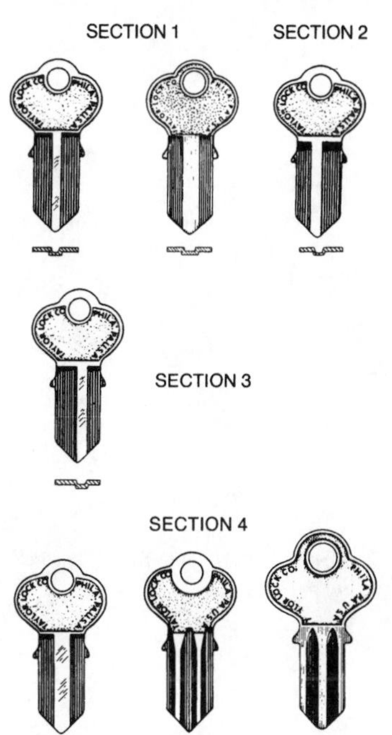

Figure 6.17 Four basic double-bitted key sections. (*Taylor Lock Company*)

Chapter

7

Pin Tumbler Locks

There are many types of pin tumbler locks. They come in various forms, including deadbolt, key-in-knob, lever, padlocks, and automobile ignition locks. The common denominator among pin tumbler locks is that they all have a pin tumbler cylinder or housing.

When a pin tumbler lock is installed on a door, you can usually see only the lock's plug or the face of its cylinder. Figures 7.1 and 7.2 show examples of common pin tumbler locks. You can quickly identify a pin tumbler lock by looking into its keyway. You'll see one or more pin tumblers hanging down (Fig. 7.3).

Construction

Although pin tumbler cylinders are simple mechanisms, some of the most secure mechanical locks rely on such a cylinder. Most pin tumbler cylinders are self-contained mechanisms. They come in a variety of shapes to fit locks of various designs. Figures 7.4, 7.5, and 7.6 show different types of pin tumbler cylinders.

The basic parts of a pin tumbler cylinder include a cylinder case (or shell), plug (or core), keyway, upper pin chambers, lower pin chambers, springs, drivers (or top pins), and bottom pins. It's easy to remember all those parts once you understand their relationships to each other.

The *cylinder case* houses all the other parts of the cylinder. Figure 7.7 shows the parts of a pin tumbler cylinder. The part that rotates when the proper key is inserted is called the *plug*. The *keyway* is the opening in the plug that accepts the key. The drilled holes (usually five or six) across the length of the plug are called *lower pin chambers*; they each hold a *bottom pin*. The corresponding drilled holes in the cylinder case directly above the plug are called *upper pin chambers*; they each hold a *spring* and a *driver*.

Figure 7.1 Pin tumbler cylinder key-in-knob lock. (*Schlage Lock Company*)

Figure 7.2 Pin tumbler cylinder deadbolt lock. (*Schlage Lock Company*)

Figure 7.3 When you look into the keyway of a pin tumbler cylinder, you can see a bottom pin hanging down.

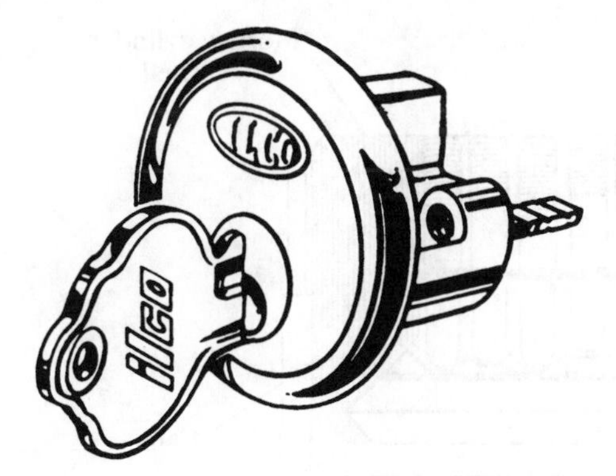

Figure 7.4 Pin tumbler cylinder for rim locks. (*Ilco Unican Corp.*)

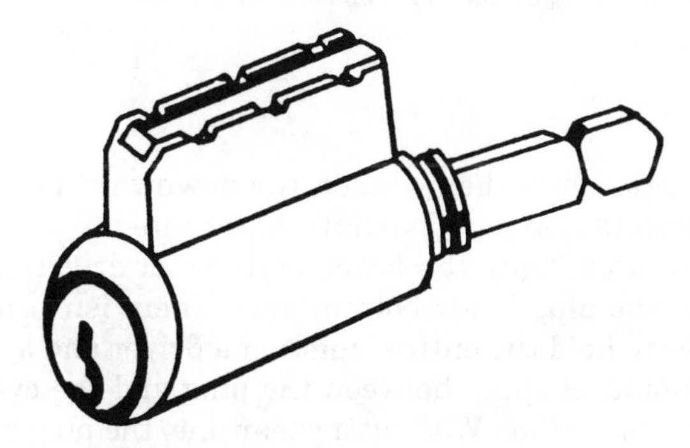

Figure 7.5 Pin tumbler cylinder for key-in-knob locks. (*Ilco*)

Figure 7.6 Pin tumbler cylinder for mortise locks. (*Ilco*)

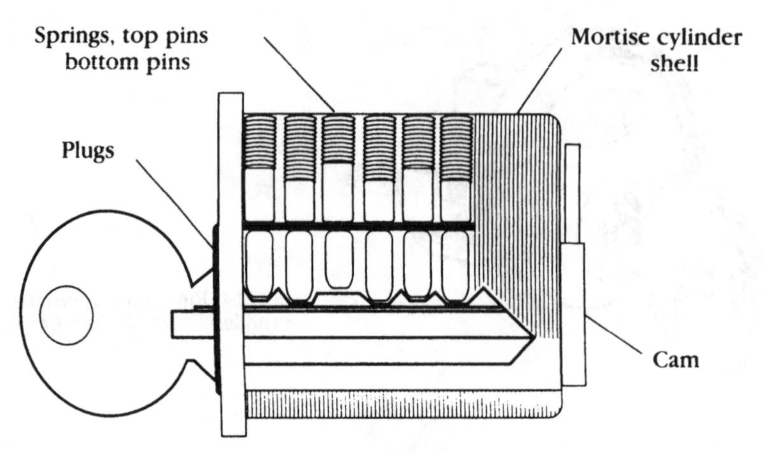

Figure 7.7 Cutaway view of a mortise pin tumbler cylinder from the side. (*The Locksmith Guild*)

How pin tumbler cylinders work

When a key is not inserted into the cylinder, the downward pressure of the springs "drives" the drivers (top pins) partially down into the plug to prevent the plug from being rotated. Only the lower portions of drivers are pushed into the plug, because the plug holds bottom pins. There isn't enough room in a lower pin chamber to hold the entire length of a driver and a bottom pin.

There is a small amount of space between the plug and the cylinder case. That space is called the *shear line*. Without a shear line, the plug would fit too tightly in the cylinder case to rotate. When a properly cut key is inserted, it causes the top of all the bottom pins and the bottom of all the drivers to meet at the shear line. While the pins are in that position, the plug is free to rotate to the open position (Fig. 7.8).

Plug. The plug contains a keyway, usually of the paracentric (off-center) type. Although they usually contain five or six lower pin chambers, some plugs have four or seven. The lower pin chambers are spaced fairly evenly along the upper surface of the plug, and they are aligned as closely as modern production techniques allow. By putting a plug with bottom pins into a plug holder and inserting a key, you can see how the plug will work in a cylinder (Fig. 7.9). The plug may be machined with a shoulder at its forward surface; this shoulder mates with a recess in the cylinder and provides

- A reference point for regulating the alignment of pin chambers in the case and the plug.
- A safeguard to prevent the plug from being driven through the cylinder, either deliberately or through resistance developed as the key enters the keyway.

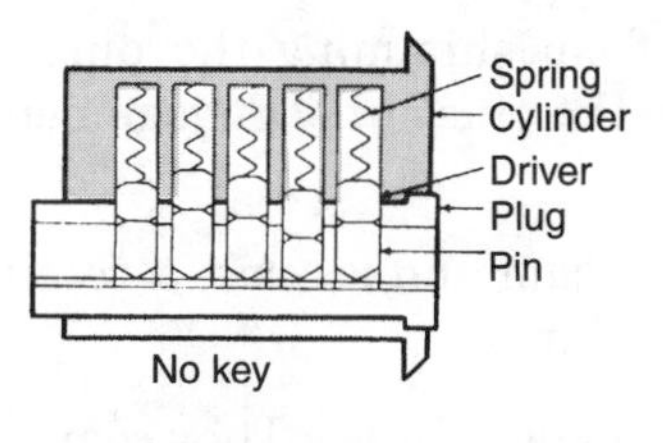

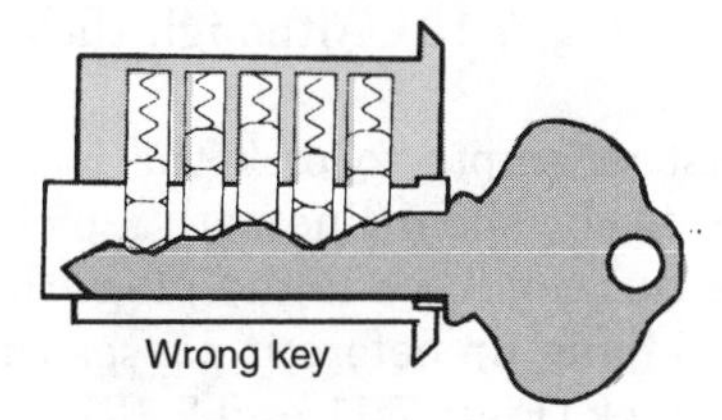

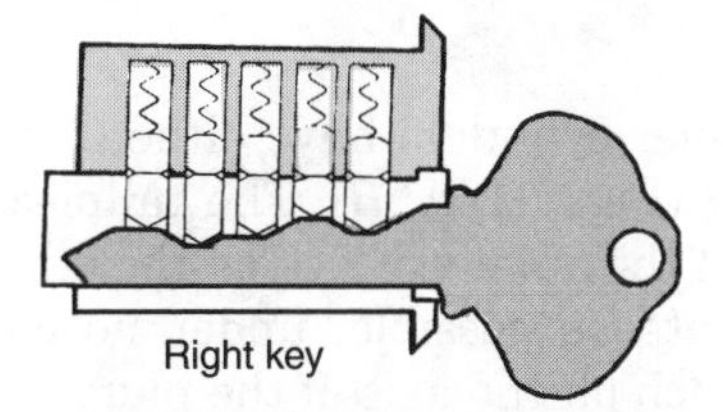

Figure 7.8 Cutaway view of side of a pin tumbler cylinder. (*The Locksmith Guild*)

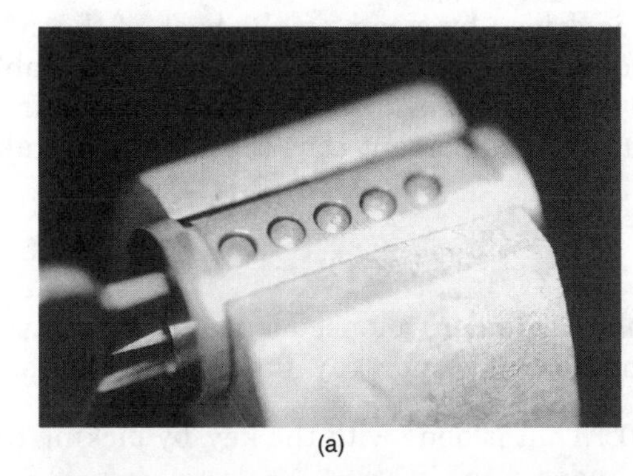

(a)

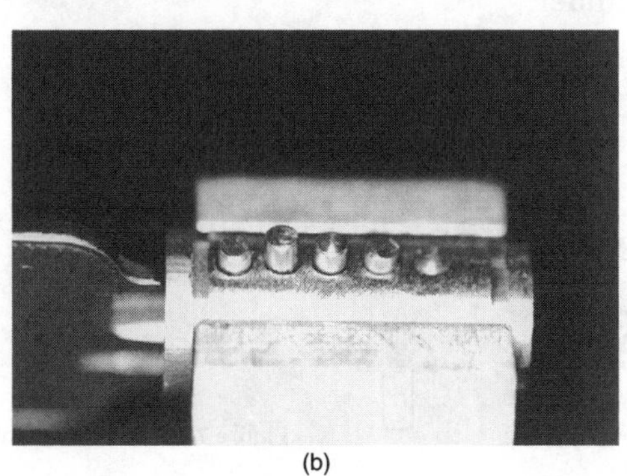

(b)

Figure 7.9 *A.* The right key brings all the bottom pins to the top of the plug, allowing it to rotate in the plug holder as shown (or in a lock cylinder). *B.* The wrong key causes some pins to rise too high or too low, not allowing it to rotate in the plug holder as shown (or in a lock cylinder).

- A safeguard to discourage a thief from shimming the pins. Without this shoulder it would be possible to force the pins out of engagement with a strip of spring steel.

The plug is retained at the rear by a cam and screws, a retainer ring, or a driver that locks the plug into the cylinder.

Pins. Top and bottom pins are usually made of brass. They come in a variety of lengths, diameters, sizes, and shapes (Fig. 7.10). Although they're small, pin tumblers are very strong.

The shape of the pins helps resist attempts to pick the lock. A standard cylindrical driver easily can be lifted to the shear line with a pick while a tension wrench is used to apply a little turning force to the plug. However, a top pin with a broken profile will tend to hang up before it passes the shear line, which makes a lock more difficult to pick (Figs. 7.11 and 7.12). (To learn more about lock picking see Chap. 14.)

Tailpieces and cams. Most pin tumbler cylinders have either a tailpiece or a cam attached to the rear of the plug (Fig. 7.13*A*, *B*). The tailpiece is loose to allow some flexibility in the location of the auxiliary lock on the other side of the door. Alignment should be as accurate as possible; under no circumstances should the tailpiece be more than ¼ inch off the axis of the plug.

On the other hand, pin tumbler cylinders for mortise locks do not have such tolerance. The best is driven by a cam on the back of the cylinder (Fig. 7.13*C*). If the lock is used on office equipment, the cam is probably a milled relief on the back of the plug; it might also be a yokelike affair, secured to and turning with the plug. It is important that these locks are aligned with the bolt mechanism.

Disassembly

To disassemble a pin tumbler cylinder, you need a small screwdriver, a *following tool* of the correct diameter, and a pin tray. Proceed as follows:

1. Turn the plug about 30°. That is done with the key, by picking the cylinder, or by shimming the cylinder.

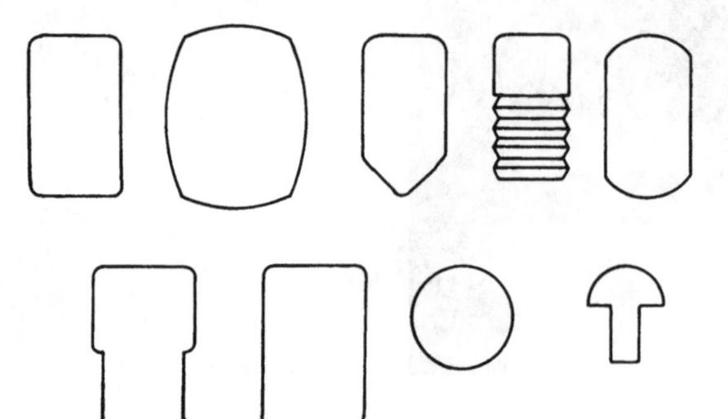

Figure 7.10 Various pin tumblers.

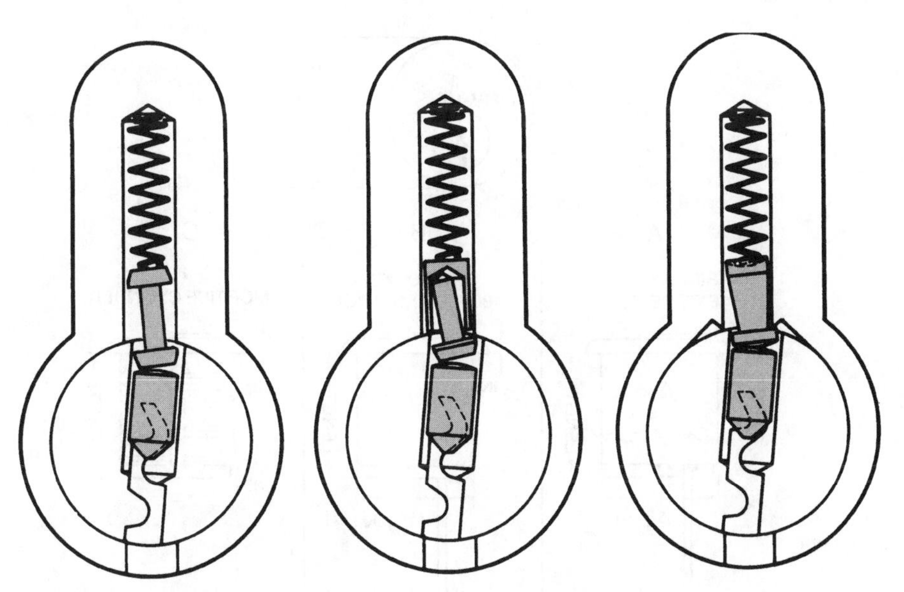

Figure 7.11 High-security top pins "hanging up" below the shear line. (*DOM Security Locks*)

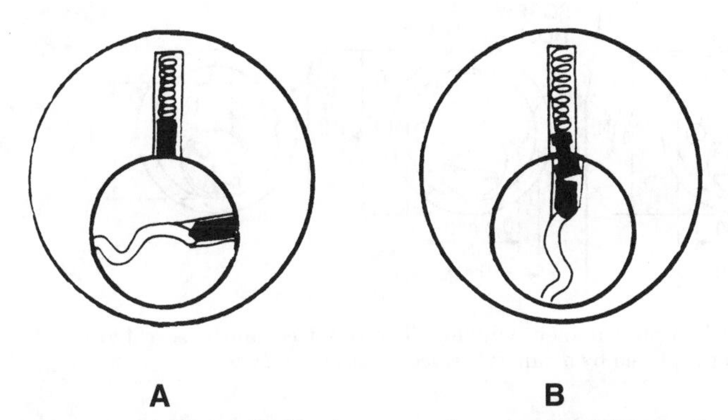

Figure 7.12 *A.* The cylinder shown employs standard drivers; it has been picked and the plug is free to rotate. *B.* The cylinder uses a spool tumbler; it has also been picked, but can be rotated only a fraction of an inch before the spool tumbler jams.

2. Remove the cam or tailpiece by removing two screws or a retainer ring. (Although they're not essential, a small pair of *snap ring pliers* can be useful for removing retainer rings.)
3. While holding the cylinder so the pins are in a vertical position, slip the appropriate size plug follower tool into the cylinder directly behind the plug. Use the plug follower tool to slowly push the plug out of the face of the cylinder. Do not create a gap between the plug and the plug follower tool or the springs and drivers will fall out.

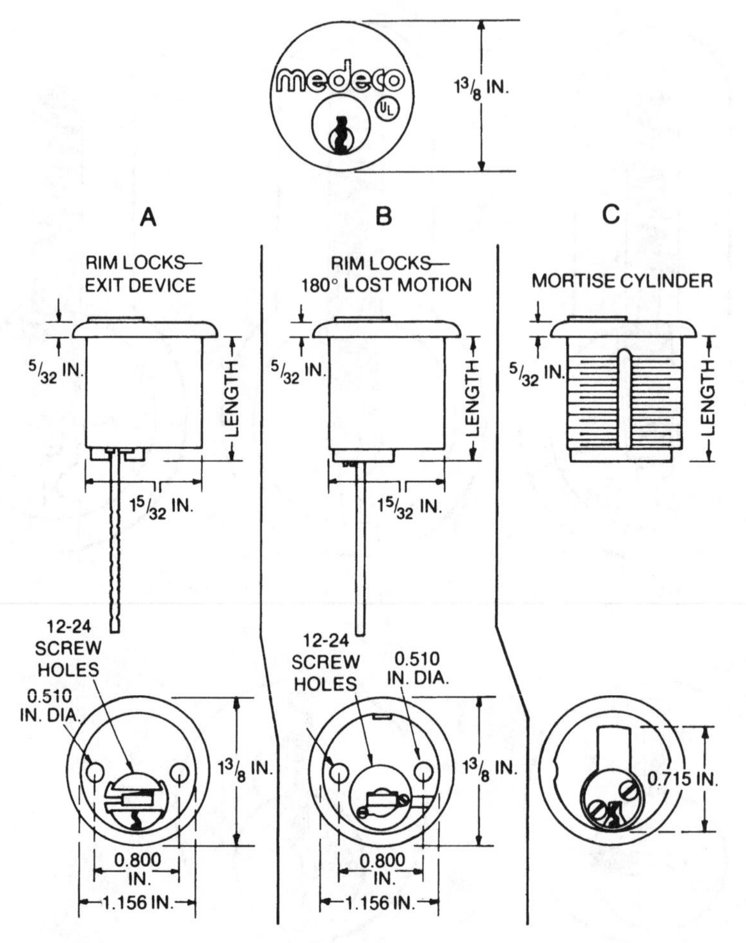

Figure 7.13 Medeco pin tumbler cylinders. The tailpiece can be serrated (*A*), solid (*B*), or replaced by a cam (*C*), (*Medeco Security Locks*)

4. Hold the plug as it leaves the cylinder; don't let the bottom pins fall out of the plug. Use your index finger to cover all pin chambers in the plug except the one closest to the face of the plug. Empty that chamber and put the pin into the first compartment of your pin tray.
5. Uncover the next pin chamber in the plug and empty the pin into the corresponding compartment of the pin tray. Continue this procedure until all the bottom pins are out of the plug and they are in proper order in the pin tray. Then you can put the plug down for awhile.
6. Hold the cylinder case and slowly pull the plug follower out the back of the cylinder. Stop each time a driver and spring fall from an upper pin chamber. Place them in the tray directly above each corresponding bottom pin.

Assembly

To assemble a pin tumbler cylinder, you need a plug follower, a small screwdriver, and pin tweezers. A *plug holder* is optional; it holds the plug in an upright position, leaving your hands free while you insert the bottom pins.

This procedure assumes that you have a key but do not know the pin combinations. (In normal reassembly, you already know which pins go into which chambers and you simply insert the springs and pins in the proper chambers.)

1. Place the plug into the plug holder and insert the key into the plug's keyway.
2. Taking one bottom pin at a time, insert it into the lower pin chamber nearest the face of the plug. If the pin stands above or below the shear line, try another pin. When a pin appears to be just at the shear line, turn the plug with the key. If the plug turns 360° in the plug holder, the bottom pin is probably the right one.
3. Repeat the procedure for each lower pin chamber until all the bottom pins just reach the shear line. *Note*: Do not remove the key or turn the plug over.
4. Move to the cylinder case and insert the plug follower through the back of the cylinder until it covers the third upper pin chamber from the face of the cylinder. Turn the cylinder case upside down so the upper pin chambers are facing the floor.
5. Using your tweezers, drop a spring into the second pin chamber from the face of the cylinder. Then, use your tweezers to insert a driver on the spring. By pushing the driver into the pin chamber while applying slight forward pressure on the plug follower, the plug follower will bind the driver into place. That will allow you to release the driver with your tweezers and use your tweezers to push the pin into the upper pin chamber while simultaneously pushing the plug follower over the driver. The plug follower will hold the driver and spring in place so they don't pop out of the cylinder. Repeat the procedure with the first upper pin chamber from the face of the cylinder.
6. At this point, you should have two upper pin chambers filled and covered by the plug follower. Push the plug follower out of the face of the cylinder to the point where your plug follower is covering only the upper pin chambers you've already filled. The remaining upper pin chambers should be exposed.
7. Insert a spring and driver in the exposed upper pin chamber nearest the plug follower. Cover that pin chamber with your plug follower. Repeat this step until all the upper pin chambers are filled.
8. Turn the cylinder case upright so the drivers and springs are in a vertical position.

9. Slowly remove the key from the plug so the bottom pins don't pop out of the plug. Insert the plug into the cylinder by pushing the back of the plug against the plug follower from the face of the cylinder. After the plug has forced the plug follower out of the rear of the cylinder case, rotate the plug until the upper pin chambers are aligned with the lower pin chambers.
10. The pins should lock the plug into place. *Note*: When inserting the plug, be sure the lower pin chambers are at least 30° out of alignment with the upper pin chambers.
11. Make sure your key works by inserting it into the plug and rotating the plug. Be careful while rotating the plug and removing the key; only the pins are holding the plug in place. With the key in the plug, all the pins should be at the shear line and the plug can easily be pulled out of the cylinder. If that happens, all the drivers and springs will fall out.
12. Attach the cam or tailpiece to the cylinder and retest your key.

Choosing a Pin Tumbler Lockset

Pin tumbler cylinder locksets are the most popular types of locks. These locksets are available in a wide variety of types and styles. Some of the general characteristics of such locksets include the following:

Security. The pin tumbler cylinder lockset provides better than average security by virtue of its internal design. Generally speaking, the more pins in a pin tumbler cylinder, the better security the lock offers. However, security also depends on the quality of the specific lock and on its application.

Quality. Quality depends on the intended service of the lockset. Light-, medium-, and heavy-duty locksets are available.

Type. Locksets are identified by their function (e.g., lavatory, classroom, residential)

Visual appeal. The lockset should match the decor of its surroundings; this is particularly important in new construction.

Hand. The location and direction of the swing of the hinges determines the hand of the door (Fig. 7.14). Taking the entrance side of the door as the reference point, there are four possible hands: left-hand, left-hand reverse (the door opens outward), right-hand, and right-hand reverse.

It is important to match the lockset to the door hand. Failure to do so can cause bolt/striker misalignment and result in an upside-down installation of the cylinder. When the cylinder is rotated 180°, the weight of the pins is on the springs. The pins weigh only a few grams, but that is enough to collapse the springs and disable the lock in time when a cylinder is installed upside down. Unless stipulated in the order, manufacturers supply right-hand mortise locksets; some of these can be modified in the field.

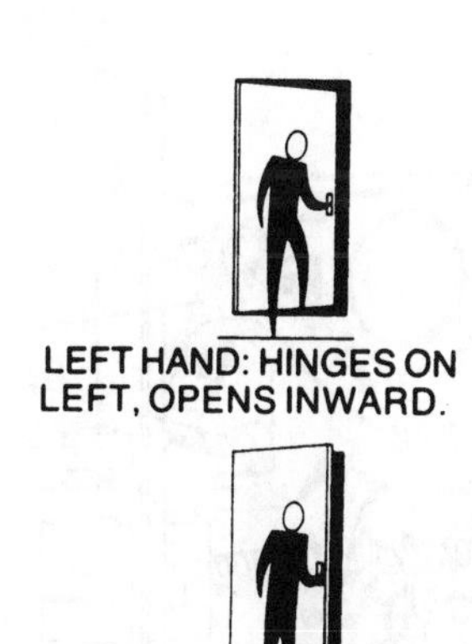

Figure 7.14 The *hand* of a door is a term that describes the location of the hinges and the direction of swing. (*Eaton Corp.*)

Pin Tumbler Cylinder Mortise Locks

The pin tumbler cylinder mortise lock is frequently used in homes, apartments, businesses, and large institutions. It's extremely popular and provides excellent security. Never mistake the pin tumbler cylinder mortise lock with its inferior cousin, the bit key mortise lock. Figure 7.15 shows two mortise-type pin tumbler cylinder locks without the cylinders in them.

Although the individual parts may vary slightly among different models, most basic pin tumbler cylinder mortise locks contain essentially the same parts. Practice disassembling and reassembling a few of these locks, and study the relationships of the various parts.

Specific pin tumbler cylinder locksets

The information presented in the following pages was supplied by various manufacturers. While it does not cover all pin tumbler cylinder locksets, it does give an overview of the current state of the art. If you need further information, contact a factory representative.

Kwikset Corporation. Kwikset's key-in-leversets are part of the company's Premium line (Fig. 7.16). The leverset is constructed of solid brass and steel. It's engineered to meet most building code requirements, including major handicapped codes, and built to absorb the punishment of high-traffic areas. It's UL listed (see Chap. 8) and backed by a 10-year limited warranty. The leverset is available in a wide range of finishes.

Kwikset offers several styles of entrance handlesets (Fig. 7.17) to complement any entrance architecture from traditional to contemporary.

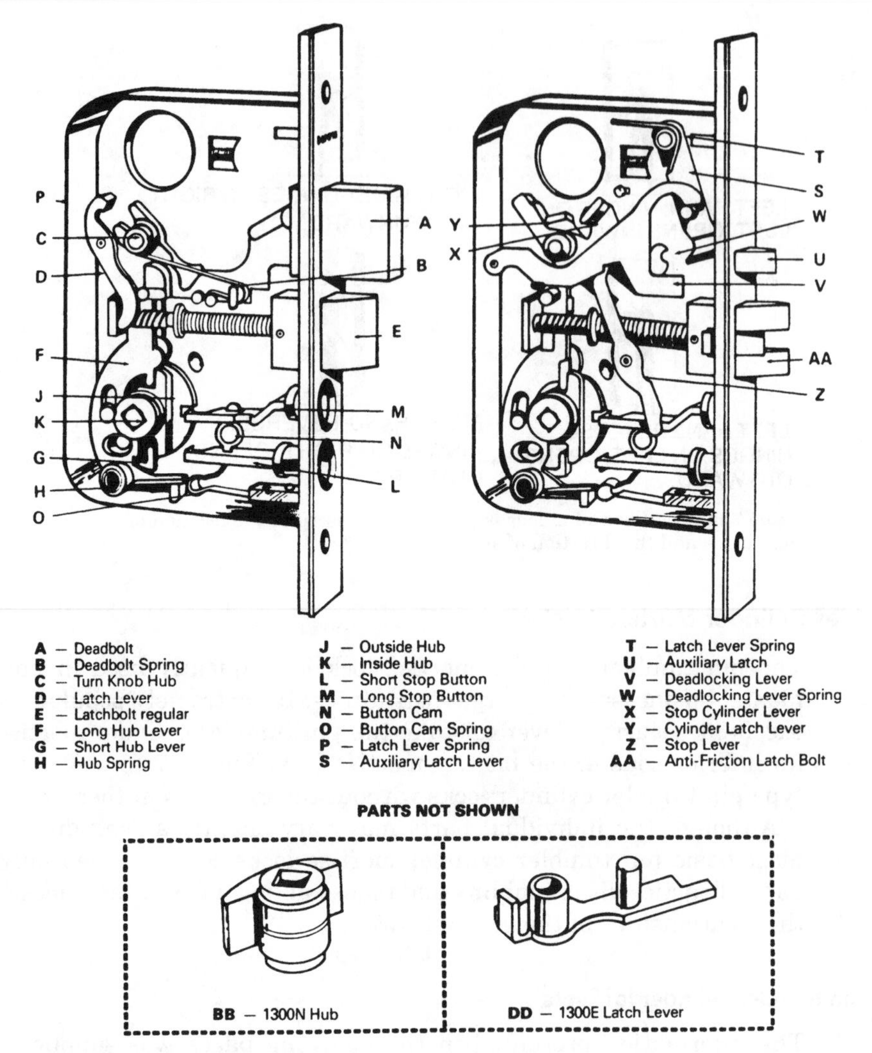

Figure 7.15 Inside views of pin tumbler cylinder mortise locks. (*Dominion Lock Co.*)

Kwikset's Series 500 Premium Entrance Lockset was designed and built in response to specific needs of the building industry for an improved, heavier-duty lockset for offices, apartments, townhouses, condominiums, and finer homes (Fig. 7.18).

The Series 500 Locksets include features such as a panic-free operation on the interior knob; a solid steel reinforcing plate to protect the spindle assembly; a reversible deadlatch with a ⅔-inch throw, solid-brass, beveled latchbolt; a solid-steel strike plate with heavy-duty screws; and a free-spinning exterior knob for protection against wrenching.

Figure 7.16 Kwikset's Premium Keyed Leverset. (*Kwikset Corporation*)

Figure 7.17 Kwikset's Entrance Handlesets. (*Kwikset Corporation*)

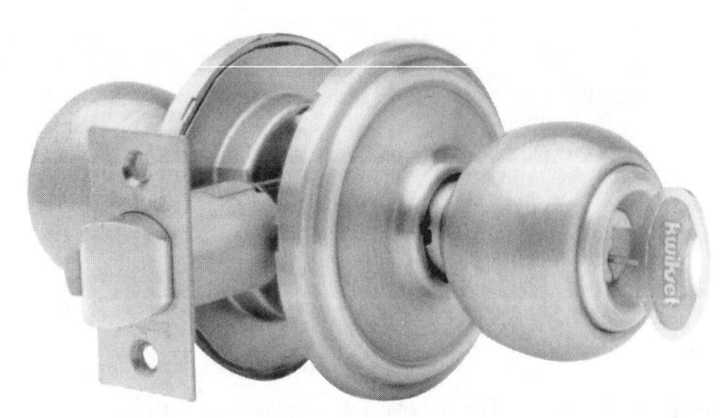

Figure 7.18 Kwikset's Premium Entrance Locksets. (*Kwikset Corporation*)

Kwikset's Protecto-Lok (Fig. 7.19) combines an entry lockset with a 1-inch cylinder deadlock. Both deadbolt and deadlatch retract with a single turn of the interior knob or lever, assuring immediate exit in case of an emergency. Protecto-Lok is UL listed when ordered and it is used with a special UL latch and strike.

Kwikset's Safe-T-Lok (Fig. 7.20) combines deadbolt security with instant exit. When you lock it from the interior, the key remains in the cylinder and cannot be removed until the deadbolt is fully retracted in an unlocked position. Safe-T-Lok has a 1-inch steel deadbolt with a heat-treated steel insert that turns with any attempted cutting.

Schlage Lock Company. Schlage's B Series deadbolt locks provide primary or auxiliary security locking for both commercial and residential applications. It includes the standard duty B100 Series for residential structures (Fig. 7.21), the heavy-duty B400 Series for commercial requirements, and the most secure B500 Series for maximum performance. All products in this series feature a full 1-inch throw deadbolt with Schlage's exclusive wood frame reinforcer to deter "kick-in" attack.

Schlage's commercial quality D Series locks (Fig. 7.22) are specified when the highest-quality mechanisms are required. Precision manufactured to exact tolerances, the D Series is best suited for commercial, institutional, and industrial use. Specialized functions satisfy many unique locking applications.

Schlage's L Series mortise lock line is a heavy-duty commercial mortise lock series containing a wide array of knob, lever, and grip handle designs (Fig. 7.23). It's available in a variety of keyed and nonkeyed functions and in most decorative finishes. It is UL listed and can be used in offices, schools, hospitals, hotels, and commercial buildings as well as residences. The L lock meets or exceeds the ANSI A156.13 specification, making it an excellent choice for any building where security, safety, and design compatibility are prime considerations.

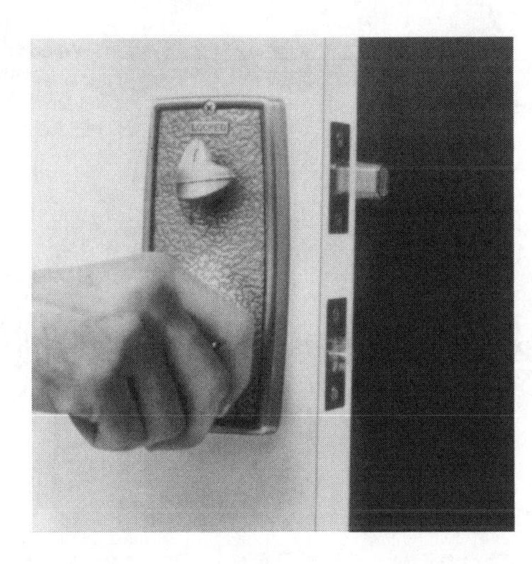

Figure 7.19 Kwikset's Protecto-Lok. (*Kwikset Corporation*)

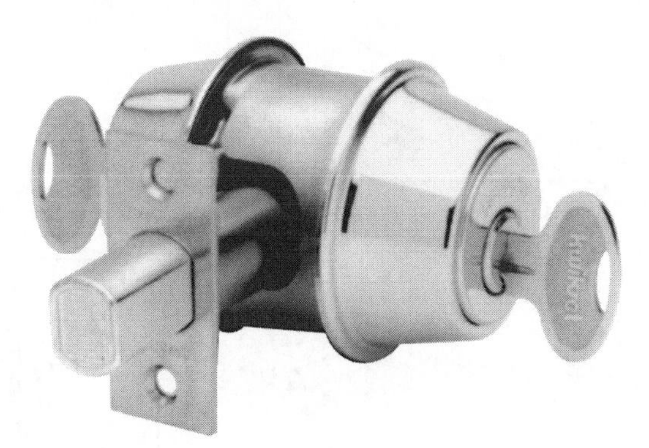

Figure 7.20 Kwikset's Safe-T-Lok. (*Kwikset Corporation*)

The L lock has one common mortise lock case for knob, lever, and grip handle trim, providing tremendous versatility. In addition, it provides excellent design flexibility in trim combinations, such as knob by lever, lever by knob, etc. All may be specified on the same L lock case.

The Cylinder Key

Figure 7.24 illustrates basic key nomenclature and Fig. 7.25 shows an assortment of key blanks. Notice the differences in the bow, blade, and length and in the width, number, and spacing of the grooves.

The bow usually has a specific shape that identifies the lock manufacturer. The blade length is indicative of the number of pins within the lock that the key operates (e.g., a five-pin key is shorter than a six-pin key). The height of the blade sometimes indicates the depths of the cuts in a series; generally, the higher the blade, the deeper the cuts.

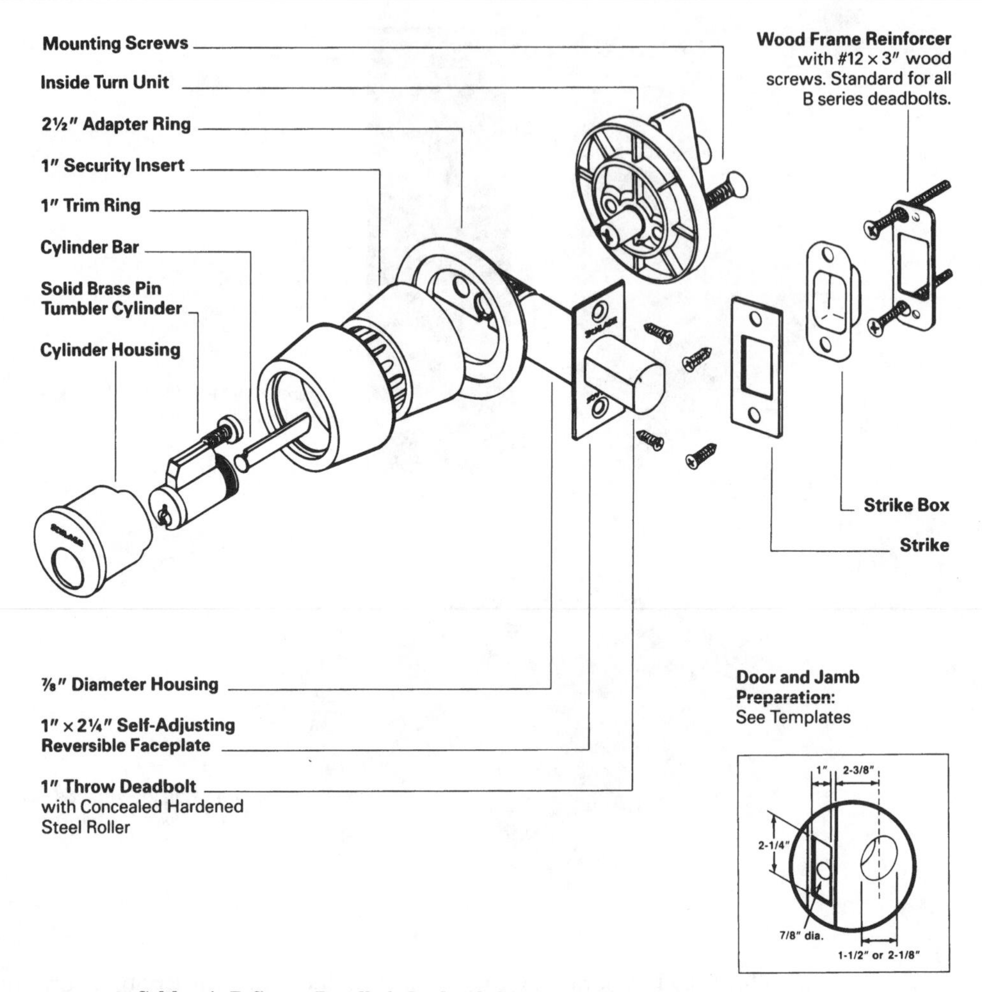

Figure 7.21 Schlage's B Series Deadbolt Lock. (*Schlage Lock Company*)

Key blanks are made of brass, nickel-brass, nickel-steel, steel, aluminum, or a combination of these and other alloys. The metal determines the strength of the key and its resistance to wear.

Duplicating a cylinder key by hand

Duplicating a cylinder key by hand isn't much different from duplicating a lever or warded key by hand. The procedure is as follows:

1. Smoke the original and mount it together with the blank in a vise. The blank must be in perfect alignment with the original.
2. Using Swiss round and warding files, begin at the tip of the key and work toward the bow. Start in the center of each cut.

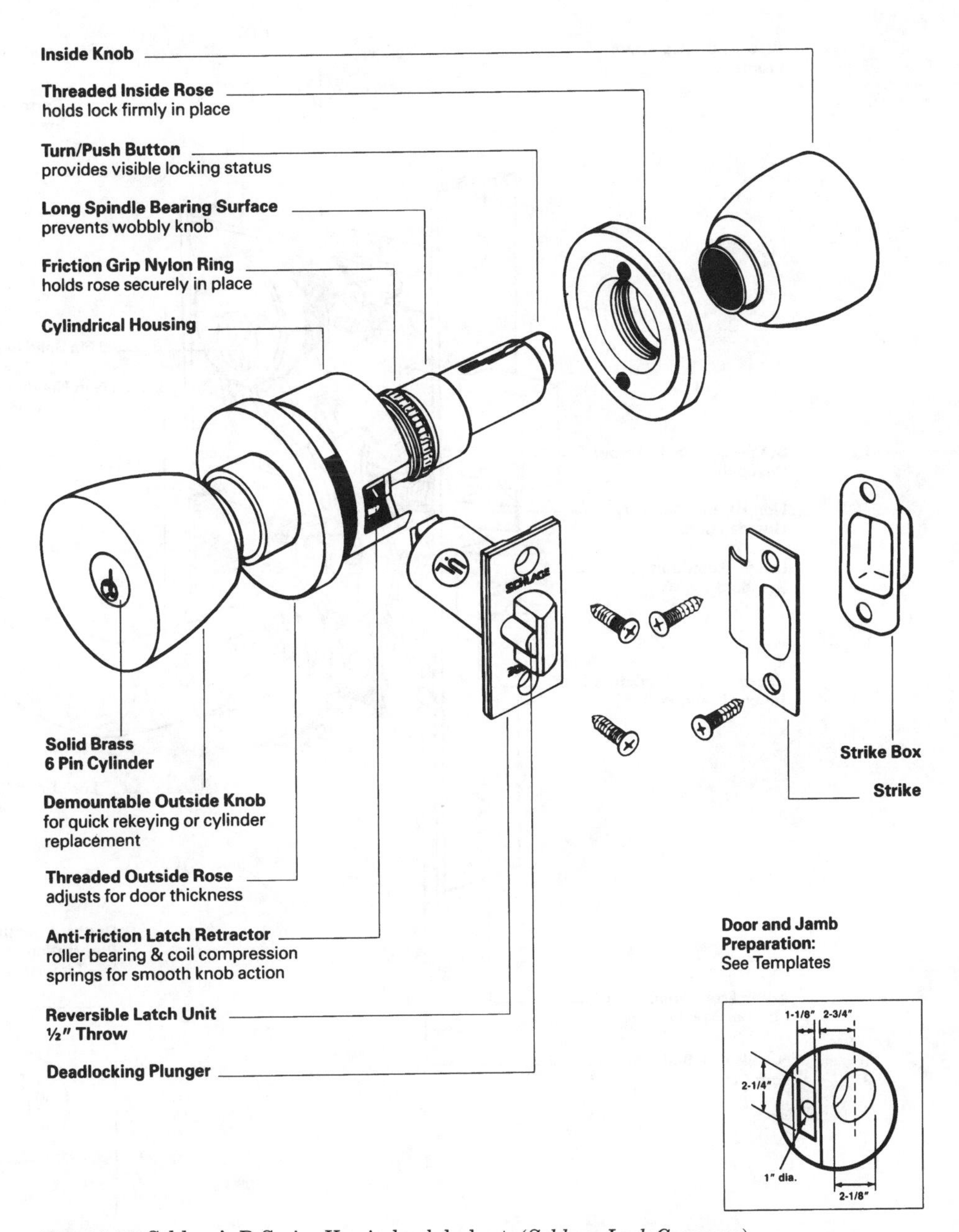

Figure 7.22 Schlage's D Series Key-in-knob lockset. (*Schlage Lock Company*)

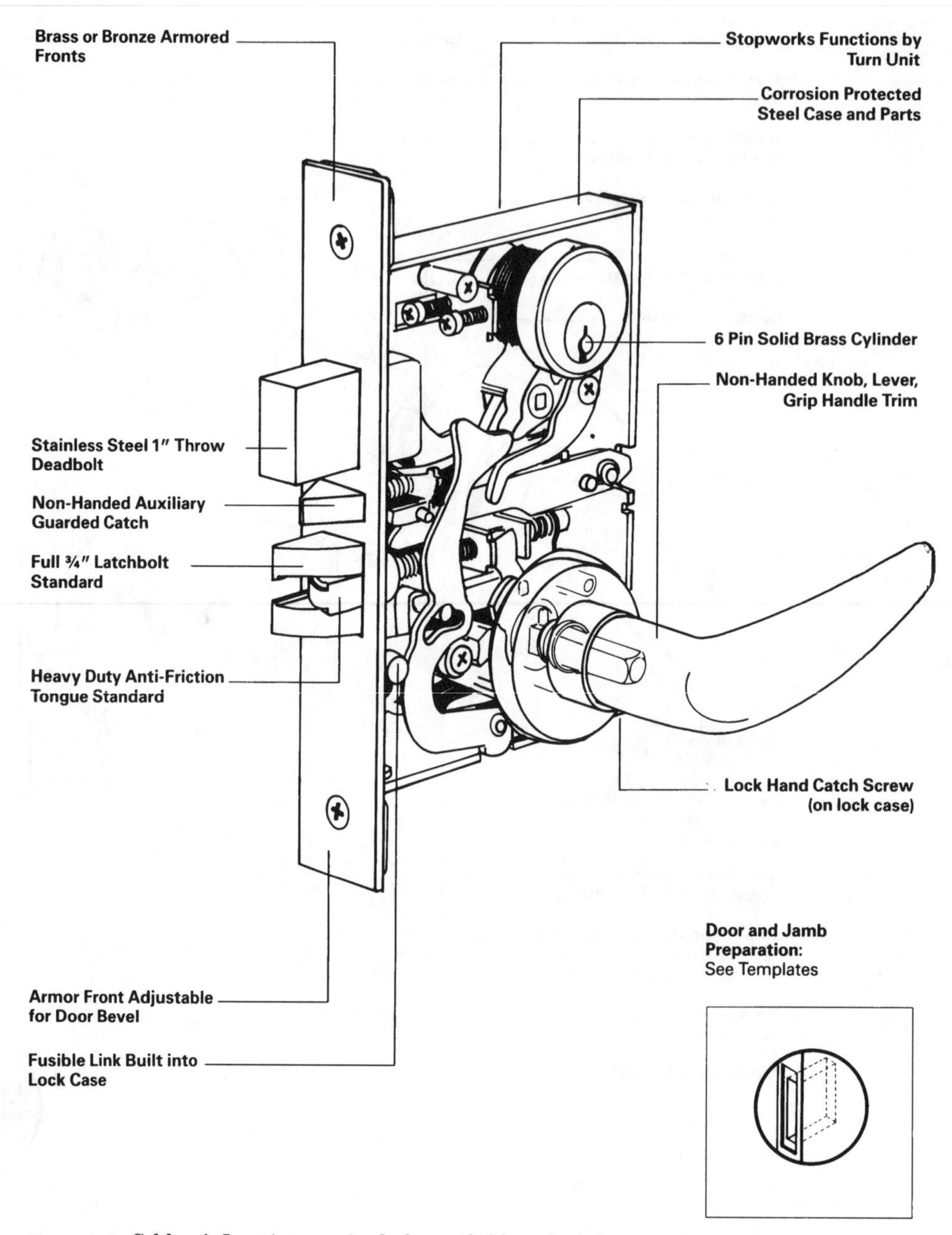

Figure 7.23 Schlage's L series mortise lockset. (*Schlage Lock Company*)

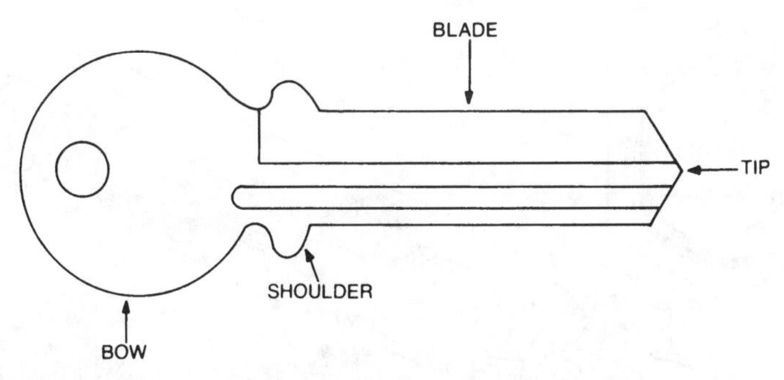

Figure 7.24 Basic cylindrical key nomenclature.

Figure 7.25 Assorted key blanks. (*Master Lock Company*)

3. Go slowly near the end of each cut, but continue to cut with firm and steady strokes. Stop when the file touches the soot on the original key.
4. Once you have finished, remove the keys from the vise. Wire-brush the duplicate to remove any burrs, then polish the key.
5. Hold the duplicate up to the light. Place the original in front of it and behind it to determine if the duplicate is accurately cut. Shallow cuts can be cut a little deeper with a file; if a cut is too deep, start over with a new key blank.

Vending machine locks

The tubular key lock has become standard on vending machines. The arrangement of the pin tumblers increases security and requires a different type of key—one that is tubular. A typical tubular key lock and key is shown in Fig. 7.26. The cam works directly off the end of the plug.

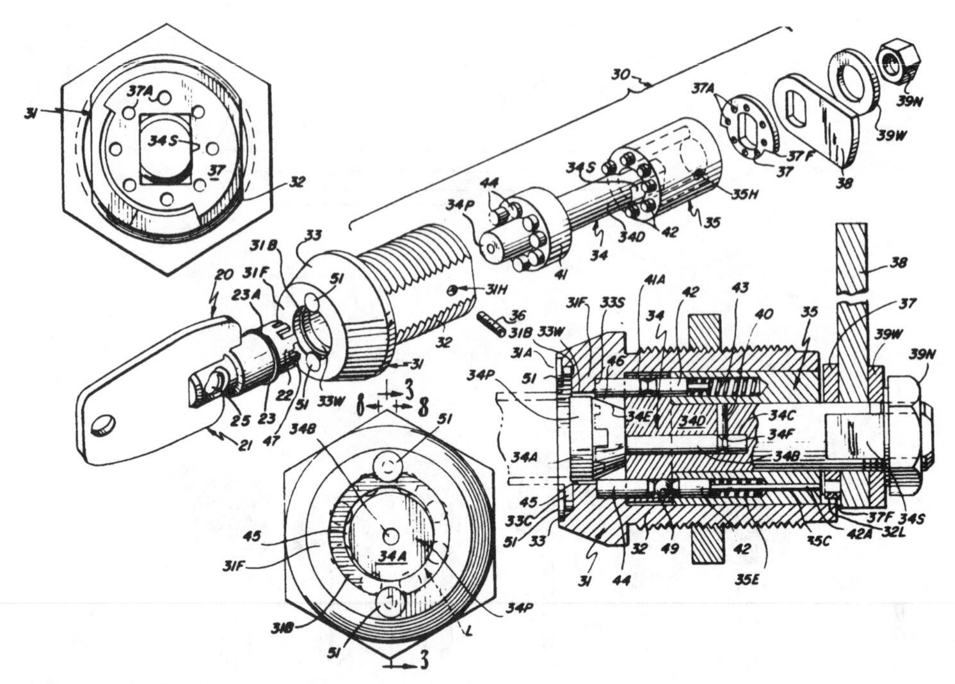

Figure 7.26 An “Ace” Tubular Key Lock.

The key has its bitting disposed radially on its end. The depth and spacing of each cut must match the pin arrangement. In addition, there are two notches, one on the inside and the other on the outer edge of the key. These notches align the key to the lock face. Otherwise the key could enter at any position.

The key bittings push the pins back, bringing them to the shear line. Once the pins are in alignment, the key is free to turn the plug and attached cam.

The pins within the lock are entirely conventional in construction, with the exception of the bottom one. A ball bearing is sandwiched between the pin and its driver. The bearing reduces friction and increases pin life. The pin in question is not interchangeable with others in the lock. Pin tolerance is extremely critical. There is no room for sloppy key cutting.

Disassembly. To disassemble the tubular key lock, follow this procedure:

1. Place the lock into its holder. Ace makes a special vise for these cylinders.
2. Drill out the retainer pin at the top of the assembly. Use a No. 29 drill bit and stop before the bit bottoms in the hole (Fig. 7.27).
3. Remove what is left of the pin with a screw extractor.
4. Insert the appropriate plug follower into the cylinder. Apply light pressure. Plug follower dimensions are length 1.50 inches, outside diameter 0.373 inch, inside diameter 0.312 inch.

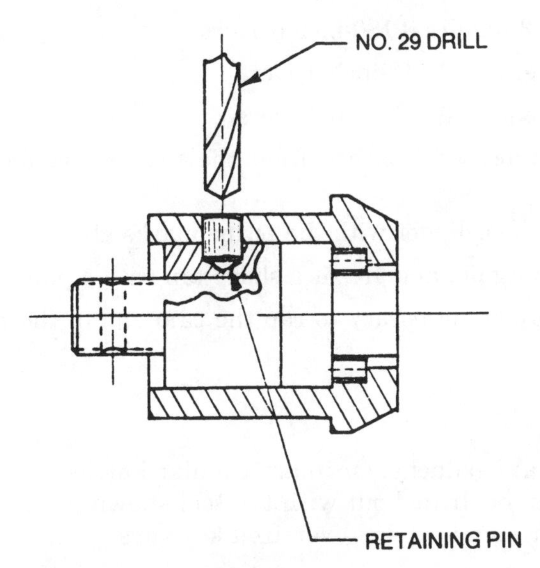

Figure 7.27 Use a No. 29 drill bit to remove the retaining pin. Alternately, use a No. 42 bit, thread a small metal screw into the hole, and extract the pin and screw by prying upward on the screw head. (*Desert Publications*)

5. Lift off the outer casing from the bushing assembly.
6. Scribe reference marks on the plug sections as an assembly guide.
7. Remove the follower from the cylinder. *Note*: Perform this operation carefully. The pins are under spring tension and must be kept in order. If not, you will have a monumental job sorting the pins. There are 823,543 possible combinations!

Assembly. Assemble the lock in the reverse order of disassembly. Replace the retaining pin with an Ace part, available from locksmith supply houses.

Rekeying. Rekeying is not difficult if you approach the job in an orderly manner.

1. Cut the key. There are seven bit depths, ranging from 0.020 to 0.110 inch in 0.015-inch increments.
2. Remove the pins with tweezers. *Note*: Pin lengths range from 0.025 to 0.295 inch in increments of 0.015 inch. Drivers are available in 0.125-, 0.140-, and 0.180-inch lengths.
3. Select new pins using the key combination as a guide.
4. Install the pins. The flat ends of the core pins are toward the key (Fig. 7.26). This pattern must be followed:

Core pins 1, 2, and 3 require 0.180-inch drivers.

Core pins 4 and 5 require 0.140-inch drivers.

Core pins 6 and 7 require 0.125-inch drivers.

Note: Pins are numbered clockwise from the top as you face the lock. Insert the key.

5. Install the plug in the cylinder with the scribe marks aligned.
6. Insert a new retaining pin and give it a sharp tap with a punch.
7. Test the key. It may be necessary to rap the cam end of the plug with a mallet.

Lockout

A lockout can be a real headache. Ordinary tubular key locks—those without ball bearings—can be drilled out with the tool shown in Fig. 7.28. Hole saws are available for standard and oversized keyways.

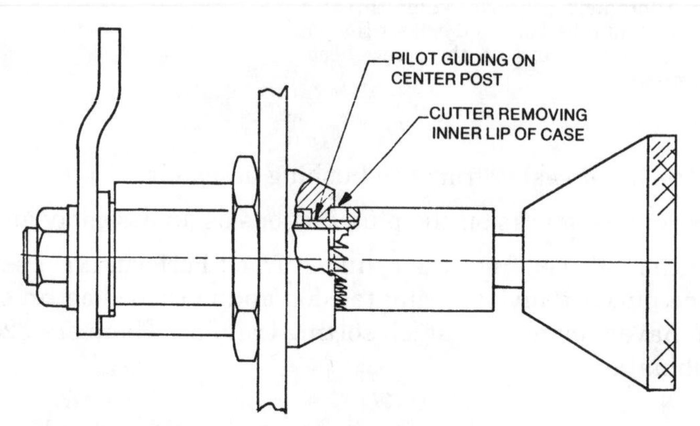

Figure 7.28 Drilling the lock requires a piloted hole saw, available from locksmith supply houses. (*Desert Publications*)

Chapter

8

High-Security Mechanical Locks

With respect to locking devices, the term *high security* has no precise meaning. Some manufacturers take advantage of that fact by arbitrarily using the term to promote their locks.

Most locksmiths would agree that to be considered *high security* a lock should have features that offer more than ordinary resistance against picking, impressioning, drilling, wrenching, and other common forms of burglary. The most secure locks also provide a high level of key control. The harder it is for an unauthorized person to have a duplicate key made, the more security the lock provides.

Underwriters Laboratories Listing

Founded in 1894, Underwriters Laboratories, Inc. (UL) is an independent, nonprofit product-testing organization. A UL listing (based on UL standard 437) is a good indication that a door lock or cylinder offers a high degree of security. If a lock or cylinder has such a listing, you'll see the UL symbol on its packaging or on the face of the cylinder.

To earn the UL listing, a lock or cylinder must meet strict construction guidelines and a sample model must pass rigorous performance and attack tests. Some of the requirements are as follows:

- All working parts of the mechanism must be constructed of brass, bronze, stainless steel, or equivalent corrosion-resistant materials or have a protective finish complying with UL's Salt Spray Corrosion test.
- Have at least 1000 key changes.
- Operate as intended during 10,000 complete cycles of operation at a rate not exceeding 50 cycles per minute.

- The lock must not open or be compromised as a result of attack tests using hammers, chisels, screwdrivers, jaw-gripping wrenches, pliers, hand-held electric drills, saws, puller mechanisms, key impressioning tools, and picking tools.

The attack test includes 10 minutes of picking, 10 minutes of key impressioning, 5 minutes of forcing, 5 minutes of drilling, 5 minutes of sawing, 5 minutes of prying, 5 minutes of pulling, and 5 minutes of driving. Those are net working times, which don't include time used for inserting drill bits or otherwise preparing tools.

Key Control

Another important factor in lock security is *key control*. The most secure locks have patented key blanks that are available only from the lock manufacturer. At the next level of key control are key blanks that can be cut only on special key machines. This type of key control greatly reduces the number of places where an unauthorized person can have a key duplicated. The least secure locks use keys that can be copied at virtually any hardware or department store.

Lock and Key Patents

One feature of a high-security lock is that it provides a high degree of key control. That is, there is restriction on who can make a copy of the key. Such locks are used in situations where you don't want just anybody to be able to run down to the local hardware store and make copies. One way of restricting copies is by having a patented lock or blank. That allows a lock manufacturer to decide who may copy its keys.

Not all patents offer a lot of key control, however. There are two relevant types of patents: design and utility. A design patent only protects how a thing looks. In 1935, Walter Schlage received a design patent for a key bow. The bow's distinctive shape made it easy for locksmiths and other key cutters to recognize the company's keys. But the patent didn't prevent aftermarket key blank manufacturers from making basically the same key blanks with slightly different bow designs. Design patents are good for 14 years.

Utility patents are more popular and provide more protection against unauthorized manufacturing. All patents expire eventually, and they aren't renewable. Utility patents are granted for 20 years, beginning on the application filing date. Medeco Security Lock's original utility patent (No. 4,635,455) was issued in 1970. Because that patent prevented companies from making key blanks that fit Medeco locks, the company was able to maintain maximum key control. As soon as the patent expired (in 1986), many companies began offering compatible key blanks. The company responded by introducing a new lock, the Medeco Biaxial, getting a utility patent (No. 4,393,673) that expired in 2004. The company's newest patented product is its Medeco3, which expires in 2021. Table 8.1 shows the patent expiration dates for several high-security locks.

TABLE 8.1 Lock Makers Patent Expiration Dates

Manufacturer/product line	Expiration date
Medeco-Medeco^3	2021
Corbin Russwin Pyramid	2018
Lori-L10	2016
Arrow Flex Core	2015
ASSA V-10	2014
Mul-T-Lock Interactive	2014
Sargent Signature	2014
Schlage Everest	2014
Schlage Primus/Everest	2014
Kaba Peaks	2008
Mul-T-Lock	2007
Kaba Gemini Peaks	2006
Medeco Biaxial	2005
Schlage Primus	2005
Abloy-Disklock Pro	2004

Just because a patent has expired doesn't mean that there are aftermarket blanks available for the lock. High-security lock makers plan for patent expirations. They make many different keyways and keep that number a secret. They also make a point of evenly distributing keyways to different geographic areas. Aftermarket blanks are made only if the blank makers can get sample keyways and if they believe that they can sell enough of the blanks to make it worth their while. If you're thinking about carrying a high-security line, ask the manufacturer how many keyways it has. A good manufacturer won't tell you.

Types of High-Security Mechanical Locks

This section provides detailed information about some of the most popular high-security mechanical locks available. Much of the information was obtained from manufacturer's technical bulletins and service manuals.

Schlage's Primus

The Schlage Primus is one of Schlage Lock Company's newest high-security mechanical locking systems. It features a specially designed, patent-protected key that operates either the UL listed #20-500 High Security Series Cylinders or the Controlled Access #20-700 Series Cylinders (not UL listed). Both series are available in Schlage A, B, C/D, E, H, and L Series locks (Figs. 8.1 and 8.2).

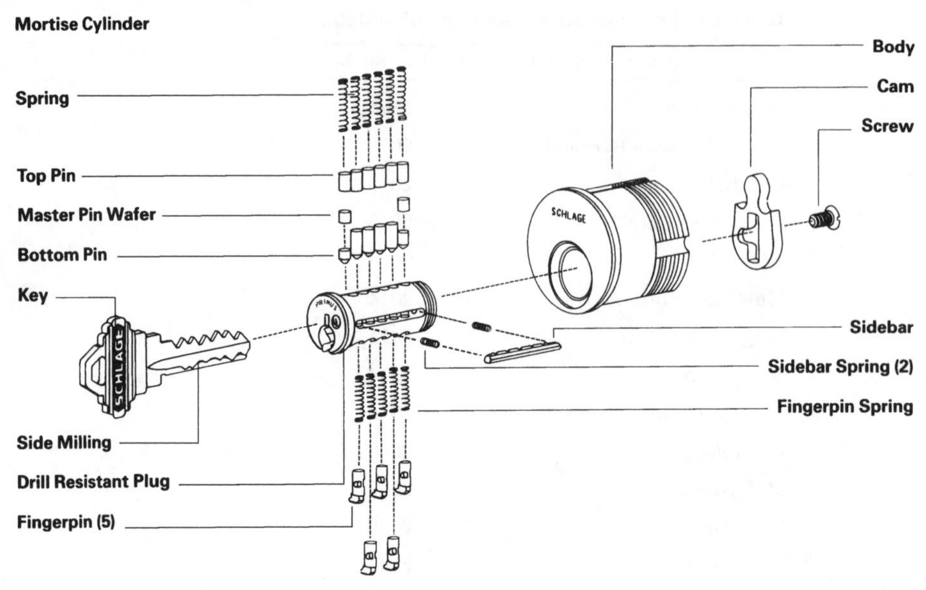

Figure 8.1 An exploded view of a Schlage Primus mortise cylinder. (*Schlage*)

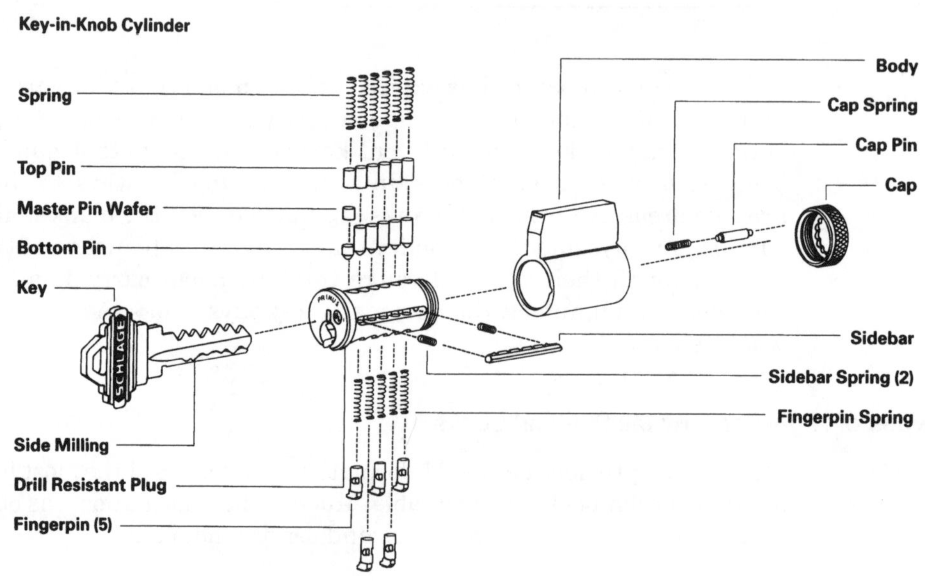

Figure 8.2 An exploded view of a Schlage Primus key-in-knob cylinder. (*Schlage*)

Both series are easily retrofitted into existing Schlage locks, and they can be keyed into the same master key system and operated by a single Primus key. The Primus key is cut to operate all cylinders, while those that operate the standard cylinders will not enter a Primus keyway.

The Primus security cylinder is machined to accept a side bar and a set of five finger pins, which, in combination with Schlage's conventional six pin keying system, provides two independent locking principles operated by a Primus key. Hardened steel pins are incorporated into the cylinder plug and housing to resist drilling attempts.

Primus security levels. The Primus system features five different levels of security. Each level requires an appropriate ID card for key duplication.

Security Level One. Primus cylinders and keys have a standard side milling allocated to Level One and are on a local stock basis. Level One also provides the flexibility of local keying into most existing Schlage key sections; it utilizes a Primus key to operate both systems. Key control is in the hands of the owner, who holds an ID card for the purpose of acquiring additional keys. Level One is serviced through qualified locksmiths who are fully trained at a Schlage Primus I Center.

Security Level One Plus. Level One Plus increases Level One security by allocating a restricted side-bit milling to a limited number of local Primus centers, which also control keying and key records.

Security Level Two. At Level Two, service is carried out solely by original Schlage Primus distributors. Also, key duplication for the exclusive side-bit milling requires the authorized purchaser's signature. Factory keying is also available.

Security Level Three. At Level Three, the Schlage Lock Company maintains control of the inventory, keying, keying records, installation data, and owner's signature. Keys for the geographically restricted, side-bit millings can be duplicated only by the factory after the owner's signature has been verified.

Security Level Four. Level Four provides all the restrictions of Level Three, plus an exclusive side-bit milling.

Kaba

Kaba locks are *dimple key* locks. There isn't just one, but an entire family of different Kaba cylinder designs. Each one is designed to fill specific security requirements (Figs. 8.3 to 8.5).

The family includes Kaba 8, Kaba 14, Kaba 20, Kaba 20S, Saturn, Gemini, and Micro. Those that use numbers in their designations generally reflect not only the order of their development, but also the number of possible tumbler locations in a cylinder of that particular design (Figs. 8.6 and 8.7).

Handing and key reading. The concept of *handing* is basic to the understanding of all Kaba cylinders. With Kaba cylinders handing is not a functional installation limitation as you might expect. A left-hand cylinder will operate both clockwise and counterclockwise and function properly in a lock of any hand. The handing of Kaba cylinders refers only to the positions where the pin chambers are drilled.

Figure 8.3 Some Kaba locks fit Adams Rite narrow stile metal doors. (*Lori Corp.*)

Figure 8.4 Double cylinder deadbolt lock with "Inner Sanctum" core. (*Lori Corp.*)

Figure 8.5 Micro-Kaba switch lock and keys. On the right is the lock with the core removed. The switch size (compare with coin) provides for a wide variety of uses with electrical and electronic circuitry. (*Lori Corp.*)

Figure 8.6 Kaba standard mortise cylinders can be used on Adams Rite doors. (*Lori Corp.*)

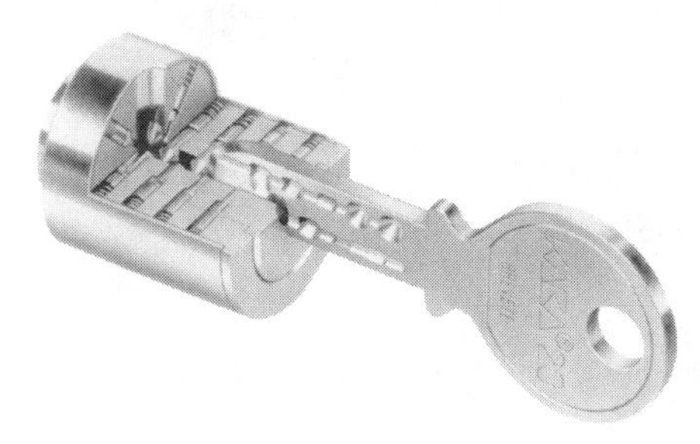

Figure 8.7 Cutaway view of a Kaba 20 cylinder. (*Lori Corp.*)

Figure 8.7 shows the orientation of the two rows of side pins in a Kaba 20 cylinder. Notice that they are staggered much like the disc tumblers in some foreign car lock cylinders.

There are two possible orientations of these staggered rows of side pins. Either row could start closer to the front of the cylinder. The opposite row would then start farther from the front. These two orientations are referred to as *right-* or *left-hand*. If the cylinder is viewed with the right side up, the hand of most Kaba designs is determined by the side whose row of pins begins farther from the front when viewing the face of the cylinder.

If a key is to do its job and operate a cylinder, obviously the cuts must be in the same positions as the pins in the cylinder. That means there are two ways to drill dimples in the keys as well.

To determine the hand of a Kaba key, view it as though it was hanging on your key board (Fig. 8.8). Notice that one row of cuts starts farther from the bow than the other. The row that starts farther from the bow determines the hand. If the row starting farther from the bow is on the left side, it is a left-hand key. If the row starting farther from the bow is on the right side, it is a right-hand key.

Occasionally, you may find a Kaba key with both right- and left-hand cuts (Fig. 8.9). This key is called *composite bitted*. It is primarily used in maison key systems (a type of master key system).

Key reading. Determining the hand is the first step in key reading. Next, you need to know the order in which the various dimple positions are read.

The positions are always read bow to tip beginning with the hand side. After the hand side comes the nonhand side, then the edge. Remember to go back

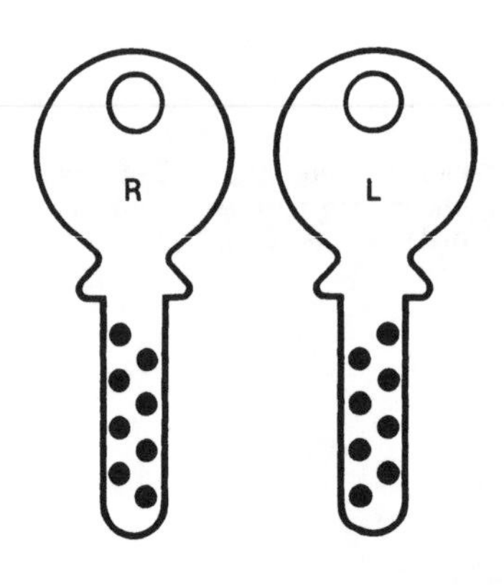

Figure 8.8 It's easy to determine the hand of a Kaba key. (*Lori Corp.*)

Figure 8.9 Examples of Kaba composite bitted keys. (*Lori Corp.*)

to the bow to start each row (Fig. 8.10). For composite bitted keys, use a Kaba key gauge to help mark the cuts of both hands.

All Kaba designs except Micro have four depths on the sides; Micro has three. Kaba's depths are numbered opposite from the way we would normally expect. Number 1 is the deepest and #4 is the shallowest. These depths can be read by eye with very little practice (Fig. 8.11). The increment for side depths of Kaba 8 and 14 is 0.0157 inch (0.4 mm). For Gemini, it is 0.0138 inch (0.35 mm).

For the edges, reading is a bit different for the various Kaba designs. The Kaba key gauge is very useful for reading the edges. Find the section of the gauge that corresponds to the design of the Kaba key you're attempting to read (e.g., Kaba 8). Place the key under the gauge so it shows up through the edge slot. When the shoulder of the key hits the stop on the gauge, you're ready to read the positions of the cuts.

Because of the nonstandard cylinder drilling for Kaba 8, the codes show two columns for the edge (Fig. 8.12). There are only two depth possibilities on the edge of Kaba 8 and Kaba 14: *cut* = #2 and *no-cut* = #4. The combinations in Fig.

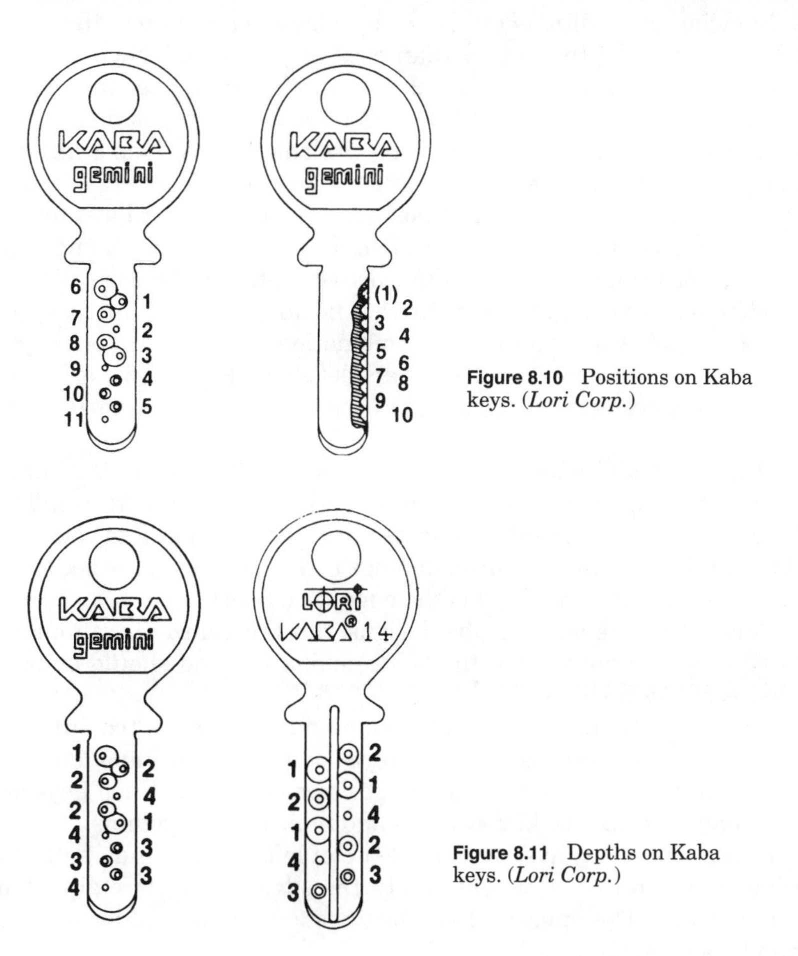

Figure 8.10 Positions on Kaba keys. (*Lori Corp.*)

Figure 8.11 Depths on Kaba keys. (*Lori Corp.*)

SIDES		EDGE	
		key	cyl.
R2214	4233	4	147
R2242	4233	4	147
R2412	4233	4	147
R4212	4233	4	147
R2244	1233	4	147
R2414	1233	4	147

Figure 8.12 Kaba 8 combinations. (*Lori Corp.*)

8.12 call for a key with a cut only in position 4, while the cylinder is drilled in positions 1, 4, and 7. This means the cylinder receives a #2 pin in position 4 for the key cut. Positions 1 and 7 require a #4 pin because there is no cut on the key.

For the Kaba 14 and new Kaba 8 using ME series codes, the edge is read differently because you know automatically which positions are involved in every case. These cylinders are all drilled with odd edges (positions 1, 3, 5, and 7).

Knowing the positions involved, the edge combination of these keys is notated in terms of depths, rather than positions. A Kaba key gauge can be used to determine the positions of the #2 edge cuts. If there is a cut in position 5 only, the combination would be 4224. If there are cuts in positions 1, 3, and 7, the combination would be 2242. If there are cuts in positions 3 and 5, the combination would be 4224, etc. We know there are four chambers in the cylinder and they are in the odd-numbered positions. Therefore, we have to come up with four bittings in the edge combination. *Remember:* cut = #2 and no-cut = #4.

For Kaba Gemini, there are three active depths on the edge, plus a high #4 cut used in masterkeying. Because there is no *no-cut* on Gemini, a key gauge should not be necessary to read the edge combination. Right-hand stock keys will always have cuts in positions 3, 5, 7, and 9 and left-hand keys for factory master key systems will almost always have cuts in positions 2, 4, 6, 8, and 10 (Fig. 8.13).

Bitting notation. Before the bitting comes an indication for the hand, "R" or "L." As mentioned, all bittings are read and notated, bow to tip. The key combination is broken up into separate parts corresponding to each row of pins in the cylinder. The first group of bittings is the hand side, the second group is the opposite side, and the third is the edge (Fig. 8.14).

This holds true for all Kaba designs, but the guard pin cut on a Saturn key (a #3 depth) is not part of the key combination and should be ignored for all phases of key reading.

Notation of composite bitted key combinations isn't much different from that of regular keys. Composite bitted keys have both right- and left-hand side cuts and often have both even and odd position edge cuts. Such keys are normally used only in selective key systems and maison key systems.

If most of the key system is made up of left-hand cylinders, that is the hand which is listed first. Conventionally, the edge bittings are all listed as part of the first line. The opposite hand bittings are written under those of the main hand, as illustrated in Fig. 8.15.

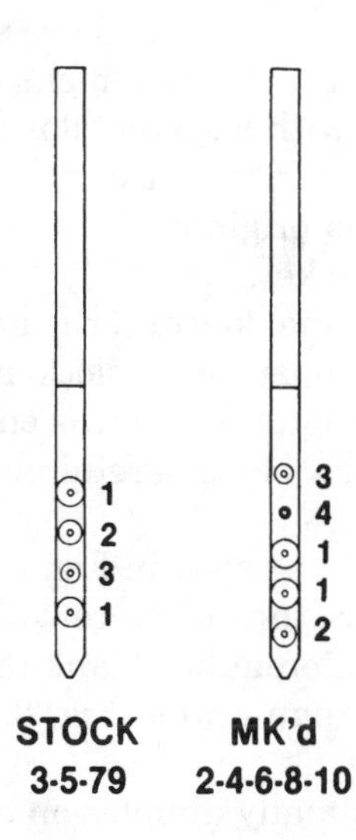

Figure 8.13 Kaba Gemini edge. (*Lori Corp.*)

Kaba 8	R 1421-2414-17
Kaba 14	L 34121-14123-2442
Kaba Gemini (Stock)	R 14214-241414-1231
Kaba Gemini (MK'd)	L 11314-442121-14321
Kaba Saturn	L 142-313

Figure 8.14 Writing key combinations. (*Lori Corp.*)

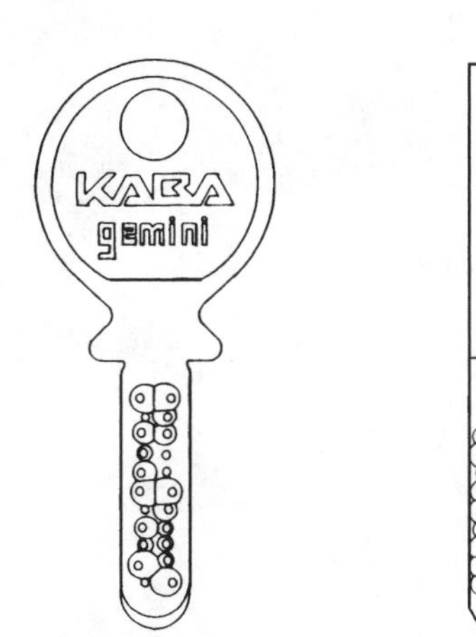

L 43121-124231-314213112
R 24133-122431

Figure 8.15 Composite bitted key. (*Lori Corp.*)

Identifying nonoriginal keys. Because the dimensions of the key blanks for K-8, K-14, and Saturn are identical, people in the field sometimes duplicate a key from one design onto another's key blank. This can lead to problems later, especially when quoting prices to a customer. If a key says "Kaba 8" but it has really been cut for a Kaba 14 cylinder, you might quote a Kaba 8 price and order a Kaba 8 cylinder only to find that you can't set it to the customer's key. To avoid confusion at a later date, a genuine key blank with a system designation such as "Kaba 8," "Kaba 14," or "Kaba Saturn" should only be used to make keys for that particular design. If you didn't sell the job originally, you should always verify the design by counting the dimples on the key.

If you need a new cylinder to match a nonoriginal key or if the key was made poorly and you must make a code original to operate the lock properly, you must also be able to determine which Kaba design the key was cut for. This is easily done by counting the dimples on the sides, being careful not to overlook any of the tiny #4 cuts.

Kaba 8 has two rows of four cuts (eight total), as seen in Fig. 8.16. Kaba 14 has two rows of five cuts (10 total) (Fig. 8.17). Saturn has a row of three and a row of four (seven total). Kaba Gemini uses a different key blank that is thicker and narrower. The dimples are oblong rather than round (Fig. 8.18). It has a row of five cuts and a row of six cuts (11 total).

Composite bitted keys have exactly twice as many dimples on a side.

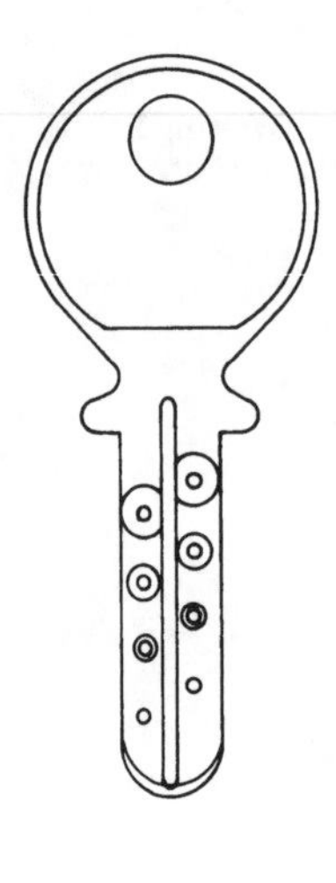

Figure 8.16 Eight bittings equal Kaba 8. (*Lori Corp.*)

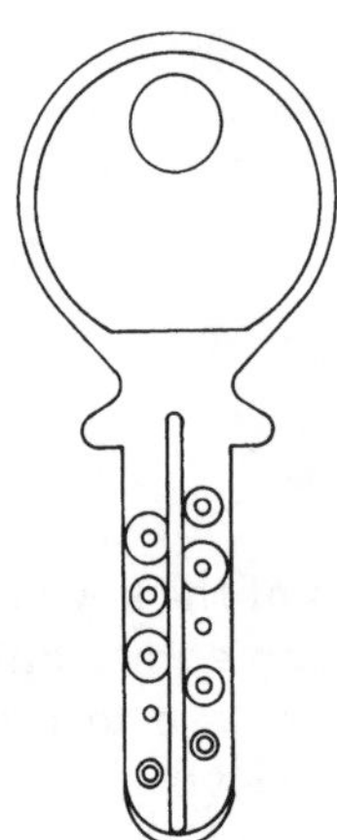

Figure 8.17 Ten bittings equal Kaba 14. (*Lori Corp.*)

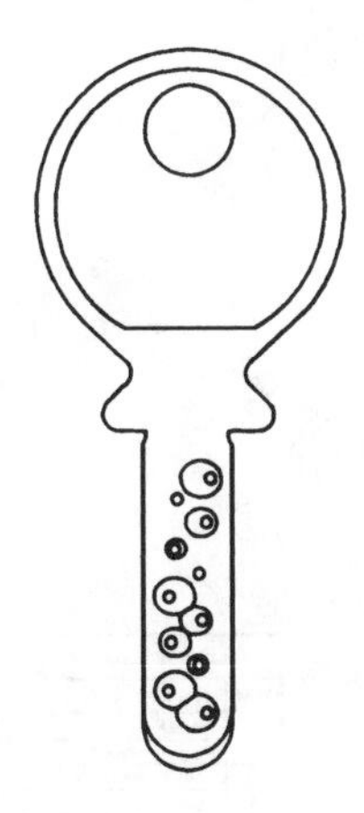

Figure 8.18 Eleven oblong bittings equal Kaba Gemini. (*Lori Corp.*)

Medeco locks

Locks manufactured by Medeco Security Locks, Inc. are perhaps the most well known high-security mechanical locks in North America. For that reason, much of this remaining section is devoted to reviewing how the various types of Medeco locks operate.

General operation. Before studying in detail the specifications of Medeco locks, you must first understand how these locks operate. Medeco's 10- through 50-series locks incorporate the basic principles of a standard pin tumbler cylinder lock mechanism. A plug, rotating within a shell, turns a tailpiece or cam when pins of various lengths are aligned at a shear line by a key. Figure 8.19 shows an exploded view of a Medeco cylinder.

The plug rotation in a Medeco lock is blocked by the secondary locking action of a sidebar protruding into the shell. Pins in a Medeco lock have a slot along one side. The pins must be rotated so that this slot aligns with the legs of the sidebar. The tips of the bottom pins in a Medeco lock are chisel pointed, and they are rotated by the action of the tumbler spring seating them on the corresponding angle cuts on a Medeco key (Fig. 8.20).

The pins must be elevated to the shear line and rotated to the correct angle simultaneously before the plug will turn within the shell (Fig. 8.21). This dual-locking principle and the cylinder's exacting tolerances account for Medeco's extreme pick resistance.

Medeco cylinders are also protected by hardened, drill-resistant inserts against other forms of physical attack. There are two hardened, crescent-shaped plates within the shell that protect the shear line and the side bar from drilling attempts. There are also hardened rods within the face of the plug and a ball bearing in the front of the sidebar. Figure 8.22 shows the shapes and positions of the inserts.

To fit within the smaller dimensions necessary in a cam lock, Medeco developed the principle of a *driveless rotating pin tumbler.* It is used in the 60- through 65-series locks. The tumbler pin and springs are completely contained within the plug diameter (Fig. 8.23).

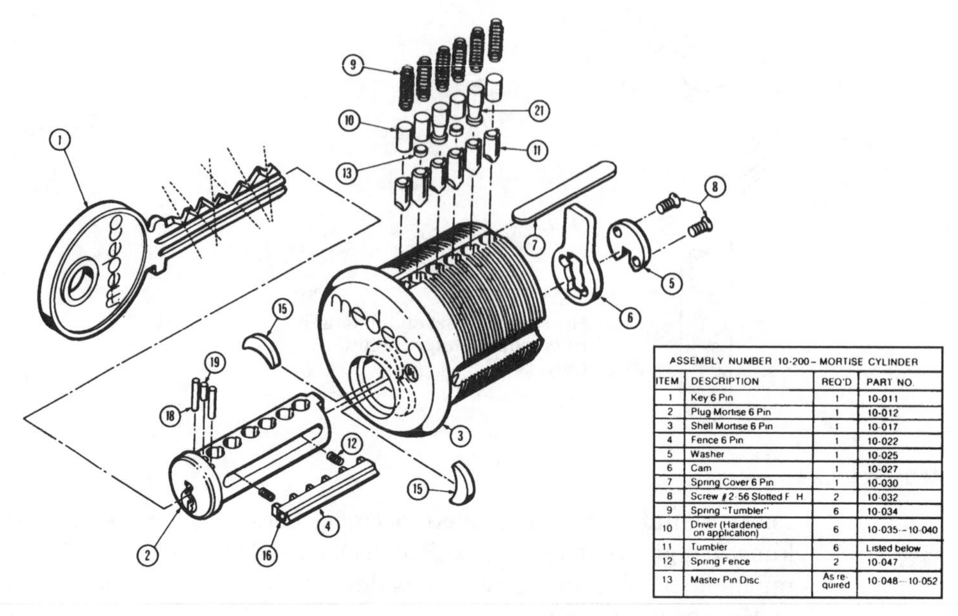

ASSEMBLY NUMBER 10-200– MORTISE CYLINDER			
ITEM	DESCRIPTION	REQ'D	PART NO.
1	Key 6 Pin	1	10-011
2	Plug Mortise 6 Pin	1	10-012
3	Shell Mortise 6 Pin	1	10-017
4	Fence 6 Pin	1	10-022
5	Washer	1	10-025
6	Cam	1	10-027
7	Spring Cover 6 Pin	1	10-030
8	Screw #2-56 Slotted F H	2	10-032
9	Spring "Tumbler"	6	10-034
10	Driver (Hardened on application)	6	10-035 – 10-040
11	Tumbler	6	Listed below
12	Spring Fence	2	10-047
13	Master Pin Disc	As required	10-048 — 10-052

Figure 8.19 Exploded view of an original Medeco mortise cylinder. (*Medeco Security Locks*)

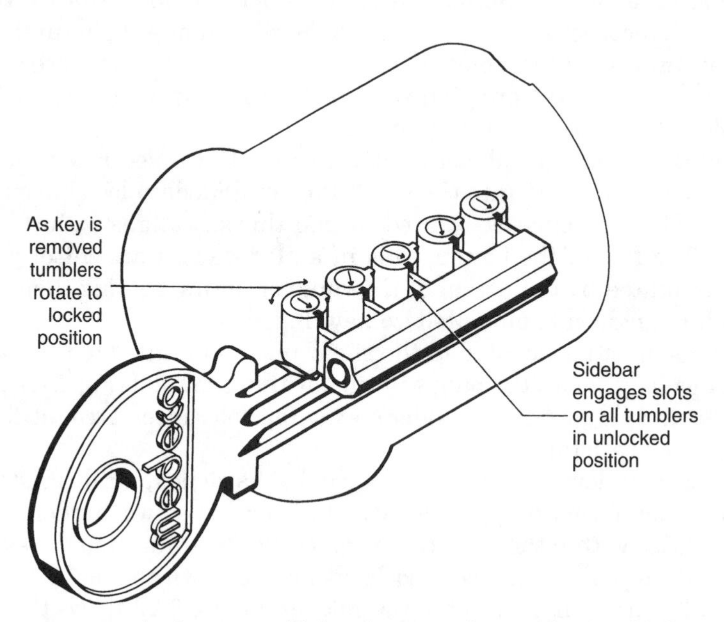

Figure 8.20 Angle cuts on a Medeco key cause the tumblers in a Medeco cylinder to be raised to the shear line while simultaneously rotating into position to allow the sidebar's legs to push into the pins. (*Medeco Security Locks*)

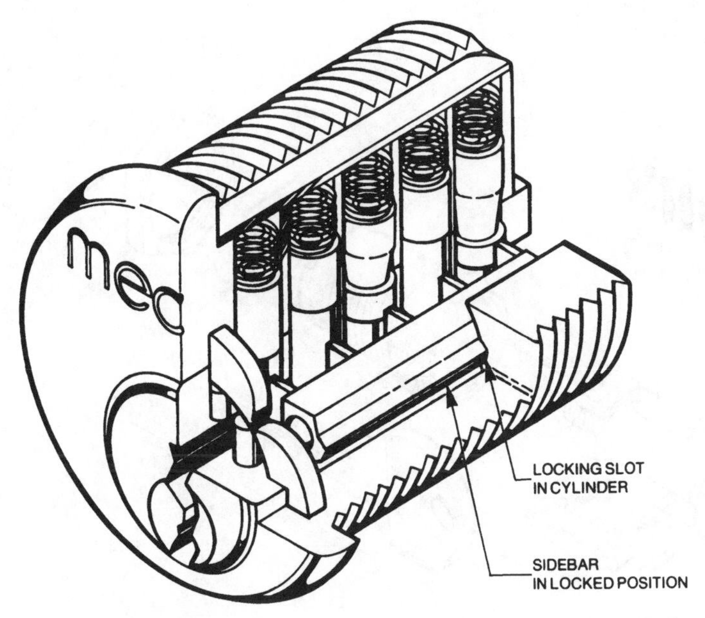

Figure 8.21 A cutaway view of an original Medeco mortise cylinder showing how both the pins and sidebar obstruct the plug from rotating. (*Medeco Security Locks*)

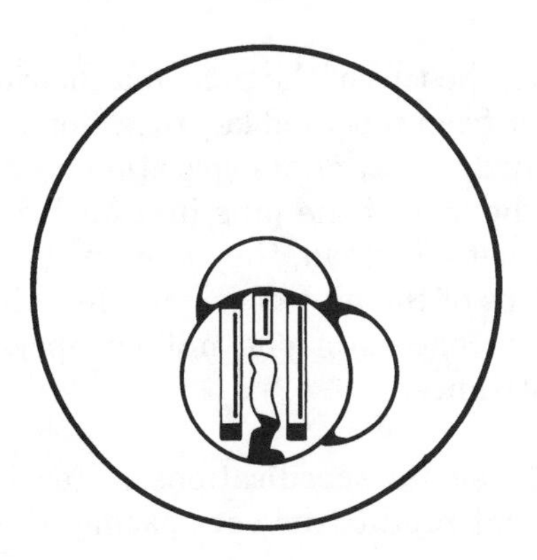

Figure 8.22 Face of a Medeco cylinder showing the positions of the hardened steel drill-resistant pins and crescents. (*Medeco Security Locks*)

The plug rotation is blocked by the locking action of a sidebar protruding into the shell. The pins are chisel pointed and have a small hole drilled into the side of them. The pins must be rotated and elevated by corresponding angled cuts on the key so that each hole aligns with a leg of the sidebar and allows the plug to rotate. In addition, the cylinder is protected against other forms of physical attack by four hardened, drill-resistant rods within the face of the plug.

Despite the exacting tolerances and additional parts, Medeco's cylinders are less susceptible to wear problems than are conventional cylinders. As in all

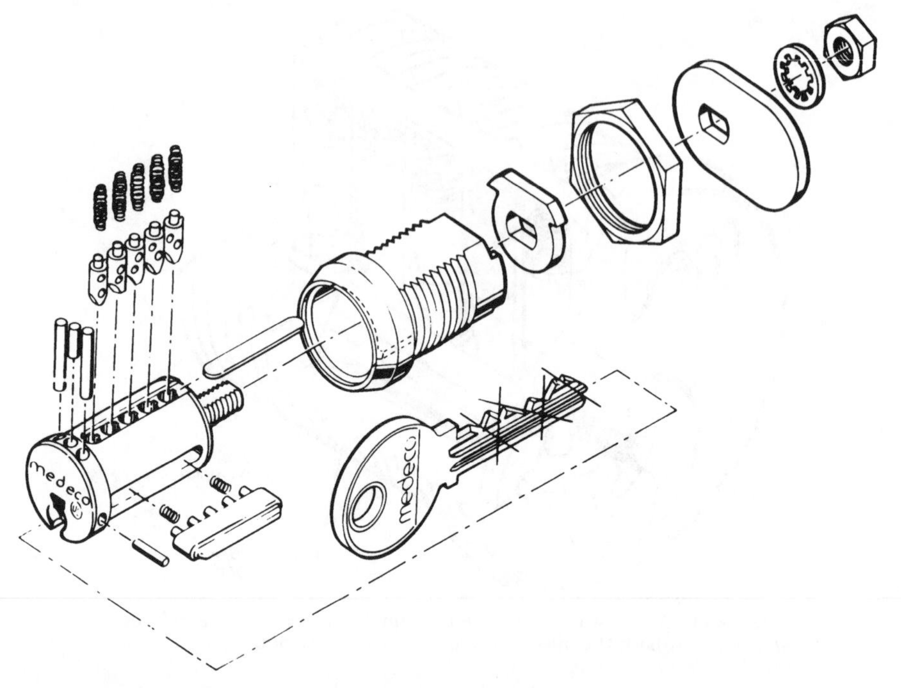

Figure 8.23 Exploded view of an original Medeco cam lock. (*Medeco Security Locks*)

standard pin tumbler cylinder locks, the tips of the pins and the ridges formed by the adjacent cuts on the key wear from repeated key insertion and removal.

In a Medeco lock, this wear has little effect on its operation because, in contrast to a standard lock cylinder, the tips of the pins in a Medeco lock never contact the flat bottoms of the key cuts. Instead, they rest on the sides of the key profile; thus, the wear on the tips of the pins does not affect the cylinder's operation (Fig. 8.24). Cycle tests in excess of one million operations have proven Medeco's superior wear resistance.

Medeco keys. There are four dimensional specifications for each cut on an original Medeco key. They are the cut profile, the cut spacing, the cut depth, and the cut angular rotation.

The profile of the cut on all original Medeco keys must maintain an 86° angle. This dimension is critical because the pins in a Medeco lock are chisel pointed and seat on the sides of the cut profile rather than at the bottom of the cut. Prior to June 1975, Medeco keys were manufactured with a perfect V-shaped profile. After this date, the keys were manufactured with a 0.015-inch-wide flat at the bottom.

Spacing of the cut on a Medeco key must be to manufacturers' specifications. For the 10-series stock keys, Medeco part KY-105600-0000 (old #10-010-0000) and KY-106600-0000 (old #10-011-0000), the distance from the upper and lower shoulder to the center of the first cut is 0.244 inch. Sub-

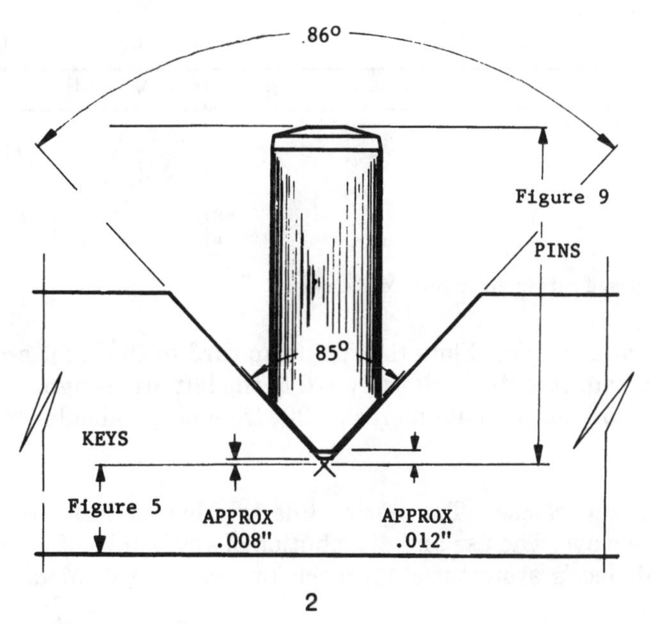

Figure 8.24 Angles of an original Medeco bottom pin and key cut. (*Medeco Security Locks*)

sequent cuts are centered an additional 0.170 inch. For the 60-series stock keys, Medeco part KY-105400-60000 (old #60-010-6000) (five pin) and KY-104400-6000 (old #60-011-6000) (four pin), the distance from the upper shoulder to the center of the first cut is 0.216 inch. Subsequent cuts are centered an additional 0.170 inch. The distance from the bottom shoulder to the center of the first cut is 0.244 inch on this and all Medeco keys. For the 60-series thick head keys, Medeco part KY-114400-6000 (old #60-611-6000) (five pin) and KY-114400-6000 (old #60-611-6000) (four pin), the distance from the shoulder to the center of the first cut is 0.244 inch. Subsequent cuts are centered an additional 0.170 inch.

Standard Medeco keys in the 10- through 50-series are cut to six levels with a full 0.030-inch increment in depths. Keys used in extensive masterkeyed systems and on restricted Omega keyways are cut to 11 levels with a half step 0.015-inch increment in depths.

Because of the size limitations, Medeco keys in the 60- through 65-series are cut to four levels with a 0.030-inch increment in depth. Keys used in extensive masterkeyed systems and Omega keyways are cut to seven levels with a 0.015-inch increment in depth.

In addition to the dimensions above, each cut in a Medeco key may be cut with any one of three angular rotations. These rotations are designated as right (R), left (L), or center (C). When you look into the cuts of a Medeco key as illustrated in Figure 8.25, concentrating on the flats of the cut, you will notice that flats that are positioned perpendicular to the blade of the key are

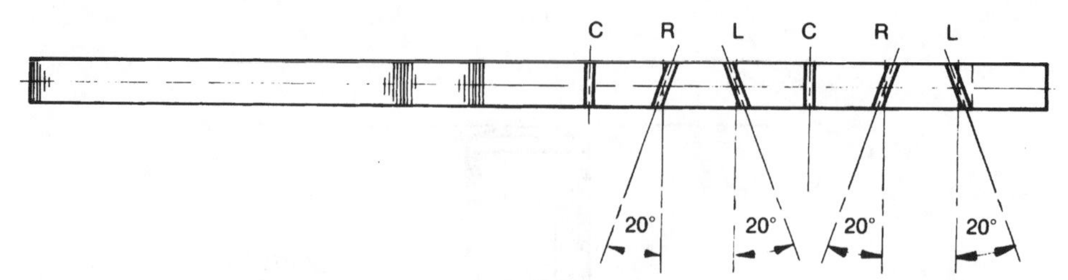

Figure 8.25 Angles of cuts in an original Medeco key.

designated *center angles.* Flats that point upward to the right are designated *right angles,* and flats that point upward to the left are designated *left angles.* Right and left angles are cut on an axis 20° from perpendicular to the blade of the key.

Keyways and key blanks. The entire line of Medeco locks is available in numerous keyways. The use and distribution of key blanks of various keyways is part of Medeco's systematic approach to key control. Medeco offers four levels of security.

Signature Program. Consumers can get new keys made simply by going to the locksmith from whom the lock was purchased. Keys for the Signature Program are restricted by Medeco to locksmiths who have contracted with the company.

Card Program. A consumer is issued a card embossed with control data and that has space for a signature. The keys are owned by Medeco and are controlled by contract. An authorized locksmith will make duplicate keys only after verifying the card data and signature.

Contract restricted. A business or institution can enter into a contract with Medeco for a specially assigned keyway. The key blanks can then be ordered through any authorized Medeco distributor with appropriate authorization of the business or institution.

Factory Program. The Factory Program is Medeco's highest level of key control for consumers, businesses, and institutions. Keys are made from factory-restricted key blanks that are never sold by the factory. Duplicate keys are available only directly from the factory on receipt and verification of an authorized signature.

10-series pins. Medeco bottom pins differ significantly from standard cylinder pins in four respects. The differences occur in the diameter, the chisel point, the locator tab, and the sidebar slot.

- Medeco pins have a diameter of 0.135 inch. That is 0.020 inch larger than the 0.115-inch diameter pin in a standard pin tumbler cylinder.
- All bottom pins are chisel pointed with an 85° angle. The tip is also blunted and beveled to allow smooth key insertion.

- The *locator tab* is a minute projection at the top end of the tumbler pin. The locator tab is confined in a broaching in the shell and the plug; it prevents the bottom pin from rotating a full 180°. A 180° rotation causes a lockout because the sidebar leg is not able to enter the sidebar slot in the pin.
- The *sidebar slot* is a longitudinal groove milled in the side of the bottom pins to receive the sidebar leg. The slot is milled at one of three locations in relationship to the axis of the chisel point.

In 1986 the patent for the original Medeco lock system expired. Now other key blank manufacturers can produce those key blanks. The original system, referred to by Medeco as "controlled," is still being used throughout the world. To allow greater key control, however, Medeco produced a new lock system called the Medeco Biaxial system, and a newer one called Medeco^3.

Medeco's Biaxial lock system. From the outside, Medeco Biaxial cylinders look similar to those of the original system. Internally, however, there are some significant differences. Figures 8.26 and 8.27 show exploded views of Medeco Biaxial cylinders.

Biaxial pins differ from original Medeco pins in three respects: the chisel point, the locator tab, and the pin length. Biaxial pins are made of CDA340 hard brass and are electroless nickel plated. They have a diameter of 0.135 inch and are chisel pointed with an 85° angle. However, this chisel point is offset 0.031 inch in front of or to the rear of the true axis or centerline of the pin (Fig. 8.28).

Fore pins are available in three angles: B, K, and Q. Aft pins are also available in three angles: D, M, and S. Pins B and D have a slot milled directly above the true centerline of the pins. Pins K and M have a slot milled 20° to the left of the true centerline of the pin. Pins Q and S have slots milled 20° to the right of the true centerline of the pin (Fig. 8.29).

The locator tab, the minute projection limiting the pin rotation, was moved to the side of the pin roughly 90°. It is now located along the centerline of the pin opposite the area for sidebar slot (Fig. 8.30).

Medeco Biaxial key specifications. There are four dimensional specifications for a Medeco Biaxial key: the cut profile, the cut depth, the cut spacing, and the cut angular rotation.

The cut profile of the Medeco Biaxial key remains at 86°. However, keys in the 10- through 50-series locks are cut to six levels with a full 0.025-inch increment in depths.

Spacing of the cut on Medeco Biaxial keys must be to manufacturers' specifications. Because Medeco Biaxial pins have the chisel point forward or aft of the pin centerline, the dimensional spacing on the Biaxial key blank can change from chamber to chamber.

From the shoulder to the center of the first cut using a fore pin (either B, K, or Q), the dimension will be 0.213 inch. From the shoulder to the center of the first cut using an aft pin (either a D, M, or S), the dimension will be 0.275 inch. Subsequent fore cuts and subsequent aft cuts are spaced at 0.170 inch (Fig. 8.31).

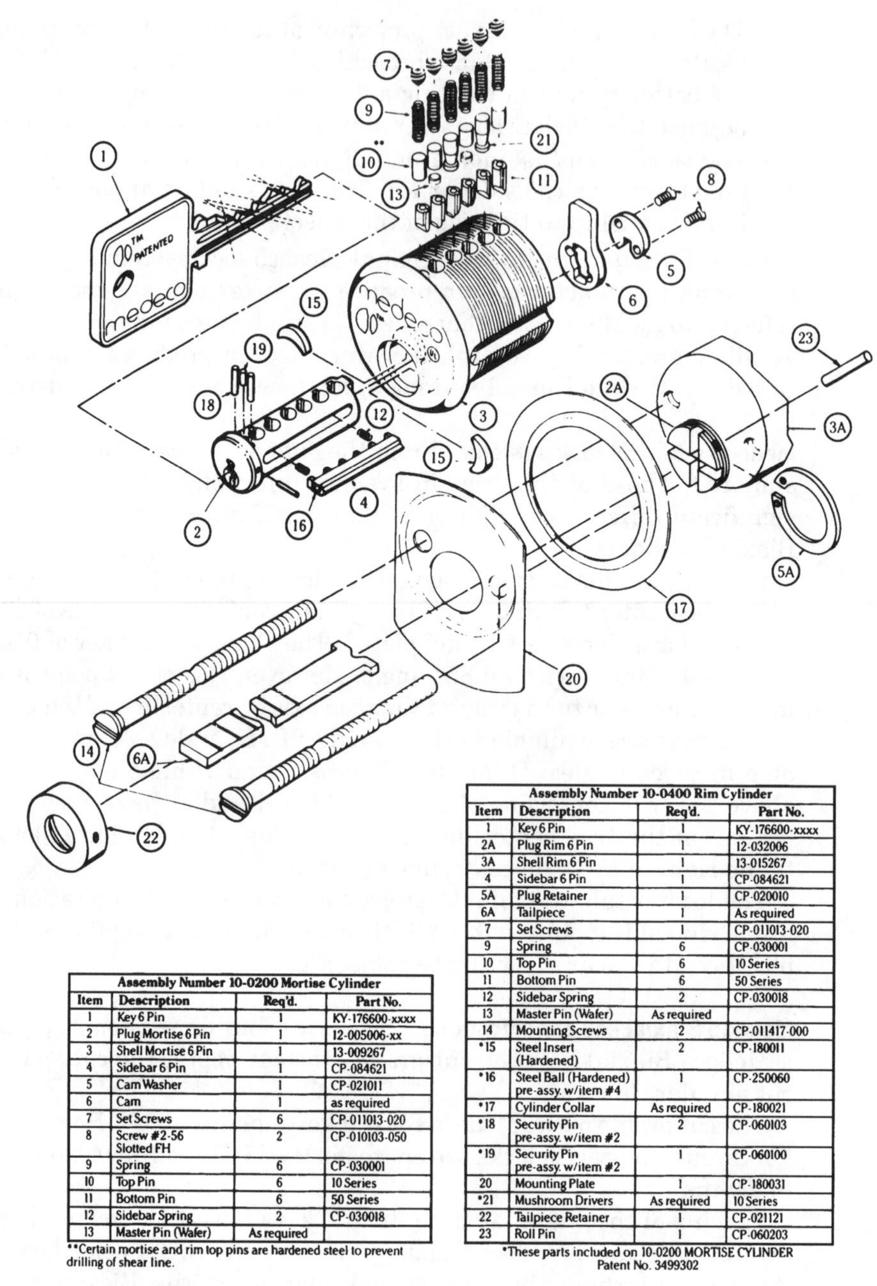

Assembly Number 10-0200 Mortise Cylinder			
Item	Description	Req'd.	Part No.
1	Key 6 Pin	1	KY-176600-xxxx
2	Plug Mortise 6 Pin	1	12-005006-xx
3	Shell Mortise 6 Pin	1	13-009267
4	Sidebar 6 Pin	1	CP-084621
5	Cam Washer	1	CP-021011
6	Cam	1	as required
7	Set Screws	6	CP-011013-020
8	Screw #2-56 Slotted FH	2	CP-010103-050
9	Spring	6	CP-030001
10	Top Pin	6	10 Series
11	Bottom Pin	6	50 Series
12	Sidebar Spring	2	CP-030018
13	Master Pin (Wafer)	As required	

**Certain mortise and rim top pins are hardened steel to prevent drilling of shear line.

Assembly Number 10-0400 Rim Cylinder			
Item	Description	Req'd.	Part No.
1	Key 6 Pin	1	KY-176600-xxxx
2A	Plug Rim 6 Pin	1	12-032006
3A	Shell Rim 6 Pin	1	13-015267
4	Sidebar 6 Pin	1	CP-084621
5A	Plug Retainer	1	CP-020010
6A	Tailpiece	1	As required
7	Set Screws	1	CP-011013-020
9	Spring	6	CP-030001
10	Top Pin	6	10 Series
11	Bottom Pin	6	50 Series
12	Sidebar Spring	2	CP-030018
13	Master Pin (Wafer)	As required	
14	Mounting Screws	2	CP-011417-000
*15	Steel Insert (Hardened)	2	CP-180011
*16	Steel Ball (Hardened) pre-assy. w/item #4	1	CP-250060
*17	Cylinder Collar	As required	CP-180021
*18	Security Pin pre-assy. w/item #2	2	CP-060103
*19	Security Pin pre-assy. w/item #2	1	CP-060100
20	Mounting Plate	1	CP-180031
*21	Mushroom Drivers	As required	10 Series
22	Tailpiece Retainer	1	CP-021121
23	Roll Pin	1	CP-060203

*These parts included on 10-0200 MORTISE CYLINDER
Patent No. 3499302

Figure 8.26 Exploded view of a Medeco biaxial mortise and rim cylinder. (*Medeco Security Locks*)

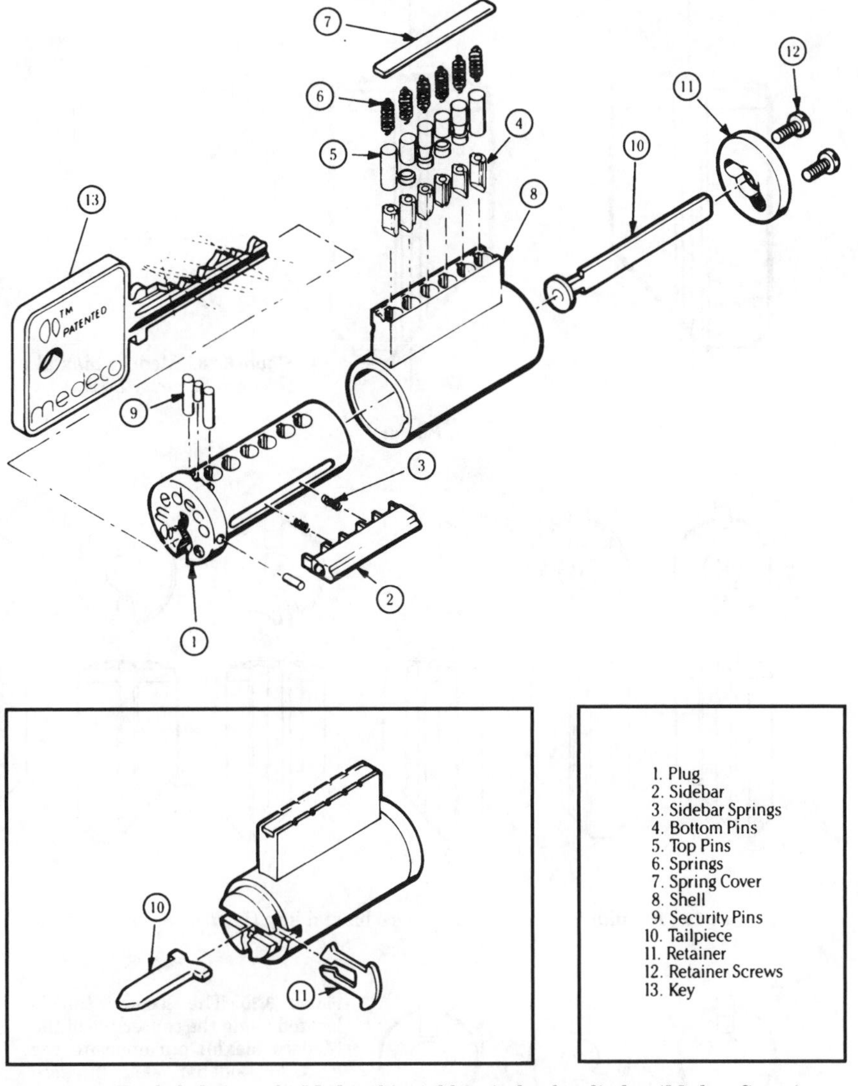

Figure 8.27 Exploded view of a Medeco biaxial key-in-knob cylinder. (*Medeco Security Locks*)

While there are only three angular rotations on a key, each rotation can be used with either a fore pin or an aft pin. Angular cuts B and D are perpendicular to the blade of the key; B is a fore cut and D is an aft cut. Angular cuts K and M have flats pointing upward to the left; K is a fore cut and M is an aft cut. Angular cuts Q and S have flats pointing upward to the right; Q is a fore cut and S is an aft cut. K, M, Q, and S angles are cut on an axis 20° from perpendicular to the blade of the key (Fig. 8.32).

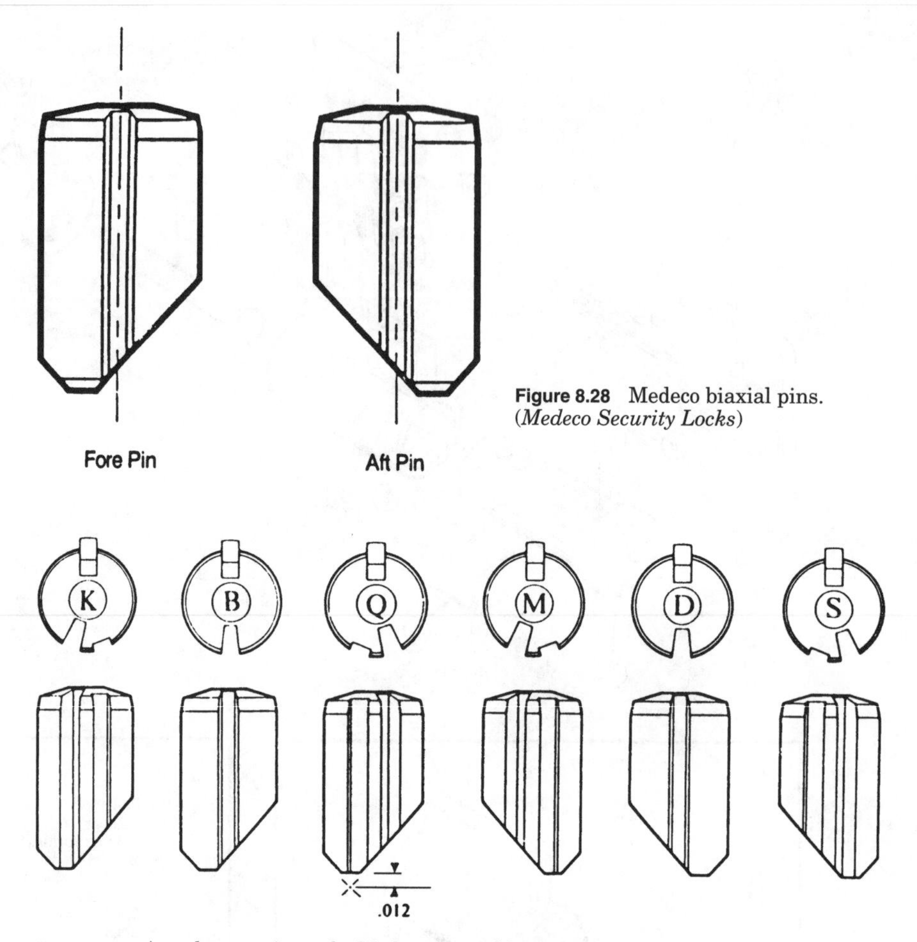

Figure 8.28 Medeco biaxial pins. (*Medeco Security Locks*)

Figure 8.29 Angular rotations of a Medeco biaxial key. (*Medeco Security Locks*)

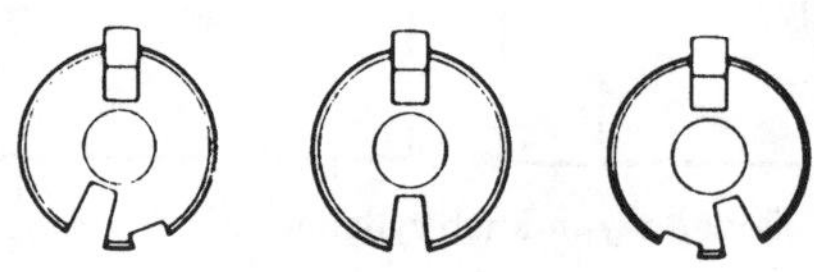

Figure 8.30 The locator tab is located along the centerline of the Medeco biaxial pin opposite the area for sidebar slot. (*Medeco Security Locks*)

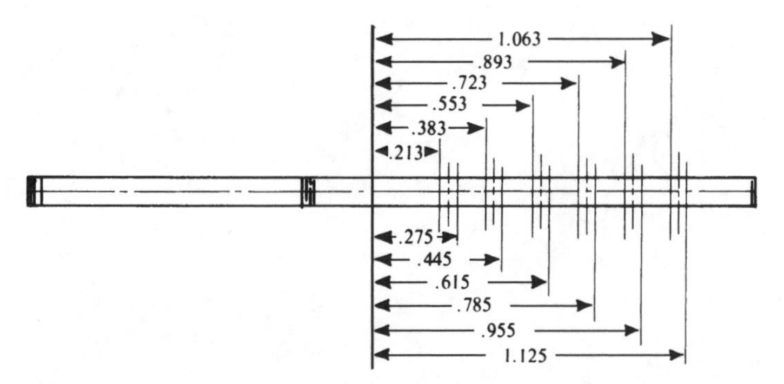

Figure 8.31 Spacing for Medeco biaxial keys. (*Medeco Security Locks*)

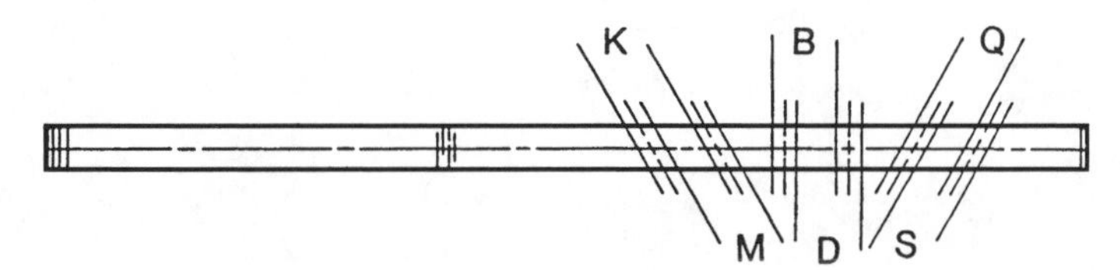

Figure 8.32 Angles for cuts in Medeco biaxial keys. (*Medeco Security Locks*)

Chapter

9

Masterkeying

Masterkeying can provide immediate and long-range benefits that a beginning locksmith can find desirable. This chapter covers the principles of masterkeying and techniques for developing master key systems.

Coding Systems

Coding systems help the locksmith distinguish various key cuts and tumbler arrangements. Without coding systems, masterkeying would be nearly impossible.

Most coding systems (those for disc, pin, and lever tumbler locks) are based on depth differentiation. Each key cut is coded according to its depth; likewise, each matching tumbler receives the same code. Depths for key cuts and tumblers are standardized for two reasons: It is more economical to standardize these depths (mass production would be impossible without some kind of standardization) and depths, to some extent, are determined by production.

Master Key Systems

In most key coding systems, tumblers can be set to any of five possible depths. These depths are usually numbered consecutively 1 through 5. Since most locks have five tumblers, each one with five possible settings, there can be thousands of combinations. Master keys are possible because a single key can be cut to match several lock combinations.

In developing codes, there are certain undesirable combinations that cannot be used. The variation in depths between adjoining tumblers cannot be too great. For example, a pin tumbler key cannot be cut to the combination 21919 because the cuts for the 9s rule out the cuts for the 1s. Likewise, a pin tumbler lock with the combination 99999 would be too easy to pick. The undesirable code combinations vary depending on the type of tumblers, the coding system, and the number of possible key variations. The more complex the system, the greater the possibility of undesirable combinations.

Masterkeying Warded Locks

Since a ward is an obstruction within a lock that keeps out certain keys not designed for the lock, a master key for warded locks must be capable of bypassing the wards. Figure 9.1 shows a variety of side ward cuts that are possible on warded keys. The master key (marked *M*) is cut to bypass all the wards in a lock admitting the other six keys.

As explained earlier, cuts are also made along the length of the bit of a warded key. These cuts correspond to wards in the lock. To bypass such wards, a master key must be narrowed.

Because of the limited spaces on a warded key, masterkeying is limited in the warded lock. The warded lock, because it offers only a very limited degree of security, uses only the simplest of master key systems. Figure 9.2 shows some of the standard master keys that are available from factories.

Masterkeying Lever Tumbler Locks

Individual lever locks may be masterkeyed locally, but any system that requires a wide division of keys would have to be set up at the factory. A large selection of tumblers is required. The time involved in assembling a large number would make the job prohibitive for the average locksmith.

There will be occasions when you are asked to masterkey small lever locks. There are two systems. The first is the *double-gate* system (Fig. 9.3); the other is the *wide-gate* system (Fig. 9.4). Double gating is insecure. As the number

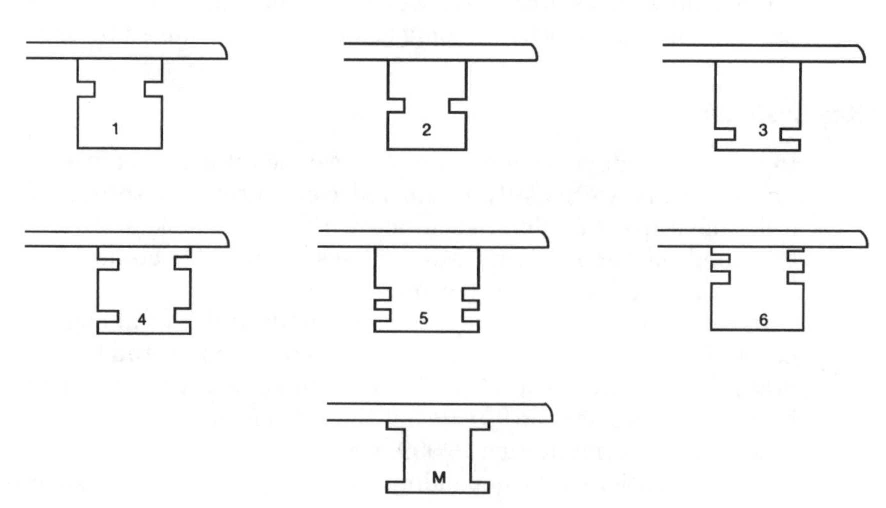

Figure 9.1 The master key at the bottom of the drawing replaces the six change keys above it.

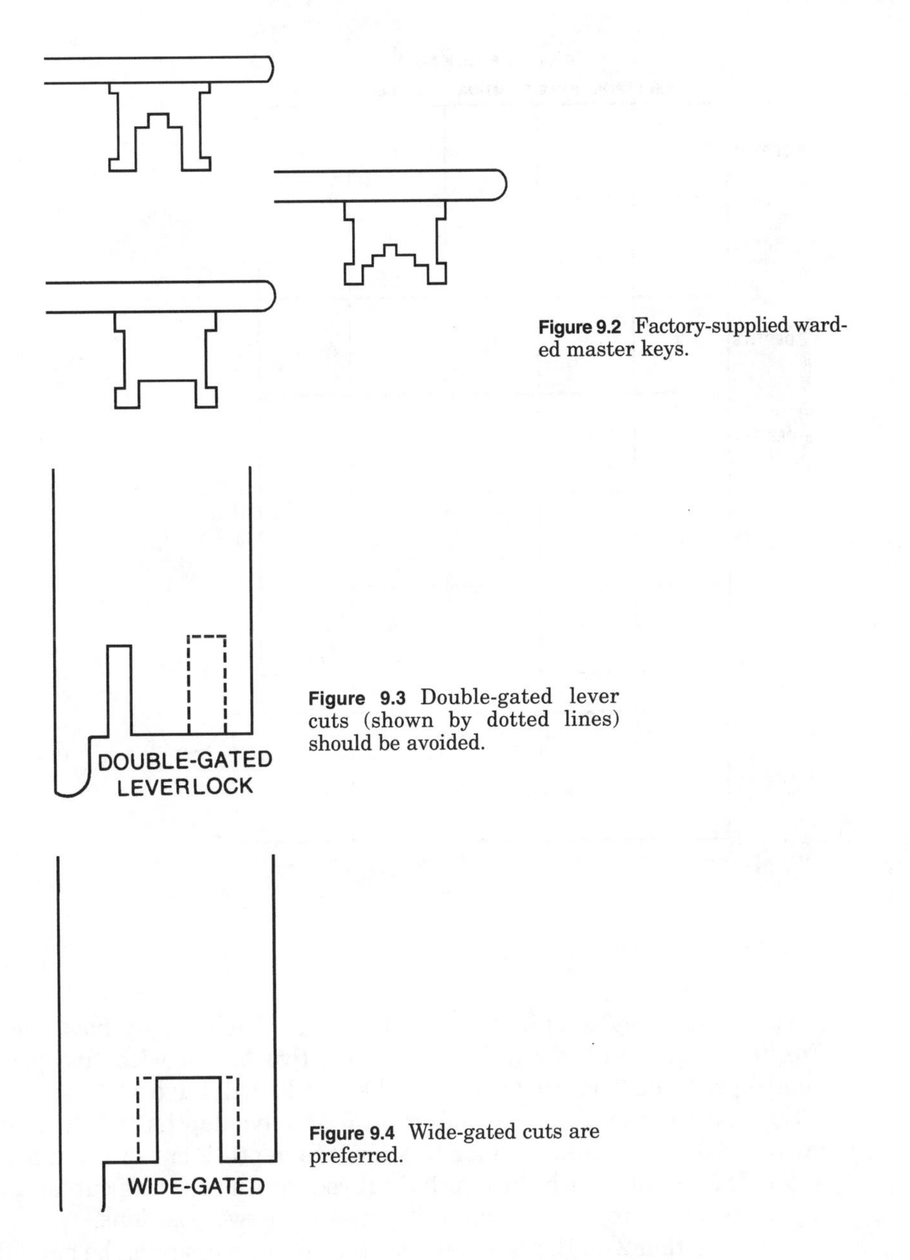

Figure 9.2 Factory-supplied warded master keys.

Figure 9.3 Double-gated lever cuts (shown by dotted lines) should be avoided.

Figure 9.4 Wide-gated cuts are preferred.

of gates in the system increases, care must be taken to prevent cross-operation between the change keys. For example, you may find a change key for one lock acting as the master key.

With either system, begin by determining the tumbler variations for the lock series in question. If the keys to all the locks are available, read the numbers stamped on the keys. Otherwise, disassemble the locks and note the tumbler depths for each one. Next, make a chart listing the tumbler variations (Fig. 9.5).

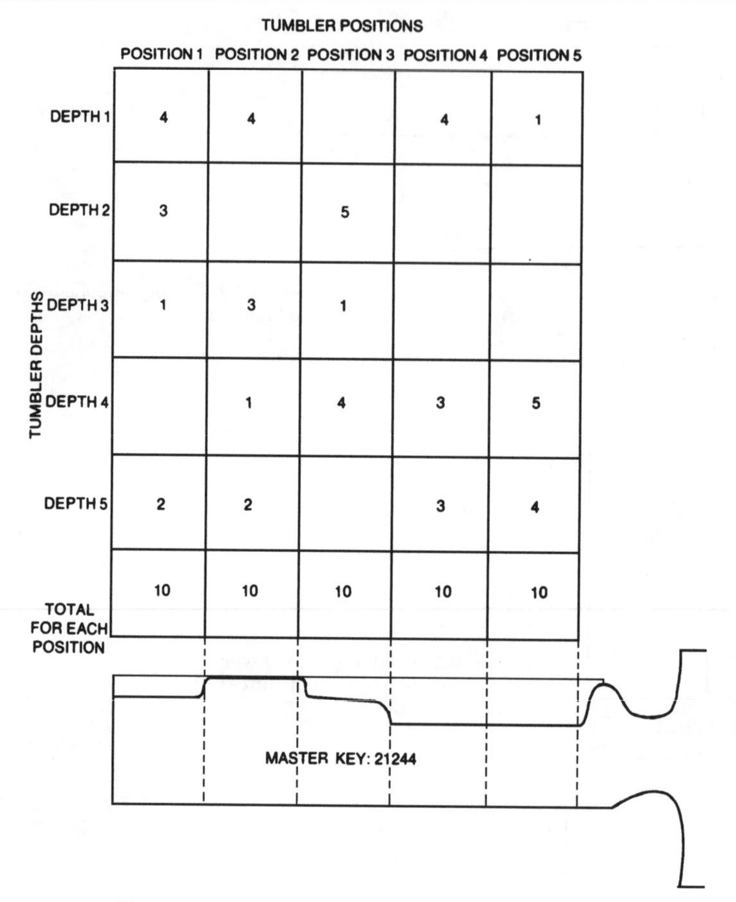

TUMBLER POSITIONS
POSITION 1 POSITION 2 POSITION 3 POSITION 4 POSITION 5
TUMBLER DEPTHS
DEPTH 1 4 4 4 1
DEPTH 2 3 5
DEPTH 3 1 3 1
DEPTH 4 1 4 3 5
DEPTH 5 2 2 3 4
TOTAL FOR EACH POSITION 10 10 10 10 10
MASTER KEY: 21244

Figure 9.5 Masterkeying by the numbers.

The master key combination can be set up fairly easily now. The chart in Fig. 9.5 is for ten-lever locks, each with five levers with five possible key depths per lever. The master key for these locks will have a 21244 cutting code.

Suppose the tumblers in the first position have depths of 1, 2, 3, and 5. You must file depths 1 and 3 wider to allow the depth 2 cut of the master key to enter. The tumbler with the depth 5 cut requires a separate cut, or double gating. It needs a cut that will align it properly at two positions.

From position 2 on the chart, we see that four levers must be cut; all of these require a double-gating cut. In position 3, four require a double-gating cut and one needs widening. In position 5, only one requires another gating and four require widening at the current gate.

Another masterkeying method is to have what is known as a *master-tumbler lever* in each lever lock. The master tumbler has a small peg fixed to it that passes through a slot in the series tumblers. The master key raises the master tumbler. The peg, in turn, raises the individual change tumblers to the proper height so the bolt post passes through the gate in each lever. The lock is open.

This system should be ordered from the manufacturer. The complexities of building one yourself require superhuman skill and patience.

Masterkeying Disc Tumbler Locks

Disc tumbler locks have as few as three discs and as many as twelve or more. The most popular models have five.

Figure 9.6 shows how the tumbler is modified for masterkeying. The left side of the tumbler is cut out for the master; the right side responds to the change key. The key used for the master is distinct from the change key in that its design configuration is reversed. The cuts are, of course, different. The keyway in the plug face is patterned to accept both keys. Since the individual disc tumblers are numbered from 1 to 5 according to their depths, it is easy to think of the master key and disc cuts on a 1 to 5 scale, but on different planes for both the key and the tumblers.

Uniform cuts are taboo. The series 11111 or 22222 would give very little security since a piece of wire could serve as the key. Other uniform cuts are out because they are susceptible to shimming. To keep the system secure, it is best to keep a two-depth interval between any two change keys. For example, 11134 is only one depth away from 11133, so 11134 should be used and 11133 omitted. The rationale is that 11134 is the more complex of the two.

The next step is to select a master key combination composed of odd numbers. At the top of your worksheet, mark the combination you have selected for the master key. Below it add a random list of possible change-key numbers. If you choose a systematic approach in developing change-key numbers, you compromise security. On the other hand, the systematic approach ensures a complete list of possible combinations. Begin systematically; then, randomly select the change-keys combinations.

A single code could be used for all customers. The main point to remember is to use different keyways.

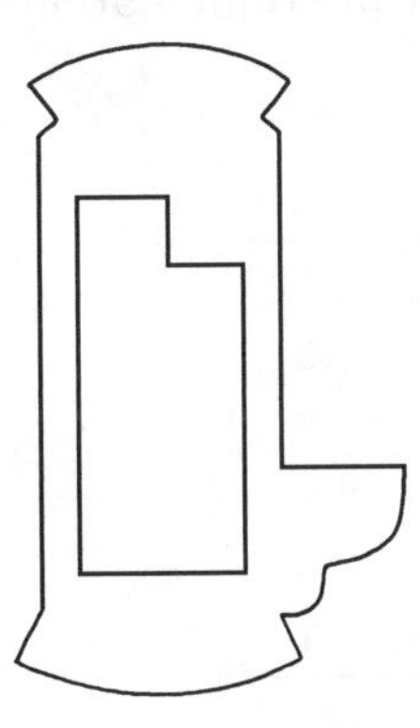

Figure 9.6 The master key operates on the left side of the tumbler.

Masterkeying Pin Tumbler Locks

Masterkeying is more involved than modifying the cylinder. It requires the addition of another pin sandwiched between a top and bottom pin. This pin, logically enough, is called the *master pin.*

The master system is limited only by the cuts allowed on a key, the number of pins, and the number of pin depths available. Since this book is for beginning and advanced students, the subject will be covered on two levels: the simple master key system for no more than 40 locks in a series and the more complex system involving more than 200 individual locks.

It is important to remember that a master key system should be designed in such a way as to prevent accidental cross keying.

Pins are selected on the basis of their diameters and lengths. Master-pin lengths are built around the differences between the individual pin lengths. Consider, for a moment, the Yale five-pin cylinder with pin lengths ranging from 0 to 9 cuts. Each pin is 0.115 inch in diameter. Lower pin lengths are as follows:

0 = 0.184 inch	5 = 0.276 inch
1 = 0.203 inch	6 = 0.296 inch
2 = 0.221 inch	7 = 0.315 inch
3 = 0.240 inch	8 = 0.334 inch
4 = 0.258 inch	9 = 0.393 inch

As an illustration, let's masterkey 10 locks with five tumblers each. Each tumbler can have any of 10 different individual depths in the chamber.

Determine the lengths of each pin in each cylinder. Mark these down on your worksheet. The master key selected for this system may have one or more cuts identical to the change keys.

Cut a master key to the required depths. In this instance, load each cylinder plug by hand.

Using the known master key depth, subtract the depth of the change key from it. The difference is the length of the master key pin. If the change key is 46794 and the master key is 68495, the master pin combination will be as follows:

Chamber position	Bottom	Master
1	4	2
2	6	2
3	7	3
4	9	0
5	4	1

Follow this procedure for each plug. Notice that not all the chambers have a master pin. Such complexity is not necessary and makes the lock more vulnerable to picking. Each master pin represents another opportunity to align the pins with the shear line. Figure 9.7 is an extreme instance, with five master pins and three master keys.

In practice, locksmiths avoid most of this arithmetic by compensating as they go along. For example, an unmasterkeyed cylinder has double pin sets. Masterkeying means that an additional pin is added in (usually) position 1. This drives the bottom pin lower into the keyway. The bottom pin must be shortened to compensate.

A grand master key or a great-grand master key adds complexity to the system. You have two choices: add master pins in adjacent chambers (the Yale approach shown in Fig. 9.8), or stack pins in the first chamber. Suppose you have a No. 4 bottom pin and the appropriate master pin (Fig. 9.9*A*). In order

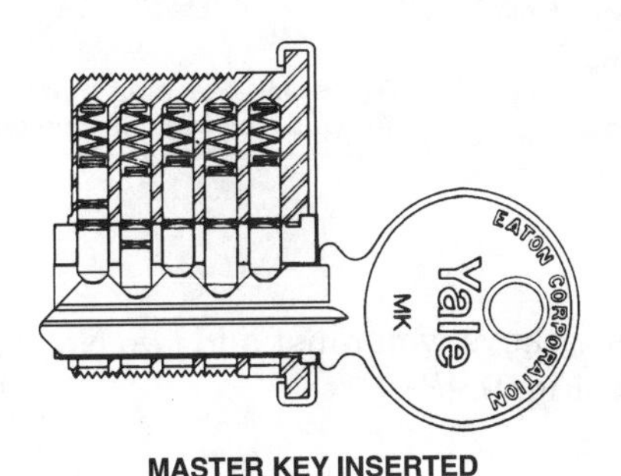

Figure 9.7 Masterkeying pin tumbler locks means a combination other than the change key will raise the pins to the shear line.

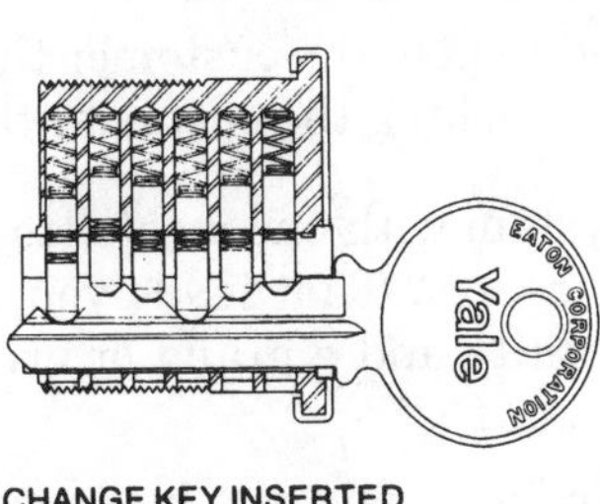

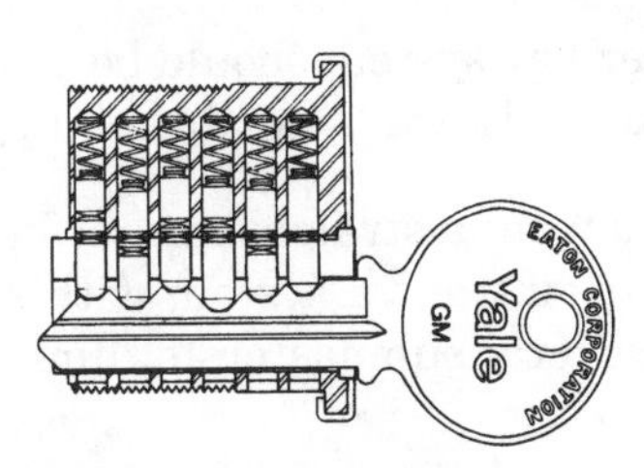

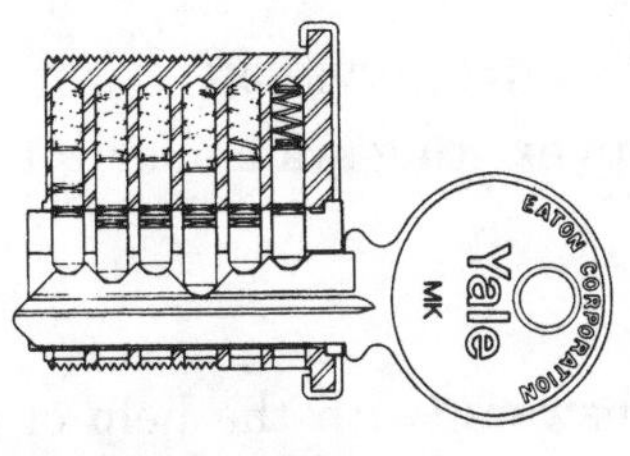

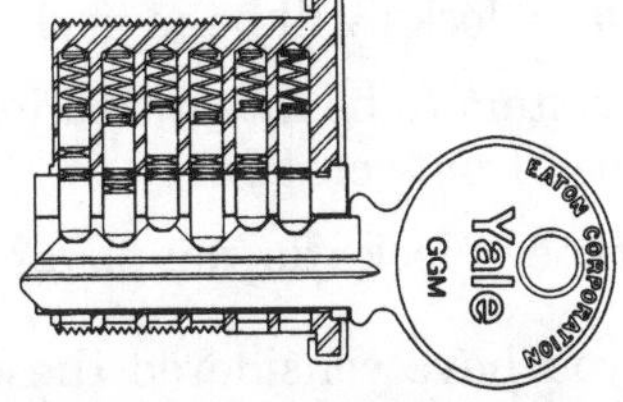

Figure 9.8 The Yale great-grand master system requires five master pins.

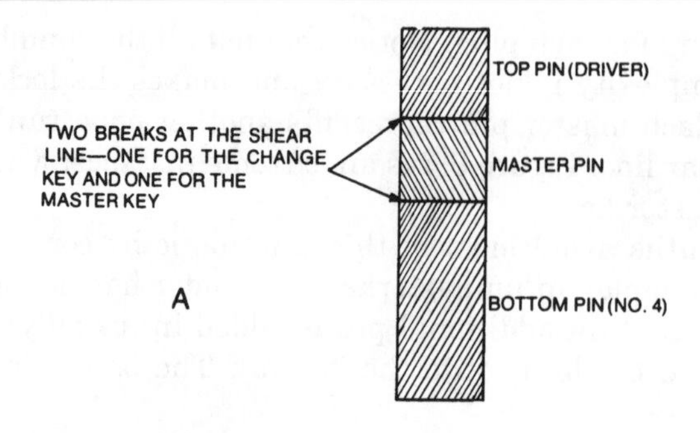

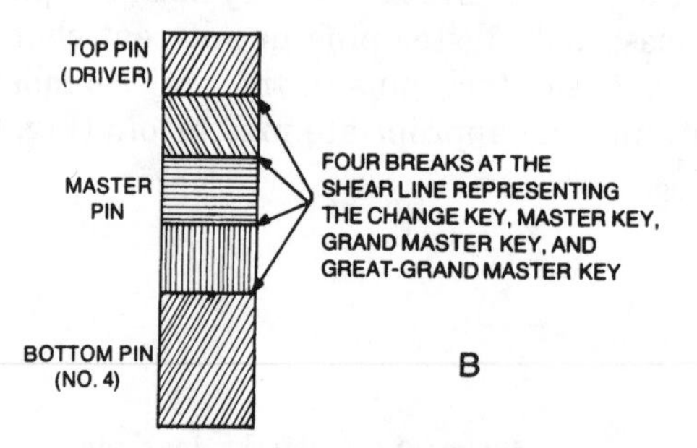

Figure 9.9 Stacking pins is an alternative way of masterkeying.

to use grand and great-grand master keys, you must add two No. 2 pins so that all four pins will operate the lock (Fig. 9.9*B*).

Developing the Master Key System

A master key system should be planned to give the customer the best security that the hardware is capable of. Begin by asking the customer these questions:

- Do you want a straight master key system with one master to open all locks? Or, do you want a system with submasters? That is, do you want a system with several submasters of limited utility and a grand or great-grand master?
- What type of organizational structure is within the business? Who should have access to the various levels?
- How many locks will be in each submaster grouping?
- Is the system to be integrated into an existing system, or will the system be developed from scratch?
- What type of locks do you have?

Once you have considered these answers (with the help of your catalogs, specification sheets, technical bulletins, and experience), you are ready to begin developing a master key system.

The purpose of masterkeying is to control access. A key may open one lock, a series of locks, a group composed of two or more series, or every lock in the system.

Master key terms

Before we go much further into the subject, it is wise to spend a few minutes defining terms. This glossary was prepared by Eaton Corporation (Yale locks) and is reprinted with their permission.

Bicentric cylinder. A pin tumbler cylinder with two plugs, which effectively make it two cylinders in one. The bicentric cylinder is recommended for large, multilevel master key systems where maximum security and expansion are required.

Building master key. A master key that opens all or most of the locks in an entire building.

Change key (or individual key). A key that will usually operate only one lock in a series, as distinguished from a master key that will operate all locks in a series. Change keys are the lowest level in a master key system.

Changes (key). See *key changes.*

Construction breakout key (CBOK). A key used by the owner to make all construction master keys permanently inoperative.

Construction master key (CMK). A key normally used by the builder's personnel for a temporary period during construction. It operates all cylinders designated for its use. The key is permanently voided by the owner upon acceptance of the building or buildings from the contractor.

Control key. A key used to remove the central core from a removable core cylinder.

Controlled cross keying. See *cross keying.*

Cross keying. When two or more different change keys (usually in a master key system) intentionally operate the same lock.

Cross keying, controlled. When two or more change keys under the same master key operate one cylinder.

Cross keying, uncontrolled. When two or more change keys under *different* master keys operate one cylinder.

Department master key. A master key that gives access to all areas under the jurisdiction of a particular department in an organization, regardless of where these areas are in a building or group of buildings.

Display room key. A special hotel change key that allows access to only designated rooms, even if the lock is in the shut-out mode. With many types of hotel locks, this key will also act as a shut-out key, making all other change and master keys inoperative, except the appropriate individual display room key and the emergency key.

Dummy cylinder. One without an operating mechanism; used to improve the appearance of certain types of installations.

Emergency key (EMK). A special, usually top-level, hotel master key that operates all the locks in the hotel at all times. An emergency key will open a guest room lock even if it is in the shut-out mode. With many types of hotel locks, this key acts as a shut-out key, making all other change and master keys inoperative, except the appropriate individual display room key and the emergency key.

Engineer's key (ENG). A selective master key used by various maintenance personnel to gain access through many doors under different master and grand master keys. The key can be set to operate any lock in a master key system and, typically, fits building entrances, corridors, and mechanical spaces. Establishing such a key avoids issuing high-level master keys to maintenance personnel. See *selective master key.*

Floor master key. A master key that opens all or most of the locks on a particular floor of a building.

Grand master key (GM or GMK). A key that operates a large number of keyed-different or keyed-alike locks. Each lock is usually provided with its own change key. The locks are divided into two or more groups, each operated by a different master key. Each group can be operated by a master key only, or by the grand only or by a master and the grand.

Great-grand master key (GGM or GGMK). A key that operates a large number of keyed-different and/or keyed-alike locks. Each lock is usually provided with its own change key. The locks are divided into two or more groups, each operated by a different grand master key. Each of these groups is further subdivided into two or more groups, each operated by a different master key. A group can be operated by the great-grand only, a grand only, a master only, or any combination of the three.

Guest room change key. The hotel room key that is normally issued to open only the one room for which it was intended. The guest room key cannot be used to set a hotel lock in the shut-out mode from the outside of the room, nor will it open a hotel lock from the outside if the lock is in the shut-out mode.

Hotel-function shut-out. When a hotel-function lock is in the shut-out mode, regular master keys of all levels and the guest room key will not open the lock from the outside of the room. Most hotel-function locks can be set in the shut-out mode with the thumbturn or pushbutton from the inside, or with the emergency or display room key from the outside.

Hotel keying. Keying for hotel-function locks.

Hotel great-grand master keys, grand master keys, or master keys. Depending on the level of the system, these keys function as they would in a normal system except they cannot be used to set a hotel lock in the shut-out mode from the outside of the room, nor can they open a hotel lock from the outside if the lock is in the shut-out mode.

Housekeeper's key. A grand master key in a great-grand level hotel system that normally operates all the guest rooms and linen closets in the hotel.

Interchange. See *key interchange.*

Key bitting. A number that represents the depth of a cut on a tumbler-type key. A bitting is often expressed as a series of numbers or letters that designate the cuts on a key. The bittings on a key are the cuts that actually mate with the tumblers in the lock.

Key bitting depth. The depth of a cut that is made in the blade of a key. See *root depth.*

Key bitting list. A list originated and updated by the lock manufacturer for every master key system established. This list contains the key bittings of every master key and change key used in the system. Each time an addition is made to the system, all new bittings are added to the list. It is essential that a complete copy of this list be furnished to any personnel servicing a master key system locally. The lock manufacturer should be informed of any changes made locally to a keying system.

Key bitting or cut position (also called spacing). The location of each cut along the length of a key blade. It is determined by the location of each tumbler in the lock. Bitting position is measured from a reference point to the center of each cut on the key. The most common reference point is the key stop, but the tip of the key is sometimes used.

Key change number. A recorded number, usually stamped on the key for identification. A key change number can be either the direct bitting on the key or a code number.

Key changes (chges). The total possible number of different keys available for a given type of tumbler mechanism. In master key work, the number of different change keys available in a given master key system.

Key interchange. An undesirable situation, usually in a master key system, whereby the change key for one lock unintentionally fits other locks in the system.

Key section (KS). The cross-sectional shape of a key blade that can restrict its insertion into the lock mechanism through the keyway. Each key section is assigned a designation or code by the manufacturer. A key section is usually shown as a cross section viewed from the bow towards the tip of the key. See *keyway.*

Key set (or set). A group of locks keyed exactly the same way. A key set is usually identified with a key symbol. See *standard key symbols.*

Keyed-alike (KA). A group of locks operated by the same change key. Not to be confused with masterkeying.

Keyed-different (KD). A group of locks each operated by a different change key.

Keying. A term, used in the hardware industry, that refers to the arrangement of locks and keys into groups to limit access.

Keying levels. The stratification of a master key system into hierarchies of access. Keying systems are available with one or more levels. The degree of complexity of the system depends on the number of levels used. Generally, the top-level master key can open all locks in the system. Each successive intermediate level of master key can open fewer locks but can open more locks than a change key.

Keying system chart. A chart indicating the structure and expansion of a master key system, showing the key symbol and function of every master key of every level.

Keyway (Kwy). The shape of the hole in the lock mechanism that allows only a key with the proper key section to enter. See *key section.*

Levels. See *keying levels.*

Maid's key. A hotel master key, given to the maid, that gives access only to the guest rooms and linen closets in a designated area of responsibility. A hotel is normally divided into floors or sections with a different maid's key for each floor or section. A maid's key will not open a guest room if the lock is in the shut-out mode.

Multiple key section system (or sectional key sections). Used to expand a master key system by repeating the same or similar key bittings on different key sections. Keys of one section will not enter locks with a different section, yet there is a master key section milled so it will enter some or all of the different keyways in the system. See *simplex key section.*

Paracentric keyway. A keyway in a cylinder lock with one or more side wards on each side projecting beyond the vertical centerline of the keyway to hinder picking. See *simplex key section.*

Pin tumblers. Small sliding pins in a lock cylinder, working against drivers and springs and preventing the cylinder plug from rotating until raised to the exact height by the bitting of a key.

Plug (of a lock cylinder). The round part containing the keyway and rotated by the key to transmit motion to the bolt or other locking mechanism.

Plug retainer. The part of a lock cylinder that holds the plug in the shell.

Privacy key. A change key set up as part of a master key system but not operated by any master keys or grand masters of any level. This key is set up for such areas as liquor-storage rooms in hotels, narcotic cabinets in hospitals, and food storage closets where valuables are kept.

Removable core cylinder. A cylinder containing an easily removable assembly that holds the entire tumbler mechanism, including the plug, tumblers, and separate shell. The cores are removable and interchangeable with other types of locks of a given manufacturer by use of a special key called the control key. See *control key*.

Root depth. Refers to the distance from the bottom of a cut on a key down to the base or bottom of the key blade. Root depth is easy to determine since it measures the amount of blade remaining, rather than the amount that was cut away (bitting depth).

Selective master key. A special, top-level master key in a grand or great-grand system that can be set to operate any lock in the entire system, in addition to the regular floor or section master key, without cross keying. Typical selective master keys include an engineer's key (ENG), nurse's key (NUR), and attendant's key (ATT). The number of selective master keys is normally limited to one or two and should be set up when the original system is established. See *engineer's key*.

Simplex key section. A single independent key section that cannot be expanded into a multiple key section system. Simplex key sections, such as the Yale "Para," are used for stock locks and small master key systems.

Spacing. See *bitting position*.

Standard key symbols. A uniform way of designating all keys and cylinders in a master key system. The symbol automatically indicates the exact function of each key or cylinder in the system, without further explanation.

Tailpiece. The connecting link attached to the end of a rim cylinder that transmits the rotary motion of the key through the door into the locking mechanism.

Top-level master key. The highest level master key in a multilevel keying system that fits most of the locks in the system.

Tumbler. One or more movable obstructions in a lock mechanism that dog or prevent the motion of the bolt or rotation or the plug and are aligned by the key to remove the obstruction during locking or unlocking.

Uncontrolled cross keying. See *cross keying*.

Visual key control. A system of stamping all keys and the plug face of all lock cylinders with standard key symbols for identification purposes. Other key and cylinder stamping arrangements are available but are not considered visual key control.

Standard key symbol code

Figure 9.10 illustrates keying levels of control and the rudiments of the standard key symbol code. Great-grand master keys are identified by the letters GGM. Grand masters carry a single letter, beginning with the first in the alphabet and identifying the hierarchy of locks that the individual grand master keys open. Master keys carry two letters; the first identifies its grand mas-

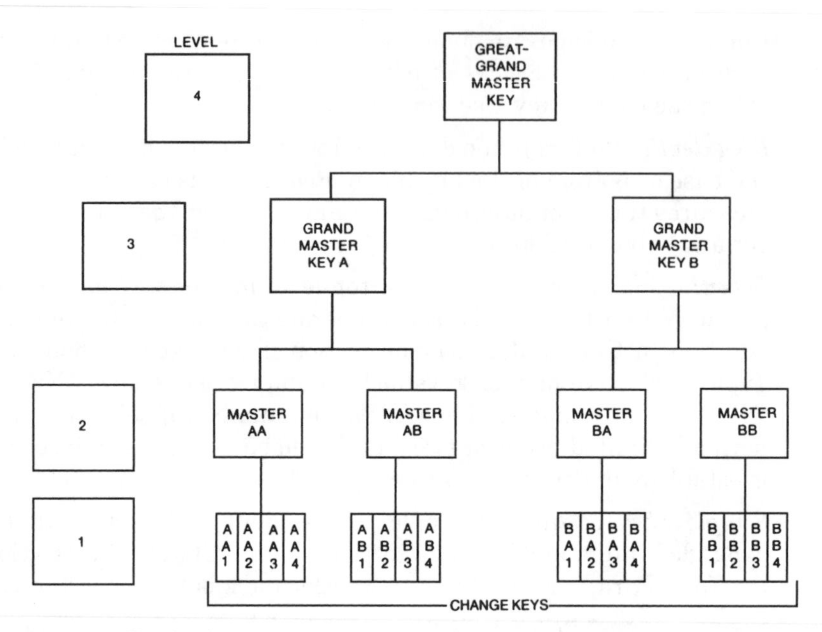

Figure 9.10 Key levels.

ter, the second identifies the series of locks under it. Thus, a master key labeled AA is in grand master series A and opens locks in master key series A. Master key AB is under the same grand master but opens locks in series B. Master BA is under grand master B and opens locks in series A. Change keys are identified by their master key and carry numerical suffixes to show the particular lock that they open. Any key in the series can be traced up and down in the hierarchy. Thus, if you misplace change key AB4, you know that master key AB, grand master A, or the great-grand master will open the lock.

There are special keys that are out of series. Some of these keys are mentioned in the glossary, together with the appropriate key symbols for them.

If cross keying is introduced into the system—that is, if a key can open other locks on its level—the cylinder symbol should be prefixed with an X. If the cylinder has its own key, it is identified with the standard suffix. For example, XAA4 is a change-key cylinder that is fourth on this level. It may be cross keyed with AA3 or any other cylinder or cylinders on this level. By the same token, master key AA, grand master A, and the great-grand master will open it. Elevator cylinders are often cross keyed without having an individual change key. It is no advantage to have a key that will operate the elevator cylinder and no other in the system. These cylinders are identified as X1X, X2X, and so on.

The symbols that involve cross keying apply to the cylinder only; all other symbols apply to the cylinder and the key. This point may seem esoteric, but ignoring it causes the factory and everybody else grief. There is, for example, no such thing as an X1X key. Nor is there an XAA4 key. Change key AA4 fits cylinder XAA4 that happens to be cross keyed with another cylinder. Key AA4 does not fit any cylinder except XAA4.

There are certain advantages to using the standard key symbol code:

- It is a standardized method for setting up the keying systems.
- It maintains continuity from one order to the next.
- It indicates the position of each key and each cylinder in the hierarchy.
- It helps to control cross keying, since each cross-keyed cylinder is clearly marked.
- It offers a method of projecting future keying requirements.
- It can be easily rendered on a chart.
- It allows better control of the individual keys within the system.
- It is a simple method of selling and explaining the keying system to an architect or building owner.

Selling the system

It is important to be able to communicate the advantages of the key symbol code and the implications it has for setting up an ordered, coherent, and secure keying system. This may take some selling on your part, since architects and building owners tend to think of keys and locks as individual entities and not part of a larger system.

Selling a comprehensive masterkeying system involves the following:

- Explaining the subject of masterkeying to the architect or owner
- Reviewing the plans of the building(s)
- Choosing the proper level of control required
- Selecting a keying system that will cover present and future expansion requirements
- Presenting the system to the architect or owner at a meeting dealing specifically with keying
- Recording all changes to the system as agreed at the meeting
- Marking plans with the proper key symbol on the side of each door where the key is to operate
- Presenting the owner with a schematic layout of the entire system and showing the layout of the masters, grand masters, etc.

Master Key System Variations

So far, the structure of the master key system has been discussed. Within this broad structure, there are many opportunities for variations. Some of these variations involve the possible range of keyways; others involve special hardware such as removable cylinders, master-ring cylinders, and rotating tumblers. Each of these variations can extend the range, flexibility, and security of the system. A locksmith must be conversant with all of them.

Keyway variations

Figure 9.11 illustrates a system of control based on keyway design. General Lock's Series 800 key section passes all cylinders in the system; 29, 30, 31, 32, and 33 are submasters, each passing four cylinders; the change keys are restricted to their own individual cylinders. In its fullest expansion, the system includes 35 different keyways on the change-key level and four master key levels.

High-security pin tumbler cylinders

Figure 9.12 illustrates features of the General lock:

- 100° key bitting for long wear (A).
- Master pins have a minimum length of 0.040 inch (B). Shorter pins tend to wedge in the chamber.
- Only two pins are standard (C).
- Pins and springs are made of corrosion-resistant alloy (D and E).
- The keyway is part of the security system (F).
- If requested, the factory supplies special identification for keys, cores, and cylinders (G).

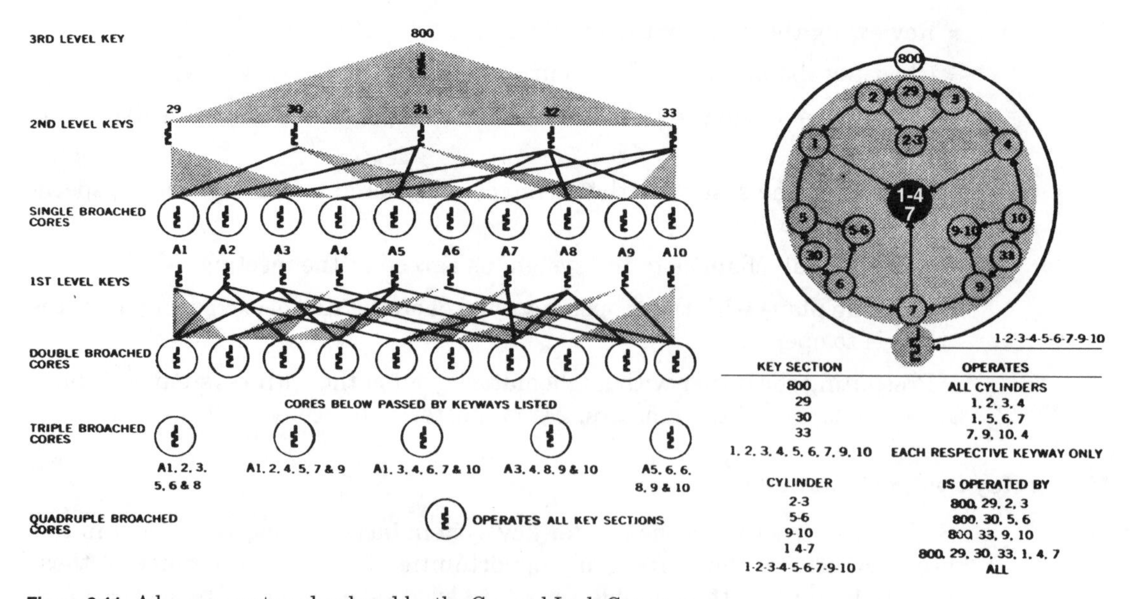

Figure 9.11 A keyway system developed by the General Lock Company.

The Emhart (Corbin) High-Security Locking System uses rotating and interlocking pins (Fig. 9.13). The pins must be raised to the shear line and, at the same time, rotated 20° so the coupling can disengage (Fig. 9.14 and 9.15). Rotation is by virtue of the skew-cut bitting on the key (Fig. 9.16). Figure 9.17 illustrates the way the cylinder is armored. The pins are protected by hardened rods and a crescent-shaped shield.

Removable-core cylinders

Removable-core cylinders are increasingly popular. Figure 9.18 illustrates the Corbin cylinder, a type typical of most. To rekey the change key, follow this procedure:

1. Obtain a Corbin rekeying kit. The kit includes the necessary pins, gauges, and tools.

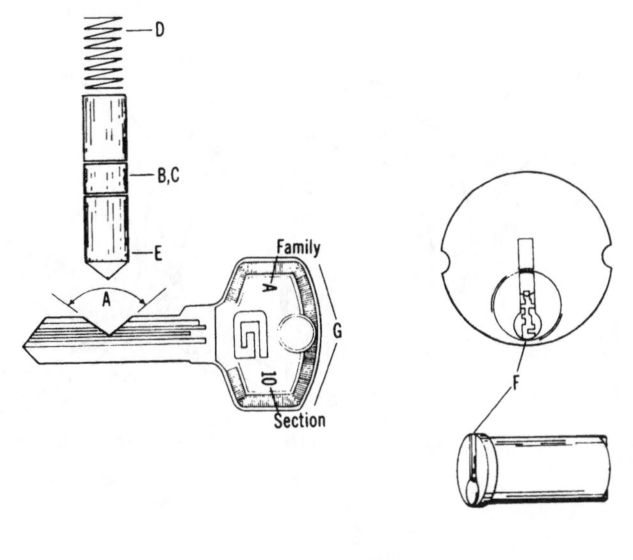

Figure 9.12 General locks have special features described in the text.

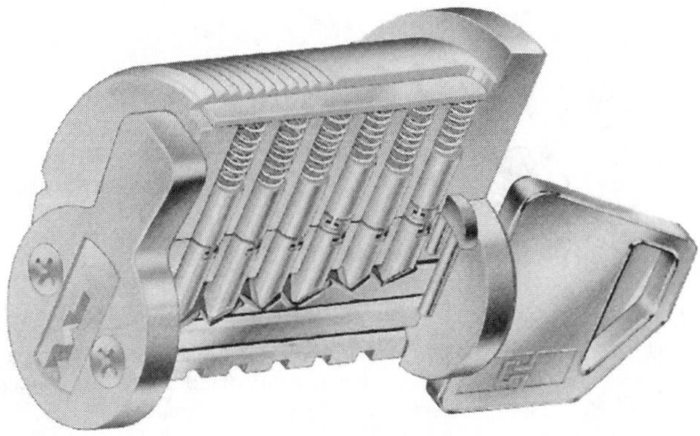

Figure 9.13 Corbin's High-Security Locking System depends on split and rotating pins.

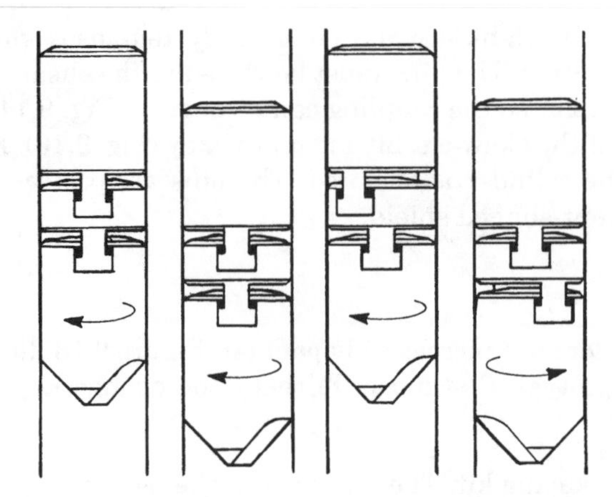

Figure 9.14 Pins are rotated in either direction to allow the joint to uncouple. (*Emhart Corp.*)

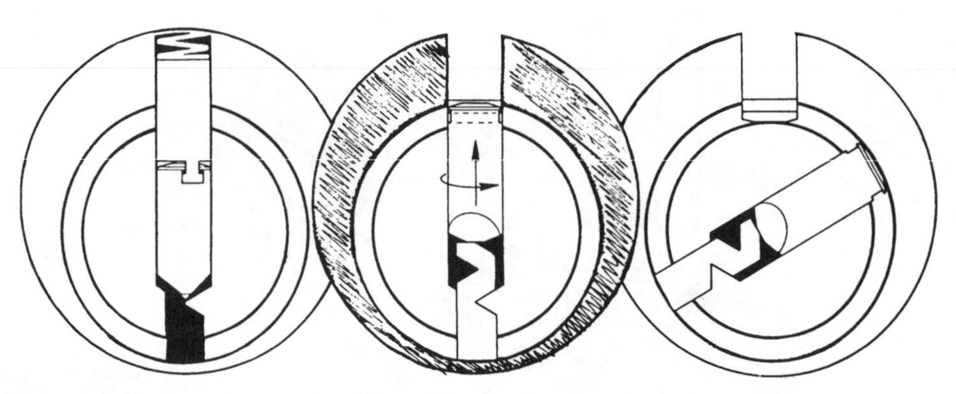

Figure 9.15 Pins must be rotated and brought to the shear line for the lock to open.

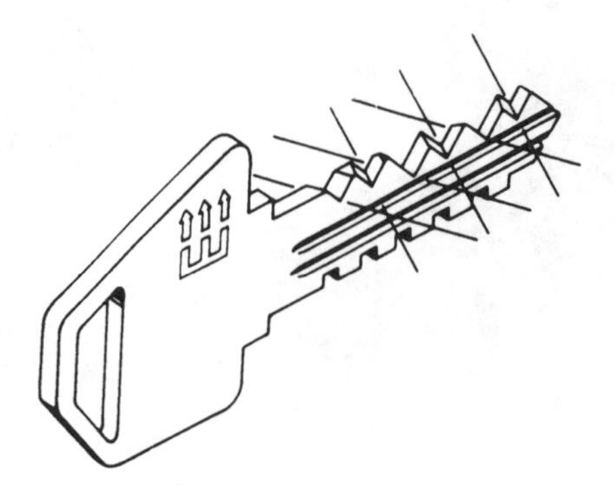

Figure 9.16 The Corbin key has a bitting cut at 20-degree angles.

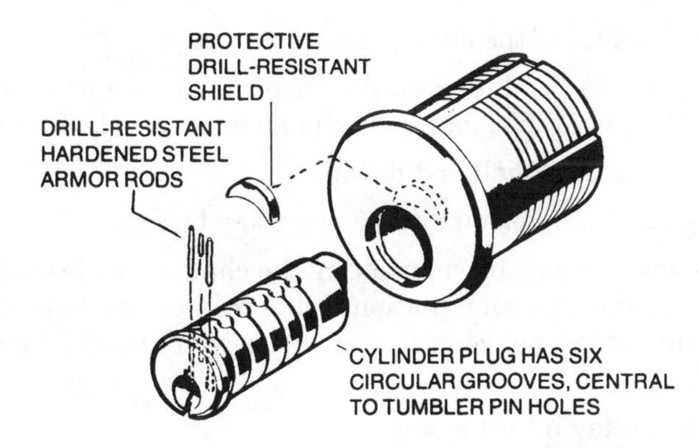

Figure 9.17 Passive defense measures include hardened steel pins and armor plate. (*Emhart Corp.*)

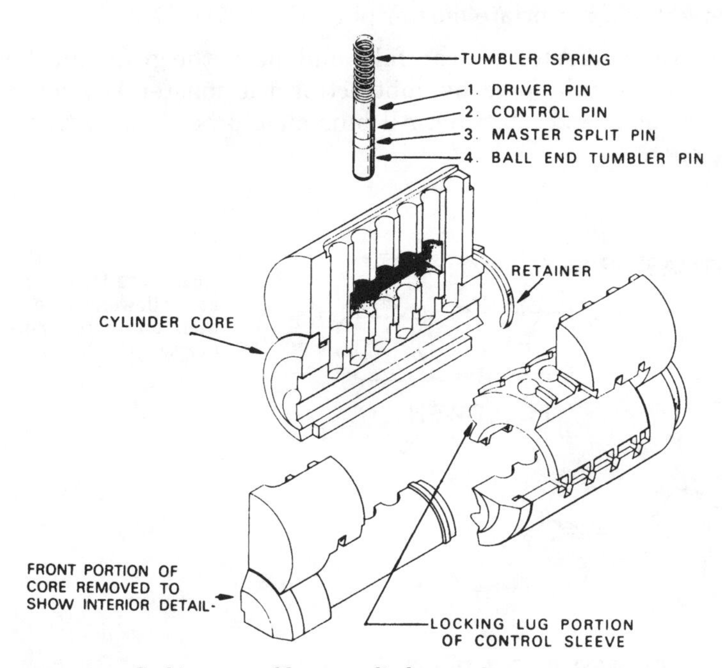

Figure 9.18 Corbin removable core cylinder.

2. Mount the cylinder in a vise.
3. Remove the plug retainer.
4. Select as the plug extractor the key with the deepest bitting. Normally, the grand master key meets this specification; however, there are instances where the engineer's key will have the deepest bitting. A shallow-cut key complicates matters by forcing the control pins, drivers, and buildup pins into the cylinder.
5 Withdraw the plug and remove all pins from their chambers. Figure 9.19 illustrates this procedure.

6. Determine the bitting of the change key.
7. Write down the new combination. As an example, suppose the original change-key bitting is 513525 and we wish to reverse it to 525315 (Fig. 9.20).
8. Install the tumbler pins, ball end down.
9. Use a depth gauge to determine the master key bitting.
10. Calculate the master pins by subtracting the change-key bitting combination from the master key combination. If the master key combination was 525763, the difference between it and the new change-key combination would be 448.
11. Insert the master key into the plug.
12. Install the appropriate master pins (Fig. 9.21).
13. Remove the master key carefully and insert the grand master key. Select the master split pins by subtracting the master key bitting from the grand master key bitting. All pins should be flush with the surface of the plug.

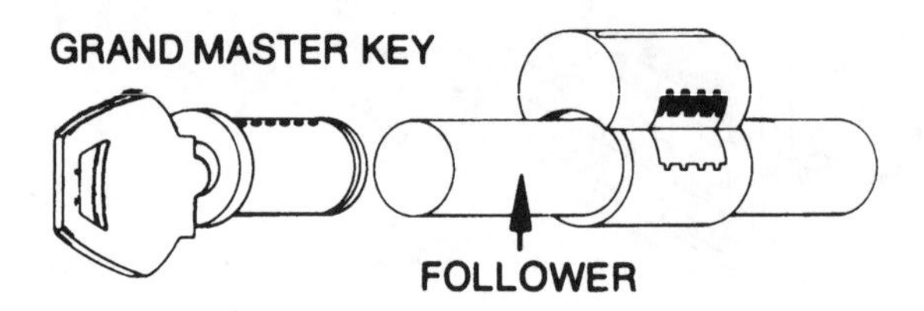

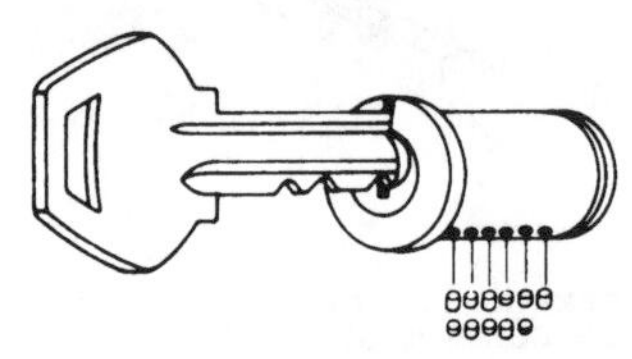

Figure 9.19 Using the appropriate follower, remove the plug, then dump the pins. (*Emhart Corp.*)

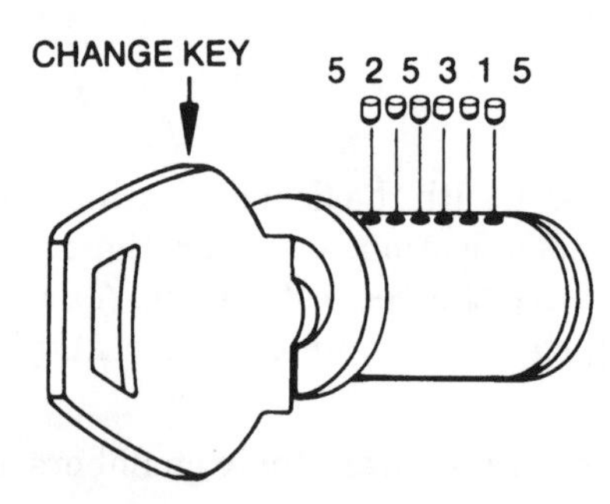

Figure 9.20 Pin length corresponds to the change-key combination. (*Emhart Corp.*)

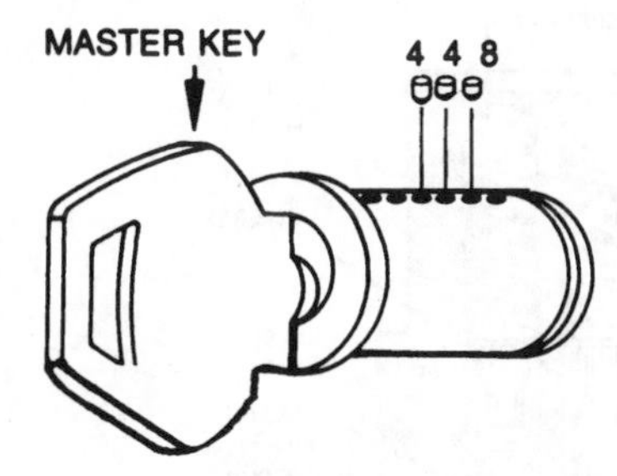

Figure 9.21 Subtract the change-key combination from the master key combination. The difference represents the length of the master pins. (*Emhart Corp.*)

14. Assemble the plug and cylinder.
15. Test all keys.
16. Lubricate the keyway with a pinch of powdered graphite.

Master-ring cylinders

The Corbin master-ring cylinder is shown in Fig. 9.22. The change key operates the plug plungers, and the master key operates the plunger in the master ring. Sometimes called *two-in-one* cylinders, these cylinders increase the range of key combinations for any given system.

To rekey the change key, follow this procedure:

1. Mount the plug in a vise and remove the cylinder slide with a pair of pliers or a small chisel.
2. Remove the springs, drivers, and pins (Fig. 9.23).
3. Ream pin holes through the shell, master ring, and plug (Fig. 9.24).
4. Assuming that the original combination was 414472, reversing the combination gives 274414. This will be the combination of the new change key.
5. Reverse the pins to conform with the new combination. That is, the pin that was first goes into the last chamber; the pin that was second goes in to the fifth chamber, and so on.
6. As you install each pin, tamp it home with a drill bit and turn the key. If the key will not turn, you have confused the pin sequence.
7. Assemble the lock and test.

To rekey the master key, follow this sequence:

1. Mount the cylinder in a vise and remove the cylinder slide with a pair of pliers or a small chisel.
2. Remove the springs, drivers, and pins.
3. Determine the master key bitting with a gauge. The combination runs from the shoulder to the tip of the key, the reverse of the usual sequence. As an example, let it be 678572.

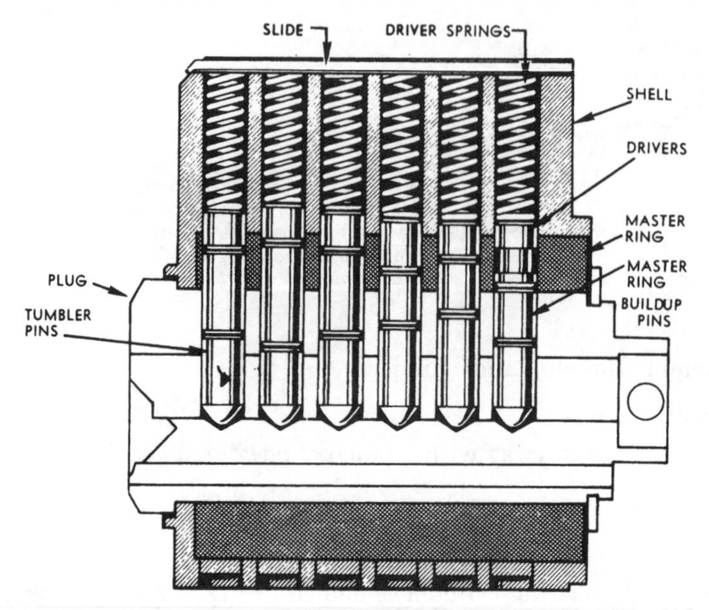

Figure 9.22 Corbin master ring cylinder.

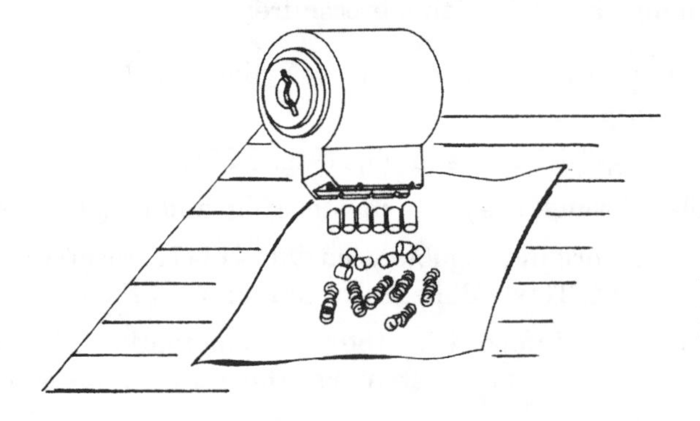

Figure 9.23 After the slide is withdrawn, remove springs, pins, and top pins. (*Emhart Corp.*)

4. Write down the change-key combination and reverse it. Suppose the combination is 275414. Reversed, it is 414572.
5. Subtract the reversed change-key combination from the master key combination (678572−414572). The difference is 264000.
6. Insert the master key into the cylinder and select the appropriate buildup pins. In this case the pins are 264. *Note*: If any number in the change-key combination is greater than the master key number above it, you must use a negative number buildup pin in the chamber. For example, a 678572 master key combination with an 814572 change-key combination requires a –2 buildup pin, together with a 6 and 4 pin.

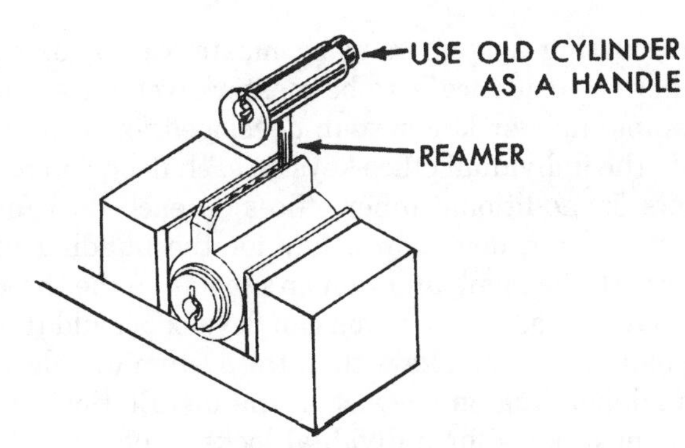

Figure 9.24 Ream the pin holes through the shell, master ring, and plug. (*Emhart Corp.*)

7. As you install each buildup pin, seat it with a drill bit and turn the key to determine that you installed the correct pin.
8. Insert the drivers into the cylinder chambers. Either of two drivers are used, depending on the lock style. Spool drivers (No. J-172) are furnished with mortise cylinders; straight drivers (No. M-099) are used on other cylinders.
9. Insert the springs into their chambers.
10. Holding the springs down with your thumb, try the change key. Do the same for the master and grand master key.
11. Mount the cylinder in the slide, hammering the slide down for a secure fit. Be careful not to damage the threads on mortise cylinders.
12. Try all the keys.

A Simple Master Key System

When you are asked to install a large number of locks for a business or a large residence wherein the customer wants the locks masterkeyed, your expertise in the realm of the master key systems is put to the test.

Probably the hardest part of developing a master key system is creating a working system that takes into consideration any variables required by the customer. For this reason, many locksmiths do not perform their own masterkeying of systems; rather, they rely on the many years of experience, expertise, and professionalism from a factory-developed system. The use of a factory system ensures that the system meets all the criteria set forth by the customer, the cylinders are correct, and the possibility of extending the system in the future can be assured. For the locksmith, time, personnel, and money are saved through this process.

On the preceding pages, you have seen various aspects of masterkeying, the types of locks that can be used, the differences in the systems, and the potential for system expansion at a later date. I will not discuss the varied intricacies of such a system but will view a small system that you may want to consider and keep in the back of your mind for small jobs that come up. These are jobs wherein an existing system may be masterkeyed, using the same cylinders, or a job where a residence is to be masterkeyed for a homeowner.

Figure 9.25 is a sample master key system developed for a three-floor office building. In addition to the individual office keys on each floor, you have the master key, extra cylinders for additional inner offices on each floor (or for use as replacement cylinders), a main door key, a key for the building maintenance shop, lavatory keys for both the men's and women's rooms on the three floors, and the start of an additional key code system you can develop should the building or other facility require more individual keys than the system was designed for.

Within this system, depending on the size of the installation you are developing, select the number of keys for individual locks to be used. In this case,

Master: 431472

First floor	Second floor	Third floor
522534	652534	752534
522544	652544	752544
522554	652554	792554
522564	652564	792564
522574	652574	792574
522594	652584	792584
522504	652594	792594
522524	652504	792504
522634	652624	792624
522644	652634	792634
522654	652644	792644
522664	652654	792654
522674	652664	792664
522684	652674	792674
522694	652684	792684
522604	652694	792694
522624	652734	792604
522724	652744	792734
522734	652754	792744
522744	652764	792754

Extra cylinders for each floor:

552784	652514	752614
522514	652524	752601
522614	604714	792602

Main door key: 346572
Bld maintenance shop key: 434672
Construction key: 466662

Note: If building were to have four or more floors, then the last three digits for the various individual key could be as above, but the first three digits would be:

Fourth floor: 514
Fifth floor: 614
Sixth floor: 714

If each floor is to have two lavatories, they could be keyed as follows:

First floor:	731573	331577
Second floor:	831573	331578
Third floor:	931573	331579

Figure 9.25 Sample master key system for a three-floor building.

we are talking about 20 individual door locks on each floor. The master key has already been designated. The progression in the development of the master key system is that the depths of the various individual keys are such that any given key can be recut to become a master key.

In actually rekeying the various cylinders, you will need the standard pins for each cylinder, and then a complete set of master pins. In this case, though, you may choose to use the lower ranking pins of the current system. Why? Because the cylinder pins are worn in addition to the interior of the cylinder itself. Also, in this case, it will save you time and money in having to order the pins.

What you are doing is determining for each key the depths that will be required for each pinning. As an example, let's take two cylinders from the first floor for rekeying. The first cylinder, 522534, will have the following pinning:

Regular Pin: 4 - 2 - 1 - 4 - 3 - 2
Master Pin: 1 - 1 - 1 - 1 - 4 - 2

With these pins in the cylinder, with the driver pins and tension spring included, either the individual or the master key will open the lock.

For the second cylinder in the series, 522534, the pinning is as follows:

Regular Pin: 4 - 2 - 1 - 4 - 4 - 2
Master Pin: 1 - 1 - 1 - 1 - 3 - 2

Again, with these pins, both the individual and the master key will operate the lock.

Now, how were these particular pins determined? The best way to figure this out is to lay out a simple chart, as shown in Fig. 9.26. As you progress in select-

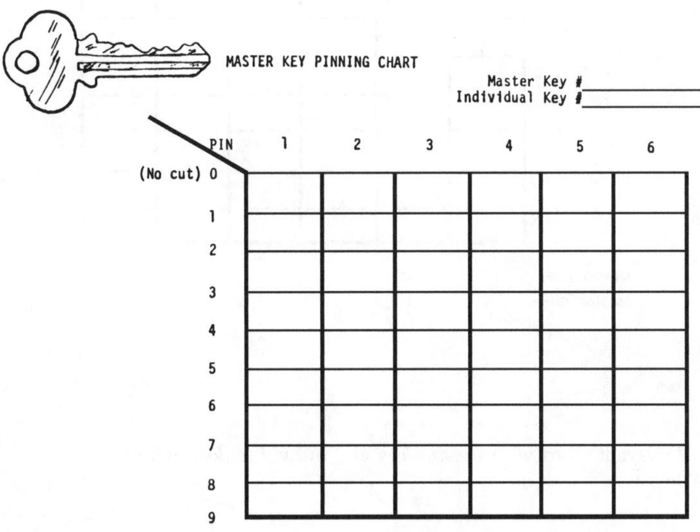

Figure 9.26 Sample chart used for arranging individual cylinder pinning for a master key system.

ing pins for the individual cylinders, first mark the appropriate points on the chart that indicate the master key number. (You can preprint this on a form and save a lot of time.) Then, as you select each cylinder to work with, put down on the chart the code number of that particular key (Fig. 9.27). The difference between the two points, if any, is the difference between the code number variations. This means that the pinning difference must be the same.

In Fig. 9.27, for column 6, notice that the individual key requires a 4 pin, but the master key requires a 2 pin. The difference of 2 is obvious. The maximum allowable height for the overall height of both pins combined must be 4, so you require a 2 pin for the master key to operate this particular pin, but also a 2 pin for the individual key. If you only had the 2 pin in the lock, only the master key would be able to open the cylinder at this point; by adding the second 2 pin, you make it possible for the individual key to also open the lock.

If you continue, you can readily see how each individual pinning is to be accomplished, what pins will be required, and the insertion sequence into the lock cylinder.

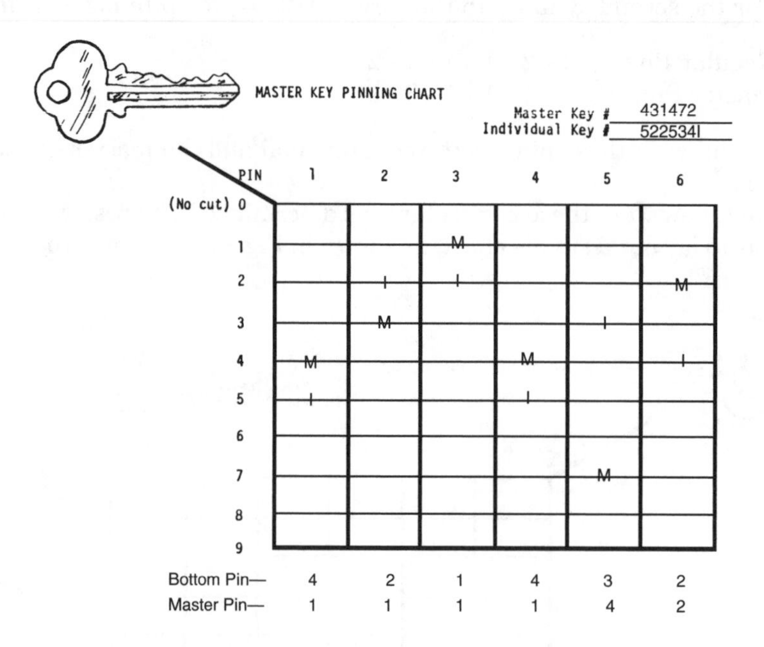

Figure 9.27 Layout chart with master key and individual key cylinder pinning indicated.

Chapter

10

Detention Locks and Hardware

The detention industry includes private, county, state, and federal sectors. It provides many opportunities for locksmiths who understand the special needs of the industry. This chapter describes the various types of detention facilities and shows how to select and install hardware for each type.

To recommend the right locks, electric strikes, and other hardware, you have to consider the security level of the facility as well as that of specific areas within the facility. Security levels include low, medium, high, and maximum. It's important to understand that a low-security facility, for example, can have low- and high-security areas. But its high-security areas might not have the same needs as a high-security area in a maximum-security facility.

Types of Locks

Mechanical and electromechanical locks are commonly used for high-security applications. Typically, the mechanical models have wafer or lever tumblers and are mortise-mounted in a 1¾-inch, hollow metal door with steel bolts and shearproof inserts. Electromechanical locks are installed in frame jambs and can be operated remotely from an electric console. They require 24 volts of direct current (Vdc) and 110 volts of alternating current (Vac).

The mogul cylinder is often used for high-security applications (Fig. 10.1). It has stainless steel pin tumblers and can be masterkeyed. Most moguls offer four keying levels: Day Key/Pass Keys, Master Key, Grand Master Key, and Great-Grand Master Key.

Standard security hardware and mortised pin tumbler locks are used for low-security applications. The locks are typically installed in 2-inch-wide jambs and come in solenoid and motor versions. (Because of its superior sideload capabilities, the motor version is specified for medium-security applications.)

Figure 10.1 The Maxi-Mogul is a popular high-security cylinder for detention facilities. (*Folger Adam Company*)

Electric Strikes

Heavy-duty electric strikes are often used in detention centers because the devices provide both mechanical and electric access control. A low- or medium-security detention facility might use them for individual confinement rooms, but they're not a good choice for individual cells in a maximum security prison.

Electric strikes are often used to create a *mantrap system*—a way of controlling access to a restricted area by using two electrically interlocked doors. A mantrap requires a person to walk through and close one door before the second door can be opened. That allows a prisoner to be inspected, searched, or detained before gaining access to a more restricted area. Highly sophisticated mantraps incorporate closed-circuit televisions (CCTV), intercoms, and security guards.

A mantrap uses a door monitor switch at each door with a connecting switch to the electric strike power leads of the opposing door. The switch acts as a power inhibitor to the opposing strike, so when one door is opened, power is cut to the second door. Only after the first door has been closed can the second be opened. While the second door is open, the first door remains locked.

Choosing the Right Equipment

In addition to knowing the security levels of the facility and its specific areas, you need to answer the following questions before choosing equipment:

- What types of doors need to be secured? (sliding, swinging, wood, metal, etc.)
- What functions must the locks perform?
- Do the locks need to operate mechanically or electromechanically?
- Do the locks need to be motor or solenoid operated?

The answers to these questions will help you choose among equipment, such as that described in the following pages.

Southern Steel 802 electric deadbolt.

The 802 is a mortise lock that provides swinging doors with auxiliary security locking and unlocking from a remote location (Fig. 10.2). It can be unlocked electrically from remote control or by key switch or pushbutton at the door. It automatically deadlocks when the door is closed, but it has no protruding deadlatch mechanism accessible to inmates when the door is open.

The lock is nonhanded and is activated by dual 115 Vac solenoids for power locking and unlocking. The case and cover are made of cold-rolled steel plate. It measures $4\frac{9}{16} \times 1\frac{3}{4} \times 9$ inches and weighs 9 pounds. The lock's stainless steel bolt is $\frac{5}{8}$ inch in diameter with a $\frac{9}{16}$-inch throw.

Special versions include:

- 802KR, which is equipped with a mogul cylinder for mechanical unlocking by key at the door and activated by dual 115 Vac solenoids for power locking and unlocking
- 802FL, which is activated by one 24 Vdc solenoid for power unlocking and has a bolt that is extended automatically when power is interrupted (fail secure)
- 802FU, which is activated by one 24 Vdc solenoid for power locking and has a bolt that retracts automatically when power is interrupted (fail secure)

Southern Steel 1010 deadlock.

The 1010 lock is a key-operated lock designed for swinging doors on access panels, plumbing chases, control cabinets, and other areas that require only

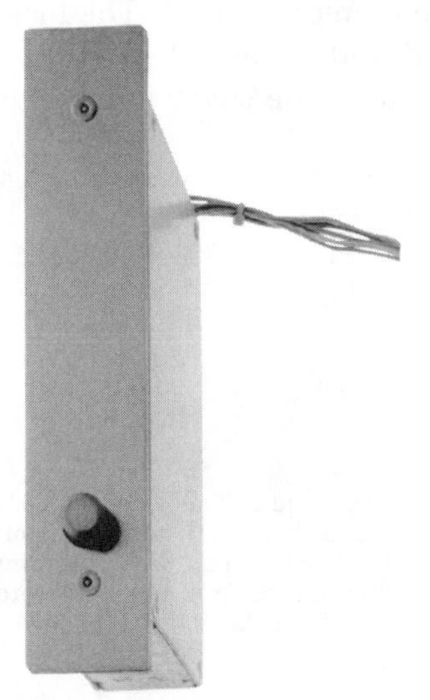

Figure 10.2 The 802 Electric deadbolt provides swinging doors with auxiliary security locking from a remote location. (*Southern Steel*)

minimum security. Figure 10.3 shows a model 1010 deadlock. According to Southern Steel's technical manuals, the lock shouldn't be used on exterior doors that directly contact the inmate population. Some jails do ignore that advice.

The lock has four holes in its corners that are used for mounting the lock to a door and securing the cover plate on the lock case. A fifth screw holds the cover plate in place; that screw hole is hidden behind a sticker on the case. You have to peel up the lower right corner of the sticker to gain access to the screw hole.

The lock measures 4¼ × 1¼ × 3 inches and weighs 4½ pounds. The cover is made of cold-rolled steel plate and the case is malleable iron. The bolt is yellow brass and measures 1½ × ¾ inches with a ⅝-inch throw.

Southern Steel 1050.

The 1050 is for swinging and sliding exterior gates. Standard mounting is for chain link fencing but can be modified for other types of gates. The lock can work by a remote pushbutton switch that activates a solenoid that raises an internal bolt, enabling the gate to open. It remains unlocked until the gate is closed, then automatically locks. The bolt can also be retracted manually with a key.

The lock cover and case are made of cold-rolled steel plate; its latchbolt is cold-rolled steel. The lock measures 7 × 3¾ × 11 inches and weighs 38 pounds. Its five tumblers are 0.097-inch spring temper brass actuated by phosphor bronze springs.

Southern Steel 1080 deadlock.

The 1080 is for swinging cell doors, corridor doors, storage room doors, and high-security doors where slam-locking isn't required. The lock is key operated and deadlocks in both the locked and unlocked position. It measures 5½ × 1½ × 3¾ inches and weighs 7 pounds. The cover is made of cold-rolled

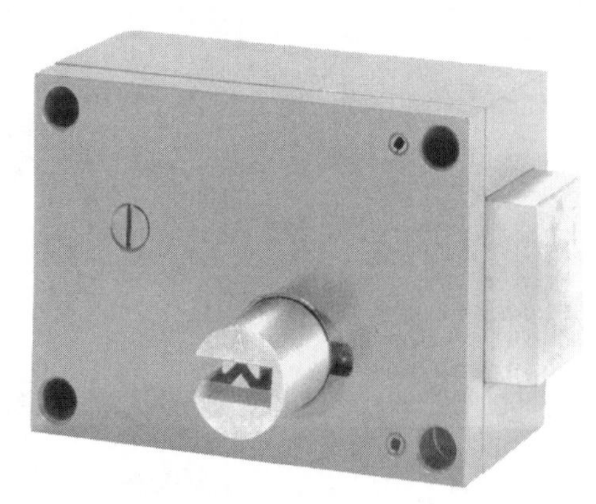

Figure 10.3 The model 1010 deadlock is for swinging doors on access panels, plumbing chassis, control cabinets, etc. (*Southern Steel*)

steel plate; the case is malleable iron. The lock's bolt is cold-rolled steel with ¼-inch diameter, case-hardened, steel insert pins; it measures 2 × ¾ inches and has a 1⅓-inch throw.

Southern Steel 10125 institutional mortise lock.

The 10125 is for swinging doors in low-security housing areas where electric or manual operation is desired. A door position switch, door closer, and power transfer hinge are recommended. The lock can work by using a remote push-button switch to activate a solenoid that locks and unlocks knobs. The lock bolt can also be retracted mechanically by mogul cylinder and knobs. The knobs can be locked out and unlocked by the mogul cylinder. When the knobs are locked out, the lock automatically deadlocks on closing.

The lock measures 4⅝ × 1¼ × 9½ inches and weighs 11 pounds. Its cover, case, face plate, and mogul cylinder are made of yellow brass. Its bolt is made of stainless steel and measures 1⅜ × ⅝ inches with a ⅝-inch throw.

Southern Steel 10195B Electrolock.

The 10195B is for electric and manual control of swinging doors in low-security inmate housing areas (Fig. 10.4). It's jamb mounted and utilizes an optional builders hardware cylinder. A door position indicator switch, door closer, and heavy-duty door pull are recommended accessories.

The lock measures 1½ × 1²³⁄₃₂ × 13⅜ inches and weighs 6½ pounds. Its bolt is made of nickel aluminum bronze. It measures ⅝- inch diameter and has a ⅝-inch throw. The lock case is made of die-cast brass and the face plate is made of brass.

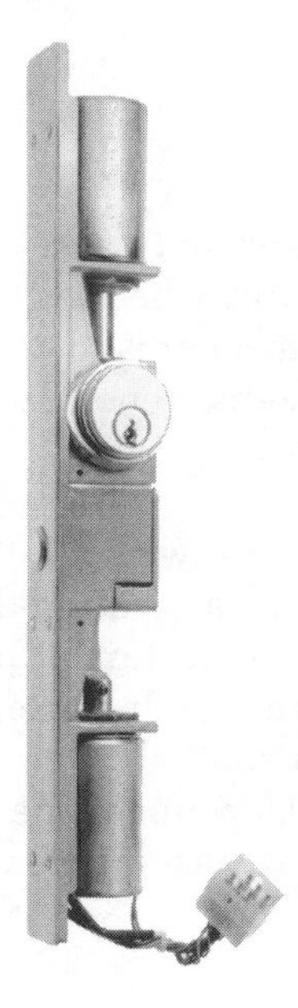

Figure 10.4 The 10195B Electrolock is for electric or manual control of swinging doors in low-security inmate housing areas. (*Southern Steel*)

In its electrical function (24 Vdc, 2½ amp), the lock can be unlocked from a remote console or by a key switch or pushbutton at the door. It automatically deadlocks when the door is closed. In its mechanical function the bolt is extended and retracted by key at the door (just like any other key-operated mechanical lock). When a key internal switch is used, however, the key will only retract the bolt.

Standard features of the lock include:

- No protruding deadlatch mechanism accessible to inmates when door is open
- Designed for mortise jamb mounting in standard door frame without visible lock pocket
- Automatic deadlocking of bolt
- Activated by power-surge dual solenoids
- During power outage, bolt remains retracted mechanically without relying on continuous electric power to its solenoid
- Bolt position indicator switch
- Nonhanded (except when internal key switch is specified)

Special features include:

- Key switch on either side or both sides
- Fail unlock version available
- Master key override—one key operates internal switch for electric bolt retraction and the other key retracts bolt mechanically
- Cylinder extenders for locks keyed stop side

Institutional hinges

202 strap hinge, 2 inches. Use 202 hinges for shutter and wicket doors where a light-duty security hinge is required. They're furnished with blank mild steel leaves for welded installation. A 202 hinge measures 4⅛ × 2 × 5⁄32 inches and weighs 0.5 pounds. Its pin is cold-rolled, case-hardened steel and measures 5⁄16-inch diameter.

203 special hinges, 3 inches. The 203 is for food passes, wicket doors, observation shutters, and other small swinging doors where a medium-duty hinge is needed. It's available with solid leaves for welded application or predrilled for fasteners. The 203FP is a version of the 203FS with a built-in stop to hold the food pass door in a horizontal position for use as a shelf.

The 203 measures 2¾ × 3 × ¼ inches and weighs 0.9 pounds. The hinge leaves are made of mild steel. The hinge pin is cold-rolled, case-hardened steel and measures 5⁄16-inch diameter.

Figure 10.5 The 204 Institutional Hinges have a nonremovable hinge pin. (*Southern Steel*)

204 institutional hinges, 4½ inches. Available in full mortise and half mortise configurations, 204 hinges can be used on swinging hollow metal or wood doors (Fig. 10.5). The hinge leaves are made of die-cast brass to provide maximum strength and durability. Each hinge features a ⅝-inch-diameter stainless steel, nonremovable pin, plus two sets of hardened steel ball bearings and races. Each unit weighs 1.6 pounds and comes with ¼-20 brass, flat-head safety screws.

204E electric power transfer hinge. Designed to supply power from door frames to electric locks on hollow metal doors, the 204E power transfer hinge contains five completely concealed and tamper-resistant Teflon-coated conductors. It's available only in full mortise configuration to be used with 204FM institutional hinges. The unit has a 1 amp capacity, 40 volts maximum. It measures 4½ × 4½ × 3⁄16 inches and weighs 1.6 pounds. Its hinge leaves are made of die-cast brass. The hinge pin is stainless steel and has a ⅝-inch diameter.

Chapter

11

Buying and Selling Safes

Virtually every consumer and business has documents, keepsakes, collections, or other valuables that need protection from fire or theft. But most people don't know how to choose a device that meets their protection needs, and they won't get much help from salespeople at department stores or home improvement centers. By knowing the strengths and weaknesses of various types of safes, you will have a competitive edge over such stores.

No one is in a better position than a knowledgeable locksmith to make money selling safes. Little initial stock is needed, they require little floor space, and safes allow for healthy price markups. This chapter provides the information you need to begin selling safes to businesses and homeowners.

Types of Safes

There are two basic types of safe: fire (or *record*) and burglary (or *money*). *Fire* safes are designed primarily to safeguard their contents from fire, and *burglary* safes are designed primarily to safeguard their contents from burglary (Fig. 11.1). Few low-cost models offer strong protection against both potential hazards. That's because the type of construction that makes a safe fire-resistant—thin metal walls with insulating material sandwiched in between—makes a safe vulnerable to forcible attacks. The construction that offers strong resistance to attacks—thick steel walls—causes the safe's interior to heat up quickly during a fire.

Most fire/burglary safes are basically two safes combined, usually a burglary safe inside a fire safe. Such safes can be very expensive. If a customer needs a lot of fire and burglary protection, you might suggest that he or she just buy two safes. To decide which type of safe to recommend, you need to know what your customer plans to store in it.

Figure 11.1 A UL-listed fire safe or chest is constructed so that its interior stays below a certain temperature during a fire. Nonrated containers might provide no more fire protection than a metal box offers. (*Sentry Group*)

Safe Styles

Fire and burglary safes come in three basic styles, based on where the safe is designed to be installed. The styles are wall, floor, and in-floor. *Wall* safes are easy to install in homes and provide convenient storage space (Fig. 11.2). Such safes generally provide little burglary protection when installed in a drywall cutout in a home. Regardless of how strong the safe is, a burglar can simply yank it from the wall and carry it out. To provide good security, a wall safe needs not only a thick steel door, but to be installed with concrete in a concrete or block wall.

Floor safes are designed to sit on top of a floor. Burglary models should either be over 750 pounds or bolted in place. Figure 11.3 shows a popular floor safe. One way to secure a floor safe is to place it in a corner and bolt it to two walls and to the floor. (If you sell a large wall safe, make sure your customer knows that the wheels should be removed from the safe.)

In-floor safes are installed below the surface of a floor (Fig. 11.4). Although they don't meet construction guidelines to earn a UL fire rating, properly installed in-floor safes offer a lot of protection against fire and burglary. Because fire rises, a safe below a basement floor won't quickly get hot inside. For maximum burglary protection, the safe should be installed in a concrete basement floor, preferably near a corner. That placement makes it uncomfortable for a burglar to attack the safe.

Figure 11.2 A wall safe can be a convenient place to store documents. (*Gardall Safe Corporation*)

Figure 11.3 A floor safe can provide more usable storage space if it has movable shelves. (*Gardall Safe Corporation*)

Make sure your customer knows that he or she should tell as few people as possible about the safe. The fewer people who know about a safe, the more security the safe provides.

Installing an In-floor Safe

Although procedures differ among manufacturers, most in-floor safes can be installed in an existing concrete floor in the following way (Fig. 11.5):

1. Remove the door from the safe, and tape the dust cover over the safe opening.
2. At the location where you plan to install the safe, draw the shape of the body of the safe, allowing 4 inches of extra width on each side. For a square body safe, for example, the drawing should be square, regardless of the shape of the safe's door.

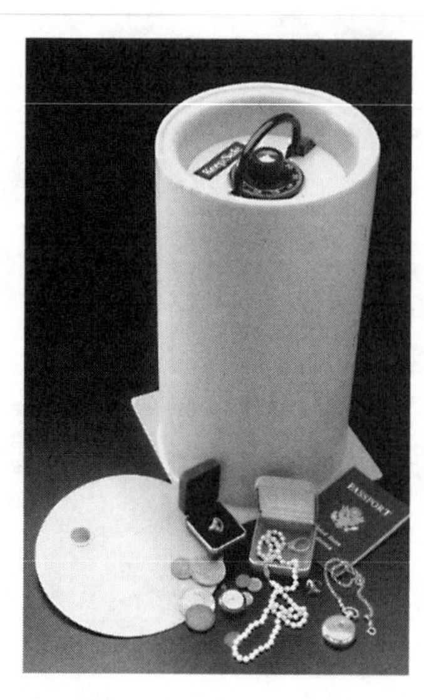

Figure 11.4 An in-floor safe can provide strong resistance to burglary and fire when properly installed in a concrete basement floor. (*Sentry Group*)

Figure 11.5 When installing an in-floor safe in concrete, make the hole the same shape as the safe walls.

3. Use a jack hammer or a hammer drill to cut along your marking.
4. Remove the broken concrete, and use a shovel to make the hole about 4 inches deeper than the height of the safe.
5. Line the hole with plastic sheeting or a weatherproof sealant to resist moisture buildup in the safe.
6. Pour a 2-inch layer of concrete in the hole, and level the concrete, to give the safe a stable base to sit on.

7. Place the safe in the center of the hole and shim it to the desired height.
8. Fill the hole with concrete all around the safe, and use a trowel to level the concrete with the floor. Allow 48 hours for it to dry.
9. After the concrete has dried, trim away the plastic and remove any excess concrete.

In-floor safes are usually installed near walls to make them hard for burglars to use tools on them.

To install an in-floor safe in concrete, use a jackhammer to cut a hole in the concrete about 4 inches wider than each side of the safe, and make the hole 4 inches deeper than the safe's height. Line the hole with plastic, and thinly line the bottom of the hole with a level layer of concrete to give the safe a solid base.

Moving Safes

Getting a heavy safe to your customer can be backbreaking unless you plan ahead. Consider having the safe drop shipped, if that's an option. Most suppliers will do that for you.

As a rule of thumb, have one person help for each 500 pounds being moved. If a safe weighs more than a ton, however, use a pallet jack or machinery mover. When moving a safe, never put your fingers under it. If the safe has a flat bottom, put three or more 3-foot lengths of solid steel rods under it to help slide the safe around, and use a prybar for leverage.

Special Safe Features

Important features of some fire and burglary safes include relocking devices, hardplate, and locks. Relocking devices and hardplate are useful for a fire safe but are critical for a burglary safe. If a burglar attacks the safe and breaks one lock, the *relocking devices* automatically move into place to hold the safe door closed. *Hardplate* is a reinforcing material strategically located to hinder attempts to drill the safe open. Never recommend a burglary safe that doesn't have relockers and hardplate.

Safe locks come in three styles: key-operated, combination dial, and electronic. *Combination dial* models are the most common (Fig. 11.6). They are rotated clockwise and counterclockwise to specific positions. *Electronic* locks are easy to operate and provide quick access to the safe's contents. Such locks run on batteries that must be recharged occasionally. For most residential and small business purposes, the choice of a safe lock is basically a matter of personal preference.

Underwriters Laboratories Fire Safe Ratings

UL fire safe ratings include 350-1, 350-2, and 350-4. A 350-1 rating means the temperature inside the safe shouldn't exceed 350°F during the first hour of a

Figure 11.6 Combination dial locks are popular for safes. (*Sargent & Greenleaf, Inc. Reprinted with permission*)

typical home fire. A safe rated 350-2 should provide such protection for up to 2 hours. Safes with a 350-class rating are good for storing paper documents because paper chars at 405°F. Retail prices for fire safes range from about $100 to over $4000. Most models sold in department stores and home improvement centers sell for under $300.

Underwriters Laboratories Burglary Safe Standard

The UL 689 standard is for burglary-resistant safes. Classifications under the standard, from lowest to highest, include Deposit Safe, TL-15, TRTL-15x6, TL-30, TRTL-30, TRTL-30x6, TRTL-60, and TXTL-60. Figure 11.7 shows a depository safe. The classifications are easy to remember when you understand what the sets of letters and numbers mean. The two-set letters in a classification (TL, TR, and TX) signify the type of attack tests a safe model must pass. The first two numbers after a hyphen represent the minimum amount of time the model must be able to withstand the attack. An additional letter and number (e.g., x6) tells how many sides of the safe have to be tested.

The TL in a classification means a safe must offer protection against entry by common mechanical and electrical tools, such as chisels, punches, wrenches, screwdrivers, pliers, hammers and sledges (up to 8-pound size), and pry bars and ripping tools (not to exceed 5 feet in length). TR means the safe must also protect against cutting torches. TX means the safe is designed to protect against cutting torches and explosives.

For a safe model to earn a TL-15 classification, for example, a sample safe must withstand an attack by a safe expert using common mechanical and electrical tools for at least 15 minutes. A TRTL-60 safe must stand up to an attack by an expert using common mechanical and electrical tools and cutting torches for at least 60 minutes. A TXTL-60 safe must stand up to an attack with

Figure 11.7 A depository safe has a slot in it that allows a store clerk to drop money in it. (*Gardall Safe Corporation*)

common mechanical and electrical tools, cutting torches, and high explosives for at least 60 minutes.

In addition to passing an attack test, a safe must meet specific construction criteria before earning a UL burglary safe classification. To be classified as a deposit safe, for example, the safe must have a slot or otherwise provide a means for depositing envelopes and bags containing currency, checks, coins, and the like into the body of the safe, and it must provide protection against common mechanical and electrical tools.

The TL-15, TRTL-15, and TRTL-30 safe must either weigh at least 750 pounds or be equipped with anchors and instructions for anchoring the safe in a larger safe, in concrete blocks, or to the premises in which the safe is located. The metal in the body must be the equivalent to solid open-hearth steel at least 1 inch thick having an ultimate tensile strength of 50,000 pounds per square inch (psi). The TRTL-15x6, TRTL-30x6, and TRTL-60 must weigh at least 750 pounds, and the clearance between the door and jamb must not exceed 0.006 inch. A TXTL-60 safe must weigh at least 1000 pounds.

TL-15 and TL-30 ratings are the most popular for business uses. Depending on the value of the contents, however, a higher rating may be more appropriate. Price is the reason few companies buy higher-rated safes. The retail price of a TL-30 can exceed $3500. A TXTL-60 can retail for over $18,000.

Such prices cause most homeowners and many small businesses to choose safes that don't have a UL burglary rating. When recommending a nonrated safe, consider the safe's construction, materials, and thickness of door and walls. Better safes are made of steel and composite structures (such as concrete mixed with stones and steel). Safe walls should be at least ½ inch thick and the door at least 1 inch thick. Make sure the safe's boltwork and locking mechanisms provide strong resistance to drills.

Selling More Safes

You'd have to cut a lot of keys to make the money you can make from selling and installing a safe. No one is in a better position than locksmiths to sell high-quality safes. Whether you're just starting to sell them or have been selling them for years, you can boost your sales.

The key to selling more safes is for you and all your salespersons to focus on the four P's of marketing: products, pricing, promotion, and physical distribution.

Products

One of the most important marketing decisions you'll make is which safes to stock and recommend. You'll need to consider quality, appearance, cost, warranty, and delivery time. Only sell good safes that you believe in. Your enthusiasm for them will make it easier for you to talk about them.

Little initial stock is needed to start selling safes. If you want to be taken seriously, however, you'll need to have a few on display. Most people want to see and touch a safe before buying it—much like when buying a car. Stock several sizes of each type of safe. This will make it easier for you to sell the customer up to a more expensive model (Fig. 11.8).

If you're just starting to sell safes, don't stock large, heavy models because they're expensive, hard to transport, and usually don't sell quickly. Consider stocking floor fire safes and in-floor safes (Fig. 11.9). In some locales, you also may want to stock gun safes. If you're planning to sell to businesses, stock TL-rated floor safes and depository safes (shown in Fig. 11.10). Square-door safes usually sell faster than round-door models.

In addition to choosing which products to sell, you'll need to choose a distributor to buy from. Some distributors have a "safe dating program," in which they'll let you stock safes without having to pay for them until you sell them. If your distributor doesn't offer such a program, ask about stocking them for 90 days before paying. And see if the distributor offers training seminars.

You and your salespeople must become familiar with what you're selling. Study literature about the safes, and attend distributor and manufacturer seminars. If you don't know much about your products, potential customers will notice. As a locksmith, you're selling your expertise as well as safes. If price were the only factor, people would buy low-end safes from department stores and home-improvement centers instead of high-quality safes from you. Major manufacturers regularly offer seminars on installation and sales.

Pricing

Buy safes at good prices, and sell them at a reasonable markup. Don't worry about not having the lowest prices in town. Marketing for peak profits involves adjusting the prices of products to meet the needs of customers and the needs of your company. Adjustments in prices mean adjustments in the customer's perception of prices. Prices should be based on perceived value. Many customers

Figure 11.8 Having multiple sizes of a safe gives you more selling options. (*Courtesy Adesco Safe Company.*)

will choose a more expensive product because they believe in the adage, "You get what you pay for."

Some customers always refuse to pay the sticker price; they feel better if they dicker the price down. You need to price your safes so that all types of customers will feel that they're getting a good deal. One way to do this is to price most items with a little room for dickering. It's a good idea to price so that you can negotiate slightly—such as by giving a 2 percent discount for cash payment.

Be careful about lowering your prices, however. Every attempt should be made to sell the safe at sticker price. If the customer objects, point out the safe's benefits and features. Keep in mind that you're selling a specialty product that will protect the valuables and keepsakes of a family or business. Make it clear that you're selling a high-quality product. One major safe dealer emphasizes the importance of quality by displaying a cheap fire safe that had been broken into.

Figure 11.9 In-floor safes are designed to be installed inside a floor. (*Courtesy Adesco Safe Company.*)

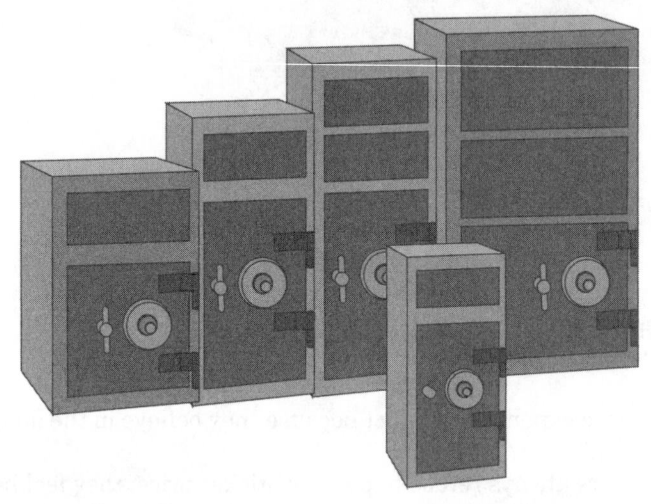

Figure 11.10 Depository safes are popular for business uses. (*Courtesy Adesco Safe Company.*)

Promotion

The quality of your promotional efforts has a lot to do with how much money a customer will be willing to pay for your safes. Promotion is mainly in the form of imaging and advertising. To design an effective imaging plan, you need to consider everything your customers see, hear, and smell during and after the selling process. Pay close attention to detail.

Your showroom needs to be pleasant for customers to be in. The safes should be displayed where they can be readily seen and touched, but not where customers will trip over them. Your customers should have to walk by safes whenever they come into your shop. The display area should have good lighting, be clean (don't let dust build on the safes), and be at a comfortable room temperature. Use racks and elevated platforms so that customers don't have to bend down to touch your smaller safes.

Use plenty of manufacturer posters, window decals, and brochures in the display area. Such materials help to educate customers about your safes. Some safe manufacturers offer display materials.

In addition to placing promotional literature near the safes, include a product label on each safe. The label should include the following information: safe brand, rating, special features, warranty, regular price, sale price (if any), and delivery and installation cost. This information will help you better describe the product to customers.

A lot of locksmith shops have a Web site. It can help you to make direct sales, as well as to promote all your products and services. You can even include a map to make it easier for customers to find you. The key to having a successful Web site is to get an easily remembered domain name. If it isn't already being used on the Internet, you can use your shop name. To find out if a domain name is in use, go to www.networksolutions.com.

After getting a domain name, you can use one of the many Web site creation programs to make your Web site, or you can hire someone to do it. Expect to pay at least a few hundred dollars for someone to make a basic Web site. To get ideas for creating one, go to an Internet search engine, such as www.hotbot.com, and enter "lock and safe" or "safe and lock." You'll find lots of locksmiths' Web sites. Once you have a Web site you'll need to promote it by including your Web address on your letterhead, business cards, service vans, and Yellow Pages ads.

The most important advertisement you can have is a listing in your local Yellow Pages. When people are looking for a safe, they don't read the newspaper; they reach for the phone book. Consider a listing under "Locks and Locksmithing" and "Safes and Vaults."

The larger your ad, the more prominent your company will seem. (And the more costly the ad will be.) To determine the right size ad to get, look at those of your competitors. If no one else has a display ad, for instance, then don't get a display ad for that Yellow Pages heading. Instead, consider getting a bold-type listing. If lots of your competitors have full- and half-page ads, however, you should have one too if you can afford it. If not, get the largest you can afford.

Some large safe dealers find television ads to be useful. Although advertising on national television can be expensive, advertising on local and cable television can be cost-effective. To do successful television advertising, you need to create professional-quality commercials and run them regularly. You can't just run them for a month or two and expect long-term results.

The most successful safe dealers take every opportunity to talk up their safes. Whenever someone buys something at your store, ask if they need a safe. And ask during every service call. Be prepared to talk about the benefits of buying one of your safes—convenience, protection, and peace of mind. Explain that you install and service your safes. Even if the person isn't ready to buy one now, he or she will remember you when he or she is ready to buy.

Physical distribution

The sale of a safe isn't the end of the transaction. Delivery of a safe should be done as soon as possible after the sale. Slight paranoia is a natural symptom in a customer who has just purchased a safe. If it takes too long to deliver it, the customer may want to cancel the order. Only work with distributors who stock a lot of safes and who can get them to you quickly.

Delivery should be done professionally and discreetly. Some safe retailers use unmarked vehicles to deliver safes. If you use an unmarked vehicle, be sure to point that out to the potential customer when you're trying to sell the safe.

By taking a little time to evaluate your current marketing strategy, you'll find ways to make it better. Just remember to carefully coordinate your decisions about the four P's of marketing, and you'll improve the fifth P—profits.

Chapter

12

Keyed Padlocks

Padlocks are used for securing outbuildings, bicycles, buildings under construction, tool boxes, paint lockers, and even automobile hoods. Because of this wide use, a locksmith needs a good working knowledge of padlocks.

While padlock exteriors vary, the functional and operational differences are few and are similar to those of other locks. Padlocks might use pin tumblers, wards, wafers, levers, or a spring bar. Some must be shackled closed before the key can be removed; this feature is made possible by a spring-loaded coupling. Key security is improved and there is less likelihood of leaving the lock open.

Choosing a Padlock

Ask the customer if he or she has a brand preference, then ask the following questions:

- Are width, case length, and shackle clearance crucial (Fig. 12.1)?
- Where will the lock be used?
- How often will the lock be opened? The price of the lock usually has a direct relationship to its wearing qualities.
- Is the lock intended to secure valuable property? An inexpensive lever lock is adequate to keep children from straying into the backyard but inappropriate for a boat trailer.
- Will the lock be used indoors or outdoors? If it is an outdoor lock, will it be protected from the elements?

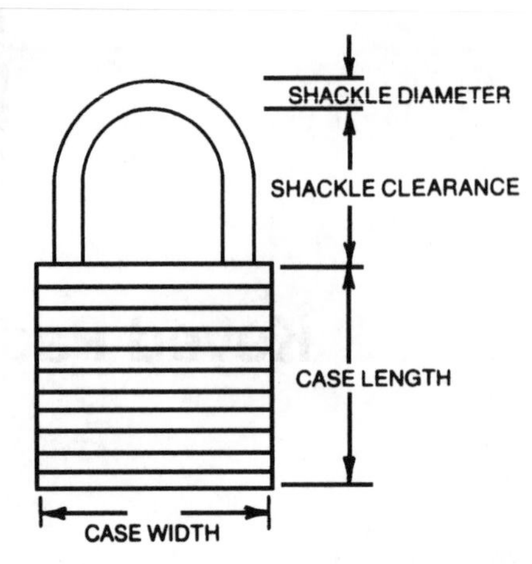

Figure 12.1 Padlock nomenclature.

Warded Padlocks

Warded padlocks locks have limited life spans, particularly when used outdoors. The cheapest locks of this type are only good for a few thousand openings and, when locked, give minimal security.

Most of these locks have three wards, although the cheaper ones have only two. Figure 12.2 illustrates the principle. The key must negotiate the wards before it can disengage the spring bar from the slot in the shackle end. Keys are flat or corrugated. The latter is a mark of the Master Lock Company.

Repairs to the lock itself are out of the question, because it would be cheaper to buy a new one.

Pass keys

Figure 12.3 shows how a pass key is cut to defeat the wards. The broad tip of the key opens the spring bar; locks with two spring bars require a key cut as shown in Fig. 12.4.

Some corrugated keys can be reversed to fit other locks by filing on the back of the blades. Alternately, you can file all the unnecessary metal off, converting the key into a pass key.

In many localities, pass keys are illegal and, unless you are a locksmithing student from an accredited school, a locksmith trainee, or a licensed locksmith, possession of such a key is a criminal offense. If you have the need for a pass key, keep it in a safe place in your home or office.

Key cutting

A warded padlock key is simple to duplicate. Follow this procedure:

1. Using the original key as a guide, select the appropriate key blank.

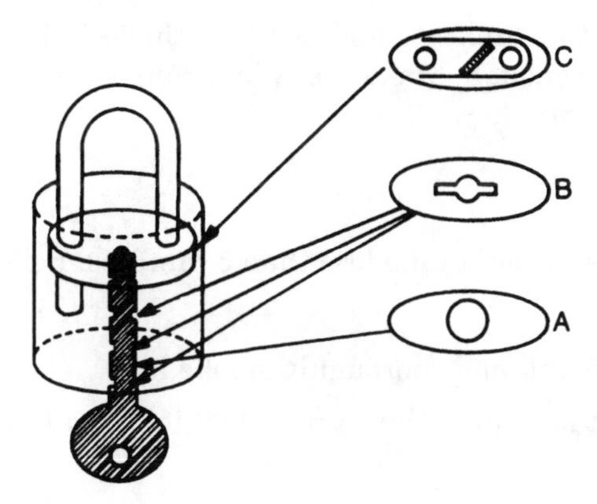

Figure 12.2 A warded padlock. Section A represents a rotating disc keyway; section B one of the three wards; and section C the spring bar.

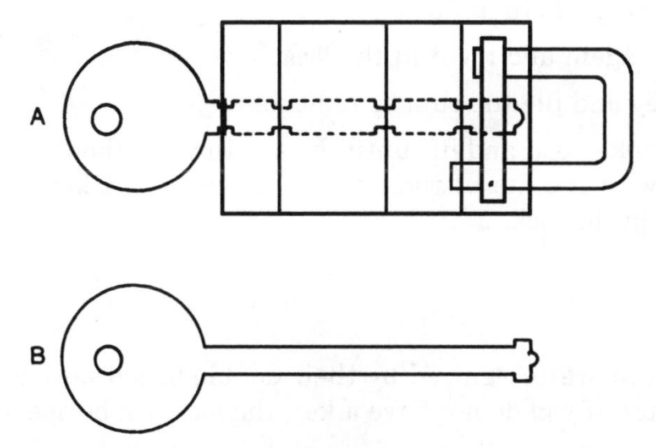

Figure 12.3 Wards and their limitations. View A shows the ward arrangement and the necessary key bitting; view B illustrates a pass key.

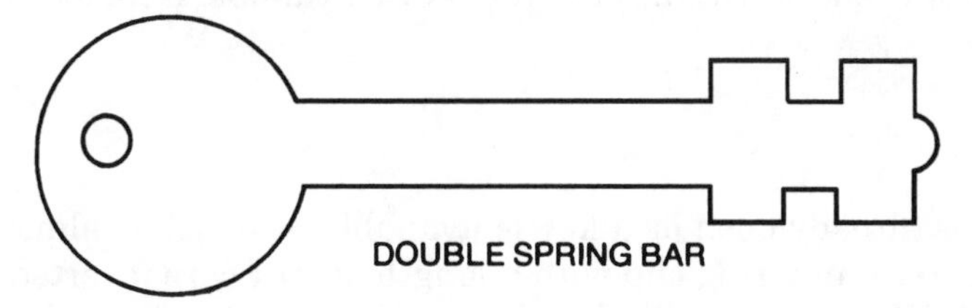

Figure 12.4 A pass key for a lock with two spring bars.

2. Smoke the original key and mount the original and the blank in your vise.
3. Using a 4-inch warding file, cut away the excess metal until the blank is an exact copy of the original.
4. Turn the keys over and repeat the operation.

5. Remove the burrs from the duplicate and test it in the lock. If it sticks, the cuts are not deep or wide enough. Make the appropriate alterations and you will have a perfect duplicate key.

Key impressioning

Impressioning a warded key should take less than 5 minutes. Follow this procedure:

1. Select the appropriate blank and thoroughly smoke it.
2. Insert the key and twist it against the wards. Do this several times to get a clear impression.
3. Mount the key in the vise.
4. Make shallow cuts where indicated.
5. Smoke the key again and try it in the lock.
6. Remove the key and file the cuts as indicated.
7. Continue to smoke, test, and file until the key turns without protest. Do not go overboard with the file. If the bits are too deep, the key will work but may break off in the lock.

Wafer Disc Padlocks

Wafer disc padlocks are recognized by their double-bitted keys and can be a headache to service. If you do not have a key, the lock can be opened by either of two methods:

- Picking is possible, but it takes practice. Purchase a set of lock picks for these locks and spend a few hours learning the skill.
- You can try the lock with a set of test keys, available from locksmith supply houses.

Disassembly

If the lock is already open or a key is available, your job is almost half done. Release the retaining ring clip with a length of stiff wire inserted into the toe shackle hole. Remove the cylinder plug and lay out the parts.

Keys

Duplicate keys can be cut by hand or with a machine designed for this purpose. These machines are expensive. Locksmiths get around the problem by stocking cylinder inserts and precut keys. The insert replaces the original wafer mechanism. A precut key requires that the discs be realigned to fit. You will need some blank discs and should have access to the cutting tool described in the chapter on double-bitted locks and keys.

In extreme cases, you can insert the key and file the disc ends for shell clearance. This is not the sort of thing that a real professional would do, but it works.

Pin Tumbler Padlocks

There are hundreds of pin tumbler locks on the market, but most fall into two categories, based on the case construction: laminated or extruded.

Laminated padlocks

Slaymaker pin tumbler padlocks are made up of a series of steel plates held together by four rivets at the casing corners. Follow these service procedures:

1. Use a hollow mill drill to shear off the rivet heads on the case bottom; this technique allows reuse of the rivets.
2. Remove the bottom plate.
3. Remove the entire cylinder section in one piece.
4. Two cylinder-housing types are used. The most popular requires a follower tool to keep the pins intact.
5. Make the necessary repairs.
6. Insert the cylinder into the casing and replace the bottom plate over the four rivet ends.
7. After checking the action, use a ball-peen hammer and repeen the rivet ends.

Extruded padlocks

Extruded locks are made from a single piece of metal, usually brass. You will need the following tools to service these locks:

- Small nail with the point cut off
- Key blanks
- Pliers
- Small punch
- Hammer
- Small light (overhead lights are too bright; a flashlight is acceptable)

To disassemble the extruded lock, first locate the plug pins or cover on the bottom of the case. Since the case is highly polished, this is sometimes hard to do. Tilting the lock under the light will show a faint outline of the cover cap or plug pins.

To assemble extruded locks held together by exposed pins, follow this procedure:

1. Determine the diameter of the plug pins.
2. Select a drill bit smaller than the diameter of the pin.
3. Drill slowly and only a fraction of an inch deep—the pin does not extend very far into the lock. If you drill through it, you can damage the spring.
4. Extract the plug.
5. Remove each spring and pin.
6. Remove the cylinder. The cylinder is held by a pin that may be covered by a small pin plug. Rapping the lock against a hard surface may shock the plug loose. If not, dislodge the pin plug with a small screwdriver.
7. Make the necessary repairs.
8. Assemble the lock.
9. Fill the pin holes with brass wire.
10. File the wire flush with the case and polish.

Extruded locks with a retaining plate over the pins require a slightly different procedure. If the lock is open, drive the plate out with a punch inserted at the shackle hole. If it is locked, follow this procedure:

1. Drill a small hole in the plate off-center.
2. Using an ice pick or other sharp instrument, drive the plate down toward the shackle. This will buckle the plate and cause the edges to rise.
3. Pry the plate loose by working a sharp instrument around the edges.
4. If the plate is not severely damaged, it can be reused. Bow the plate slightly so that the center bulges outward when the plate is installed.
5. Peen the edges of the plate to form a tight seam.

Other locks mount the plug and cylinder by means of horizontal pins running across the width of the lock. These pins can be seen under strong light.

Key fitting for pin tumbler padlocks

Follow this procedure:

1. Insert the blank into the plug.
2. Place the No. 1 pin in its chamber.

3. Using a blunted nail as a punch, drive the pin into the blank. A single light hammer tap is enough to impression the blank.
4. Remove the blank.
5. File the blank at the impression.
6. Insert the key and try to turn the cylinder. File as necessary.
7. Move to the next pin and repeat the process.

Major Padlock Manufacturers and Their Products

American (Junkunc Brothers)

All American (Junkunc Brothers) padlocks share the same patented locking device—two hardened steel balls fitting into grooves in the shackles. This arrangement is one of the best ever devised. Applying force on the shackle wedges the balls tighter. The H10 model, for example, requires more than 5000 pounds to force; test locks have been stressed to 6000 pounds and still worked. All H10 locks use a 10-blade tumbler. (The exceptions to these statements are those locks with a deadlock feature. The key must be turned to lock the shackle. These padlocks are made on an entirely different principle and are not discussed here.)

American padlocks have a removable cylinder that simplifies servicing. If a customer wants keyed-alike locks, the modification takes only a few minutes. Locksmiths can also key-alike different models.

These locks have three basic subassemblies: cylinder assembly, locking mechanism assembly, and the shackle assembly.

Cylinder removal and installation. Follow this procedure:

1. Open the lock, exposing the *retaining screw* at the base of the *shackle hole*.
2. Remove the retaining screw (I in Fig. 12.5). If the screw is stubborn, use penetrating oil and let it set a few moments before attempting to turn the retaining screw.
3. Strike the side of the lock with a leather mallet. The purpose is to force the *retaining pin* (C) into the space vacated by the retaining screw. Referring to the drawing, note how the pin retains the cylinder in the slot shown.
4. Withdraw the key together with the cylinder. Place a new cylinder in the case.
5. Assemble the lock in the reverse order of disassembly. *Note*: The H10 series uses a plated silver cap dimensionally identical to the brass caps on the other models.

The locking mechanism. The brass *retainer* is the heart of the assembly (E in Fig. 12.5). Retainers vary in size according to the lock model. A30 and AC20

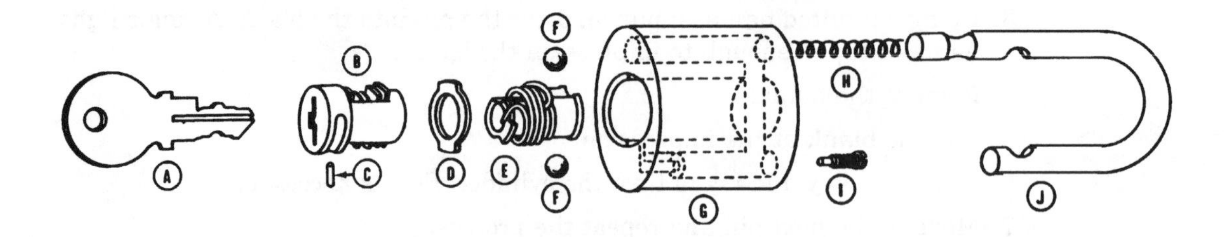

A. Key
B. Cylinder
C. Cylinder-Retaining Pin
D. Retainer Washer
E. Retainer Assembly With Coiled Spring
F. Hardened Steel Balls
G. Case
H. Shackle Spring
I. Retainer Screw
J. Shackle or "Staple"

Figure 12.5 A popular American padlock.

retainers have a groove milled in the body. L50, K60, and KC40 retainers are plain. The H10 is much larger than the others.

The retainer assembly has three parts: a washer, coil spring, and retainer body. The *washer* (D) fits over the retainer. One end is hooked to accept the coil spring. The other end of the spring is moored in a hole on the retainer.

To remove the retainer, follow this procedure:

1. Grasp the retainer body (E) with needle-nosed pliers.
2. Rotate the retainer 45° and pull.
3. Remove the steel balls.

To replace it, follow these steps:

1. Grease the balls so they will stay put.
2. Replace the balls and spread them apart.
3. Exert pressure against the shackle to hold the balls.
4. Determine that the washer is correctly aligned with the spring. The free end of the spring has to be in line with the retaining-pin hole in the case.
5. Place the retainer assembly about halfway in the case with the washer riding in the grooves provided. The end of the spring also rides in the groove.
6. Twist the retainer about a quarter turn to the right and down.

Shackle. The various models have different shackle lengths and diameters. The spring (H in Fig. 12.5) must match. For example, an L-shaped spring is used on the L50 lock. An extra pin guides the long spring on the K60. The shackle in the H10 is secured with a hardened steel pin; other models secure the shackle with the retainer mechanism.

Pin tumbler padlocks. The American five-pin tumbler padlock was designed so the cylinders can be changed quickly (Fig. 12.6). Late production 100- and 200-series use the same cylinder assembly and keyway, reducing the inventory load.

Assembly and disassembly of the pin tumbler padlock is only slightly different from that of the other American padlocks. To change the cylinder, follow this procedure:

1. Open the padlock to expose the retaining screw at the base of the shackle.
2. Remove the retaining screw with a small screwdriver.
3. Pull the cylinder out of the case. *Note:* Leave the lock unlocked; do not depress the shackle.
4. Insert a new cylinder in the case. Replace the screws, bringing the cylinder almost flush with the case.

To assemble the locking mechanism:

1. Assemble the retainer (D in Fig. 12.6) and the retainer washer (C) with the free end of the spring tightly against the left side projection on the washer.
2. Place the shackle spring (H) in the hole on the end of the shackle (I).
3. Insert the shackle spring into the deep well of the lock body.
4. Drop the balls (E) into the case bottom and move them into the pockets with a small screwdriver. There must be room for the retainer assembly.

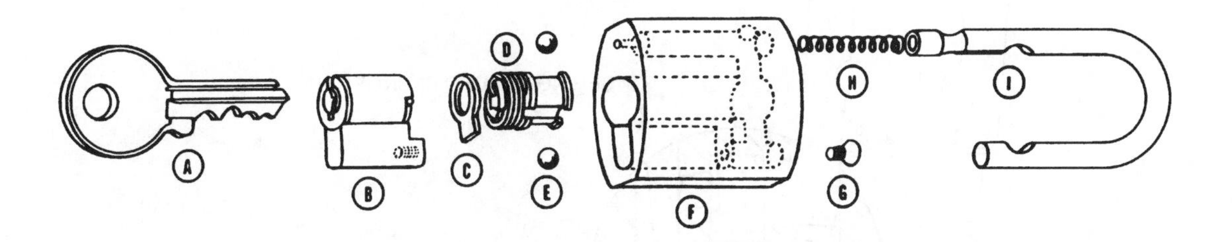

A. Key
B. Cylinder Assembly
C. Retainer Washer
D. Retainer Assembly
E. Hardened Steel Balls
F. Case or Body
G. 6/32 Brass Screw
H. Shackle Spring
I. Shackle or "Staple"

Figure 12.6 An American pin tumbler padlock.

5. Depress the shackle so the balls cannot slide back into the center of the case.
6. Insert the retainer assembly (C and D). The retainer washer (C) fits into the elongated hole in the case.
7. Make sure the assembly is bottomed in the hole.
8. Insert a retainer-assembly tool into the cylinder hole so the step on the tool engages the retainer step. Turn the tool clockwise until the lock opens.
9. Install the cylinder assembly. Holding it flush with the case bottom, insert the 6 × 32 brass screw (G) into the shackle hole and tighten it snugly.

Note: The American Lock Company will provide the retainer-assembly tool free for the asking.

Because of the close tolerances of American padlocks, they will not operate properly in extreme cold weather unless Kerns ML3849 lubricant is used. Graphite is acceptable in milder climates.

To remove the cylinder in the 600 series pin tumbler padlocks, unlock the padlock, and turn the shackle as shown in Fig. 12.7. Depress the spring-loaded plunger (A) with a small screwdriver until the plunger is flush with the wall of the shackle hole. At the same time, pull on the key. Withdraw the cylinder.

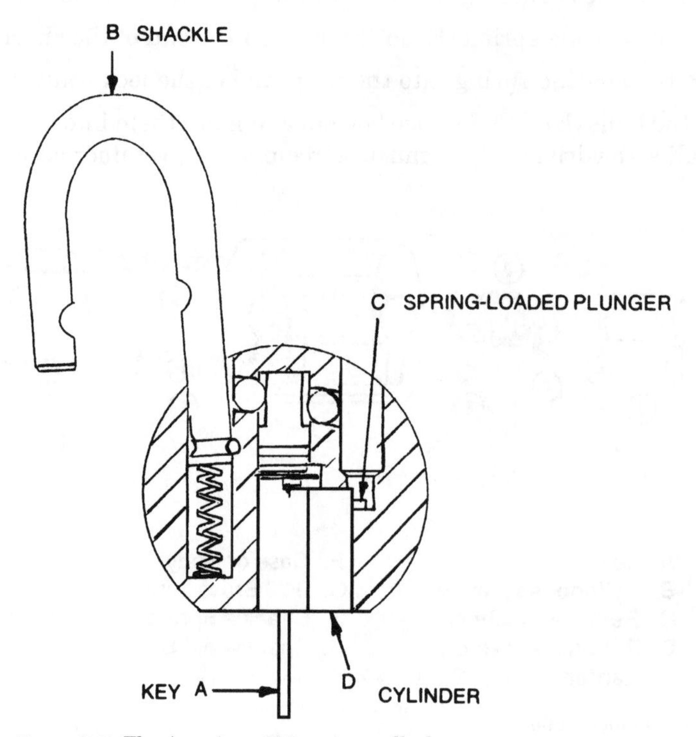

Figure 12.7 The American 600 series padlock.

Master padlocks

The Master Lock Company makes a variety of padlocks, ranging from simple warded locks to sophisticated pin tumbler types.

The Super Security Padlock. An improved version of the familiar Master padlock was recently introduced—the Master Super Security Padlock (Fig. 12.8). It was designed to give greater security than standard locks and is recommended for warehouses, storage depots, and industrial plants, as well as the home and yard. Figure 12.9 is a cutaway view of the mechanism. Salient features include:

- A patented dual-lever system to secure the shackle legs. Each lever works independently of the other and is made of hardened steel.
- The long shackle leg is tapered to align the locking levers.
- For weather protection, the case laminations are made of hardened steel and are chrome-plated.

Figure 12.8 The Master Super Security Padlock.

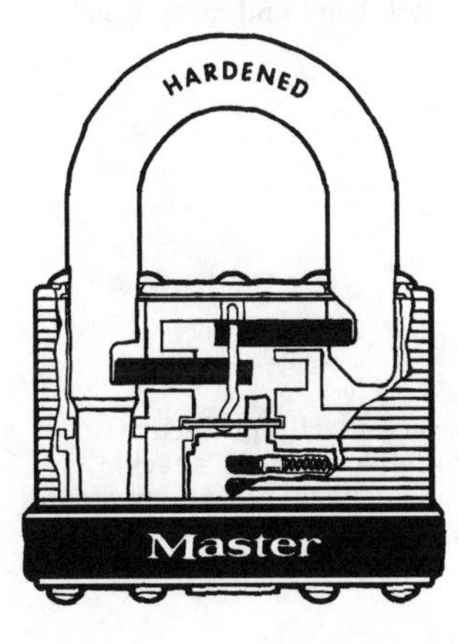

Figure 12.9 A cutaway of the Master Super Security Padlock.

- For better protection, the case is larger than that of standard locks.
- A rubberoid bumper prevents the lock body from scratching adjacent surfaces.
- Tension tests show that this high-security lock can tolerate without damage a force on the shackle of 6000 pounds.

When servicing the Super Security lock, follow this procedure:

1. Drill out the bottom rivets.
2. Remove the plates, one at a time, and keep them in order.
3. Remove the lock plug.
4. Assemble the lock using new rivets.

Master Lock's Pro Series locks. Master Lock Company's Pro Series is a complete line of high-security padlocks that are especially for commercial use. The line includes "Weather Tough," "High-Security Shrouded," and a round, solid steel lock (Figs. 12.10 through 12.12). All have black bodies and shackles made from

Figure 12.10 Pro Series "Weather Tough" padlocks are protected with "Xenoy" thermoplastic covers and feature boron alloy shackles and special drain channels that move water and debris through the lock body and away from the locking mechanism. (*Master Lock Company*)

Figure 12.11 Pro Series High Security shrouded padlocks have a steel shackle guard that protects the shackle from saws, bolt cutters, and prying. (*Master Lock Company*)

Figure 12.12 One type of Pro Series High-Security padlock comes in a round, solid steel nonshrouded version. (*Master Lock Company*)

a hardened boron steel that provides up to 15,000 pounds of cut resistance. Inside, the shackles are locked with dual ball-bearing locking mechanisms that resist up to 13,000 pounds of pulling and prying. In addition, they are shrouded in solid iron.

The cylinders of all Pro Series locks feature special spool pins to make them pick-resistant. All Pro Series padlocks are rekeyable with cylinder and shackle options that are quick and easy to change.

The Pro Series also features a line of Weather Tough locks especially for outdoor applications. These locks have the same boron alloy shackles and dual ball-bearing locking mechanisms found on the High-Security models. A xenoy thermoplastic cover protects against key and cylinder jamming from sand, dirt, and other contaminants. Special drain channels inside move water and debris through the lock body and away from the locking mechanism for trouble-free opening.

Removing and replacing the laminated and Pro Series cylinder

1. Unlock the padlock and remove the key (Fig. 12.13). Turn the shackle away from its opening. *Note*: If rekeying WO (without cylinder) padlocks, use a screwdriver to open the shackle. Insert the tip of the screwdriver through the opening in the bottom plate. Turn the mechanism as shown to trip the shackle release. (For Pro Series 6000 locks simply pull up on the shackle. Cylinder extension and ball bearings are included in a separate polybag.)
2. The cylinder retainer screw is located inside the shackle opening. Using an appropriately sized hex wrench, rotate the screw counterclockwise (left) until the cylinder retainer door separates from the padlock body.
3. Remove the retainer door and nut. Slide the cylinder out. Be sure to cradle the padlock in your hand to catch parts.
4. Place the new cylinder rear-end first into the padlock body. Replace the retainer door and nut.
5. Tighten the retainer screw with a hex wrench.

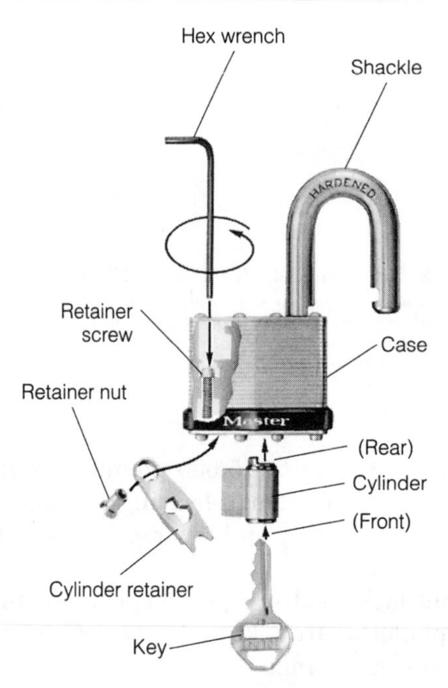

Figure 12.13 Exploded view of a laminated Master padlock. *(Master Lock Company)*

Removing and replacing the cylinder in solid-body padlock (Nos. 220, 230)

1. Follow Steps 1 and 2 of "Removing and Replacing the Laminated and Pro Series Cylinder."
2. Slide the cylinder retainer out (Fig. 12.14). Turn the retainer upside down to remove the cylinder.
3. Place the new cylinder with the rear end facing up into the retainer. Slide the retainer into the padlock body.
4. Tighten the retainer screw with a hex wrench.

Rekeying the Pro Series cylinder

1. Examine the rear of the cylinder and note the flat indent on the otherwise round rim of the cylinder shell. This is the *index mark*. While holding the cylinder, insert the key and rotate it 90° toward you (counterclockwise) so the bottom of the keyway lines up with the index mark. *Caution*: Don't rotate the keyway 180° or the pins may jam.
2. Line up the cylinder assembly tool shoulder to shoulder with the rear end of cylinder.
3. Carefully push the tool into the cylinder until the plug comes out. Leave the tool inserted in the cylinder shell to prevent top pins and springs from falling out.

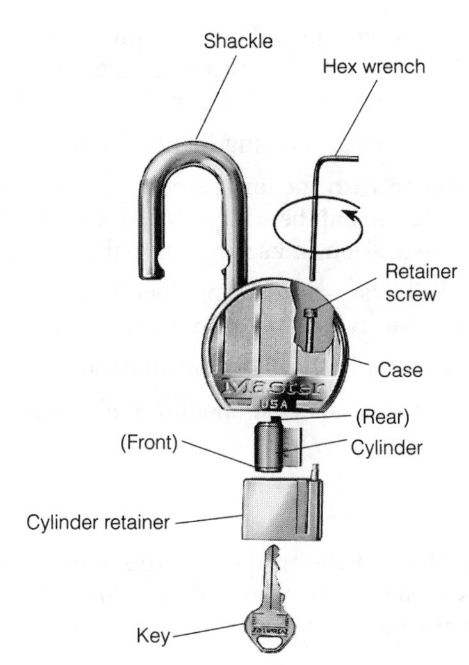

Figure 12.14 Exploded view of a solid-body Master padlock. (*Master Lock Company*)

4. Remove the key and turn the plug upside down so that bottom pins drop out.
5. Using a key cut gauge, determine the bottom pin sizes to match the new key cut.
6. Insert pins with pointed side down into the plug.
7. Insert the new key. Each pin should be flush with the plug surface.
8. Slowly push the plug back into the cylinder shell to force the assembly tool out the opposite end.
9. Carefully rotate the key 90° away from you (clockwise) back to the starting position. *Caution*: Don't rotate the keyway 180° or pins may jam.

Removing and replacing the Pro Series shackle

1. Follow steps 1 and 2 of "Removing and Replacing the Laminated and Pro Series Cylinder."
2. Remove the retainer door and nut. Slide the cylinder out.
3. Close the shackle.
4. Strike the bottom of the lock against the palm of your hand. The shackle extension and ball bearings will drop out.
5. Pull up on the shackle to remove it.
6. Insert the new shackle into the lock body.

7. To replace the ball bearings, turn the padlock upside down. Drop one ball bearing into the opening. Add a small amount of grease to lubricate if necessary. Use a hex wrench to carefully push the ball into the left channel.
8. Repeat with the other ball bearing, pushing it into the right channel.
9. Replace the cylinder extension with the large steel end in first, lining it up with the opening (make sure the ball bearings stay in either channel). Use a hex wrench to push the extension in as far as possible.
10. Place the cylinder—rear end first—into the padlock body (remove the key from the cylinder). Replace the cylinder retainer door and nut.
11. Use the key to open the lock and pull up on the shackle.
12. Insert the retainer screw into the shackle opening and tighten with a hex wrench.

ILCO

ILCO secures the cylinder with what the trade calls the "loose-rivet" method. A single brass rivet extends through the case and into the cylinder. The rivet is headless (hence the term *loose*) and is secure by virtue of its near invisibility (Fig. 12.15).

To remove the pin, follow these steps:

1. Hold the case up to the light and look for the shadow created by the pin. The pin is located ⅛ inch from the bottom of the case and is centered on the side.
2. Using a No. 48 bit, drill a hole in the end of the pin.
3. Thread a small screw into the hole.
4. Gently pry up on the screw and remove it together with the pin.

To insert a new pin, follow these steps:

1. Select a piece of brass from stock that exactly matches the diameter of the pin hole.
2. Use the next size smaller rod (to make removal easier if it is ever necessary) and cut it so that 1/32 inch stands out.

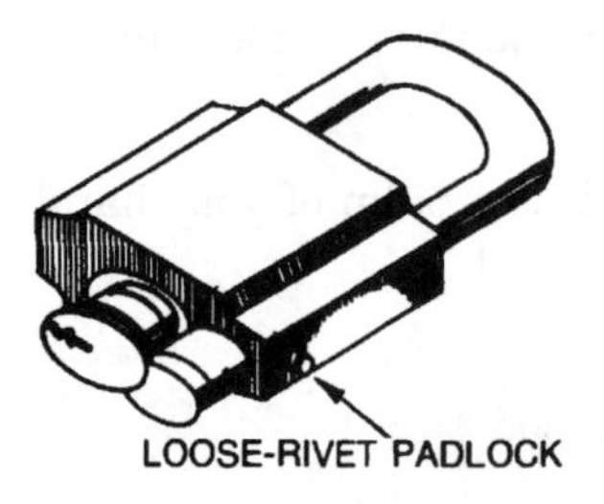

Figure 12.15 ILCO loose-rivet padlock.

3. Peen over the end of the rod.
4. Burnish the case to camouflage the pin.

Another ILCO variant is a *cylinder retaining plate* on the bottom of the case. The rivet that holds the assembly together is cunningly disguised in the maker's name stamped on the plate.

To remove the retaining plate, follow this procedure:

1. Draw a line under the ILCO name.
2. At a right angle to the first line, draw a second line (on the outside edge of the O in ILCO).
3. Drill a shallow hole to accept the tip of a 6 × 32 machine screw. Fit the screw with a washer and thread it into the hole.
4. Using the screwhead for purchase, pry the retainer out of the case.
5. Withdraw the cylinder.
6. When the cylinder has been serviced and replaced into the case, mount the retainer in its original position, tapping it home with a flat-ended punch. Fill the hole with brass stock.

Kaba security padlocks

The Kaba KP 2008 security padlock (Fig. 12.16) is a good-quality, heavy-duty lock with an interior dust cover which, when coupled with the stainless steel body, means it can be used in any weather conditions. The dimensions are in Fig. 12.17.

The KP 9 cable padlock has a plastic-coated steel cable. It is well suited to securing bulky items such as equipment on building sites, bicycles, etc. (Fig. 12.18).

Figure 12.16 Heavy-duty padlock with the Kaba 20 cylinder. (*Kaba Locks, Ltd.*)

Figure 12.17 Specification data for the heavy-duty padlock. Notice the two different shackle sizes available. (*Kaba Locks, Ltd.*)

Figure 12.18 Cable padlock suitable for securing small pieces of equipment and bicycles. (*Kaba Locks, Ltd.*)

The KP 10 padlock, which has a shackle load of over 2 tons (Fig. 12.19), is designed for strength. Its dimensions are shown in Fig. 12.20.

The Epilok security hasp and staple (Fig. 12.21) is a unique unit. A strong hasp and staple of hardened steel construction uses the unique Epilok spigot lock with its advanced Mini-Kaba locking cylinder. This unit will stand up to the most strenuous attack.

Figure 12.22 breaks down the unit a little more. The rawbolt staple (on the right) is separate from hasp unit and locking spigot lock. After the hole is

Figure 12.19 The KP 10 has a large shackle into a solid locking body for increased strength. (*Kaba Locks, Ltd.*)

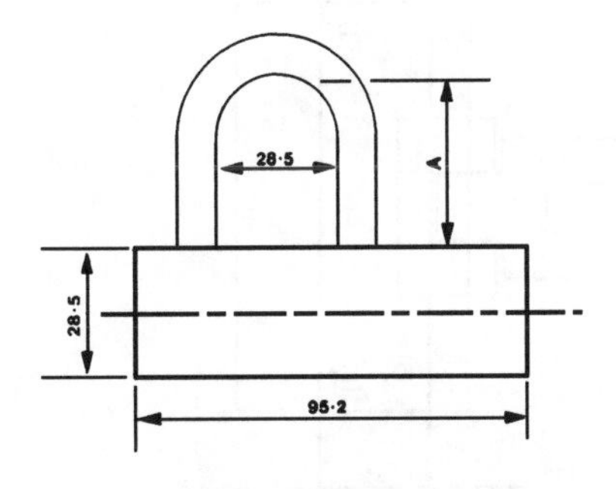

Figure 12.20 KP specification data. (*Kaba Locks, Ltd.*)

Figure 12.21 The Epilock high-security hasp and staple unit used the Mini Kaba cylinder. (*Kaba Locks, Ltd.*)

drilled and the rawbolt inserted, the staple is screwed on tight, making removal exceptionally difficult. The security of this system is much greater than that of the several screws one would expect with a security hasp. The dimensions are provided in Fig. 12.23.

The 45201 security window/patio door lock has a universal locking mechanism that can be used for any application where a sliding door or window

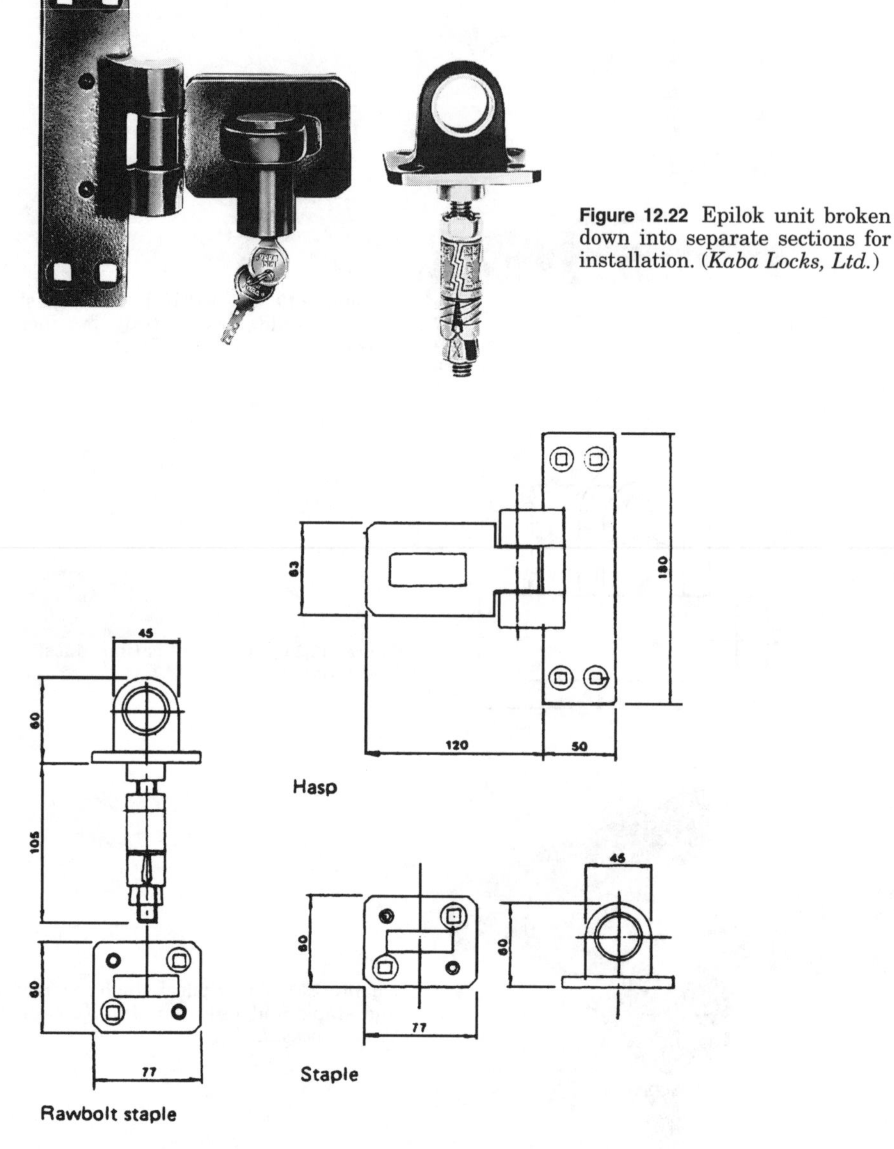

Figure 12.22 Epilok unit broken down into separate sections for installation. (*Kaba Locks, Ltd.*)

Figure 12.23 Specification data for the Epilock. (*Kaba Lock Ltd.*)

requires additional locking on the top and/or bottom part of the frame. It is also very suitable for hinged windows or pivoting windows (Fig. 12.24).

The lock is operated by a Kaba-20 cylinder mechanism. The same key can be linked to an unlimited number of locks with the same key code, including rim locks, padlocks, or mortise locks. All windows, patio doors, and front and

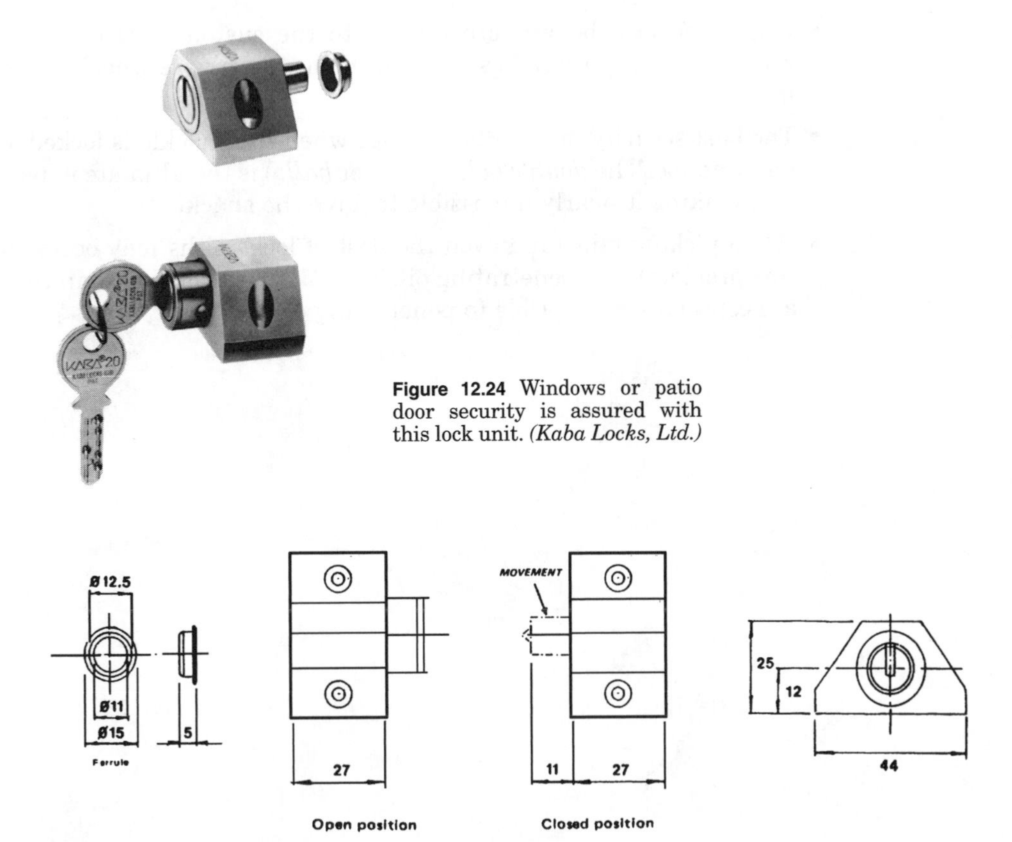

Figure 12.24 Windows or patio door security is assured with this lock unit. *(Kaba Locks, Ltd.)*

Figure 12.25 Specification data for the window/patio lock. (*Kaba Lock, Ltd.*)

back doors can be quickly locked, every time. There is no need to search for individual keys on a large key ring. The dimensions for the window/patio door lock are in Fig. 12.25.

These last two locks are excellent counter displays for the locksmith because they practically sell themselves to the customers. There is no need to provide any form of sales pitch for the product. They are top-of-the-line, with proven quality, reliability, and durability, and will give your customer years of valuable protective service.

Helpful Hints

- It's easier to work with padlocks that are open. If the lock is not open, pick it open. Sometimes the shackle bolt can be disengaged with the help of a hatpin inserted through the keyway. However, locks are getting better, and this technique doesn't work as well as it used to.

- Polish the lock before returning it to the customer. Use emery paper to remove the deep scratches, then burnish with a wire wheel. Finish by buffing.
- The best security in a padlock comes when the shackle is locked at both the heel and toe. The *double bolt action* (or *balls*) is the ultimate in padlock security, making it nearly impossible to force the shackle.
- When picking fails (and even the best of locksmiths may occasionally have this problem), use penetrating oil. Yale, Corbin, and ILCO pin tumbler locks are especially susceptible to penetrating oil.

Chapter

13

Home and Business Services

Over the years a locksmith acquires a great deal of knowledge, much of which is learned from correcting his or her own mistakes. This is a hard school. The best way to solve problems is to never let them arise in the first place.

This chapter includes some basic techniques for dealing with door problems. Some of this material might seem obvious and you might wonder why it's in here. These techniques have been tested and approved by the experts, and they can save you time, money, and the embarrassment of callbacks.

The following few pages are supplied courtesy of Corbin (Emhart Corporation). The first section applies to all locksets, regardless of make. Other sections detail service and troubleshooting procedures for specific Corbin locksets as well as other makes of locks.

Common Problems and Troubleshooting

If you encounter difficulties in the operation of a lockset, first review this checklist of common problems and solutions to see if you can clear up the difficulty.

- Is the door locked (Fig. 13.1)?
- Are you using the right key (Fig. 13.2)?
- If the latchbolt or deadbolt does not engage or disengage the strike or binds in the strike, it is usually due to bolt-strike misalignment.
- Has the door warped (Fig. 13.3)?
- Is the door binding? Frames that are out of plumb are frequently the cause of faulty operation of locksets and binding of bolts in the strike (Fig. 13.4).
- Are the hinges loose? Tighten the screws, filling holes if necessary, or rehang the door if the screws will not hold (Fig. 13.5).

Figure 13.1 Is the door locked?

Figure 13.2 Are you using the right key?

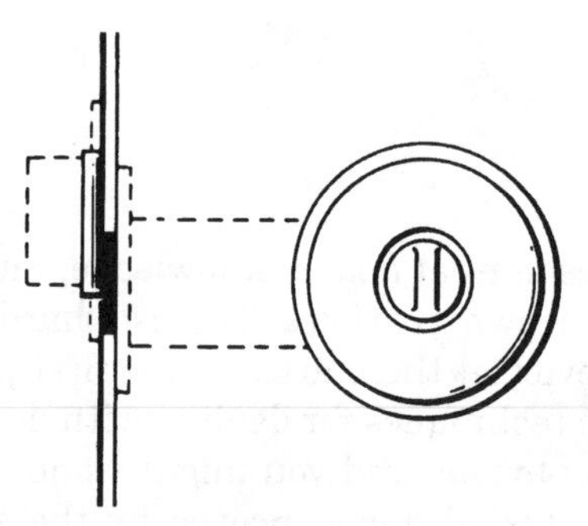

Figure 13.3 Has the door warped?

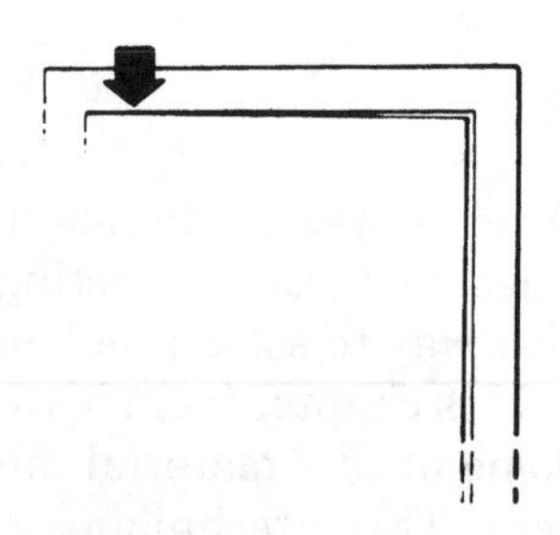

Figure 13.4 Is the door binding?

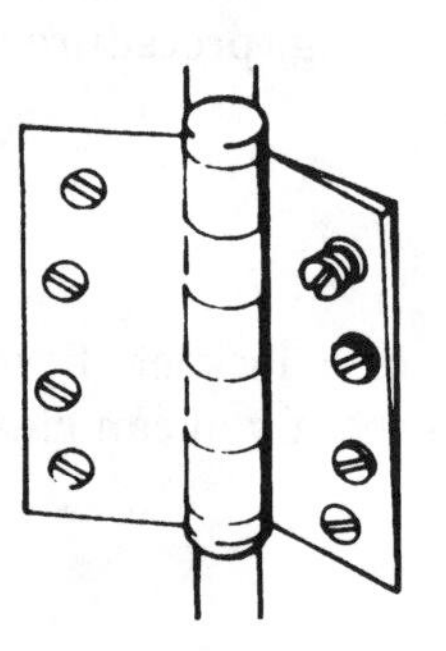

Figure 13.5 Are the hinges loose?

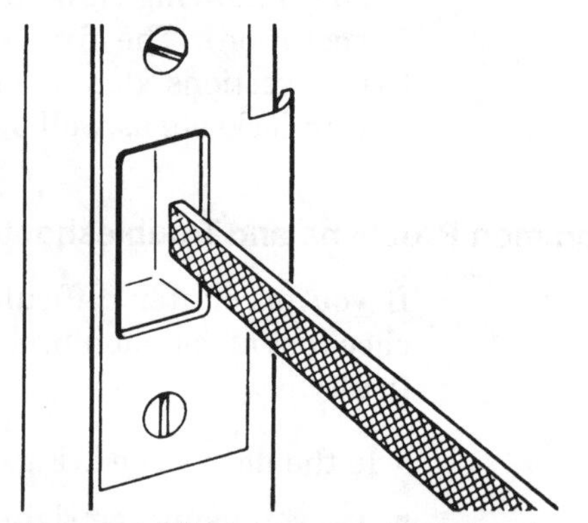

Figure 13.6 File to correct bolt-strike misalignment.

- Are the hinges worn? If excessive wear has occurred on the hinge knuckles, the door will not be held tightly. Replace the hinges.
- Is the frame sagging? If sag cannot be corrected and the door and frame returned to plumb relationship, planing or shaving the door and repositioning or shimming the strike may relieve this condition.
- When a key operates the latchbolt or deadbolt with difficulty, it is usually due to bolt-strike misalignment (Fig. 13.6).

Corbin Cylindrical Locksets

Common problems with Corbin cylindrical locksets include the following:

Problem 1: Latchbolt will not deadlock. Caused by deadlocking latch going into strike. Either the strike is out of line or the gap between the door and jamb is too great. Realign the strike or shim the strike out towards the flat area of the latchbolt (Fig. 13.7).

Problem 2: Latchbolt cannot be retracted or extended properly. Caused by latchbolt tail and latchbolt retractor not being properly positioned (Fig. 13.8). Remove the lockset from the door. Reinsert the latchbolt in the door. Looking through the hole in the door, the tail should be centered between the top and bottom of the hole. Remove the latchbolt and insert the lockcase. Looking through the latchbolt hole in the lock face of the door, the latchbolt retractor should be centered in the hole. Adjust the outside rose for proper position. Rebore the holes, if necessary, to line up the retractor and tail.

Problem 3: Latchbolt will not project from the lock face. Latchbolt tail and retractor may be misaligned (Fig. 13.9). See Problem 2. If this is not the cause, the spring is probably broken.

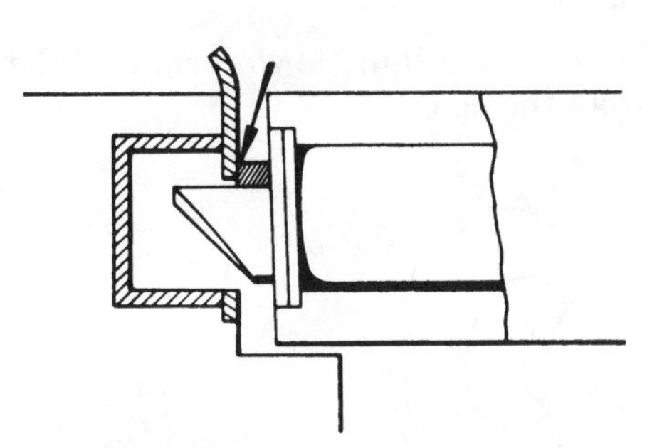

Figure 13.7 Latchbolt will not deadlock.

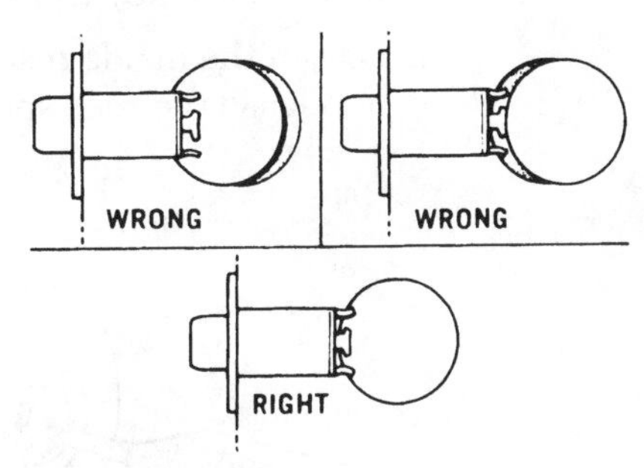

Figure 13.8 Latchbolt cannot be retracted or extended properly.

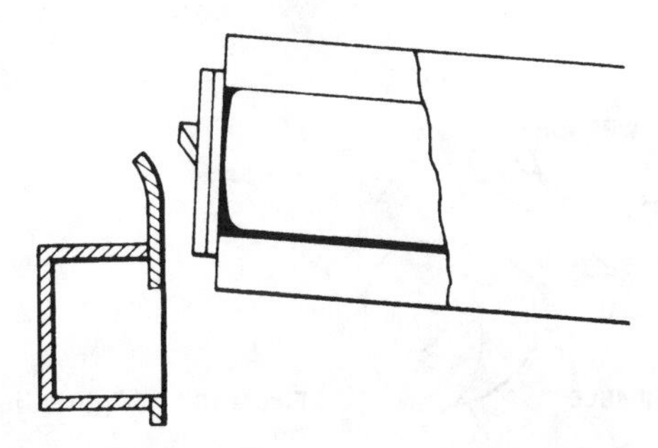

Figure 13.9 Latchbolt won't project from lock face.

Problem 4: Key works with difficulty. Lubricate the keyway (Fig. 13.10). Do *not* use petroleum products. Spray powdered graphite into the cylinder or place powdered graphite or lead pencil shavings on the key. Move the key slowly back and forth in the keyway. Bitting (notches) on the key may be worn.

Corbin heavy-duty cylindrical locksets: Service procedures

Servicing these locks is not difficult if the task is approached methodically.

Tightening the locksets. Tighten the inside rose thimble with a wrench. If the thimble needs to be taken up a great deal, tighten the *outside* rose at the same time to prevent possible misalignment and binding of both (Fig. 13.11).

If the lockset is still not tight, back off the thimble. Using a screwdriver, push down the knob retainer and remove the knob and rose. If the spurs on the back of the rose are bent, straighten and reposition the rose so the spurs are embedded in the door. If the spurs are broken off, insert a rubber band or nonmetallic washer under the rose. Tighten the thimble.

Removing and installing the locksets

1. Remove the key from the knob.
2. Loosen the inside rose thimble with a thimble wrench. Pull the lock slightly to release the rose spurs from the door.

Figure 13.10 Lubricating keyway.

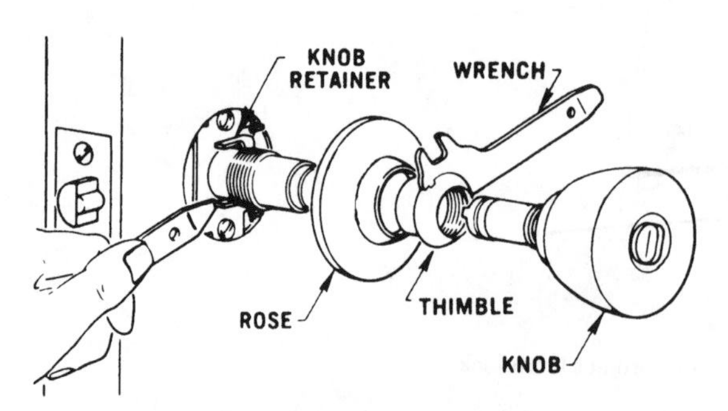

Figure 13.11 Tightening the lockset.

3. Disengage the inside knob retainer with a thimble wrench and pull out the inside knob.
4. Slide the rest of the lockset from the outside of the door.
5. Remove the bolt faceplate screws and slide the latch unit from the lockface of the door.

To reinstall the lockset, reverse the procedures above. When inserting the case and keyed knob from the outside of the door, be sure the bolt retractor in its case properly engages the latchbolt tail.

Removing and replacing the cylinder in the locksets

1. Follow the procedure in the "Removing and Installing the Locksets" section.
2. Remove one case screw and slightly loosen the other.
3. Swing out the outside knob retainer (it pivots on the slightly loosened case screw).
4. Using a screwdriver, pry the knob filler cover off the outside knob.
5. Using a special Waldes No. 3 ring pliers, remove the large Waldes ring from the groove and withdraw the shank and cylinder (Fig. 13.12).

To reinstall the cylinder, reverse the procedure. Be sure the knurled side of the Waldes ring is face up.

Changing the hand of the locksets. Remember that the lock cylinder should always be in position to receive the key with the bitting (notches) facing upward.

1. Follow the procedure in the preceding section.
2. Slightly loosen *one* case screw and back off the other (Fig. 13.13).
3. Swing out the outside knob retainer (it pivots on the slightly loosened case screw).
4. Lift the knob out and rotate it 180°. Replace it in the case. Swing the knob retainer back into place.

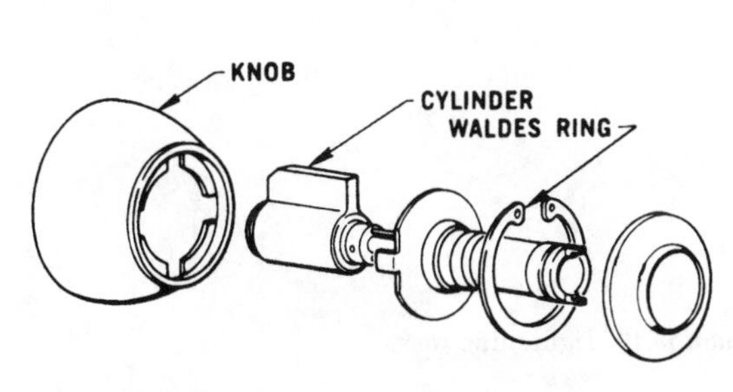

Figure 13.12 Removing cylinder.

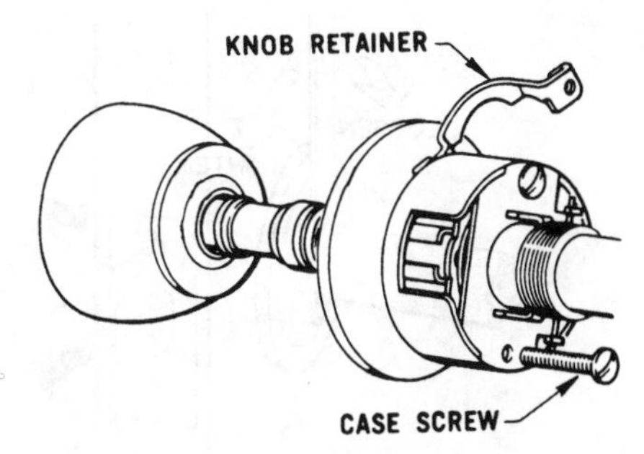

Figure 13.13 Changing lockset hand.

5. Insert the case screw and tighten both screws.
6. Reinstall (Fig. 13.14).

Corbin standard-duty cylindrical locksets: Service procedures

Tightening the locksets

1. Depress the knob retainer and remove the inside knob.
2. Unsnap the rose from the rose liner (Fig. 13.15).
3. Place the wrench into the slot in the rose liner and rotate it clockwise until it's tight on the door. If the lockset is extremely loose, tighten the outside rose and inside liner *equally* (Fig. 13.16).
4. Replace the inside rose; depress the knob retainer and slide the knob onto the spindle until the retainer engages the hole in the knob shank.

Removing locksets

1. Depress the knob retainer and remove the inside knob.

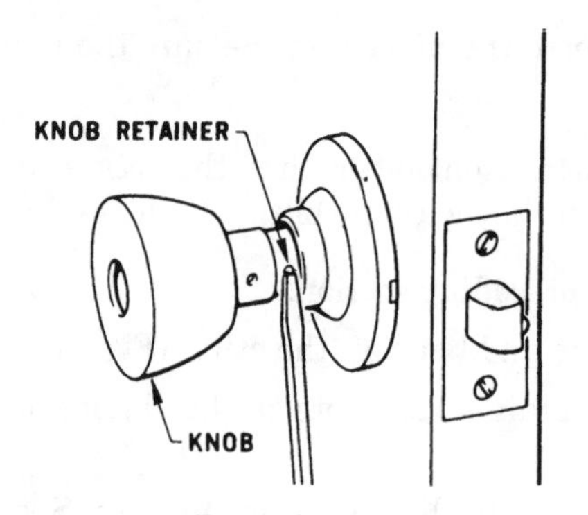

Figure 13.14 Reinstalling knob.

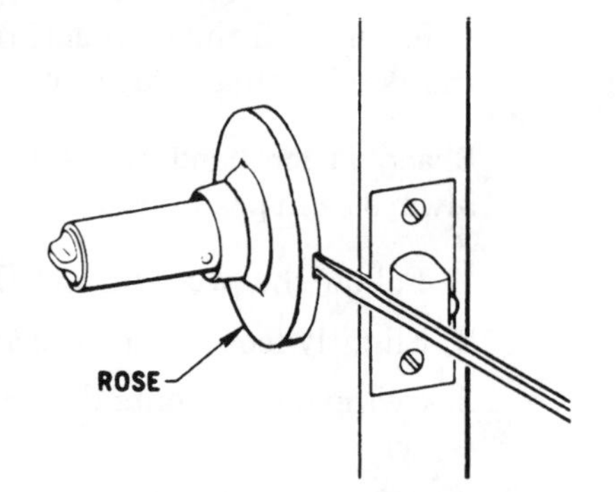

Figure 13.15 Unsnapping rose from rose liner.

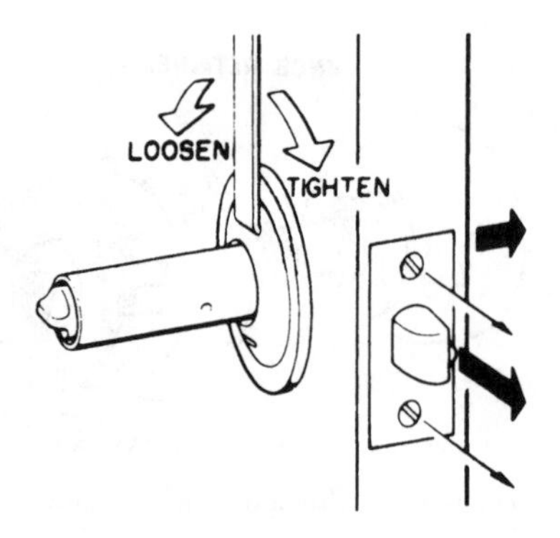

Figure 13.16 Tightening rose.

2. Unsnap the rose from the rose liner.
3. Place the wrench into the slot in the rose liner and rotate counterclockwise until it's disengaged from the spindle.
4. Remove the lock by pulling the outside knob.
5. Remove the bolt after removing the screws in the lockface of the door.

Reinstalling locksets

1. Adjust the lock for the door thickness by turning the outside rose until the edge of the rose matches one of the lines marked on the shank of the knob. (The first line is for a 1⅜-inch door; the second is for a 1¾-inch door.) When using trim rosettes, increase the adjustment to compensate for the thickness of the metal.
2. Install the latchbolt in the face of the door.
3. Install the lock from the outside of the door so the case engages the slot in the latchbolt and the tail interlocks the retractor.
4. Replace the rose liner, rose, and knob on the inside of the door.

Reversing the knobs

1. Lock the exterior knob (Fig. 13.17).
2. Hold the latchbolt in the retracted position with the key (Fig. 13.18). Depress the knob retainer with a screwdriver through the slot in the knob shank.
3. Pull the knob off the spindle (Fig. 13.19). Rotate the knob so the bitting on the key will be *up*.
4. Replace the knob by lining up the lance in the neck of the knob with the slot in the spindle.
5. Push the knob on the spindle until it hits the retainer button. Depress the retainer button and push the knob until it snaps into position.

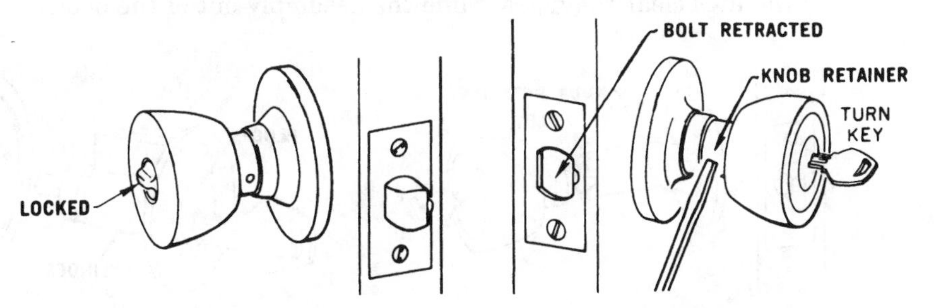

Figure 13.17 Locking exterior knob.

Figure 13.18 Retract latchbolt.

Removing the cylinders

1. Remove the outside knob. Turn the key in either direction until it can be partially extended from the plug (Fig. 13.20).
2. Hold the knob and turn the key to the left, pulling slightly on the key until the cylinder disengages.

Corbin unit locksets (300 and 900 series)

Troubleshooting

Problem 1: Latchbolt binds or rattles in strike. Adjust the strike. Because the lockset is preassembled as one unit, there are no internal adjustments to make. Check that the cutout for the lockset is square and at the right depth so the face of the lockset is flush with the face of the door. Then, adjust the nylon adjusting screw in the strike (Fig. 13.21).

Problem 2: Key does not activate the knob or latchbolt. Check for a worn key. If bitting (notches) in the key is worn down or the key is bent, the locking mechanism will not operate properly. If the key is not worn, spray powdered graphite into the keyway or put graphite or lead pencil filings on the key and move it back and forth slowly in the keyway (Fig. 13.22). Never use petroleum products. Check that a binding latch is not the cause.

Problem 3: Lockset is loose in the door. Tighten the *escutcheon* screws (Fig. 13.23). Be sure to tighten them evenly.

Problem 4: Lockset has the wrong bevel for the door. Reverse the lockset. Unit locksets with horizontal keyways may be changed from right- to left-hand regular bevel, or vice-versa by merely turning the lock upside down (Fig. 13.24).

Service procedures. Usually the unit locksets are easier to service than the cylindrical locksets. To service this series of locksets, proceed as follows:

Removing the locksets from a door. First, remove the through-bolts on the inside of the door (Fig. 13.25). Push the outside escutcheon away from the door so the lugs clear the holes. Slide the assembly out of the door.

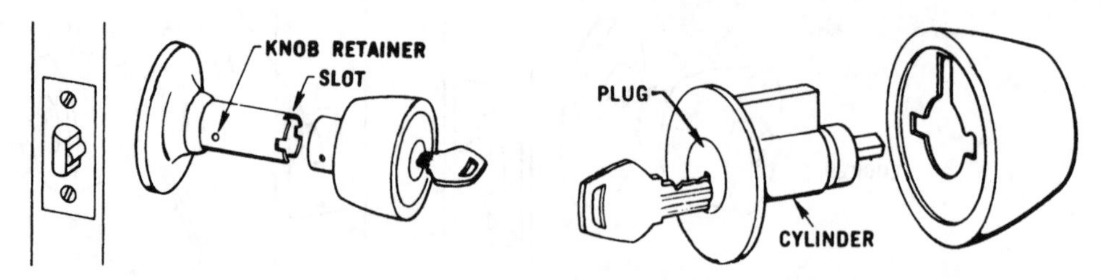

Figure 13.19 Pull knob off spindle.

Figure 13.20 Remove outside knob.

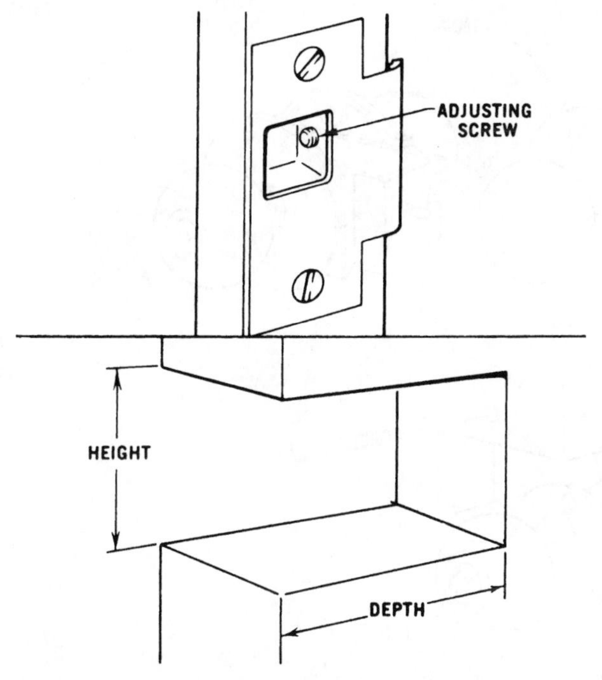

Figure 13.21 Adjusting screw in strike.

Figure 13.22 Lubricating keyway.

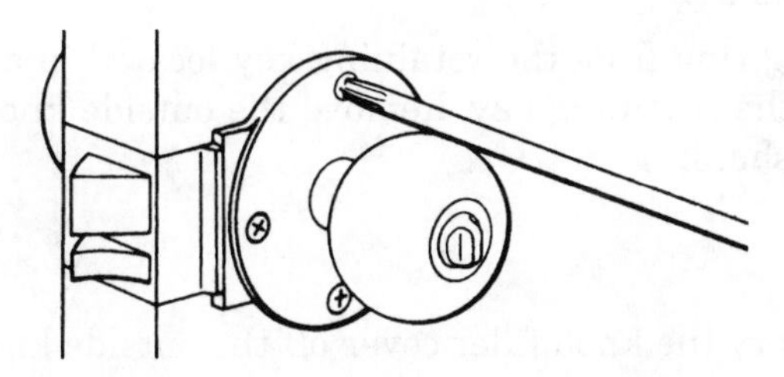

Figure 13.23 Tightening escutcheon screws.

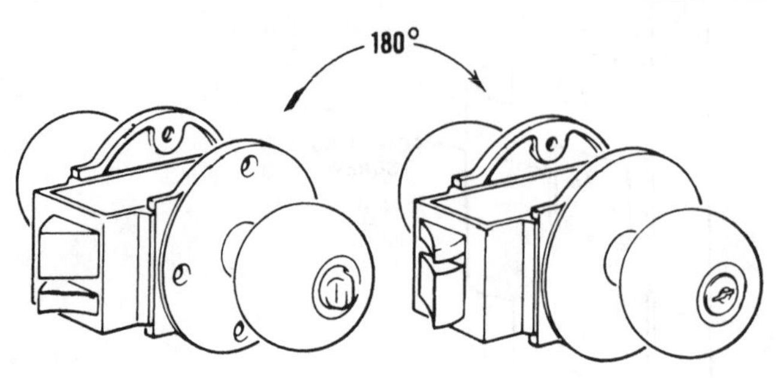

Figure 13.24 Revising lockset.

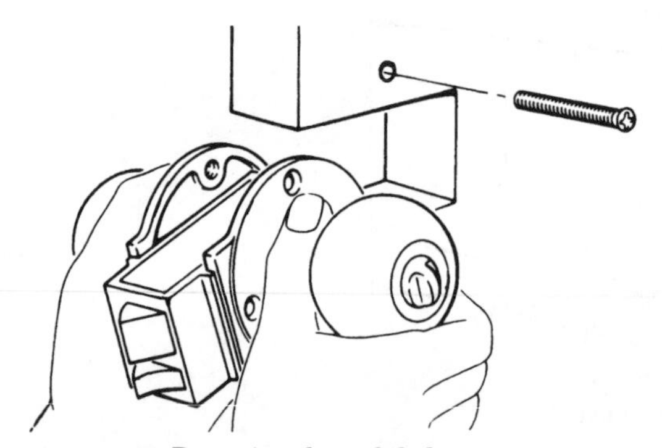

Figure 13.25 Removing through-bolts.

Removing 900 series knobs

1. Remove the attaching screws (Fig. 13.26).
2. Snap off the dust cover.
3. Pry the wire retaining ring from the knob retaining key located in a slot in the frame tube. Remove the retaining key. Remove the inside knob, which is fastened to the knob shank.
4. Remove the inside escutcheon.
5. Loosen the escutcheon on the outside of the lock by inserting a screwdriver through the access hole from the inside of the lock frame; remove the screw and escutcheon fastener.
6. Pry the wire retaining ring from the retaining key located in a slot in the frame tube. Remove the retaining key. Remove the outside knob, which is fastened to the knob shank.

Removing cylinders

1. Using a screwdriver, pry the knob filler cover off the outside knob.
2. Use No. 103F92 Waldes retaining pliers to remove the Waldes ring.

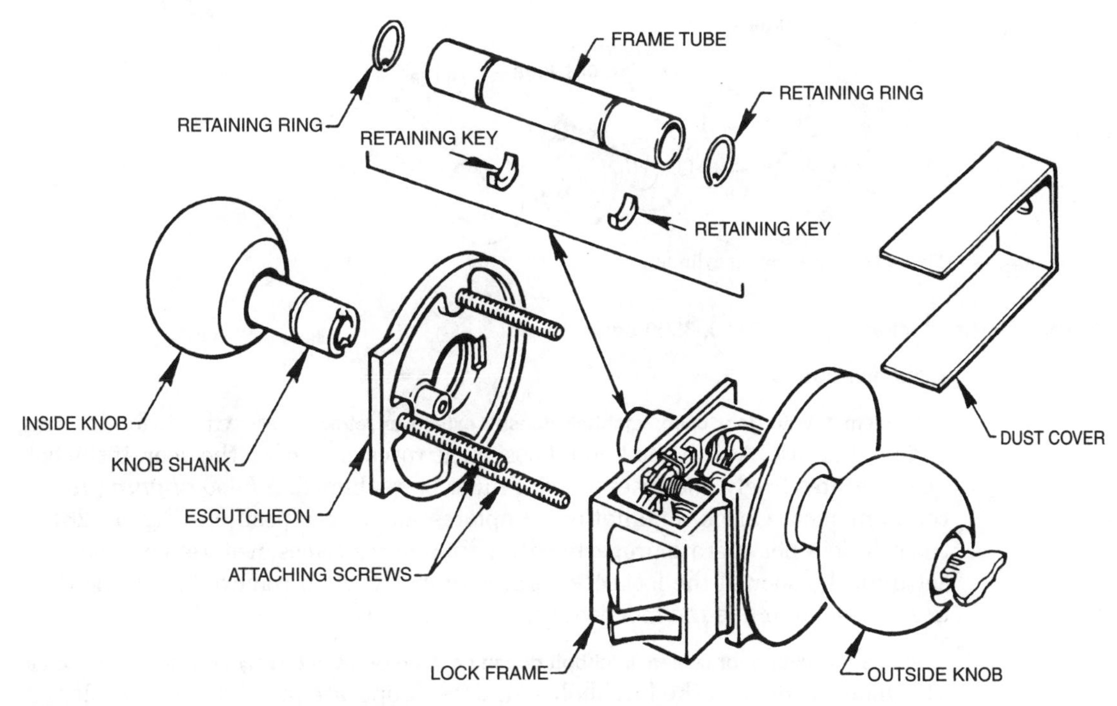

Figure 13.26 Removing attaching screws.

3. Remove the shank and lock cylinder. Rekey the lock.

Reverse the procedure to reinstall. Be sure that the beveled edge of the Waldes ring faces away from the knob and that the ring is properly seated in the groove (Fig. 13.27).

Removing 300 series knobs

1. Remove the key from the lock (lock should be unlocked). Remove the three attaching screws (Fig. 13.26).
2. Snap off the dust cover.
3. On the inside knob side, pry the wire retaining ring away from the knob retaining key located in the slot in the frame tube. Remove the retaining key.
4. Remove the knob that has the shank assembled to the knob.
5. Remove the inside escutcheon.
6. Loosen the escutcheon on the outside of the lock by inserting a screwdriver through the access hole from the inside of the lock chassis; remove the screw and washer.
7. Repeat step 3 on the outside knob and shank.

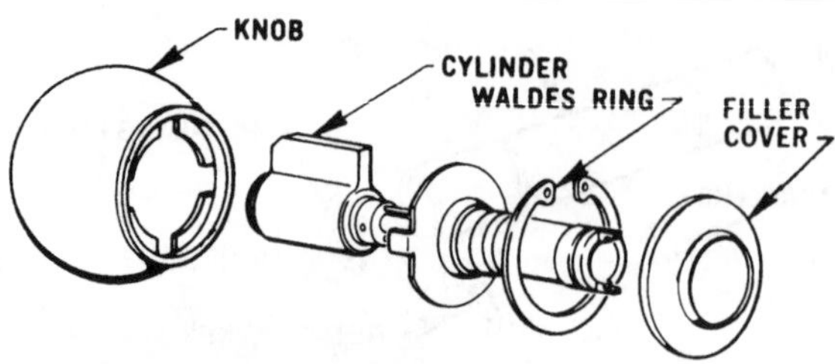

Figure 13.27 Removing cylinder.

Corbin mortise locksets (7000, 7500, 8500 series)

Troubleshooting

Problem 1: With door open, latchbolt doesn't extend or retract freely. Check for binding against the rose. Adjust the knob. Loosen the roses or trim on the door. If the bolt now operates freely, the roses or trim must be realigned. A *knob aligning tool* is recommended. Check installation templates for proper position (Fig. 13.28). If the bolt does not operate properly with trim and roses loosened, remove the lockset from the door. If the lockset operates properly when removed from the door, use a chisel to make the mortise larger so the lockset enters freely.

Problem 2: With door closed, latchbolt doesn't extend or retract freely or door won't latch at all. Open the door. If the latchbolt still doesn't operate properly, see Problem 1.

Problem 3: Latchbolt "stubs" on the strike lip. Bend the strike slightly back toward the jamb. Wax or paraffin makes an excellent lubricant, as does silicone spray (Fig. 13.29).

Problem 4: Deadbolt doesn't enter the strike. This is probably due to misalignment of strike and bolt, particularly in cases where door sag has taken place. Both latchbolt and deadbolt holes in the strike must be filed or the strike repositioned. Do *not* force the thumbpiece if the deadbolt doesn't extend and retract in the strike freely (Fig. 13.30).

Problem 5: Removing the cylinder. Loosen the cylinder locking screws in the face of the lockset. If the scalp covers the set screws, remove the scalp. Unscrew the cylinder. When replacing, be sure the locking screws are firmly seated (Fig. 13.31).

Problem 6: Key does not operate the latchbolt or deadbolt. Loosen the cylinder locking screws. The cylinder is in the wrong position in the door so the cylinder cam does not engage the locking mechanism properly. Turn the cylinder a whole turn to the left or right until it works properly. The keyway must always be in position to receive the key with bitting (notches) up. Tighten the cylinder locking screws.

Problem 7: Key turns hard when retracting the deadbolt. If the bolt operates freely with the door open, check the bolt-strike alignment. Check the scalp to be sure it is not binding the bolt or that paint over the bolt is not causing the bind.

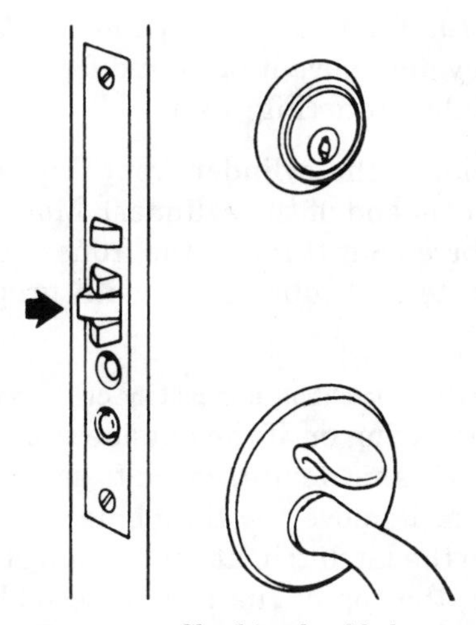

Figure 13.28 Checking latchbolt.

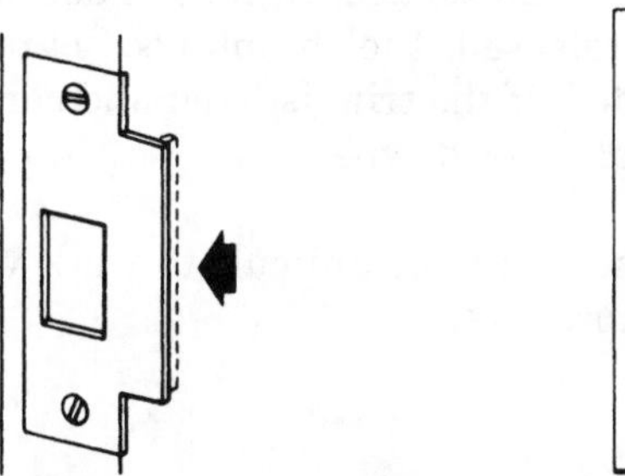

Figure 13.29 Bending strike.

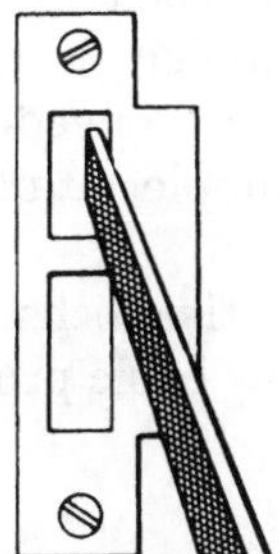

Figure 13.30 Filing strike.

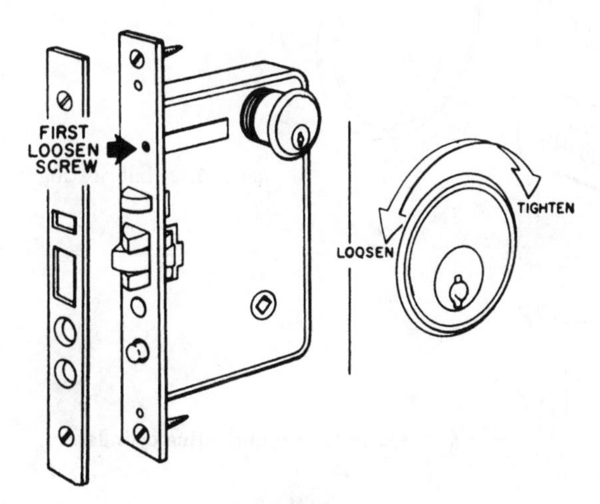

Figure 13.31 Removing cylinder.

Problem 8: Key works hard in the cylinder. Lubricate the cylinder with powdered graphite or place graphite on the key and move it back and forth slowly in the keyway. Never use any petroleum lubricants (Fig. 13.32).

Problem 9: Key breaks in the lock. Remove the cylinder from the lock. Insert a long pin or wire into the back end (cam end of the cylinder). Move it back and forth until the broken key stub is forced out through the front of the cylinder. Clean the cylinder with ethyl acetate and lubricate it with graphite before reinstalling (Fig. 13.33).

Problem 10: Thumbpiece trim doesn't retract latchbolt completely or doesn't extend bolt completely. Check for binding at inside trim; or, if the outside thumbpiece trim is used in conjunction with the inside panic device, check to see if it is operating properly and is not dogged down. Remove the thumbpiece and check the position of thumbpiece in relation to the latch trip at the bottom of the mortise lock case. When properly installed, the top of the thumbpiece should be up against the bottom of the latch trip but not lifting it (Fig. 13.34). If the bolt doesn't retract fully when the thumbpiece is pushed down, the thumbpiece is too low on the door. Move the trim up as needed. If the bolt doesn't extend completely when the thumbpiece is released, the thumbpiece is too high on the door. Move the trim down as needed. If the trim is fixed and cannot be moved, carefully bend the thumbpiece tail up or down.

Service procedures. Mortise locksets are not difficult to work with. Servicing them involves only a few basic procedures.

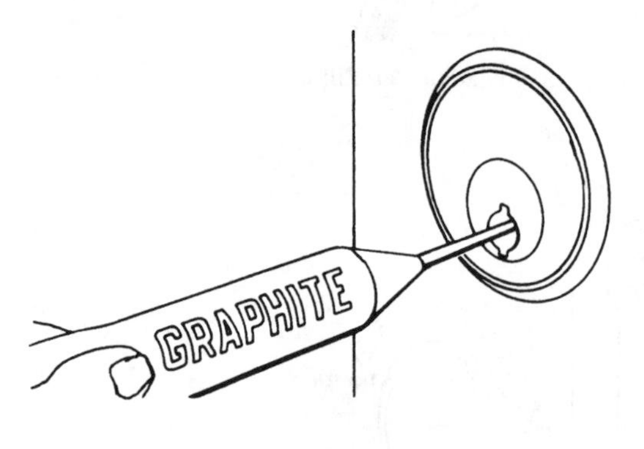

Figure 13.32 Lubricating keyway.

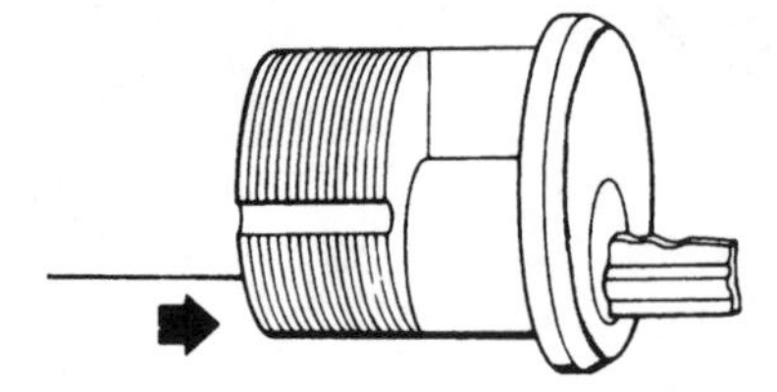

Figure 13.33 Lubricating cylinder.

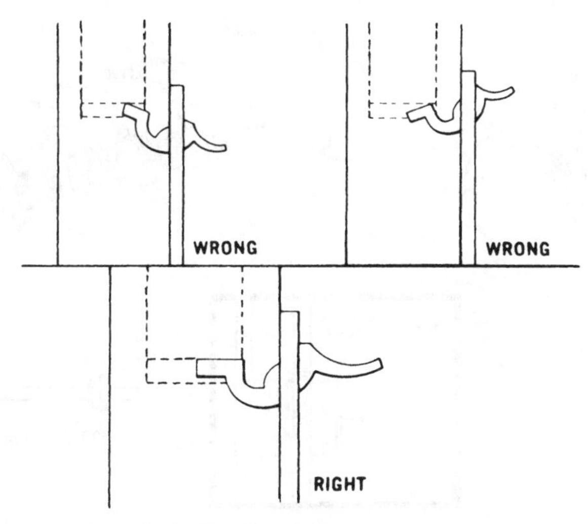

Figure 13.34 Installing thumbpiece.

Installing and adjusting the working trim. Unscrew the spindle. Remove the mounting plates, roses, and thimble from the spindle. Unscrew the sleeve from the inside knob. Remove the spindle from the knob (Fig. 13.35).

Install the rose-attaching plates. On wood doors, install flange washers through the mortise, turning the attaching plates into them. It may be necessary to mortise for the washer to permit the mortise lock to clear it. Install the mortise lock.

Align the plates using a No. 028 aligning tool. If this tool is not available, assemble the spindles allowing a $\frac{1}{16}$-inch gap between halves of the swivel spindles with mounting plates, sleeve spacer, and adjusting nut in position as shown in Fig. 13.35. Tighten the adjusting nut with your fingers. Mark screw holes through both mounting plates. Install the screws.

Reassemble the spindle, tightening the adjusting nut, and back off one quarter-turn to line up with the flats of the spacer. The spindle should have a slight end chuck. Try the knob. If the latch binds, back off the adjusting nut another quarter-turn. Disassemble the spindle.

Assemble the roses and thimble. Reassemble the spindle complete with knobs. Tighten the rose thimbles and inside adjusting plate with furnished spanner wrenches. Knobs should turn freely in either direction from either side.

Installing and adjusting the lever handle trim assembly

1. Check that the handle hole is $1\frac{1}{8}$ inch wide. On wood doors, remove the mortise lockset and make additional mortise for retainer on the inside of the case mortise. Replace the locksets and secure with screws.

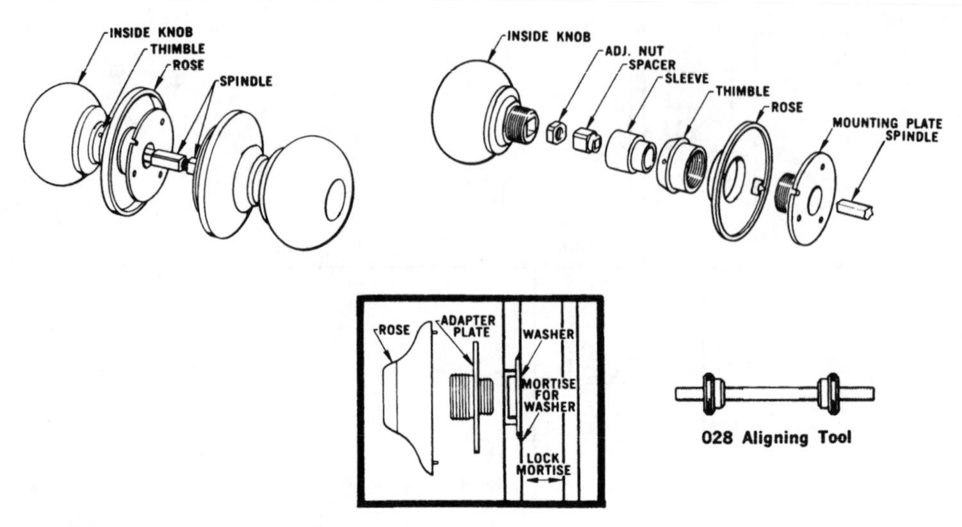

Figure 13.35 Removing spindle from knob.

2. Place the rose assembly spindle into the lock hub. The label shows the top and latch edge. Spot four screw holes through the assembly plate (Fig. 13.36). Make sure that the latchbolt operates freely. Drill holes very carefully.
3. Attach the rose assembly to the door. (With a wood door, attach through the door to the retainer.)
4. Put on the cover plates. Place the lever handle on the assembly and secure with an Allen set screw.
5. Adjust. If the latchbolt binds, remove the lower plate and slightly change the rose assembly position.

Reversing the hand. As you will remember from Chap. 7, the hand refers to the position of the hinge on the door and to the direction of swing. Some locksets are universal and fit all four hands. Others must be modified in the field.

1. Refer to Figure 13.37. Unfasten the two cap screws (A) and remove the cap.
2. Reverse the latchbolt (B). If the lock has an antifriction latch, first remove the L-shaped pin (C) and reinstall it on the opposite side after reversing the bolt. Hold the short leg against the side of the case with the pressure-sensitive decal.
3. Reverse the auxiliary latch (D) if an auxiliary latch is furnished. To do this, first remove the auxiliary latch lever and spring, then invert the latch so

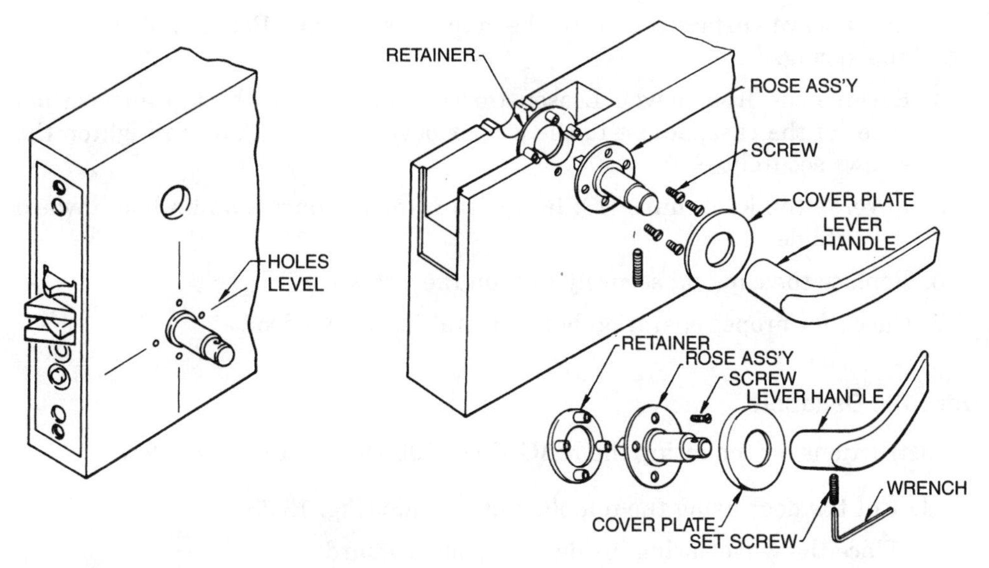

Figure 13.36 Installing lever handle.

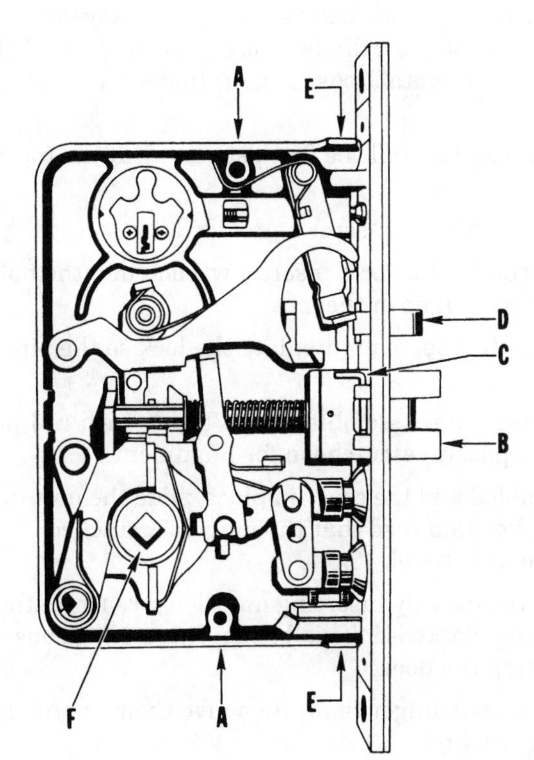

Figure 13.37 Reversing the hand.

the concave surface is toward the stop in the frame. Reassemble the lever and spring.

4. Reverse the front bevel. Loosen three screws (E) in the top and bottom edges of the case, adjust to the proper bevel (or to flat), and retighten the screws securely.
5. Reverse the knob hubs (F): hub with ⅜-inch square spindle hole toward the outside.
6. Replace the cap and securely tighten the screws.
7. Check for proper operation before installing in the door.

The MAG Ultra 700 Deadbolt

Instructions for mounting the MAG Ultra 700 Deadbolt are as follows:

1. Cut the door using the template as a guide (Fig. 13.38).
2. Place the wave spring inside the cylinder guard.
3. Insert the cylinder through the cylinder guard with the wave spring *under* the head of the cylinder.
4. Thread the cylinder into the lock body securely. The keyway must be horizontal and on the side of the cylinder closest to the bolt. If the keyway does not come to the horizontal, loosen the cylinder as necessary (always less than one turn).
5. Remove the faceplate and install the four set screws at the top and bottom of the lock.
6. Secure the cylinder with the set screws provided.
7. If you have to disassemble the lock, insert a wrench into the hole and turn counterclockwise. This will open the lock.
8. With the lock near or full open, manipulate the lock so the pins slip out of one escutcheon.
9. Reverse this procedure to reassemble. *Note:* The ³⁄₁₆-inch roll pin must fit into the hole in the opposite escutcheon for alignment.
10. With the lock assembled and the cover in place, slide the unit into the door cutout (Fig. 13.39). Position it so that the longest escutcheon is flush with the high edge of the door bevel.
11. Tighten the lock to the door by inserting an Allen wrench in the appropriate holes and turning clockwise. This brings the escutcheons of the lock together so they clamp the door.
12. Turn the mounting screws finger tight; then give them approximately one-half turn more (Fig. 13.40).
13. Install the faceplace with the screws provided (Fig. 13.41).

Figure 13.38 Cut door with template.

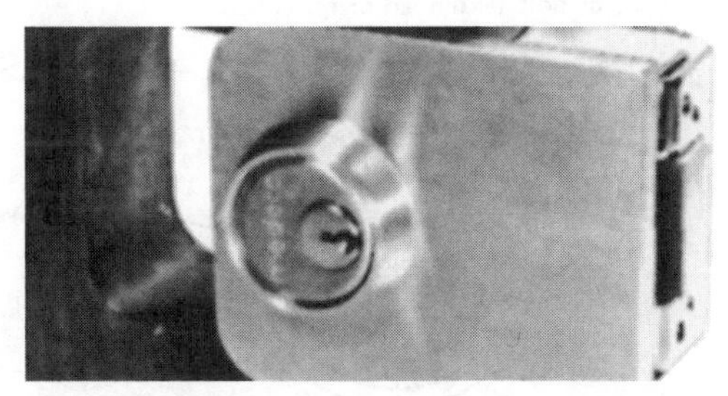

Figure 13.39 Slide lock into cutout.

Figure 13.40 Tightening mounting screws.

Figure 13.41 Installing face plate.

The Ultra 700 lock can be adapted for a single cylinder on the exterior side and a thumbturn on the interior side of the door. For the thumbturn to act as an indicator, the long portion (the actual turning mechanism) must point in the direction of the bolt (extended or retracted position), and the cam must be properly positioned on the thumbturn before the entire assembly is installed in the lock.

Check the cam position. If necessary, remove the drive pin and cam, rotate one-half turn (180°), and replace the pin. Figure 13.42 shows the thumbturn as viewed from the interior door side. When installing the lock, be sure the cam is inserted between the mechanism arms; then, tighten the cylinder set screws on the top and bottom of the lock.

The instructions are simple and easy to follow from a locksmith's viewpoint, and the time is well spent in terms of customer relations and business income. In addition, if you are replacing the interior cylinder with the thumbturn, you might be able to keep the cylinder that was removed, thus adding various parts to your supply stock. (*Note:* Never resell a used lock; the various individual parts of the lock can be used for repair parts, for study, and for various practical training exercise for new employees.)

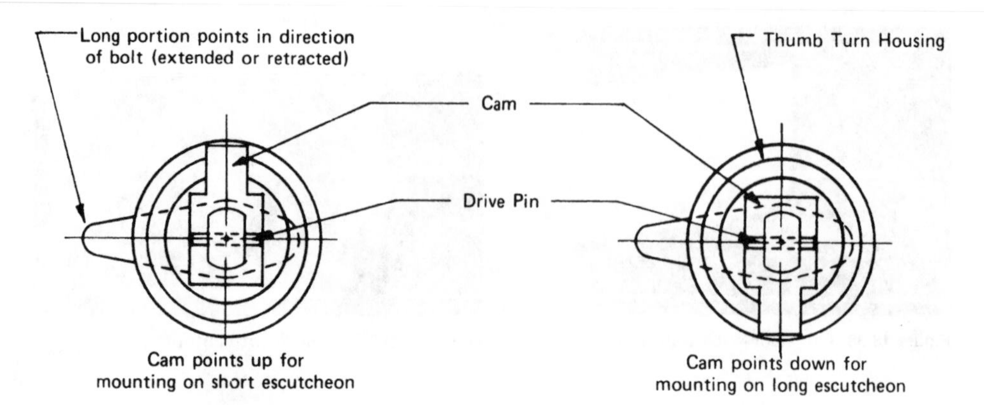

Figure 13.42 Cam direction position when mounting is dependent on whether a short or long escutcheon is used. (*M.A.G.*)

M.A.G. Engineering and Mfg., Inc.

The MR series, new from M.A.G., is an adaptation of the original Ultra 700 deadbolt, and one well worth investing in. The adaptation is for use with the Corbin Master Ring cylinder and is now available from locksmithing distributors nationwide.

The adaptation (Fig. 13.43) enables locksmiths to use the same masterkeying system but still provides the high-security lock compatible with the present system.

Like the regular Ultra 700, this new series includes no attaching screws on either side of the lock, and the key cannot be withdrawn from the lock unless the locking bolt is either fully extended or fully retracted. Also, the cylinder guard is free-turning and recessed into the escutcheon casting for maximum protection.

A new Series A deadbolt Ultra 700 is also now available for the locksmith to add to potential sales. The Series A is designed for use where one side entrance only is desired (Fig. 13.44). The blank side of the lock is a soldering casting. The key side of the lock is available in stainless steel and uses a 1⅛-inch mortise cylinder.

The Ultra 800, another M.A.G. lock, is also a deadbolt. The lock has a $^{15}/_{32}$-inch diameter × 1⅛-inch-long mortise cylinder with a standard Yale cam. Figure 13.45 illustrates the lock in an exploded view. Notice the bolt; it is a cylindrical deadbolt instead of the standard rectangular bolt mostly associated with deadbolt locks. Take note of the strike; while it looks like the average mill-run strike, M.A.G. has gone further and ensured that it is ably secured to the frame and supporting stud. In the illustration you can readily see that the interior side of the door will take either a mortise cylinder or thumbturn, providing your customer with an option feature. The lock is set by the factory for 1¾-inch doors, but instructions are included for other door thicknesses and are also below.

Figure 13.43 The Ultra 700 deadbolt adapted for use with the Corbin Master Ring Cylinder. (*M.A.G. Engineering and Mfg., Inc.*)

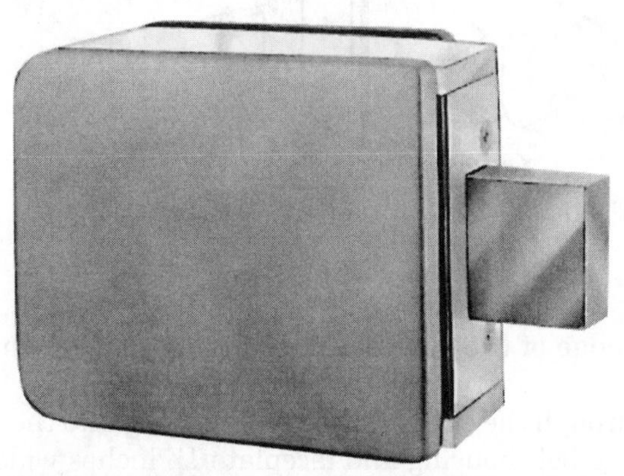

Figure 13.44 Ultra 700 deadbolt Series "A" used when one side entrance only is desired. (*M.A.G. Engineering and Mfg., Inc.*)

First, you must determine the proper setting for the lock if the door is not 1¾ inches thick. Figure 13.46 details adjusting the exterior lock housing piece for use on doors of varying thicknesses. The ring referred to is the circular ring on the reverse side (inside portion) of the exterior lock housing. Look at the reverse side, at the 8 o'clock position, to locate the pin hole; insert the pin. With this done, proceed to the door preparation using the template furnished with the lock unit.

Wood doors

Door preparation. Follow the dimensions outlined on the template. Drill a ⅛-inch pilot hole through the door face, then drill a 2½-inch hole exactly 2¾

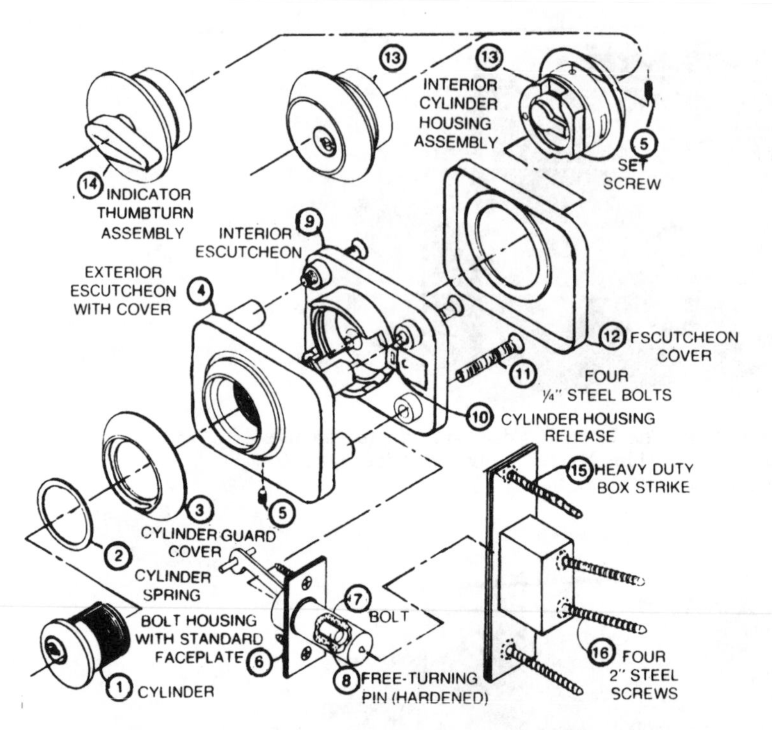

Figure 13.45 Exploded view of the Ultra 800 deadbolt. (*M.A.G. Engineering and Mfg., Inc.*)

inches from the high edge of the door bevel, referring to the template as you do so.

Drill a 1-inch hole through the edge of the door, breaking into the 2⅛-inch hole. Cut out the wood for the bolt housing and faceplate 1½ inches wide × 2¼ inches high × 5⁄32 inch deep. This faceplate hole must be centered on the 1-inch hole.

Drill the four ⅛-inch pilot holes for the four corner mounting bolts. With the pilot holes drilled, replace the ⅛-inch drill bit with a 5⁄16-inch bit and put the four 5⁄16-inch holes through the door. *Note:* M.A.G. has a drill fixture kit available for fast and accurate installation. If you expect to perform numerous installations, then the drill fixture kit is a must-have item for rapid installation.

Prior to mounting the lock, you should assemble the mortise cylinder to the exterior escutcheon of the lock, as follows. (The numbers refer to the various lock part components in Fig. 13.45.)

Put the cylinder spring (2) over the cylinder (1), then insert the cylinder into the cylinder guard cover (3) and install the escutcheon cover (12).

Thread the cylinder in until the cylinder is tight (the keyway should be toward the rear of the lock and horizontal). Tighten the set screw. The cylinder guard cover should be tight against the escutcheon cover. The ring should rotate freely one-half turn back and forth. If the cam on the cylinder touches inside the mechanism cover, back off the cylinder one turn.

NOTE: PRESS PIN IN TO DEPTH SHOWN IN CHART CORRESPONDING TO DOOR THICKNESS LOCK WILL BE MOUNTED ON. STANDARD PIN DEPTH SET BY FACTORY FOR 1 3/4" DOORS.

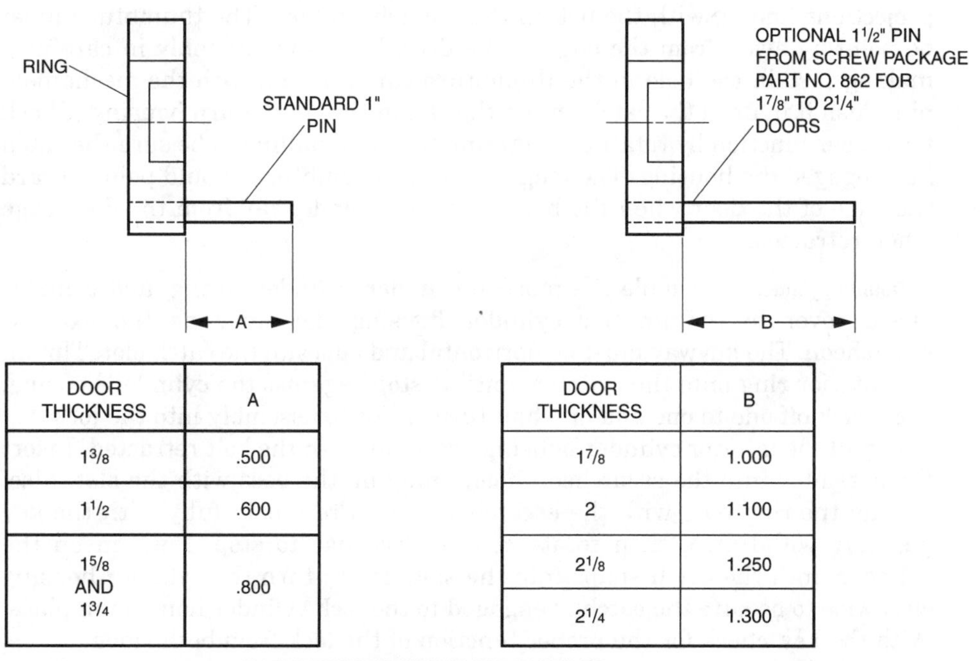

DOOR THICKNESS	A
1 3/8	.500
1 1/2	.600
1 5/8 AND 1 3/4	.800

DOOR THICKNESS	B
1 7/8	1.000
2	1.100
2 1/8	1.250
2 1/4	1.300

ALL DIMENSIONS SHOWN ARE IN INCHES

Figure 13.46 Pin adjustment settings for doors with thickness other than 1¾-inch. (*M.A.G. Engineering and Mfg., Inc.*)

Put the key in the cylinder and rotate one complete turn, first in one direction and then the opposite direction. This is to check the function of this assembly.

If you haven't adjusted the lock for other than a 1¾-99-inch door, remove the mechanism cover by first removing the spring assembly. Pry up the cover and remove the ring; then, adjust the pin for the appropriate door thickness, referring to Fig. 13.46 to ensure the details are correct.

Mounting the lock. Insert the bolt assembly into the 1-inch hole.

Insert the exterior escutcheon (with the mechanism ring in the retracted position) into the 9/16- and 2⅛-inch holes. When inserting, make sure the bolt link fits over the mechanism pin and that the *large tab* of the bolt link is toward the hole in the mechanism cover.

Place the interior escutcheon on the door and secure it to the exterior escutcheon with the four ¼-inch-diameter bolts furnished with the lock unit.

Put the cover on the interior escutcheon.

Single cylinder. Install the thumbturn with the *bolt retracted.* Place the circular wave spring over the thumbturn housing. Be sure the projection on the thumbturn cam and the projection on the housing are together. Place the thumbturn assembly into the interior escutcheon. Make sure the alignment projections line up with the bolt in the escutcheon hole. The thumbturn must be pointing away from the edge of the door. Push the assembly in carefully, making certain the hole in the thumbturn cam lines up with the mechanism pin. Push it in until the catch enters the slot in the thumbturn housing. Check for proper function by rotating the thumbturn and pulling to be sure the catch has engaged the housing. The long side of the thumbturn should point toward the edge of the door when the bolt is extended and away from the door edge when retracted.

Double cylinder. Assemble the mortise cylinder, cylinder spring, and cylinder guard cover to the interior cylinder housing the same as the exterior escutcheon. The keyway must be horizontal and opposite the catch slot. Thread the interior ring onto the cylinder until it stops against the cylinder housing, then back off one to one and one-half turns prior to assembly into the lock.

Install the interior cylinder housing assembly with the bolt retracted. Insert the assembly into the escutcheon hole, lining up the boss with the slot. Also, line up the ring boss with the escutcheon slot. Press it in fully with the key partially withdrawn, then rotate counterclockwise to stop. Push in on the cylinder until the catch snaps into the slot. Try to turn the cylinder housing clockwise to be sure the catch is engaged to the lock cylinder housing in place. With the key, check for the proper function of the lock from both sides.

Insert the bolt housing assembly into the 1-inch bolt hole. *Caution:* Make certain the opening side of the bolt housing is *up* on left-hand doors and *down* on right-hand doors.

Install the faceplate with the screws provided. Make sure the hole in the faceplate fits over the projection of the bolt housing assembly to center it properly.

Metal doors. Door preparation and procedures for mounting the lock to hollow metal doors—for doors that have no seam or a hole in the center of the door edge—are as follows.

1. Follow the dimensions outlined on the template. Drill a $\frac{1}{8}$-inch pilot hole through the face of the door located $2\frac{3}{4}$ inches away from the high edge of the door bevel. Be sure the pilot hole is the same distance from the high edge of the door on each side. Bore a $2\frac{1}{8}$-inch hole through.
2. Drill four $\frac{9}{16}$-inch holes the same as for wood doors.
3. Using the template, make three centerpunch marks on the edge of the door: one in the center, one $\frac{13}{16}$ inch above, and one $\frac{13}{16}$ inch below center. Drill $\frac{1}{4}$-inch holes.
4. Open the center hole with a $\frac{3}{4}$-inch hole saw. Remove the burr on the inside edge. Countersink top and bottom holes for 12-24 flathead machine screws.

5. Install clip nuts to the bolt housing with the long side toward the rear of the housing.
6. Insert the bolt housing through the 2⅛-inch hole in the door and place it against the inside front edge of the door, making certain the projection of the bolt housing assembly fits into the ¾-inch hole. Put the opening on the side of the housing *up* on a left-hand door and *down* on right-hand doors.
7. Secure the bolt housing to edge of the door with the two 12-24 screws.
8. Insert the bolt assembly into the bolt housing.
9. Follow steps 2 through 5 for wood doors to complete the installation.

For hollow metal doors that have a seam or fold on the edge that would interfere with drilling of holes for mounting the bolt housing, use this procedure: Cut out the door edge as shown (Fig. 13.47) and mount Install-A-Lock. (Depending on whether the door is a 1⅜- or 1¾-inch door, you will require a special order; check with your distributor to see what the Install-A-Lock special order number is for this unit.)

Finish the door preparation for hollow metal doors as previously described. Mount the lock per procedures for mounting to wood doors.

Strike installation

Wood jambs

1. With the Ultra 800 properly installed, close the door fully and throw the bolt out (by either the key or thumbturn) so that the point on the end of the

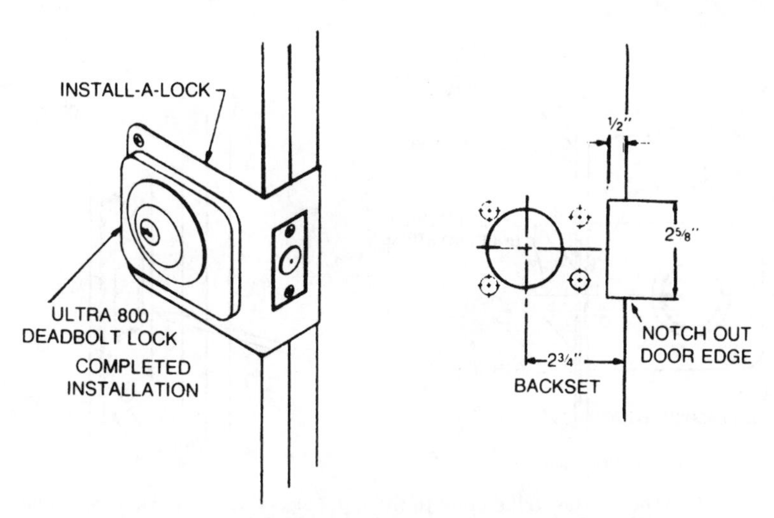

Figure 13.47 Door edge cutout required for metal door with center seam on the door edge. (*M.A.G. Engineering and Mfg., Inc.*)

bolt will make a mark on the jamb. This will be the center mark for the strike mounting.

2. Open the door. Place the strike plate against the door, ensuring that it lines up with the bolt mark centered on the strike plate hole (Fig. 13.48).
3. Mark the center hole required to be drilled; also mark around the strike plate edge (Fig. 13.49).
4. Drill two 1-inch diameter holes 1⅜ inches deep, one ⅝ inch above and one ⅝ inch below the center mark (Fig. 13.50).
5. Chisel out the wood in between.
6. Cut out the wood 1¼ inch wide × 4⅞ inches high × ⅛ inch deep for the strike box and strike plate.
7. Be sure to drill a ⅛-inch pilot hole for all four screws to aid installation.
8. Insert the strike box. Place the finished strike plate over the strike plate box and secure it with screws provided (Fig. 13.51).
9. Within the strike box, drill two ⅛-inch-diameter pilot holes on approximately a 25° angle toward the center of the jamb. This will provide for maximum holding power when the two screws are attached to the stud (Fig. 13.52). Figure 13.53 is a side view of the strike box positioning.

Metal jambs not prepared for a strike

1. Mark the jamb with the point on the end of the bolt as described for wood jambs.
2. Drill a ⅞-inch hole only. No strike plate is required.

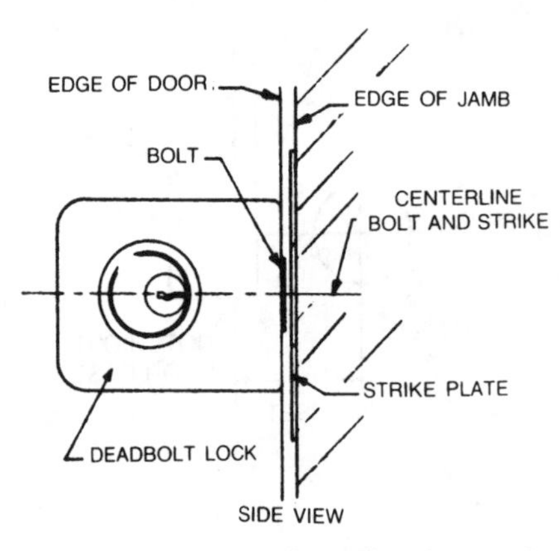

Figure 13.48 Line up the strike plate to the locking bolt. (*M.A.G. Engineering and Mfg., Inc.*)

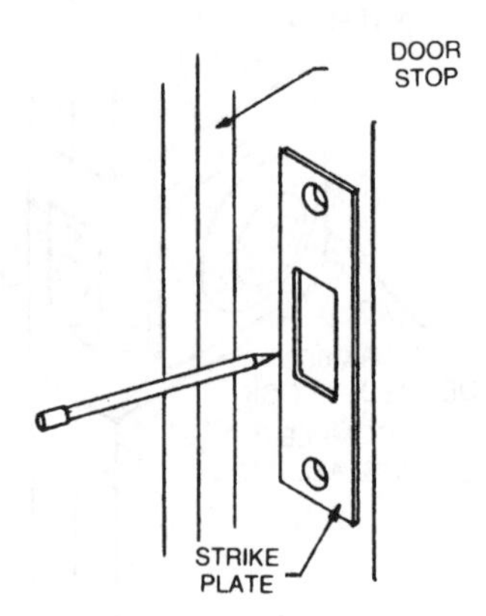

Figure 13.49 Mark strike plate. (*M.A.G. Engineering and Mfg., Inc.*)

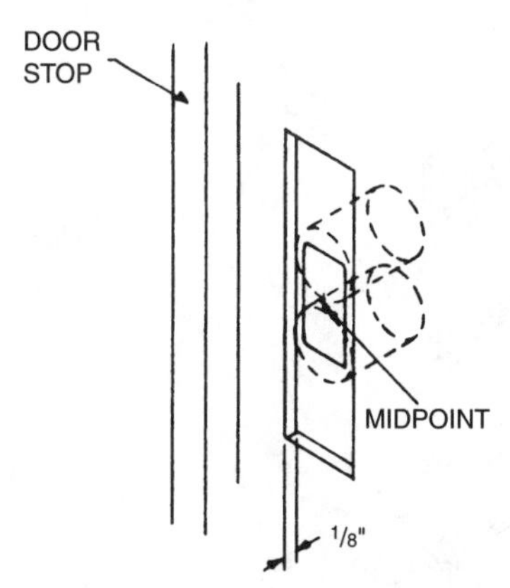

Figure 13.50 Ensure the midpoint is centered in the middle of the strike plate hole. (*M.A.G. Engineering and Mfg., Inc.*)

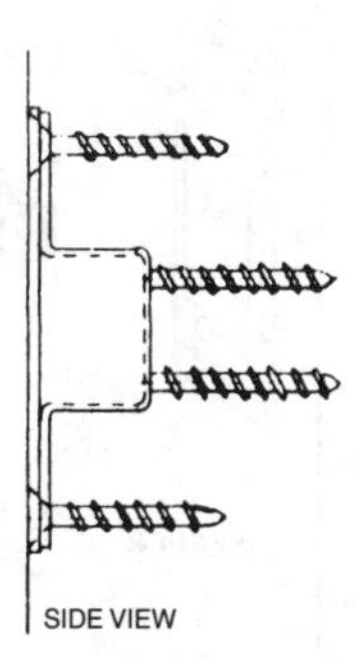

Figure 13.51 Strike box into the door frame. (*M.A.G. Engineering and Mfg., Inc.*)

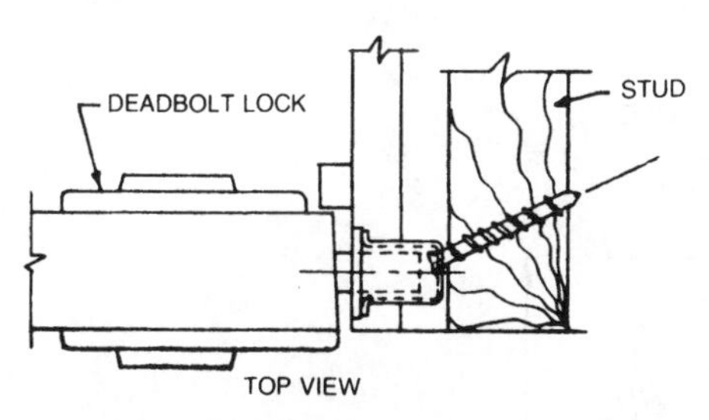

Figure 13.52 Ensure the screws go into the frame stud for maximum holding power. (*M.A.G. Engineering and Mfg., Inc.*)

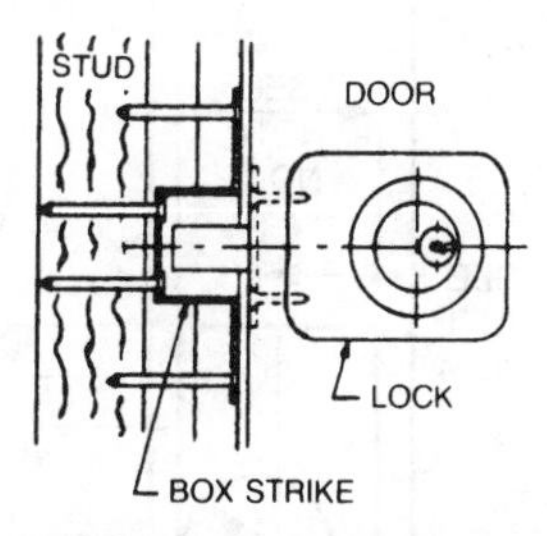

Figure 13.53 Side view of the strike box. (*M.A.G. Engineering and Mfg., Inc.*)

3. If desired by the customer, an Adjust-A-Strike may be installed with the unit. Several models are available. Figure 13.54 refers to the models 450/451: 1¼-inch × 2¾-inch lip and deadbolt strike jamb preparation. (Figure 13.55 provides positioning of parts and the strike installation completed for all models described here.)

For 1¾-inch doors, start cutout 3/16 inch away from the door stop; for 1⅜-inch doors, start the cutout on the edge of the door stop.

1. Cut out 1¼ inches × 2¾ inches, 5/32 inch deep.
2. Cut out 1½ inches, 31/32 inch deep.

For 1¾-inch door, 31/32 inch; for 1⅜-inch door, 25/32 inch.
For latches (lip strike), drill ⅝ inch deep; for deadbolts, drill 1 inch deep.

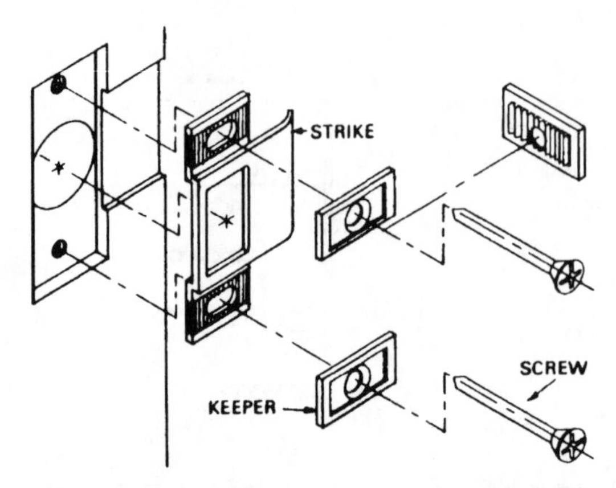

Figure 13.54 Exploded view illustrating parts relationship for the complete installation of the Adjust-A-Strike. (*M.A.G. Engineering and Mfg., Inc.*)

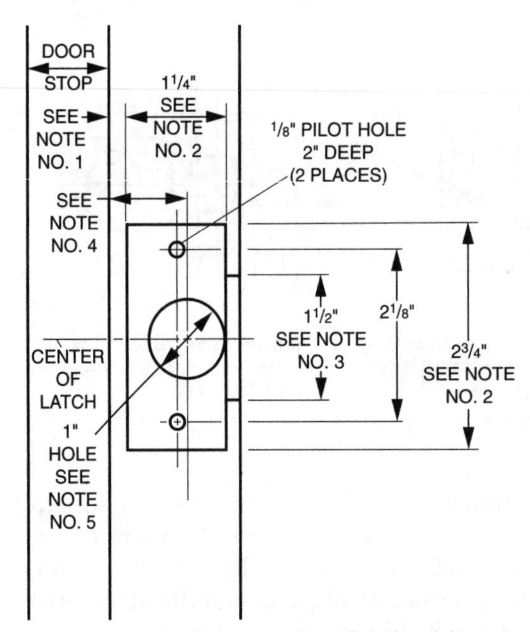

Figure 13.55 Jamb preparation for the Adjust-A-Strike models 450 and 451. (*M.A.G. Engineering and Mfg., Inc.*)

Install the strike and keepers in the center of the cutout. Close the door to check fit. Adjust if necessary.

For the model 452, for a 1¼ × 4⅞-inch lip strike, the jamb preparation is as follows:

1. For 1¾-inch door, start the cutout ³⁄₁₆ inch away from the door stop; for 1⅜-inch door, start cutout on the edge of door stop (Fig. 13.56).

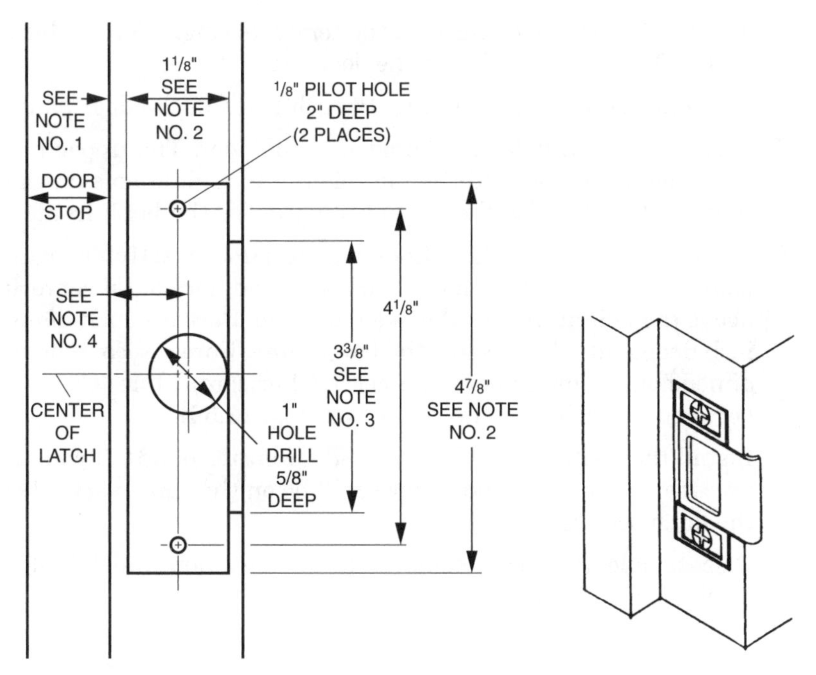

Figure 13.56 Jamb preparation and completed installation of Model 452 for a 1¾-inch door. (*M.A.G. Engineering and Mfg., Inc.*)

2. Cut out 1¼-inch × 4⅞-inch, ⅛ inch deep.
3. Cut out 3⅜ inches, 1/16 inch deep.

For 1¾-inch door, 3/32 inch. For 1⅜-inch door, 25/32 inch:

1. Install the strike and keepers in the center of the cutout. Close the door and check for proper fit, adjusting if necessary.

The Schlage G Series Lockset

The mechanics of this lockset were discussed in an earlier chapter.

Installation

These instructions, developed from material supplied by Schlage, apply for wood doors from 1⅜ to 2 inches thick.

1. Mark the height line (the centerline of the deadlatch) on the door (Fig. 13.57). *Suggestion:* Mark the line 38 inches from the floor.
2. Mark the centerline of the door thickness on the door edge.
3. Position the template (supplied with the lockset) on the door with the lower deadlatch hole on the height line. Mark center points for one 2⅛-inch hole

and two 1½-inch holes through the template (Fig. 13.57). Mark the centers for two 1-inch latch holes on the door edge.

4. Bore one 2⅛-inch hole and two 1½-inch holes in the door panel.
5. Bore two 1-inch holes in the edge of the door. The upper hole should be extended ⅜ inch beyond the far side of the middle hole on the door panel (Fig. 13.58). Mortise the edge of the door for the latch front.
6. Mark vertical and height lines on the jamb exactly opposite the center point of the lower latch hole. Mark a second horizontal centerline 2 inches above the height line for the deadbolt hole. Bore a 1-inch-diameter hole 1⅛ inch deep and ¼ inch below the height line. Bore a second hole to the same dimensions ¼ inch above the second horizontal line. Clear out the area between the holes for the strike box (Fig. 13.59).
7. Install the strike (it can be reversed for either hand). If you have done the work correctly, the strike screws will be on the same vertical centerline as the latch screws.
8. Disassemble the lock. Remove the inside knob and lift off the outside mechanism (Fig. 13.60).

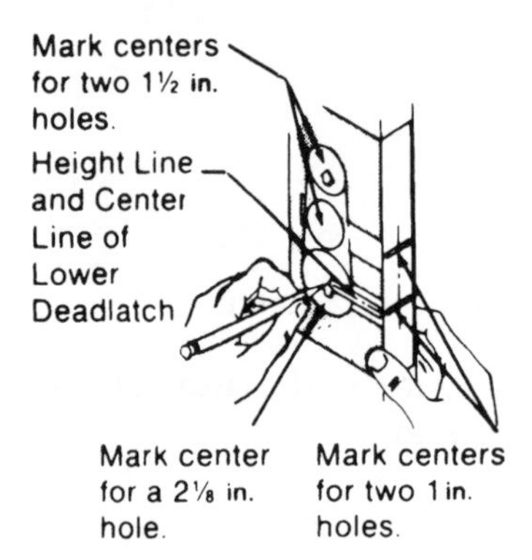

Figure 13.57 Marking holes.

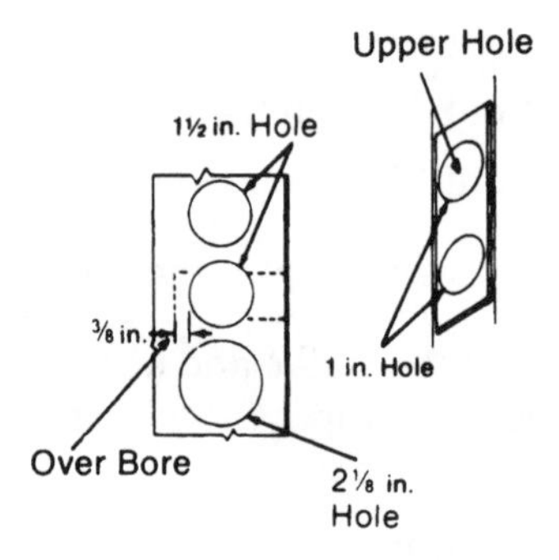

Figure 13.58 Boring holes.

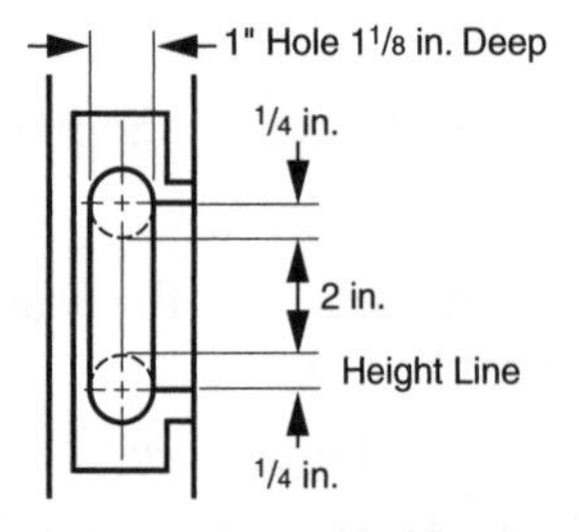

Figure 13.59 Clear area between holes.

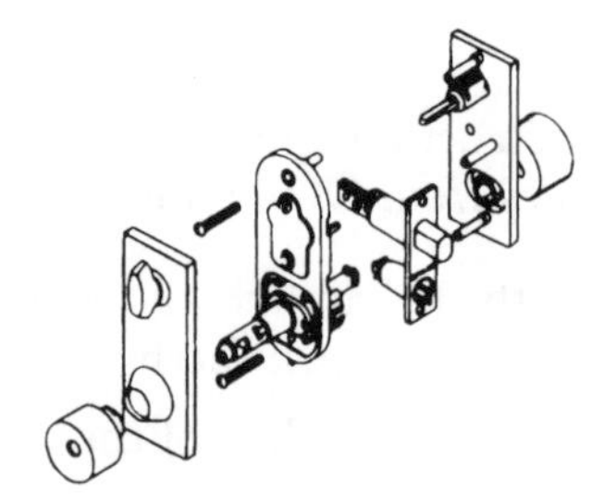

Figure 13.60 Disassemble lock.

9. With the deadbolt thrown and the crank slot in the vertical position, insert the latch unit and secure it with the screws provided (Fig. 13.61). The deadlatch can be rotated to match the door hand (Fig. 13.62). The beveled edge of the deadlatch should contact the striker as the door swings shut.
10. Install the inside mechanism with the knob button released and the crank bar in the vertical position. Insert the end of the crank bar in the slot on the deadbolt (Fig. 13.63). Engage the jaws of the slide with the deadlatch bar, and engage the deadlatch housing with the ears on the inside mechanism. This sounds more confusing than it is; see the insert for Fig. 13.63.
11. From the outside of the door, turn the cylinder bar slot with a screwdriver to check the action of the deadbolt and deadlatch. Once you are satisfied that the lock works properly, retract the deadbolt and release the knob button. Turn the cylinder bar slot to the horizontal position (Fig. 13.64). The clutch plate should be situated as shown.
12. Rotate the cylinder bar to the horizontal position with the knob spindle as shown (Fig. 13.65).

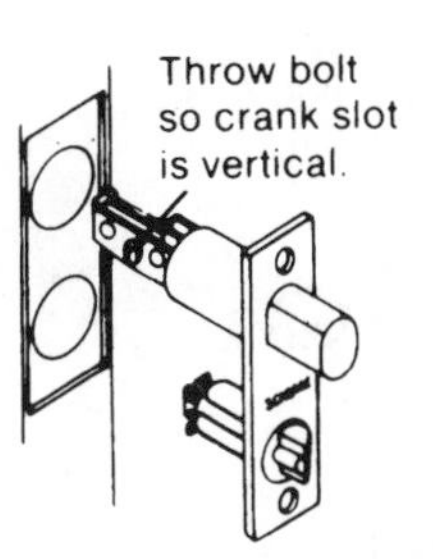

Figure 13.61 Insert latch unit.

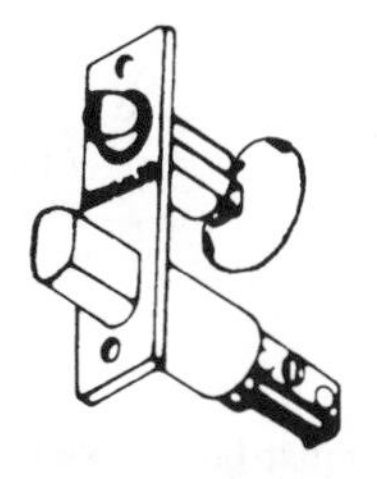

Figure 13.62 Rotate deadlatch.

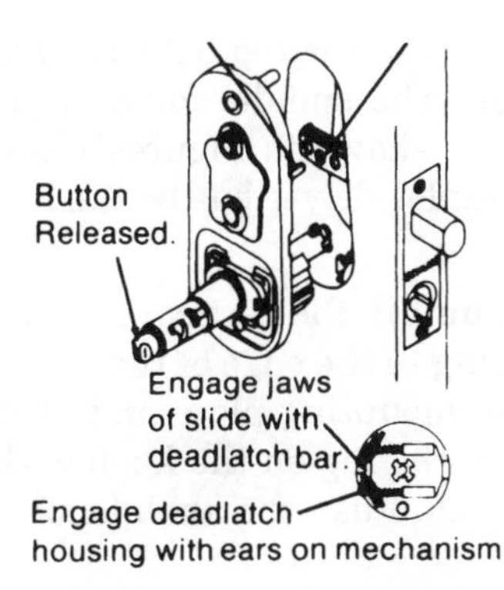

Figure 13.63 Install mechanism.

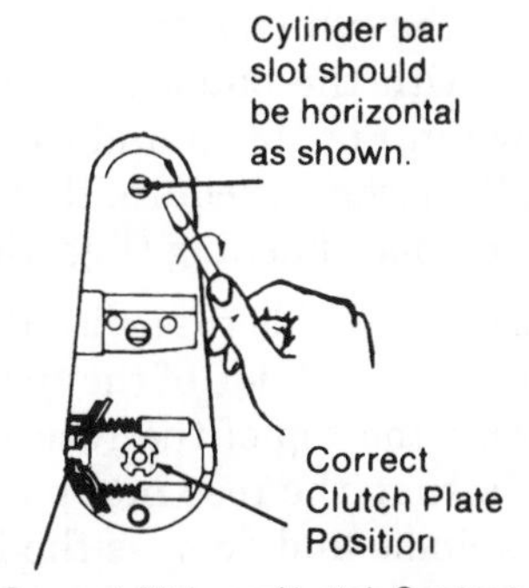

Figure 13.64 Testing lock.

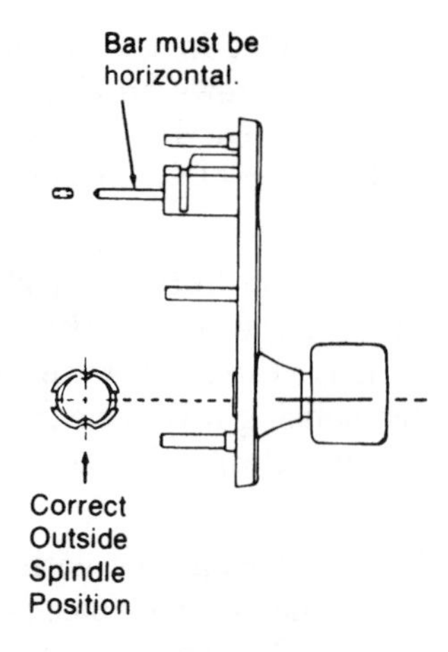

Figure 13.65 Rotate cylinder bar.

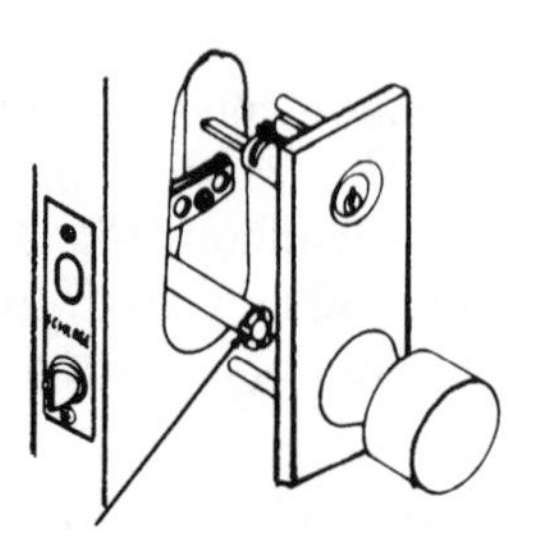

Figure 13.66 Positioning clutch.

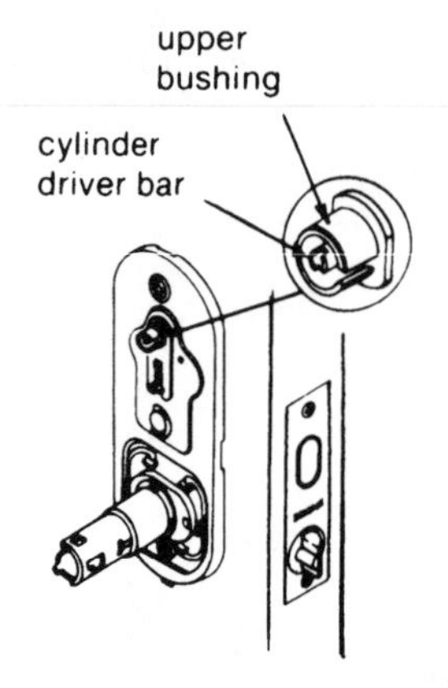

Figure 13.67 Driver bar must be vertical.

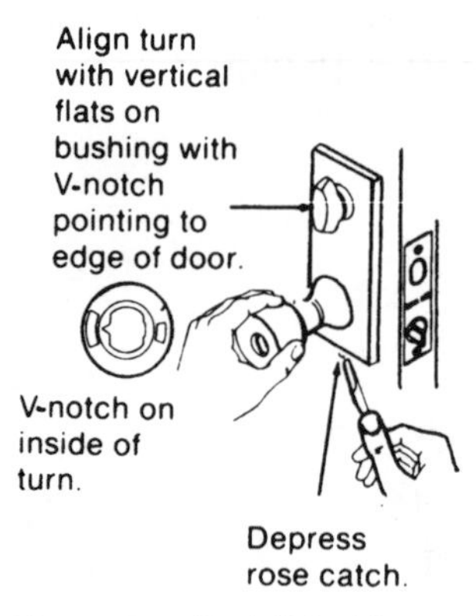

Figure 13.68 Install inside rose.

13. Insert the cylinder bar into the bar slot in the top hole. Engage the clutch plate with the outside knob spindle. The spindle must be positioned as shown in Fig. 13.65 and the clutch as shown in Figures 13.64 and 13.66. G85PD locksets require that the cylinder driver bar be flush with the end of the upper bushing (Fig. 13.67).

14. Install the inside rose with the turn unit in the vertical position and the V-notch on the inside of the turn pointing to the edge of the door (Fig. 13.68). Engage the top of the rose with the mounting plate and, depressing the catch, snap the rose into place. Align the lug on the knob with the slot in the spindle and depress the knob catch. Push the knob home.

15. Test the lock.

Changing the hand

G series locksets are available in right- and left-hand versions. However, the hand can be changed in the field.

Converting from right-hand to left-hand. Figure 13.69 is an assembled view of a right-hand unit. To convert it to left-hand operation, follow these steps:

1. Disengage the retaining rings (Fig. 13.70).

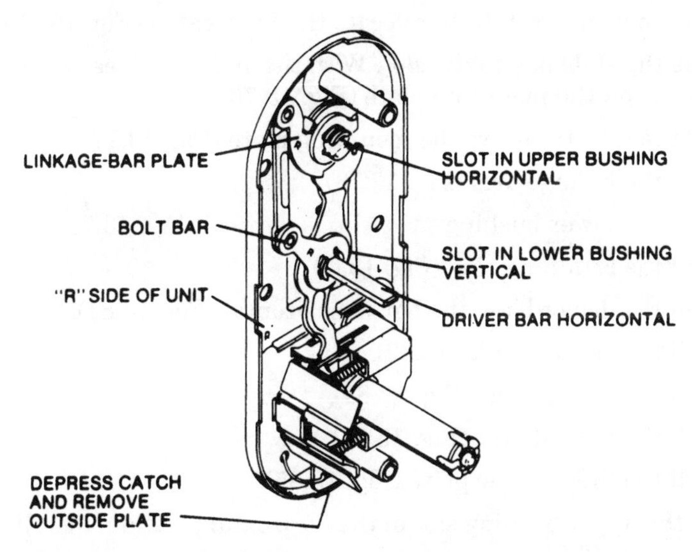

Figure 13.69 Right-hand unit.

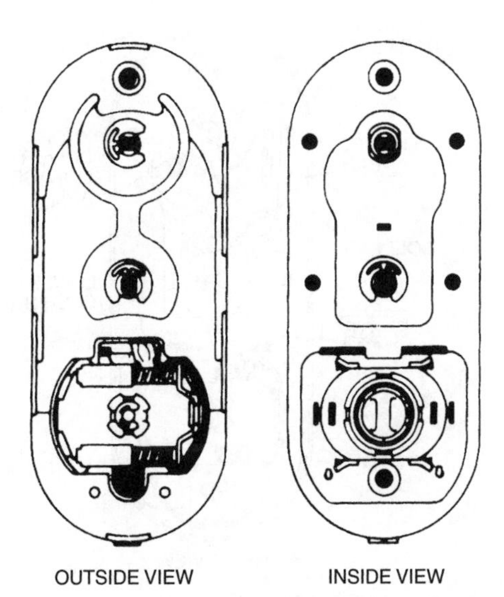

Figure 13.70 Changing hand, Step 1.

2. Lift off the mounting-plate cover (Fig. 13.71).
3. Remove the linkage-bar plate, linkage bar, bolt bar, driver bar, knob-driver linkage arm, and lower bushing (Fig. 13.72).
4. With a factory-supplied wrench or long-nosed pliers, straighten the tabs (Fig. 13.73).
5. Compress the slide with a screwdriver and pull the assembly out (Fig. 13.74).
6. Rotate the spindle 180° to relocate the knob catch (Fig. 13.75).
7. Rotate the slide assembly 180°. With the slide compressed, insert the lugs into slots on the mounting plate (Fig. 13.76).
8. Bend the tabs to secure the mounting plate (Fig. 13.77).
9. Install the linkage arm (Fig. 13.78).
10. Install the lower bushing—the slot is vertical (Fig. 13.79).
11. Install the retaining ring (Fig. 13.80).
12. Install the driver bar—the bar is horizontal (Fig. 13.81).
13. Install the bolt bar (Fig. 13.82).
14. Install the linkage bar (Fig. 13.83).
15. Install the knob driver (Fig. 13.84).
16. Install the linkage-bar plate (Fig. 13.85).
17. With the upper bushing slot in the horizontal position, install the cylinder driver (Fig. 13.86).

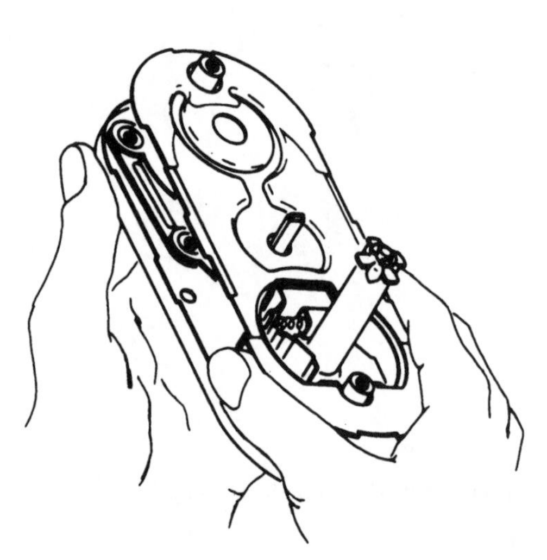

Figure 13.71 Changing hand, Step 2.

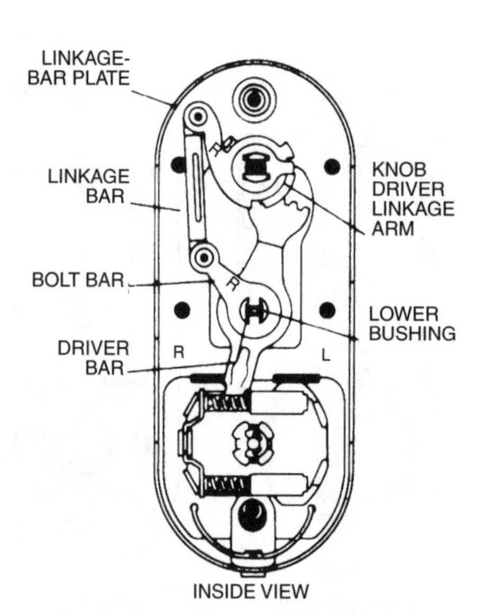

Figure 13.72 Changing hand, Step 3.

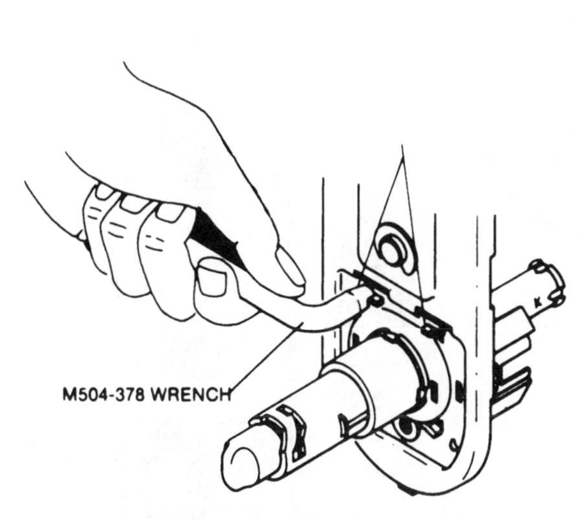

Figure 13.73 Changing hand, Step 4.

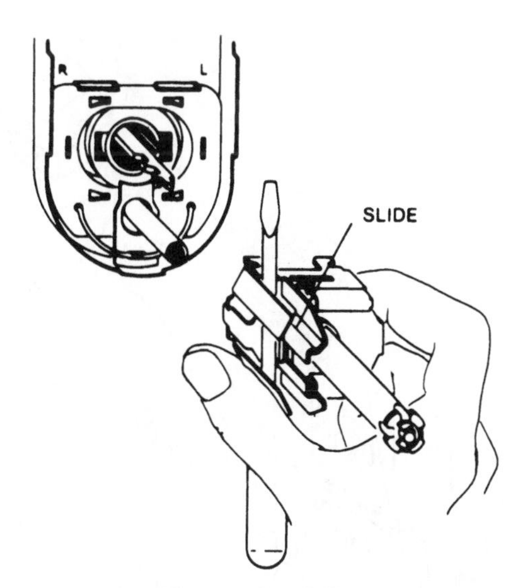

Figure 13.74 Changing hand, Step 5.

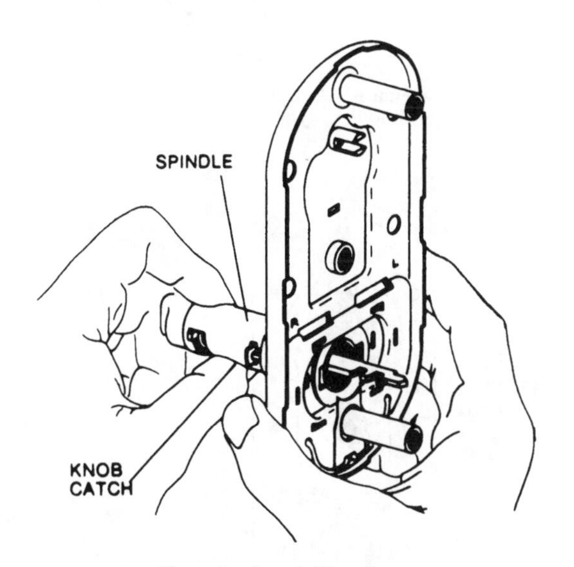

Figure 13.75 Changing hand, Step 6.

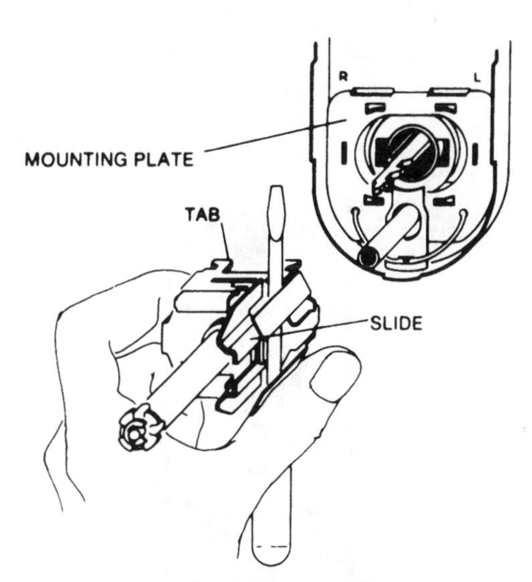

Figure 13.76 Changing hand, Step 7.

18. Replace the mounting-plate cover (Fig. 13.87).
19. Install the inside retaining rings (Fig. 13.88).
20. Turn the slot in the G85 unit to face the same direction as the slide (Fig. 13.89).
21. Turn the deadlatch 180° (Fig. 13.90).
22. Check your work (Fig. 13.91).

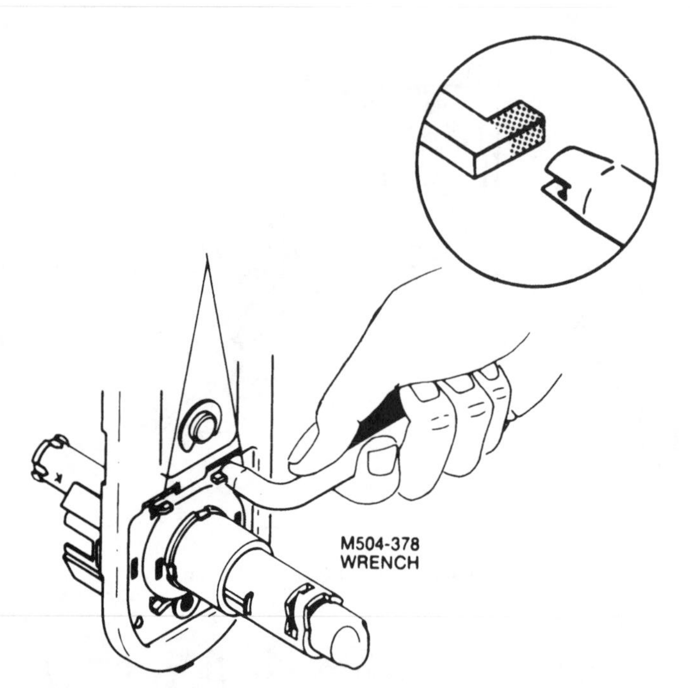

Figure 13.77 Changing hand, Step 8.

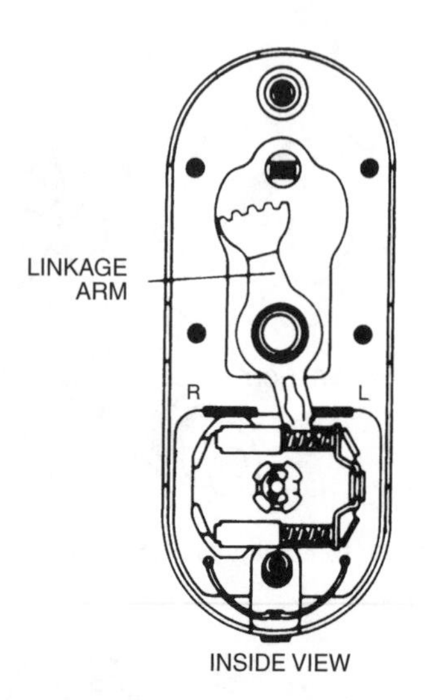

Figure 13.78 Changing hand, Step 9.

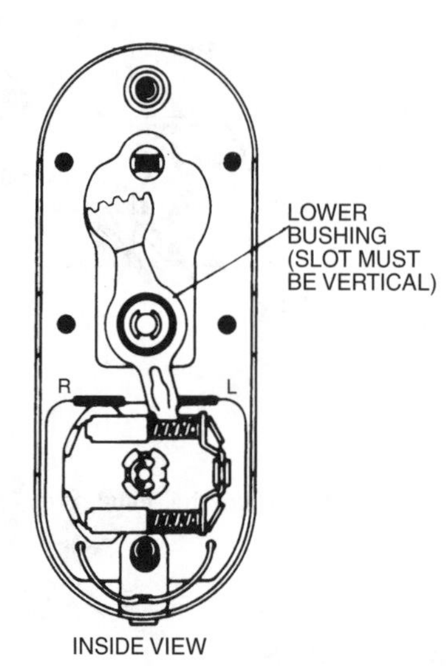

Figure 13.79 Changing hand, Step 10.

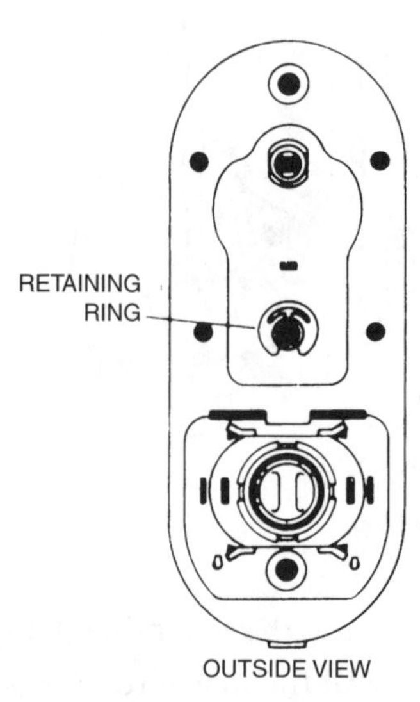

Figure 13.80 Changing hand, Step 11.

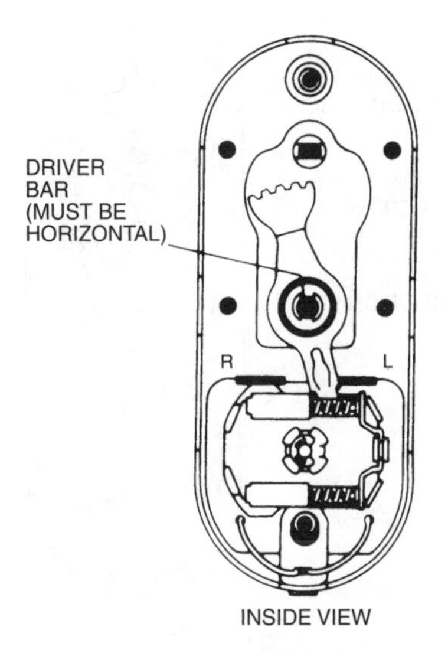

Figure 13.81 Changing hand, Step 12.

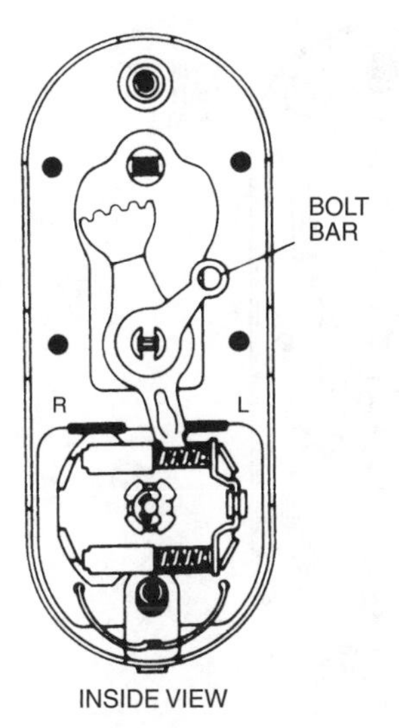

Figure 13.82 Changing hand, Step 13.

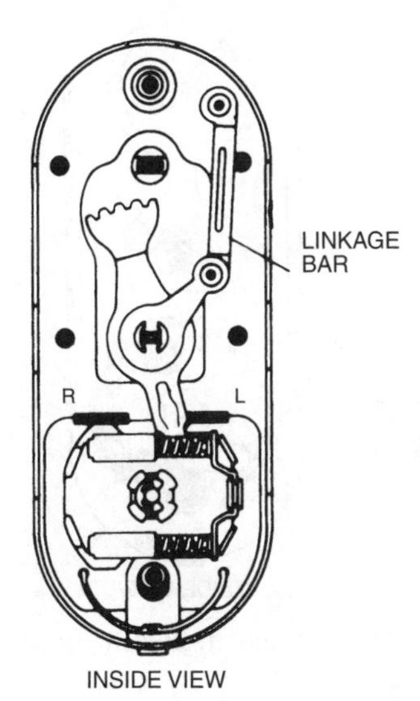

Figure 13.83 Changing hand, Step 14.

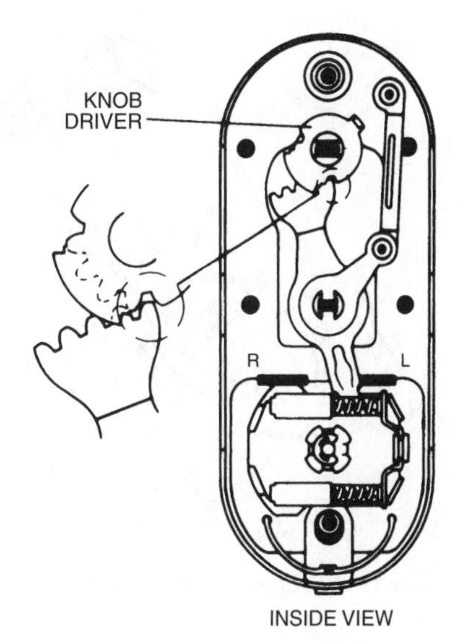

Figure 13.84 Changing hand, Step 15.

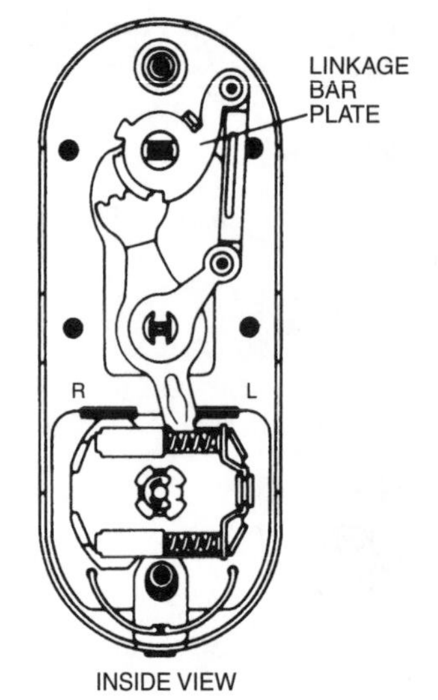

Figure 13.85 Changing hand, Step 16.

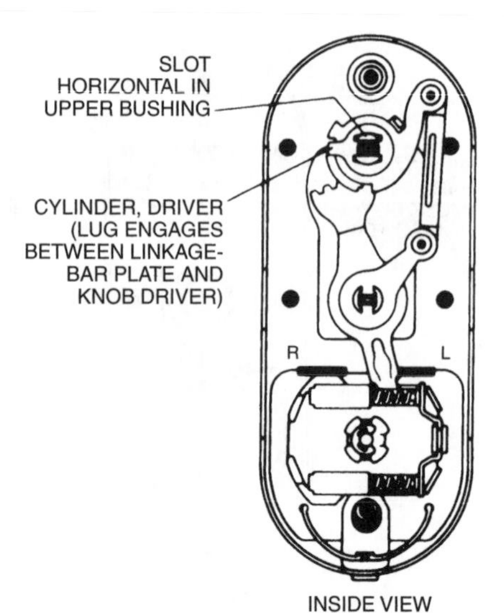

Figure 13.86 Changing hand, Step 17.

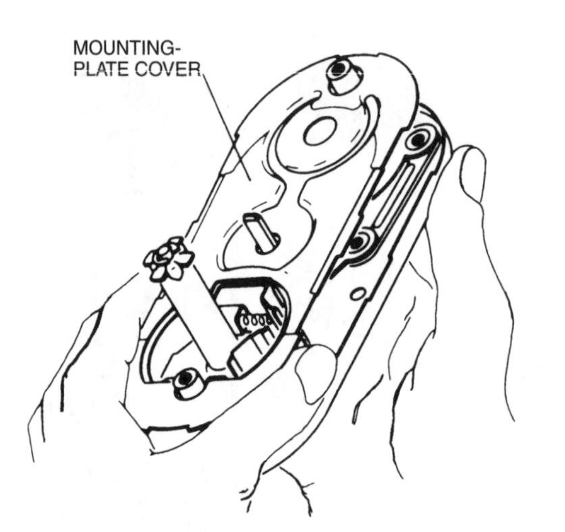

Figure 13.87 Changing hand, Step 18.

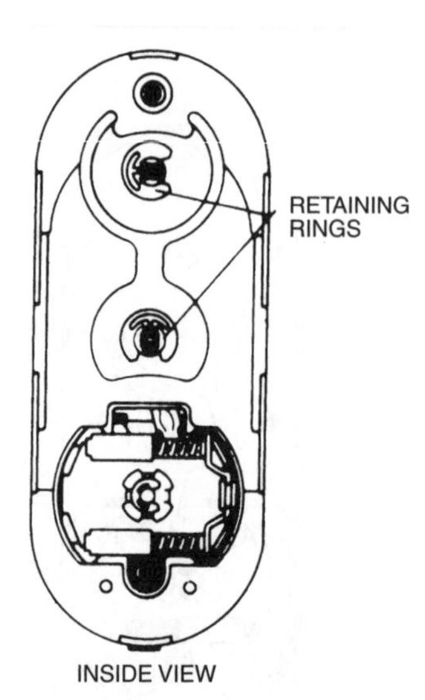

Figure 13.88 Changing hand, Step 19.

Converting from left- to right-hand. Figure 13.92 is an assembled view of a left-hand unit. To convert it to right-hand operation, follow these steps:

1. Disengage the retaining rings (Fig. 13.93).
2. Lift off the mounting-plate cover (Fig. 13.94).

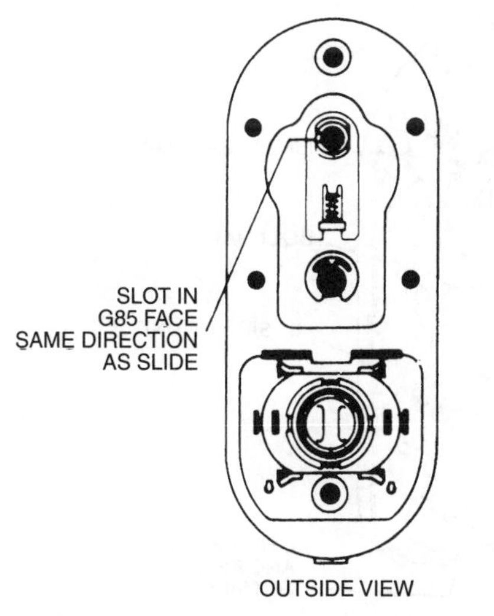

Figure 13.89 Changing hand, Step 20.

Figure 13.90 Changing hand, Step 21.

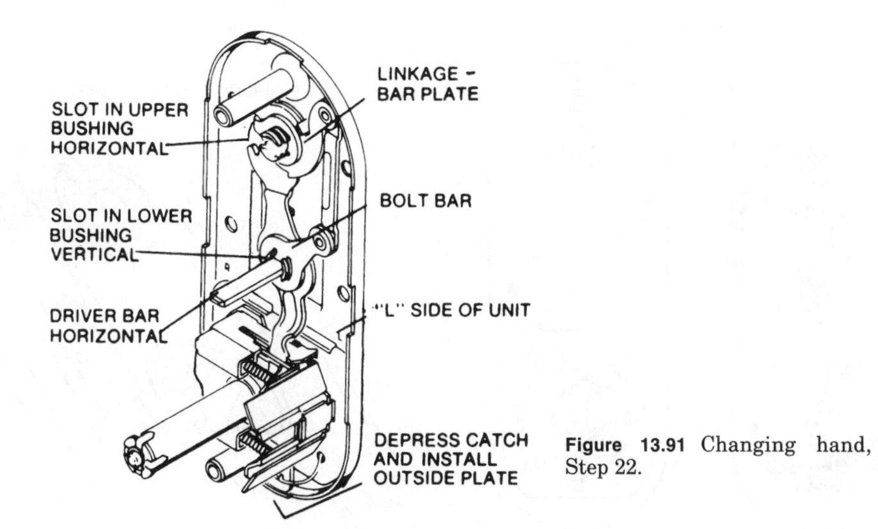

Figure 13.91 Changing hand, Step 22.

3. Remove the linkage-bar plate, linkage bar, bolt bar, driver bar, knob-driver linkage arm, and lower bushing (Fig. 13.95).
4. With a factory-supplied wrench or long-nosed pliers, straighten the tabs (Fig. 13.96).
5. Compress the slide with a screwdriver and pull the assembly out (Fig. 13.97).

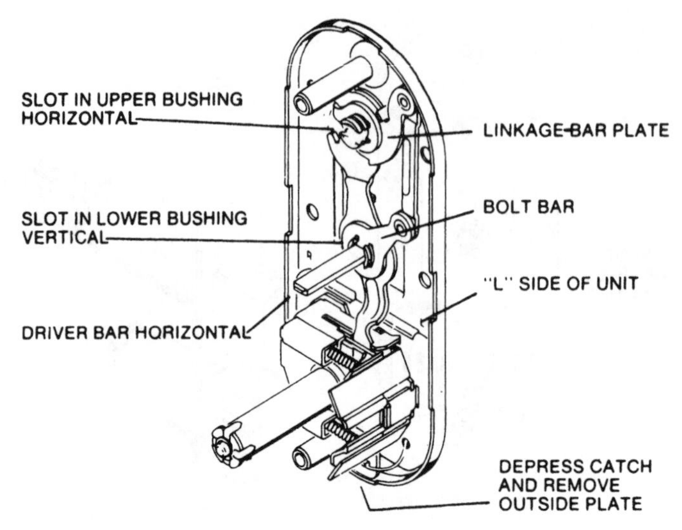

Figure 13.92 Assembled left-hand unit.

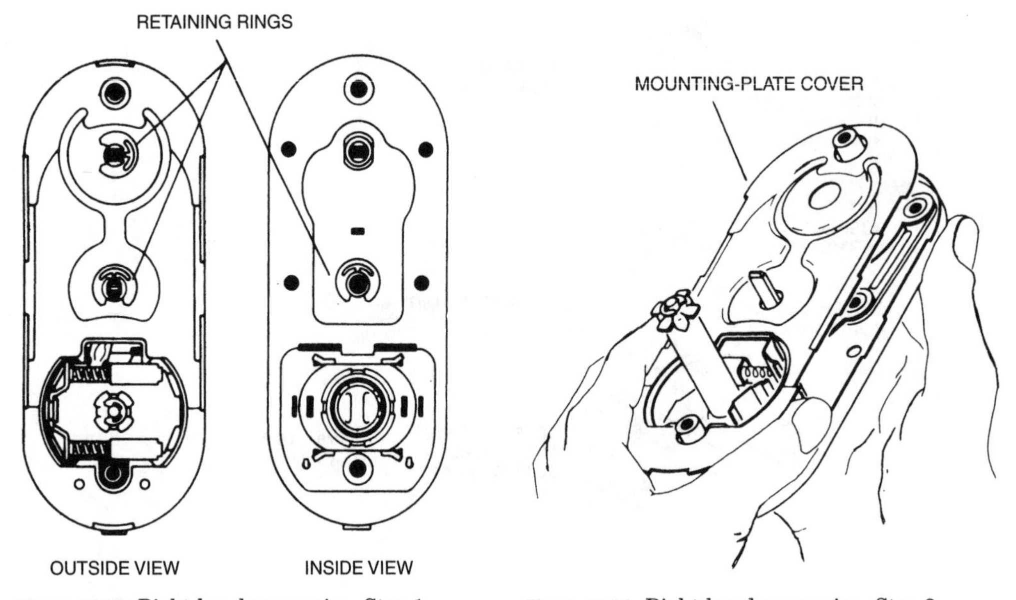

Figure 13.93 Right-hand conversion, Step 1.

Figure 13.94 Right-hand conversion, Step 2.

6. Rotate the spindle 180° to relocate the knob catch (Fig. 13.98).
7. Rotate the slide assembly 180°. With the slide compressed, insert the lugs into slots on the mounting plate (Fig. 13.99).
8. Bend the tabs to secure the mounting plate (Fig. 13.100).

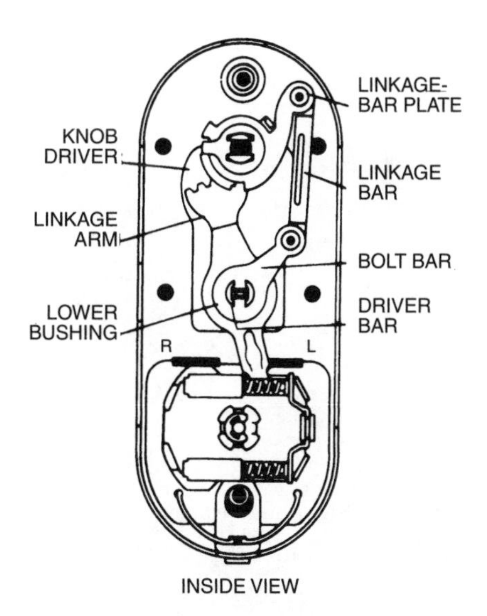

Figure 13.95 Right-hand conversion, Step 3.

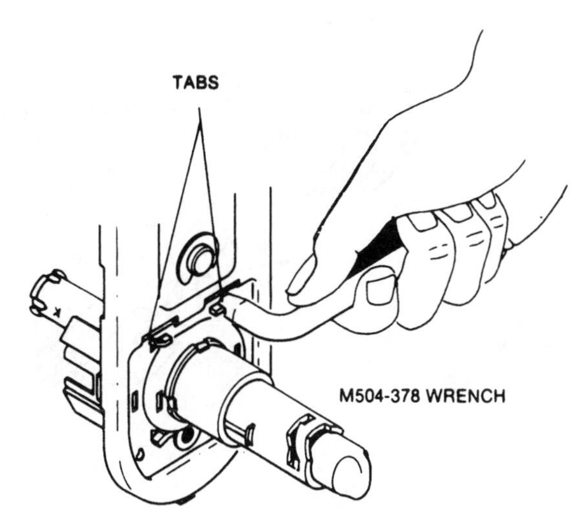

Figure 13.96 Right-hand conversion, Step 4.

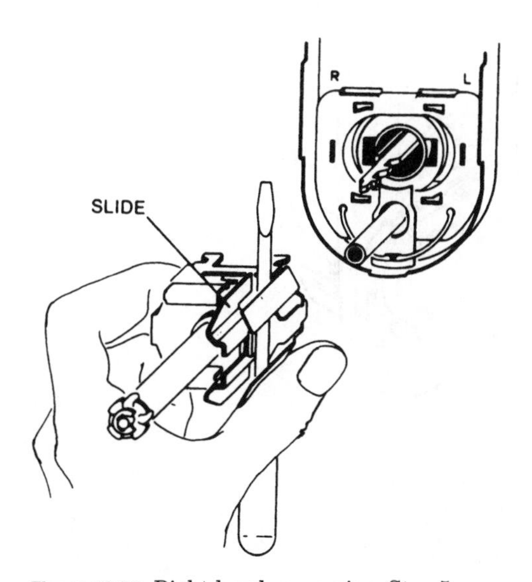

Figure 13.97 Right-hand conversion, Step 5.

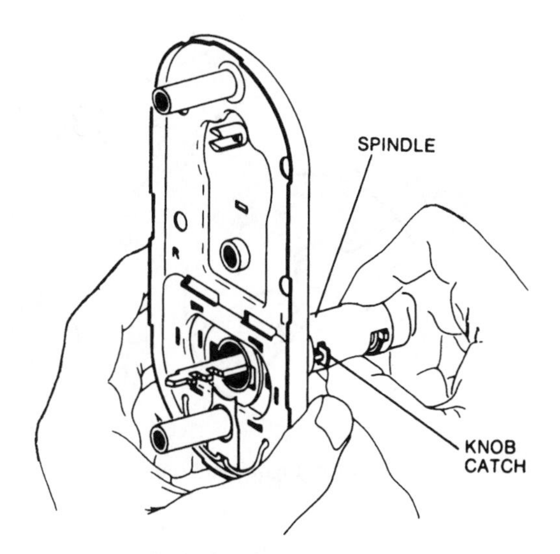

Figure 13.98 Right-hand conversion, Step 6.

9. Install the linkage arm (Fig. 13.101).
10. Install the lower bushing (Fig. 13.102).
11. Install the retaining ring (Fig. 13.103).
12. Install the driver bar (Fig. 13.104).
13. Install the bolt bar (Fig. 13.105).

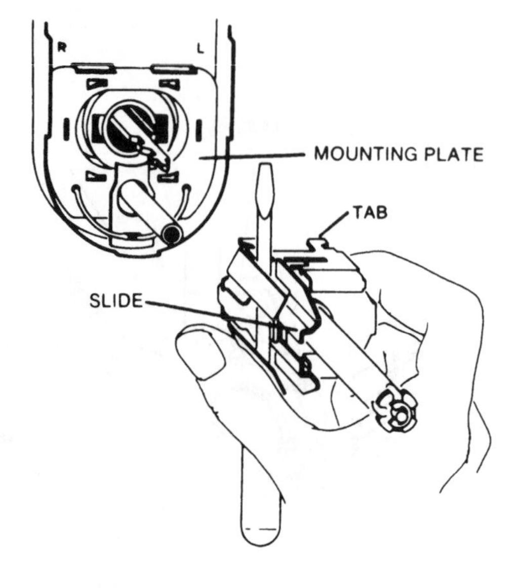

Figure 13.99 Right-hand conversion, Step 7.

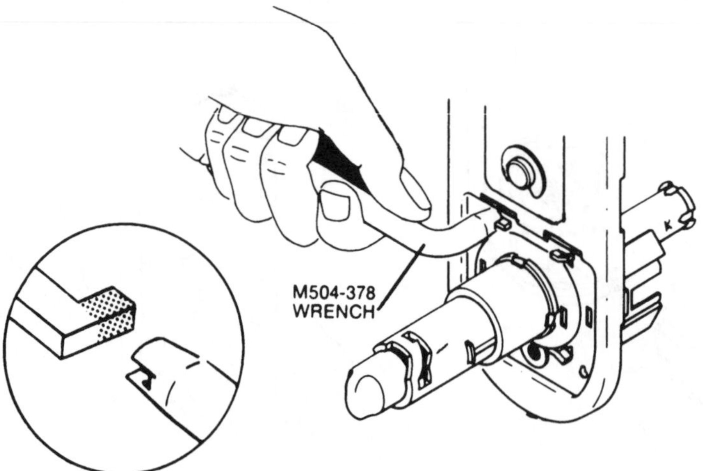

Figure 13.100 Right-hand conversion, Step 8.

14. Install the linkage bar (Fig. 13.106).
15. Install the knob driver (Fig. 13.107).
16. Install the linkage-bar plate (Fig. 13.108).
17. With the upper bushing slot in the horizontal position, install the cylinder driver (Fig. 13.109).
18. Replace the mounting-plate cover (Fig. 13.110).
19. Install the inside retaining ring (Fig. 13.111).

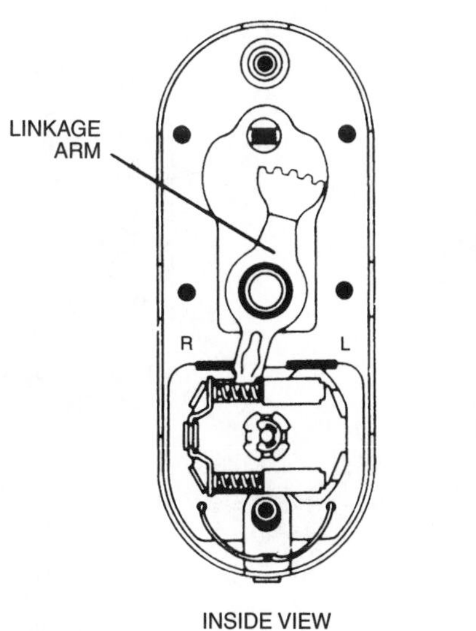

Figure 13.101 Right-hand conversion, Step 9.

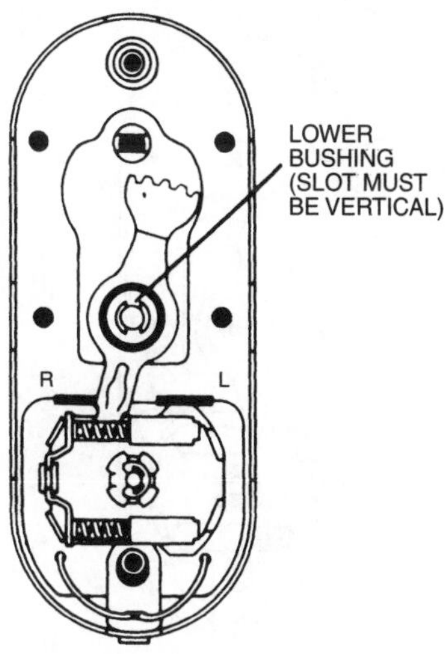

Figure 13.102 Right-hand conversion, Step 10.

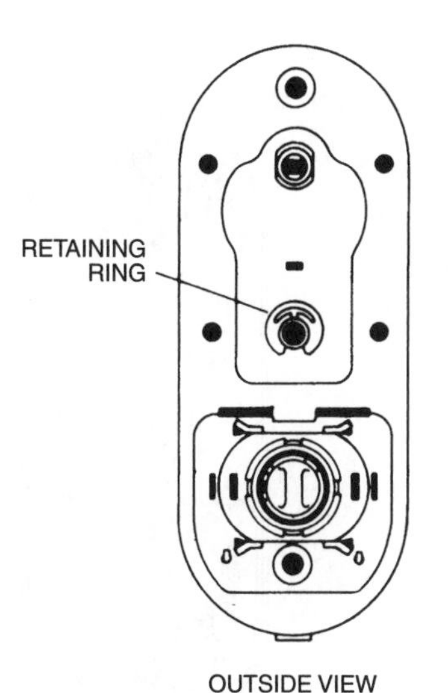

Figure 13.103 Right-hand conversion, Step 11.

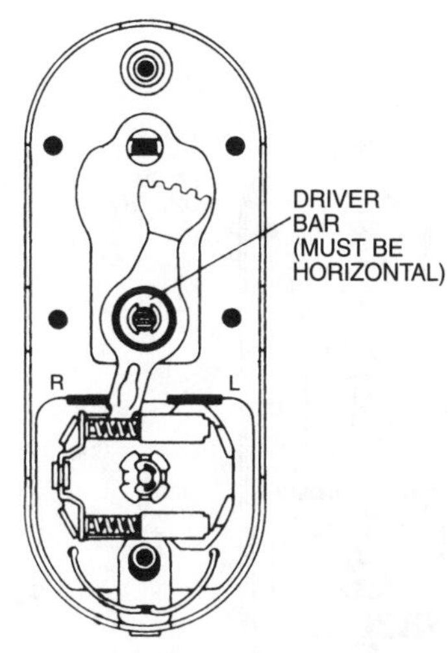

Figure 13.104 Right-hand conversion, Step 12.

20. Turn the slot in the G85 unit to face the same direction as the slide (Fig. 13.112).
21. Turn the deadlatch 180° (Fig. 13.113).
22. Check your work (Fig. 13.114).

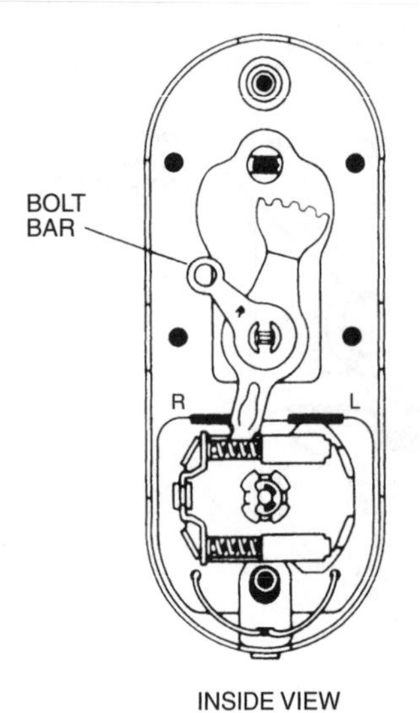

Figure 13.105 Right-hand conversion, Step 13.

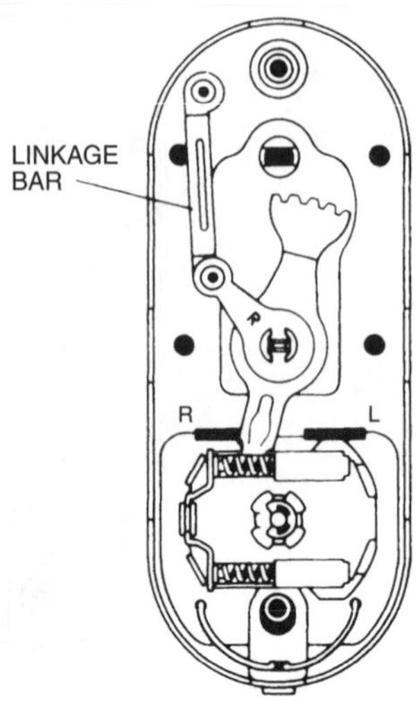

Figure 13.106 Right-hand conversion, Step 14.

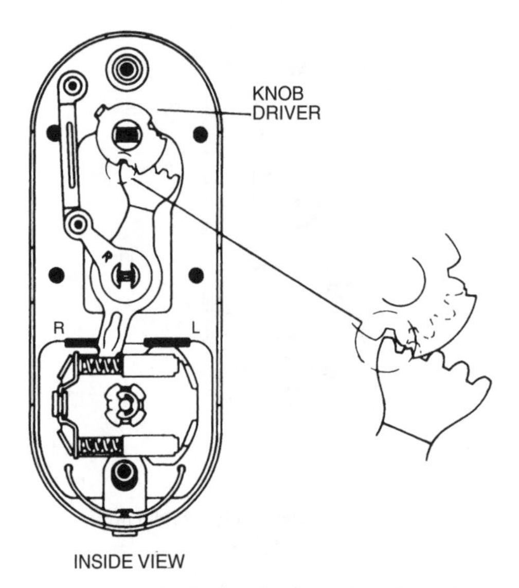

Figure 13.107 Right-hand conversion, Step 15.

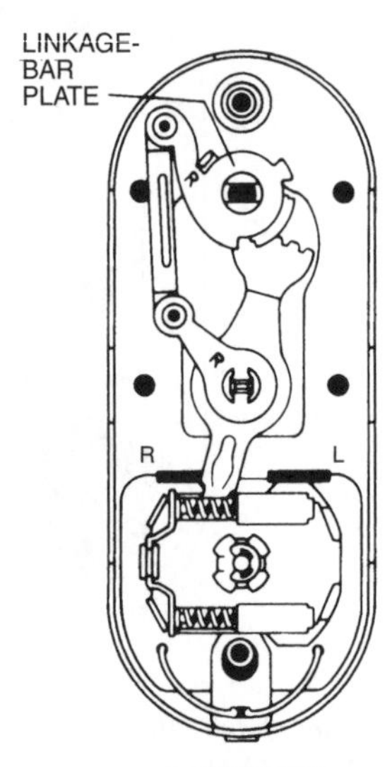

Figure 13.108 Right-hand conversion, Step 16.

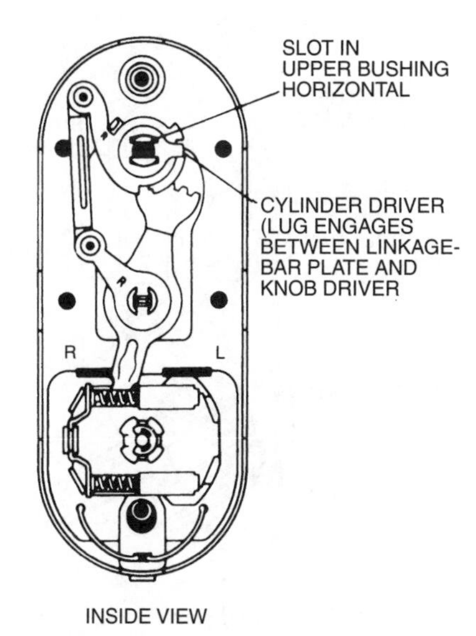

Figure 13.109 Right-hand conversion, Step 17.

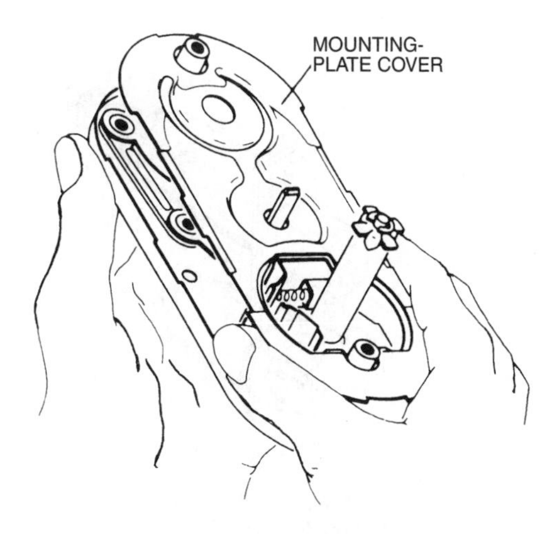

Figure 13.110 Right-hand conversion, Step 18.

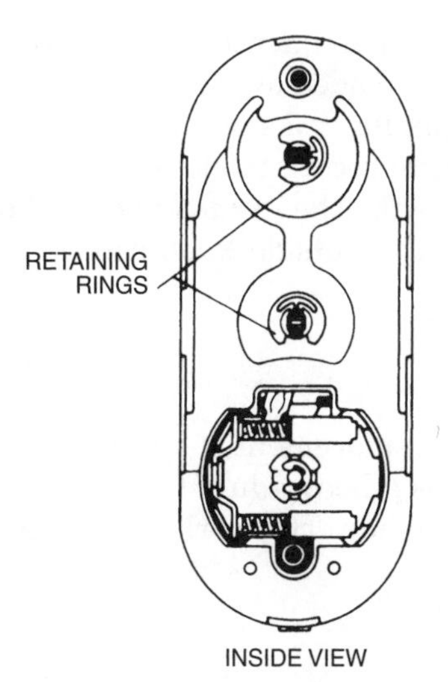

Figure 13.111 Right-hand conversion, Step 19.

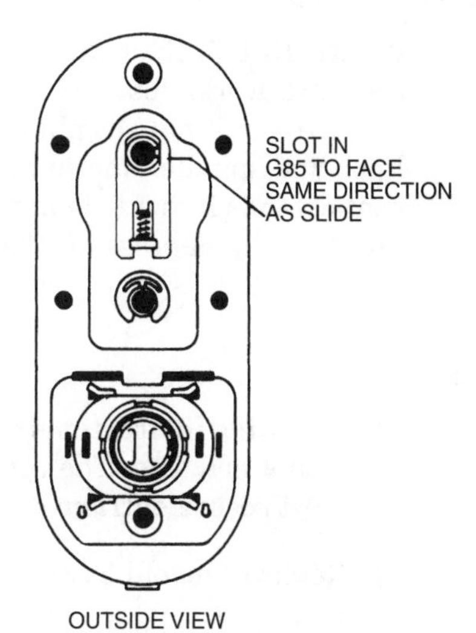

Figure 13.112 Right-hand conversion, Step 20.

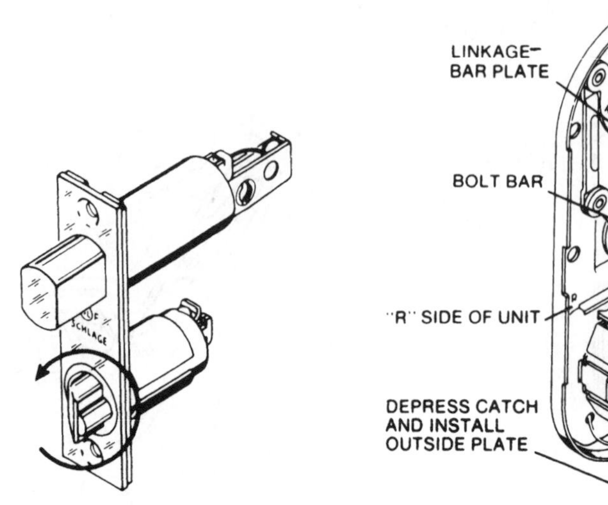

Figure 13.113 Right-hand conversion, Step 21.

Figure 13.114 Right-hand conversion, Step 22.

Attaching and removing knobs

Figure 13.115 illustrates two of the most frequently encountered methods of securing doorknobs. The knob may be threaded over the spindle and held by a set screw, or it may be pinned to the spindle by a screw that passes through the knob and spindle. Another approach is to secure the knob with a retaining lug that extends into a hole in the knob shank. The inside knob can be removed at any time by depressing the retainer; the outside knob can be removed only when the lock is open (Fig. 13.116).

Updating a lockset

The following instructions concern the replacement of a worn or outdated lockset with a new one. The replacement is a heavy-duty G series Schlage, one of the most secure entranceway locksets made. The work is not difficult.

1. Remove the old lockset (Fig. 13.117).
2. Remove the latch (Fig. 13.118).
3. If a jig is available, use it as a guide for cutting the door; if not, use the template packaged with the lockset (Figs. 13.119 and 13.120).
4. Mortise the edge of the door to receive the strike plate (Fig. 13.121).
5. Install the double-locking latch and deadbolt (Fig. 13.122).
6. Install the inside lockface assembly (Fig. 13.123).
7. Install the internal mechanism.

8. Couple the outside lockface assembly with the internal parts (Fig. 13.124).
9. Secure the inside lockface assembly with the screws provided (Fig. 13.125).
10. Snap the inside cover into place (Fig. 13.126). The lockset is now installed (Fig. 13.127).

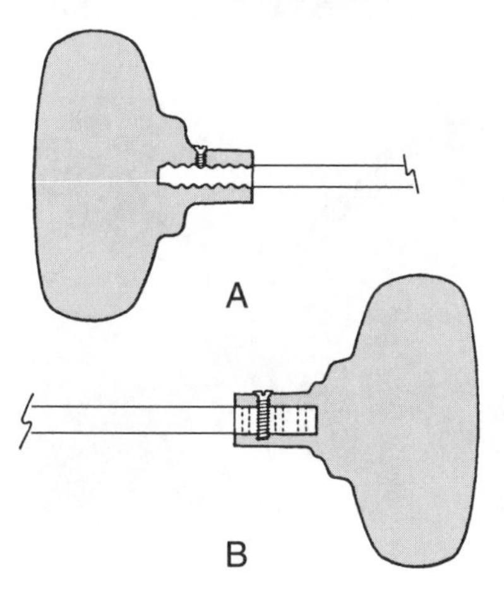

Figure 13.115 Securing door knobs.

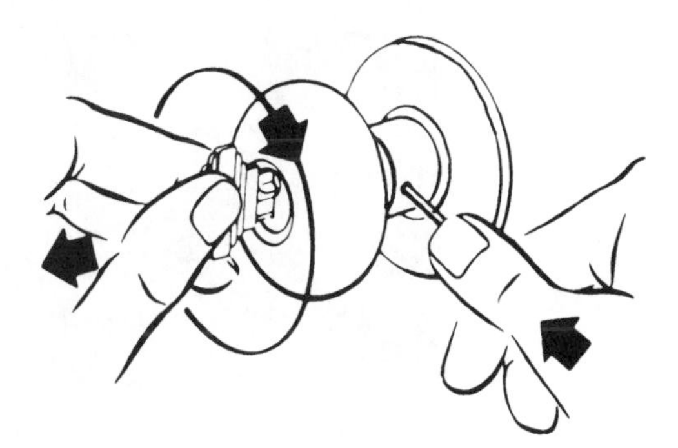

Figure 13.116 Removing knob.

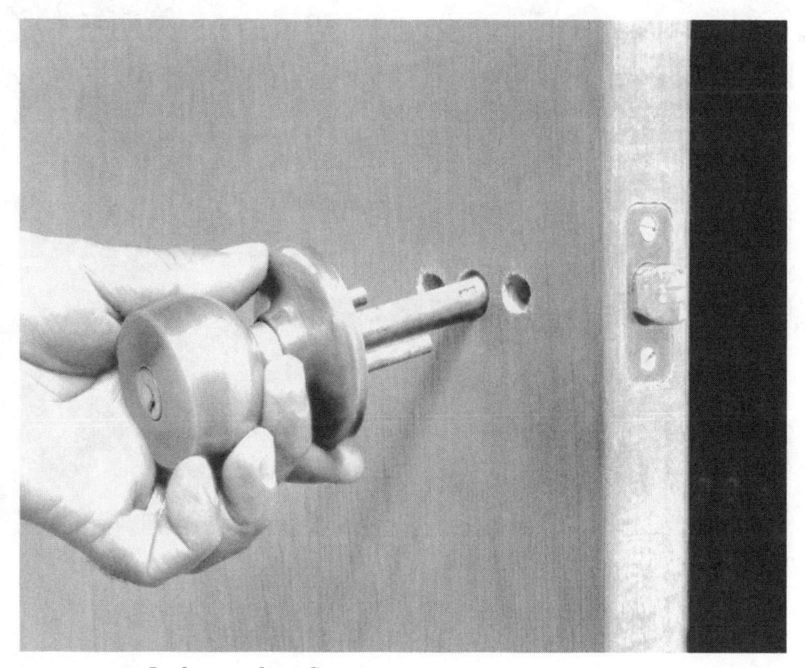

Figure 13.117 Lockset update, Step 1.

Figure 13.118 Lockset update, Step 2.

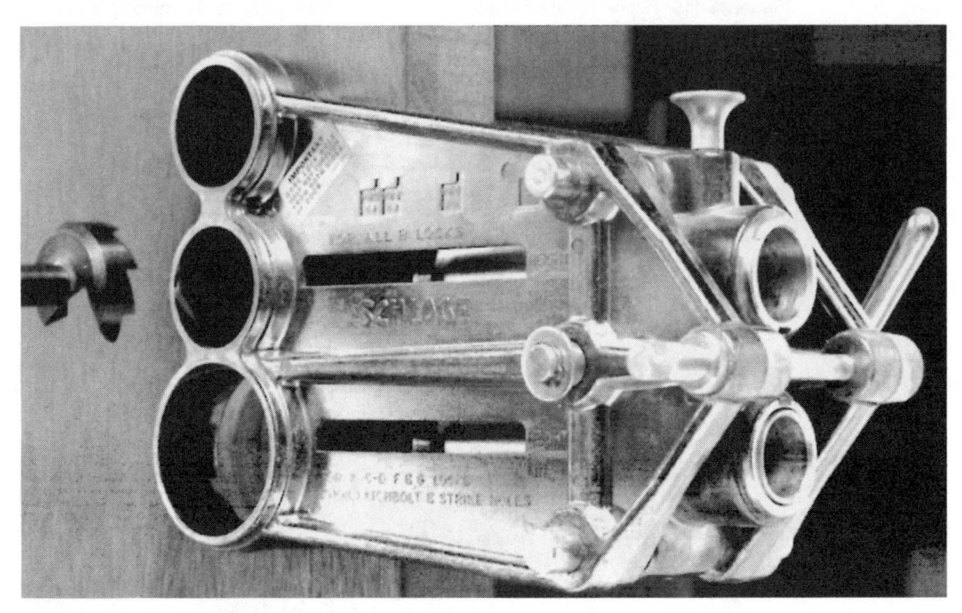

Figure 13.119 Lockset update, Step 3.

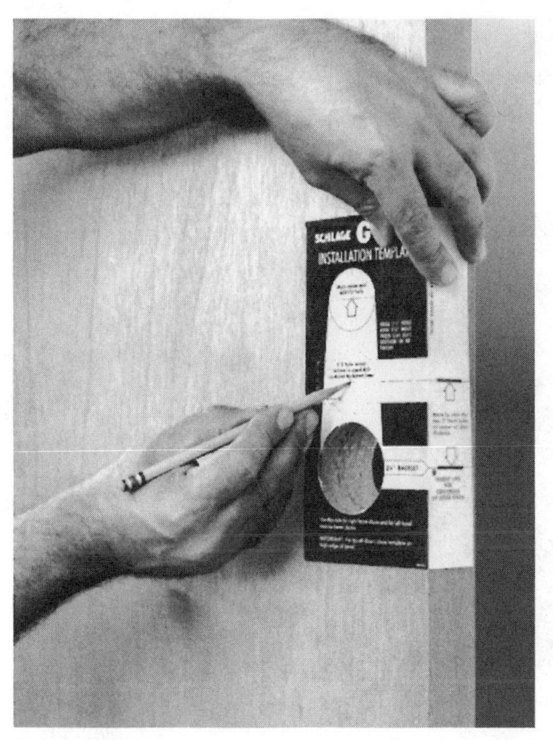

Figure 13.120 Lockset update, Step 4.

Figure 13.121 Lockset update, Step 5.

Figure 13.122 Lockset update, Step 6.

Figure 13.123 Lockset update, Step 7.

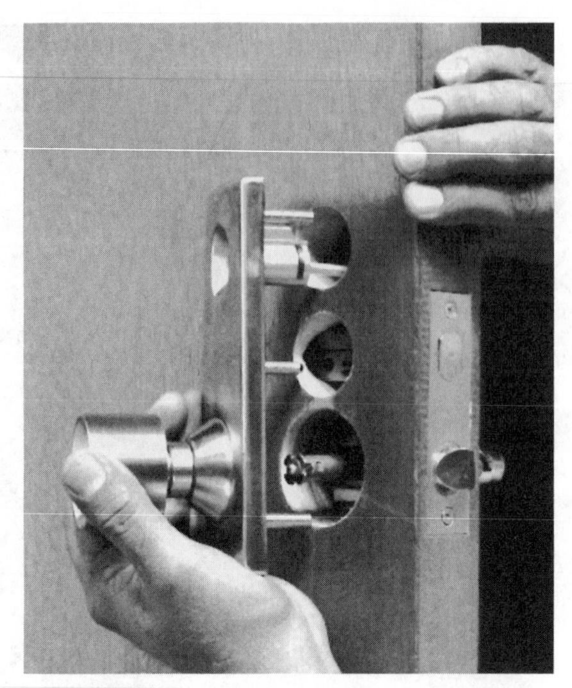

Figure 13.124 Lockset update, Step 8.

Figure 13.125 Lockset update, Step 9.

Figure 13.126 Lockset update, Step 10.

Figure 13.127 Lockset update, Step 11.

Strike plates

Quality locksets are equipped with a deadlocking plunger on the latchbolt. Otherwise, the bolt could be retracted with a knife blade or a strip of celluloid. Nevertheless, it is important to leave very little space between the door edge and the strike plate. A bolt with a ½-inch throw should have no more than ⅛ inch visible between the door and jamb. If necessary, mount the strike plate over a steel spacer. This moves the strike plate closer to the bolt.

The length of the lip is also critical, since a short lip will increase wear on the latchbolt and may frustrate the automatic door-close mechanism, leaving the door unlatched. Mortise strike plates are mounted on the same vertical centerline as the bolt (or should be). Measure the distance from the centerline of the latchbolt to the edge of the jamb and add ⅛ inch for flat strike plates and ¼ inch for curved types. Figure 13.128 illustrates a collection of Schlage strike plates.

Entrance-door security

Security is a function of the door, cylinder, bolt, and strike plate.

- There is *no* security with a hollow wooden door.
- Masterkeying should be kept as simple as possible. Each split pin increases the odds in favor of the lock picker.
- Resist demands for extensive cross keying. These demands originate with customers who insist on single-key performance for executive keys.
- Security begins with proper key-control procedures.
- It is wise to supply extra change-key blanks during the installation phase of the system. Without readily available blanks, the convenient thing to do would be to cut duplicate change keys from master key blanks. Security could be compromised since the change key could fit other locks.
- Double-cylinder locks should be used wherever possible.
- Automatic deadlatches should be specified for all locksets without a deadbolt function. Otherwise the bolt could be retracted by *loiding*, that is, by slipping a flat object between the bolt and the strike.
- Use the longest bolts available to frustrate attempts to gain entry by spreading the door frame. A strike plate cover can also have the same function.
- Shield the bolts with armored inserts.
- Reinforced strike plate mounts, particularly with metal door frames, add security by increasing the area of contact between the strike plate and frame.

Sentry Door Lock Guards

Integral to improved lock security for home, apartment, or business doors is the choice of the door strike. All locks come with a strike that is especially con-

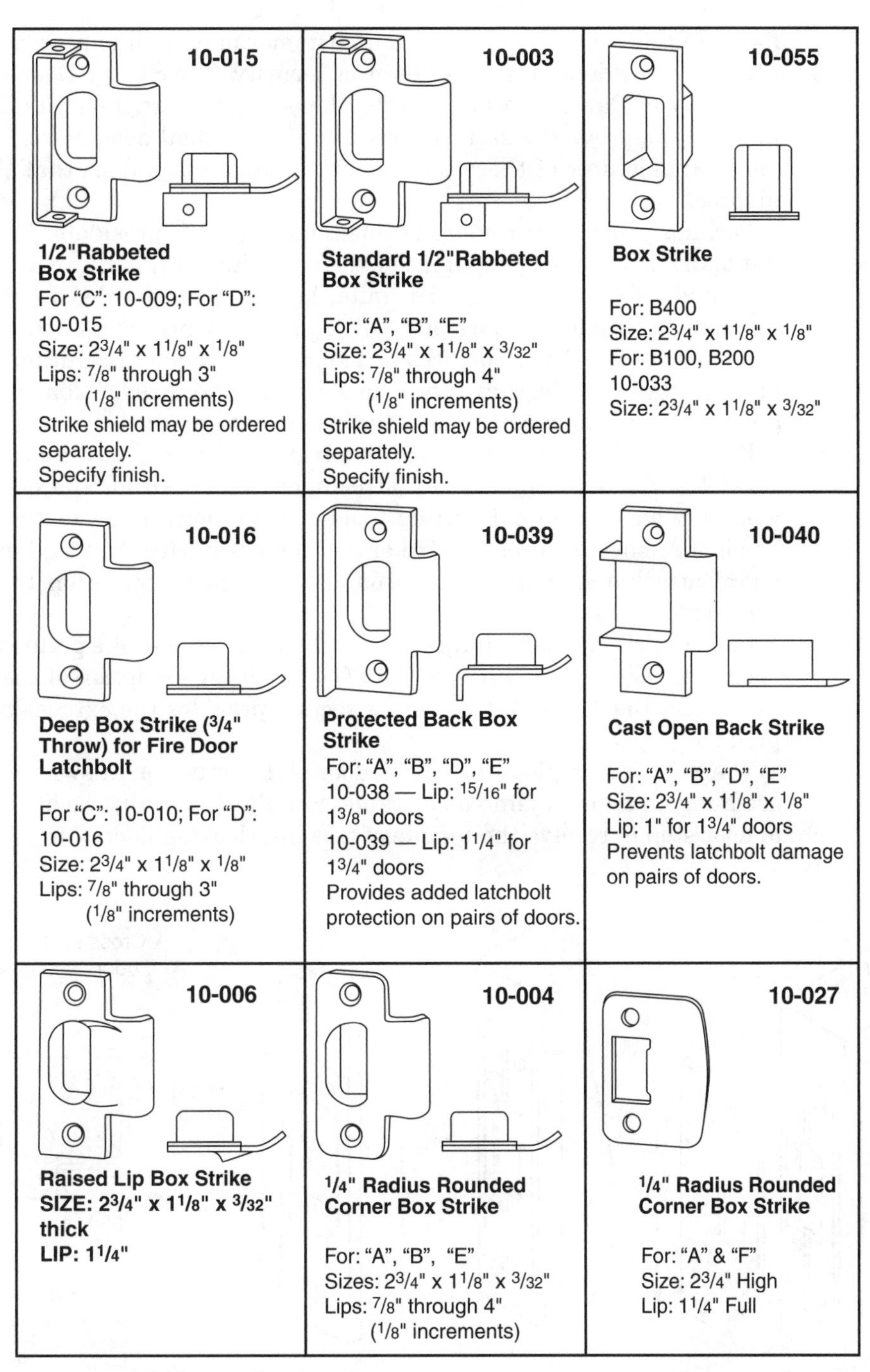

10-015
1/2"Rabbeted Box Strike
For "C": 10-009; For "D": 10-015
Size: 2 3/4" x 1 1/8" x 1/8"
Lips: 7/8" through 3" (1/8" increments)
Strike shield may be ordered separately.
Specify finish.
10-003
Standard 1/2"Rabbeted Box Strike
For: "A", "B", "E"
Size: 2 3/4" x 1 1/8" x 3/32"
Lips: 7/8" through 4" (1/8" increments)
Strike shield may be ordered separately.
Specify finish.
10-055
Box Strike
For: B400
Size: 2 3/4" x 1 1/8" x 1/8"
For: B100, B200
10-033
Size: 2 3/4" x 1 1/8" x 3/32"
10-016
Deep Box Strike (3/4" Throw) for Fire Door Latchbolt
For "C": 10-010; For "D": 10-016
Size: 2 3/4" x 1 1/8" x 1/8"
Lips: 7/8" through 3" (1/8" increments)
10-039
Protected Back Box Strike
For: "A", "B", "D", "E"
10-038 — Lip: 15/16" for 1 3/8" doors
10-039 — Lip: 1 1/4" for 1 3/4" doors
Provides added latchbolt protection on pairs of doors.
10-040
Cast Open Back Strike
For: "A", "B", "D", "E"
Size: 2 3/4" x 1 1/8" x 1/8"
Lip: 1" for 1 3/4" doors
Prevents latchbolt damage on pairs of doors.
10-006
Raised Lip Box Strike
SIZE: 2 3/4" x 1 1/8" x 3/32" thick
LIP: 1 1/4"
10-004
1/4" Radius Rounded Corner Box Strike
For: "A", "B", "E"
Sizes: 2 3/4" x 1 1/8" x 3/32"
Lips: 7/8" through 4" (1/8" increments)
10-027
1/4" Radius Rounded Corner Box Strike
For: "A" & "F"
Size: 2 3/4" High
Lip: 1 1/4" Full

Figure 13.128 Schlage strike plates.

figured for the lock, but you or your customer may wish to have a better strike that will provide a greater amount of security. The Sentry Door Lock Guards Company, of Dania, Florida, has developed a line of strike units that will dramatically improve the security posture of individual door units. The designer and manufacturer of these unique lock guards guarantees that they are jimmyproof.

Perhaps you've never heard of this company or its president, Frank Sushan, but many people and companies have. As the inventor of these unique, impressive, and effective Lock Guard units, he has been profiled in several major newspaper articles. Numerous police departments across the nation now demonstrate the Sentry devices as part of local crime prevention lectures and demonstrations. Further, the National Crime Prevention Institute has approved and endorsed the installation of these products.

Figures 13.129 through 13.133 illustrate the various lock guards available.

LG-1 is a maximum-security attachment produced for the do-it-yourselfer who may have no metal cutting tools or experience in security applications. You just remove the present striker and replace it with the fitted unit. The lock guard attachment is installed around the latch faceplate on the door when such doors open out.

LG-2 is a single plate that is used with an existing strike plate for doors that open out. LG-2A is identical to the LG-2, with the exception of the closing section under the T. The WDG-1 cross section provides the extrusion view of the product.

The LG-3 is a single interlocking door plate for use on in-swinging doors that are hung on a metal jamb door frame. Simply, a slot must be cut into the metal door stop to receive the tongue of this interlocking door plate. Any door pro-

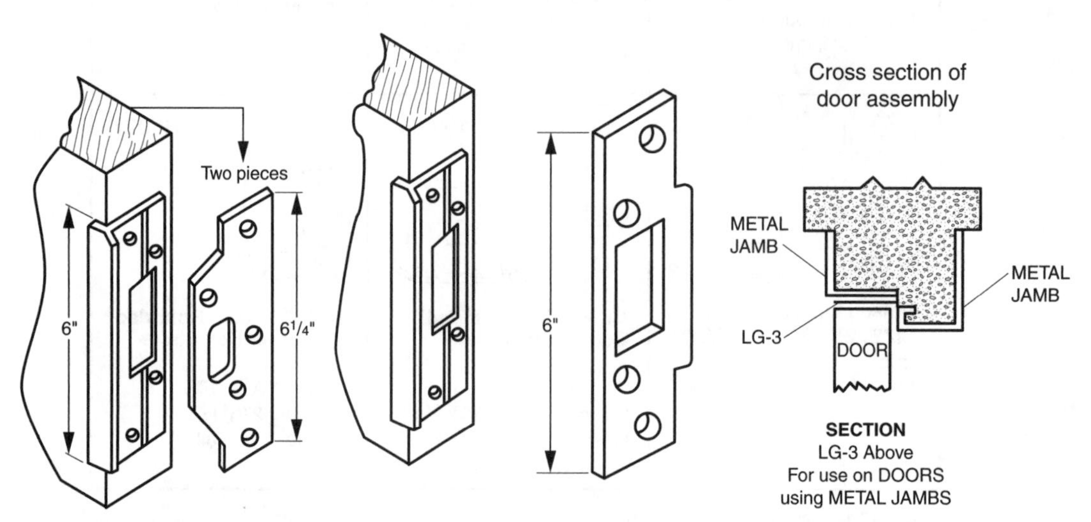

Figure 13.129 The LG-1 and LG-2 units for door units opening outward. (*Sentry Door Lock Guards Co.*)

Figure 13.130 LG-3 unit is a single, interlocking plate for in-swinging doors hung in a metal frame. (*Sentry Door Lock Guards Co.*)

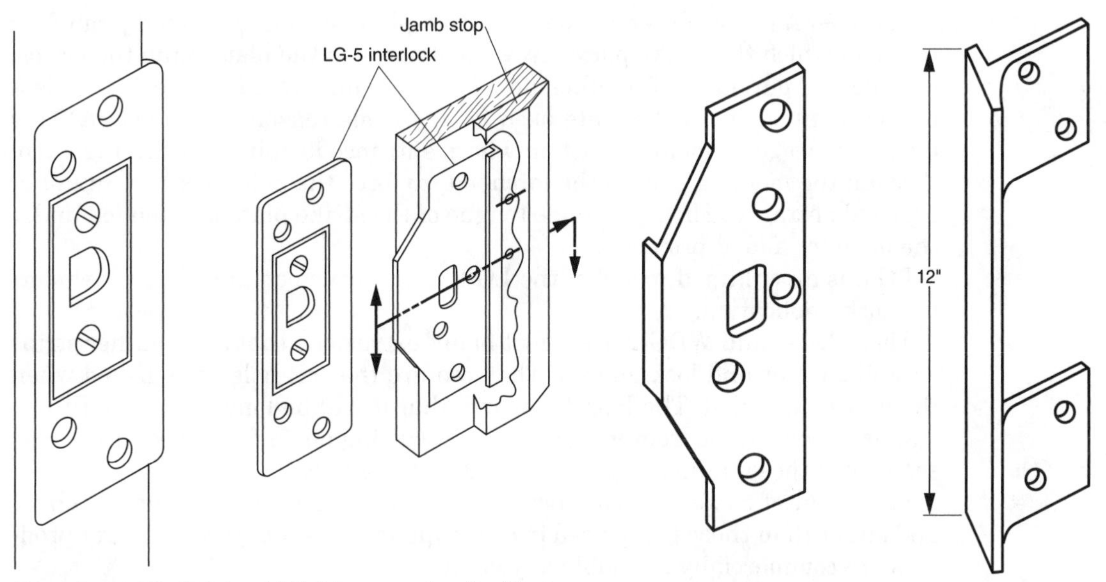

Figure 13.131 The LG-4 and LG-5 door guard units. (*Sentry Door Lock Guards Co.*)

Figure 13.132 LG-5A and LG-6 lock guards. (*Sentry Door Lock Guards Co.*)

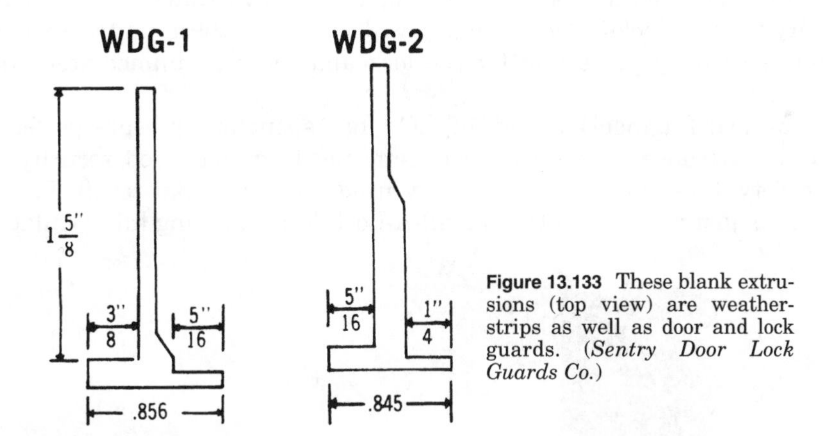

Figure 13.133 These blank extrusions (top view) are weatherstrips as well as door and lock guards. (*Sentry Door Lock Guards Co.*)

tected by this unit does not require locking with a key from the outside after the door is closed. The lock cannot be loided, jimmied, or jacked apart. Figure 13.133 shows a cross section of the installation.

The LG-4 is used similar to the LG-3; when used alone, the difference is that a slot must be made in a wood door jamb to receive the tongue of the unit.

The LG-5 is a two-piece unit for installation on wood doors that open inward only. As indicated for the LG-3 and LG-4, such protected locks cannot be jimmied or jacked open. The doors and jambs will remain together and will not separate.

The LG-5A is a replacement safety strike plate for use on a wood jamb door frame in which the strike plate screws will not hold the plates onto the jambs, making the purchase of another new lock unit unnecessary. The LG-5B is a replacement safety strike plate used for the same reason as the LG-5A. This unit provides additional protection where entrance is gained by driving a tool through the molding behind the latchbolt, to force the bolt back to gain unauthorized entrance. The right angle tongue of the strike plate is extended under the mold for added protection.

LG-6 is an enlarged model of the LG-2 unit, for use on any old or new mortise lock application.

The WDG-1 and WDG-2 are 7-foot blank extrusions that are weather strips as well as door and lock guards. They require that all holes are drilled when the unit is installed. The length ensures that it will be long enough to fit any standard door frame, requiring only slight cutting down for perfect fits. Cross sections of these units are provided in Fig. 13.133.

As a point of customer reference, these lock guard units are much stronger and larger than those being used in other applications today. No stronger products are commercially available anywhere.

M.A.G.'s Install-A-Lock

Lock changing and door reinforcement kits provide an extra measure of security for door locks. For old and new doors alike, these units provide the ultimate security protection for the lock and the area immediately around the locking unit.

M.A.G. Engineering and Mfg. Co. has spent many years perfecting these units. Because of the ever-increasing need for upgraded security in today's society, these products are widely used and beneficial to all. Figure 13.134 shows just one type of the Install-A-Lock lock changing kits available.

Figure 13.134 Install-A-Lock lock changing kit component parts. (*M.A.G. Engineering and Mfg., Inc.*)

These kits can be applied to cylindrical locks, deadlocks, deadbolt and key-in-knob locksets, and mortise locksets (Figs. 13.135 and 13.136). The lock changing kit enables you to install a lock on any door within minutes. It is made of heavy gauge stainless steel or brass, and the assembled unit brings the ultimate protection to new or old doors; it also covers damaged portions on burglarized doors. Even worn, old, mistreated doors can take on a new glow when graced with an Install-A-Lock kit. Designed to cover jimmy marks, splits, and other defacements around old locks, it also provides new strength in the locking area. With the latch plate secured metal-to-metal—*not* wood—the completed installation is extremely solid and secure.

The one-piece construction provides a great savings in labor time. The doors are then protected, after only a few minutes' installation time, right up to the leading edge on both sides.

Figure 13.135 Install-A-Lock with a key-in-knob unit, plus a deadbolt locking unit added for extra security. (*M.A.G. Engineering and Mfg., Inc.*)

Figure 13.136 Install-A-Lock for a mortise lockset. (*M.A.G. Engineering and Mfg., Inc.*)

The security cost is economical and comes in a variety of sizes to meet varying requirements. The cylindrical lock kit will accept locksets with a 2- to 2⅛-inch bore and the deadbolt and key-in-knob allow for two backsets of 2⅜ or 2¾ inches. Simply put, these are the finest remodeling kits available.

Installation instructions

Deadlock or night latch

1. Remove the old lock and all screws from the door.
2. Bore holes for the new lock, if necessary, using the template furnished with the lockset.
3. Chisel out the front edge of the door 1¼ inch wide × 2⅝ inches high × ½ inch deep in the center of the latch hole.
4. Slide the Install-A-Lock unit and position the large hole in the unit to line up with the hole in the door for the lockset.
5. After the unit is in position, install the lock in the door using the 8-32 machine screws that come with the kit. These secure the latch to the unit.
6. Install the four screws and washers to secure the unit to the door. Be sure that the face of the unit is flush with the door face.
7. Install the striker in the door jamb.

For new installations of mortise locks, follow the manufacturer's installation instructions, then follow the instructions below for the Install-A-Lock for mortise locks. For preparing a door in which a mortise lock is currently mounted, follow the instructions below only.

1. Remove the lock from the door.
2. Cut the door edge ⅝ inch deep and 8½ inches high centered on the lock.
3. Install two screw fasteners into the recessed area of the Install-A-Lock with the lip facing down.
4. To determine which side is long, place the unit on a flat surface. The unit will tilt slightly in the direction of the short side. Place the long side on the high edge of the door bevel.
5. Install the lock in the unit with the cylinder at the top.

Sectional trim

1. Select the proper Install-A-Lock insert. Place it between the door and the under side of the Install-A-Lock with the raised area out, fitting it on the large rectangular hole.
2. Install the cylinder and knobs to hold the Install-A-Lock unit in place.

3. Check that the lock matches the strike correctly.
4. Ensure that the Install-A-Lock unit is tight against the door.

Escutcheon trim

1. Place the unit on the door.
2. Install the cylinder and knob assembly to hold the Install-A-Lock unit in place. Do not install the escutcheon retaining screws as yet.
3. Check that the lock matches the strike correctly.
4. Ensure that the unit is tight against the edge of the door.
5. Mark and drill the top mounting hole closest to the edge of the door, using a 13⁄32-inch drill, and install through bolt. Repeat the same procedures on the bottom outside hole closest to the door edge; tighten the bolt securely.
6. Drill and install the remaining three through bolts, installing the center bolt first.
7. For escutcheon, drill and install the escutcheon screws as supplied by the lockset manufacturer.

Double-lock units

1. Remove the lock, latch, and all screws (Fig. 13.137).
2. See Fig. 13.138. Mortise out the front edge of the door (1) 2⅝ inches high × ½-inch deep in the center of the lock hole (2).
3. Using the template furnished with the deadbolt lockset, mark and drill holes (3) and (4) 3⅝ inches from the center of the bottom hole.

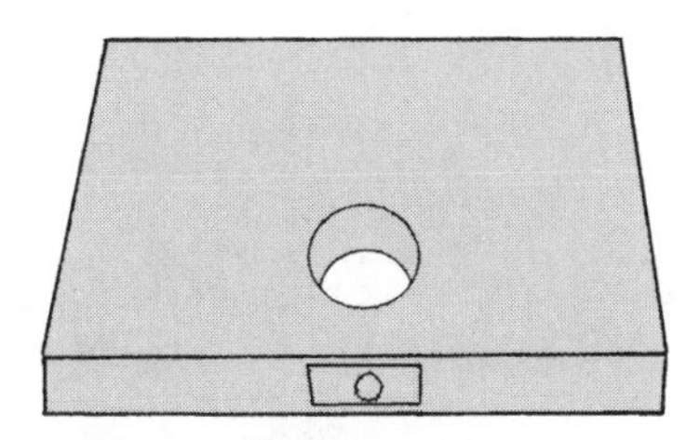

Figure 13.137 Door section with the old lock and latch removed. (*M.A.G. Engineering and Mfg., Inc.*)

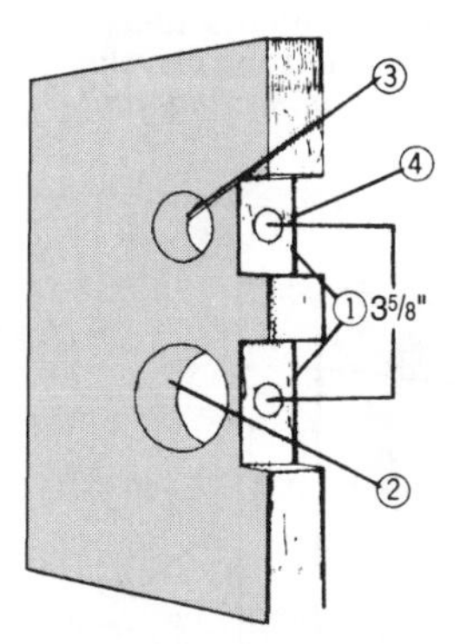

Figure 13.138 Two mortise cuts are required in the door edge in addition to the required holes. (*M.A.G. Engineering and Mfg., Inc.*)

4. Slide the unit (5) over the door and position both holes to line up with the holes in the door (Fig. 13.139).
5. Install both locksets (6) (before installing the mounting screws). Install both striker plates (7) 3⅝ inches between centers. The strike plate installation should be per the specific instructions for the installed lockset (Fig. 13.140).
6. Close the door and check that both the latch and deadbolt (8) enter the strikes (7). If they do not enter, move the Install-A-Lock assembly slightly up or down until the latches enter (Fig. 13.141).
7. Install the mounting screws (9) with finishing washers, being sure that the unit is flush with the door.

Exit Alarm Locks and Panic Bar Deadlocks

Alarmed and deadbolted exit door locks are great concerns of any locksmith's business customers. These units are used to control, monitor, or warn of access

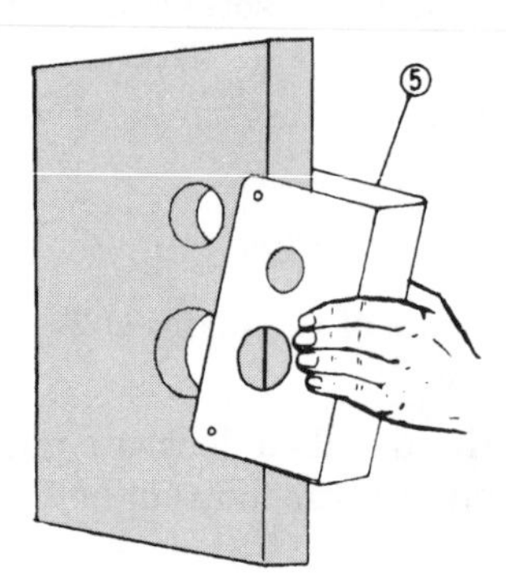

Figure 13.139 Line up the Adjust-A-Lock unit when fitting into the door. (*M.A.G. Engineering and Mfg., Inc.*)

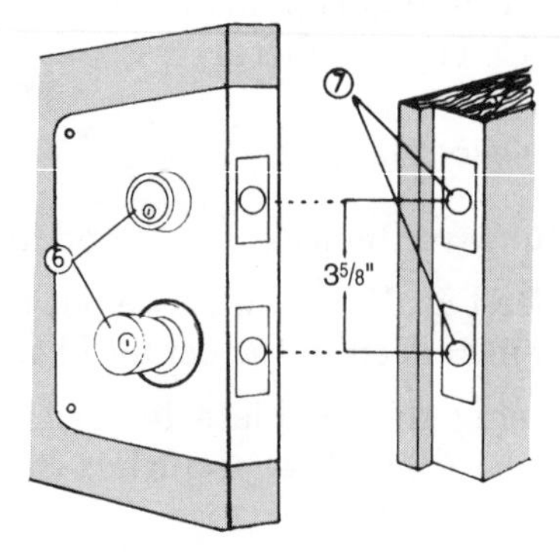

Figure 13.140 Install the locksets and strike plates. (*M.A.G. Engineering and Mfg., Inc.*)

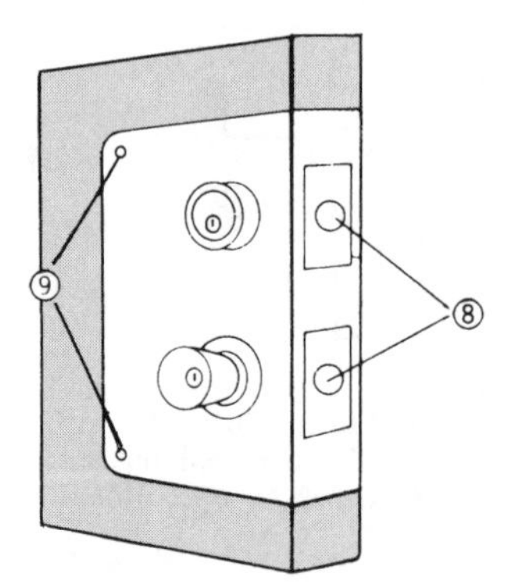

Figure 13.141 Install the strikes and mounting screws to complete the installation. (*M.A.G. Engineering and Mfg., Inc.*)

(either inward or outward) through various interior and exterior doors. One-way (emergency) doors use these locks. For controlled access areas, such as stockrooms and back rooms, these lock alarm units are a great convenience and additional security.

The Alarm Lock Corporation, which specializes in these devices, has a long list of satisfied customers including American industry, the U.S. government, major chain and department stores, large food markets, factories, hospitals, post offices, schools, and small- and medium-size businesses in your own city.

The deadbolt exit alarm lock and the double-deadlock panic bar lock evolved to meet a major problem found at government and private-industry buildings. It was (and still is) illegal to lock or bolt an exit or emergency door because it would endanger life. Also, as long as the doors were unlocked, they became an invitation to unauthorized intrusion from the outside. These intrusions led to the theft every year of many millions of dollars from the buildings that were illegally entered.

The solution to the problem is a locking device that allows for emergency exit, sounds an alarm when the door is opened from either the inside or out, and ensures that nobody can enter the area unchallenged. The Safetygard 11 deadbolt alarmed security lock (Fig. 13.142) and the Safegard 70 panic bar alarm deadlock (Fig. 13.143) are only two of the products that were developed to meet the pressing needs of government and private business.

Safetygard 11 Deadbolt Alarmed Security Lock

The safety alarm lock provides secure, full-time locking of emergency exit doors while complying with standard safety, fire, and local building codes. It provides maximum day and night security and a powerful deterrent against pilferage—from within or without—and always remains panicproof.

The safety alarm lock provides instant automatic exit in case of an emergency. It combines the best features of panic devices, plus a deadbolt and a

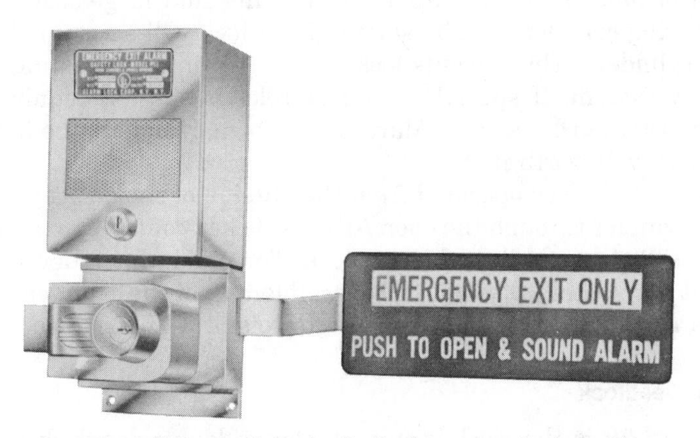

Figure 13.142 Safetygard 11 deadbolt alarm security lock. (*Alarm Lock Corporation*)

Figure 13.143 Safegard 70 panic bar alarm deadlock. (*Alarm Lock Corporation*)

warning system. The first person reaching the door applies a slight pressure to the emergency clapper plate, which automatically releases the lock, opens the door, and sounds the alarm that the door is open.

Alarm lock operation. The complete lock mechanism retracts from the edge of the door when the clapper arm depresses, which withdraws the deadbolt and sounds the powerful twin horn alarm. The alarm continues until the lock is reset to the original position with the appropriate key. The alarm can only be bypassed by an individual authorized to carry and use the bypass key. The mounting plate is already predrilled when the unit arrives from the factory, so it is easily mounted on all styles of doors. The lock is furnace mounted, and it requires no mortising.

Keying. The safety alarm lock unit arrives without the cylinders. (You can sell the cylinder along with the unit, possibly keying the cylinder to meet a certain level of master key operation in medium- and large-size business and manufacturing complexes.) The safety alarm lock will accommodate all standard rim cylinders. This permits lock integration into an existing or proposed master key system. If special keying or rekeying is not required, the unit should be factory ordered with Alarm Lock's own cylinders; the increased cost is slight, but well worth it.

The lock may be key operated from the outside of the door by mounting an extra rim cylinder through the door. An outside key control cam and cover tamper alarm switch are built into every lock. The outside cylinder, though, must have a flat horizontal tailpiece. The overall length of the cylinder and tailpiece must be a minimum of 3½ inches in length to suit a 1¾-inch door.

Safegard 70 Panic Alarm Deadlock

The Safegard 70 is the world's only available double-deadlock panic lock. It provides for antipilferage alarm security from within and deadlock security against intrusion from the outside; also, it meets all requirements as an emergency exit device.

With deadbolt locking, the Safegard 70 is designed to deter the unauthorized use of the emergency and fire exit doors. Instant emergency panic exiting is provided, though, as authorized by law. An added advantage is that a remote electrical control for this unit is available.

The unit can be operated in four different modes:

Maximum security locking. Once the door is closed, the key projects the bolt, firmly securing the door by deadbolting it. The crossbar must be depressed to release the deadbolt and deadlatch and open the door. When the crossbar depresses, the alarm activates and remains so until the door is closed and relocked by the authorized relocking key.

Outside key and pull access. By adding an outside cylinder and door pull, you have another mode of access. The key retracts the deadbolt and the pull releases the deadlatch to open the door. Entry remains unrestricted from both sides until the deadbolt is relocked by the key from either the inside or the outside.

Outside key only access. The key must retract the deadbolt and release the deadlatch momentarily to open the door. The door relatches with each closing; the key is required for each re-entry. To relock the door and rearm the alarm, the deadbolt must be projected by use of the key.

Bar dogging function. This provides for free passage from both sides of the door.

The Safegard 70 dimensions are shown in Fig. 13.144. Modes of operation for single doors, pairs of doors, and pairs of doors with center mullions are shown in Fig. 13.145.

Narrow stile glass doors

Narrow stile glass doors have evolved the need for a more secure type of locking device than is used in other applications. The architectural trend for such doors will continue for a number of years to come. As an individual concerned with security, you must be extremely aware of what is available to meet a variety of locksmithing situations, both in the home and the general business environment.

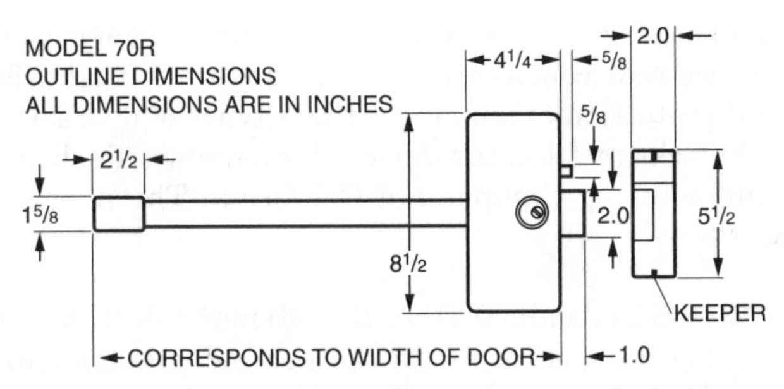

Figure 13.144 Safegard 70 specifications. (*Alarm Lock Corporation*)

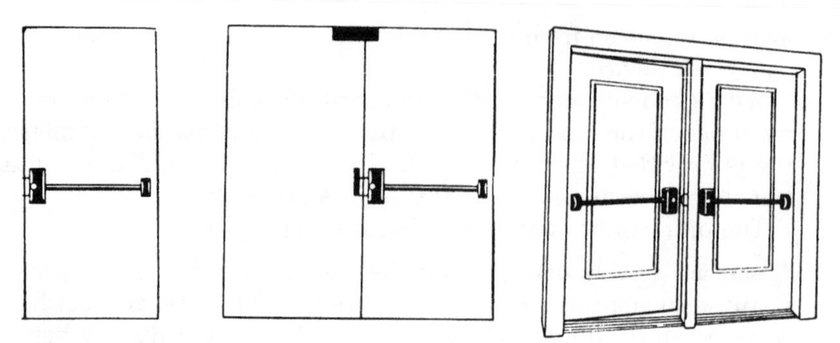

Figure 13.145 Modes of door operation with the Safegard 70 installed. (*Alarm Lock Corporation*)

Initially, realize that because of this architectural trend, the aluminum used in narrow stile glass doors has shrunk in width and material distance from that of a wooden door with a large piece of glass mounted in it. Because the aluminum is narrower and less thick, it naturally becomes slightly more flexible. At the same time, of course, the space to house the lock also becomes smaller. It is impossible to retract a long bolt in the conventional manner because there is no place for it to go. The conventional lockbolt in the aluminum frame of a glass door has a ½- to ⅝-inch throw with an effective penetration of the strike, in some instances, of as little as ¼ inch. This door can be opened by just about anyone with a couple of screwdrivers or, if the frame is weak or worn, with a well-directed frontal attack by a moving body (the intruder). This highly attractive situation means that the lock installed must be able to stop both the hammer and bar and general forced entry.

To an extent, maximum-security considerations for sliding glass doors must also be considered. These doors have spawned misconceptions, some fostered by locksmiths who really don't know what they are talking about because they haven't studied the doors, their locks, and the other applications.

This confusion has resulted in some of these doors being installed still lacking the most basic elements of protection that any door should have. Chief among these is a *real* lock—a secure lock that can be operated by a key if desired and one whose key is compatible with the system of other locks in the same building.

Convenience and security are both necessities to the owner/user of the lock, and this is where we now consider the specialized locks that offer the maximum security and protection available for these types of doors. Only one company specializes in locks and locking devices for narrow stile doors; that is the Adams Rite Manufacturing Company of California. The next several pages cover their products.

Deadlocks. The MS1850A deadlock (Fig. 13.146) is the first in a series of specialized locks developed especially with the narrow stile door in mind. This lock uses a high bolt of laminated steel nearly 3 inches long and actuated by an uncomplicated pivot mechanism. This lock has made the Maximum Security Deadlock the standard for the entire narrow stile door industry. The length of the bolt provides the maximum security for any single-leaf door, even a tall and flexible one. It is also excellent for installation applications where the gap between the door and the jamb is greater than it should be.

The lock case measures 1 × 5³⁄₁₆ inches × the desired depth. The case depth varies with the backset (Fig. 13.147). The cylinder backset can be ⅞, ³¹⁄₂₂, 1⅛, or 1½ inch. Remember that in measuring the backset, it is from the centerline of the faceplate to the centerline of the cylinder.

The bolt is ⅝ × 1⅜ × 2⅞ inches with a 1⅜-inch throw. It is constructed of a triply laminated steel, the center ply having an Alumina-Ceramic core to defeat any hacksaw attack, including rod-type "super" hacksaws.

The faceplate for the MS1850A comes in five variations, as illustrated in Fig. 13.148. Figure 13.149 shows the dimensions (in inches and millimeters), illustrates the stile preparation for the lock, lock and cylinder installation, and information about the cylinder cam.

Latch locks. The MS+ 1890 latch/lock (Fig. 13.150) is a deadlocked Maximum Security unit offering the highest type of protection after-hours for the businessperson or homeowner, in addition to traffic control convenience for business management during the business day.

A typical installation in a bank or a store could require any of three modes of door control:

- Both the lock and latchbolts retracted for unrestricted entry and exit during business hours
- Handle-operated latch for exit-only traffic
- Maximum security hookbolt for overnight lockup

Figure 13.146 The MS1850A deadlock for narrow stile doors. (*Adams Rite Mfg. Co.*)

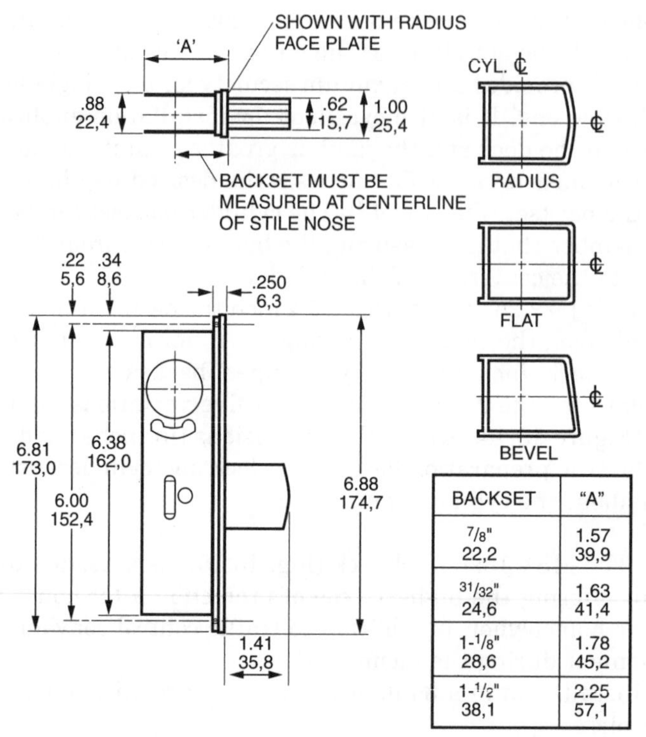

Figure 13.147 Critical lock dimensions for the SM1850A deadlock. (*Adams Rite Mfg. Co.*)

FLAT: MS1850A	
RADIUS: MS1851A	
RADIUS WITH WEATHERSTRIP: MS1851AW	
L. H. BEVEL: MS1852A(LH)	
R. H. BEVEL: MS1852A(RH)	

Figure 13.148 MS1850A face plate variations. (*Adams Rite Mfg. Co.*)

This has an operational advantage not necessarily provided by other locks that may seem similar. It requires a 360° turn of the key to throw or retract the hook-shaped deadbolt. When unlocked, turning the key 120° farther will also retract the latchbolt. When using the door handle that can also be obtained for the lock, the handle retracts the spring-loaded latchbolt only.

Figure 13.151 shows the overall dimensions of the lockset. Figure 13.152 shows the stile preparation, strike location and dimensions necessary, and the latch/lock, cylinders, and paddle installation. Latchbolt holdback information is in Fig. 13.153.

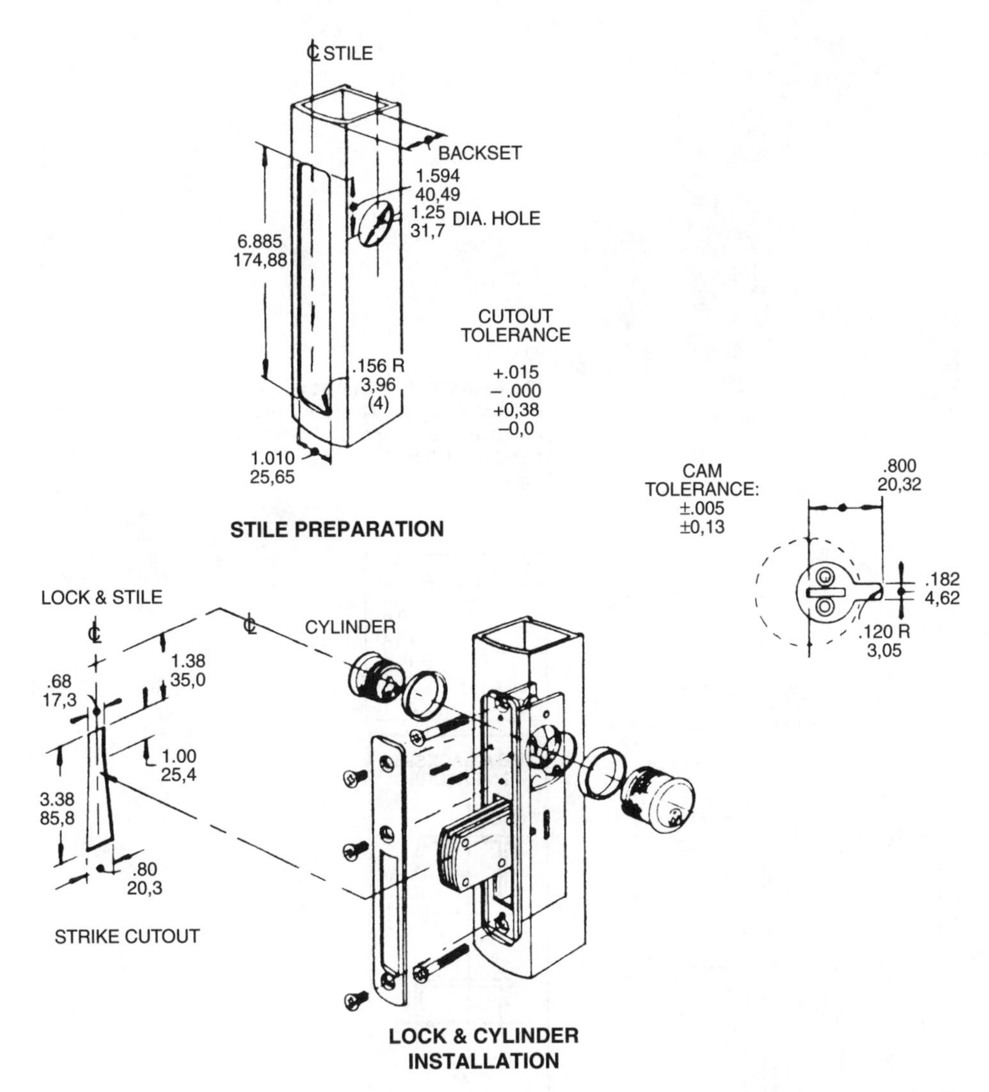

Figure 13.149 Lock and cylinder preparation, stile preparation, and critical cam dimensions for the MS1850A. (*Adams Rite Mfg. Co.*)

Threshold bolts. Along with a lock installation, you might also be required to install a corresponding threshold bolt mechanism in a narrow stile door (Fig. 13.154). The use of the threshold bolt allows maximum security for pairs of doors with the turn of a single operating key. With the simultaneous dropping of a hardened steel hexbolt into the threshold and pivoting of the doors' stile, the two-point lock secures the entire double-door entranceway. The threshold bolt is harnessed to the rear of the pivoted bolt. The standard threshold bolt rod is sufficient for a cylinder height up to 53⅞₁₇ inches maximum. It is fully threaded and can be cut off to allow for low cylinder heights.

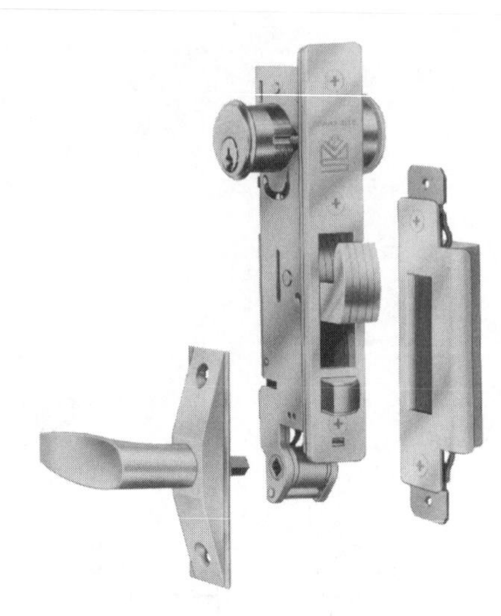

Figure 13.150 MS1890 Maximum Security Latch/Lock unit, with an armored strike and a lever handle. (*Adams Rite Mfg. Co.*)

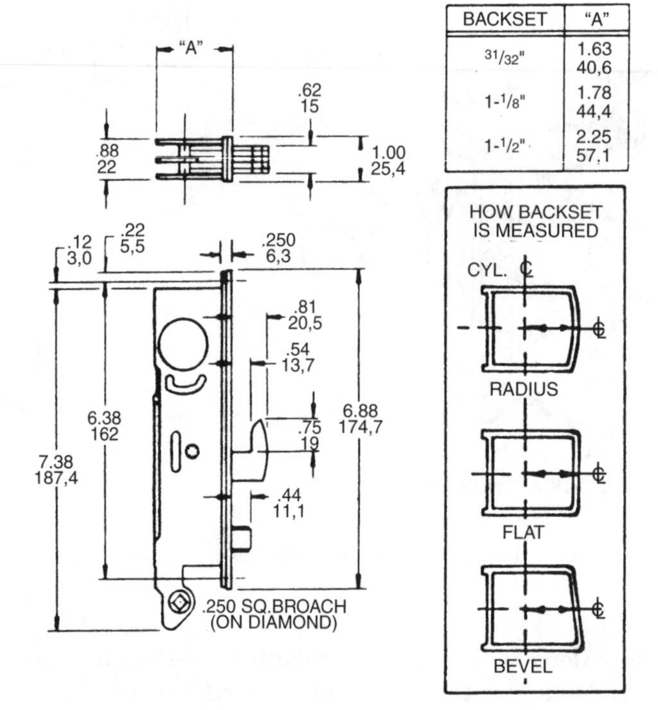

BACKSET	"A"
31/32"	1.63 40,6
1-1/8"	1.78 44,4
1-1/2"	2.25 57,1

Figure 13.151 Latch/lock specifications. (*Adams Rite Mfg. Co.*)

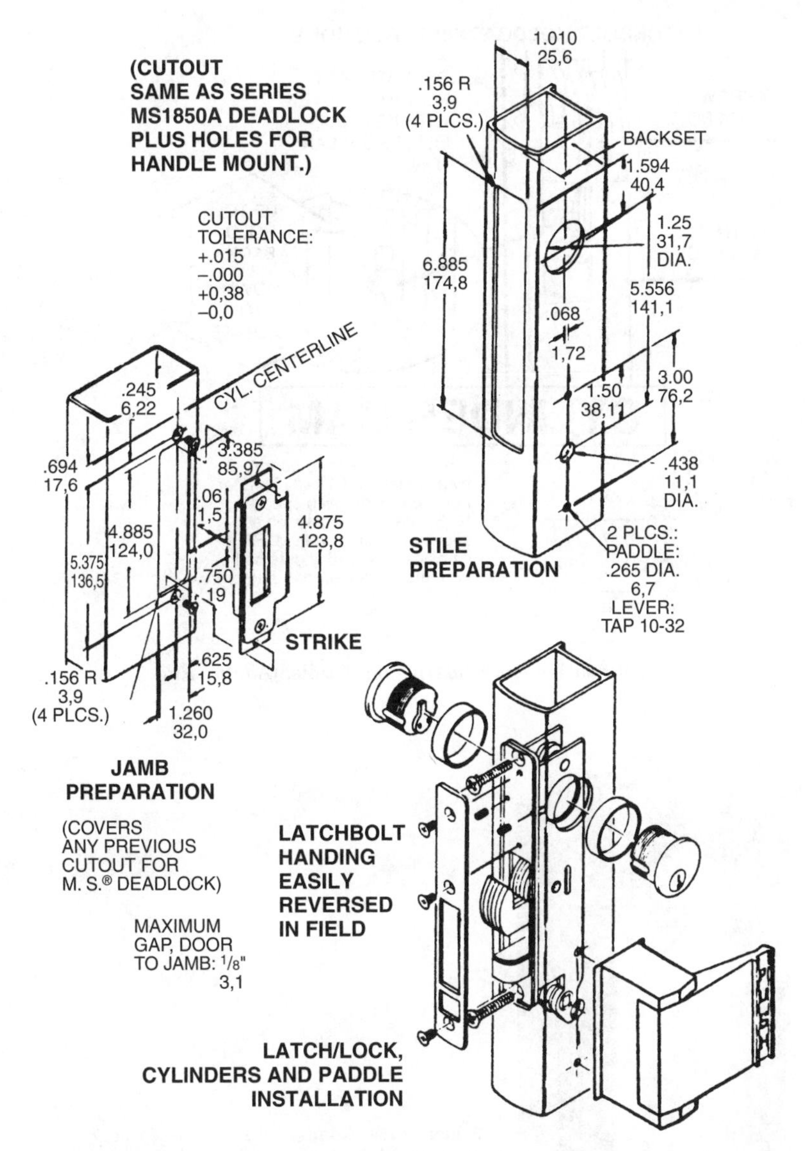

Figure 13.152 Metal stile jamb and paddle/cylinder installation procedure in an exploded view. (*Adams Rite Mfg. Co.*)

A full turn of the key (or a thumbturn if one is installed on the interior side of the door) throws the counterbalanced bolt into the opposite door and the drop bolt into the threshold. The key can only be removed when the bolts are in a positive locked or unlocked position—not halfway. Figure 13.155 shows the dimensions, Fig. 13.156 demonstrates the installation of the threshold bolt, and Fig. 13.157 illustrates the header bolt installation.

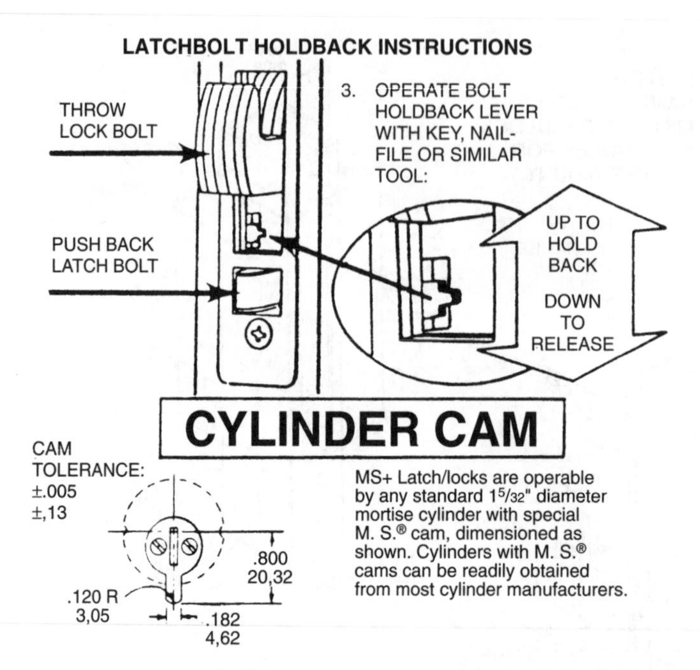
LATCHBOLT HOLDBACK INSTRUCTIONS
THROW LOCK BOLT
PUSH BACK LATCH BOLT
3. OPERATE BOLT HOLDBACK LEVER WITH KEY, NAIL-FILE OR SIMILAR TOOL:
UP TO HOLD BACK
DOWN TO RELEASE
CYLINDER CAM
CAM TOLERANCE:
±.005
±,13
.800
20,32
.120 R
3,05
.182
4,62
MS+ Latch/locks are operable by any standard 1 5/32" diameter mortise cylinder with special M. S.® cam, dimensioned as shown. Cylinders with M. S.® cams can be readily obtained from most cylinder manufacturers.

Figure 13.153 Latch bolt holdback instruction details. (*Adams Rite Mfg. Co.*)

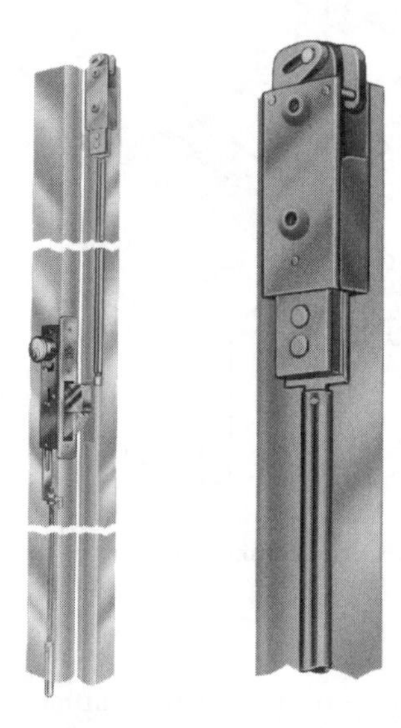

Figure 13.154 Adams Rite threshold bolt (left) and header bolt (right); the unit has the MS1850A maximum security lock installed. (*Adams Rite Mfg. Co.*)

Narrow stile door deadlock strikes. Three basic strikes are available. The majority of Adams Rite deadlocks are installed in metal construction where the strike cutout can simply be a slot. For aesthetic reasons or, in the case of certain locks, for added security, specific strikes should be used.

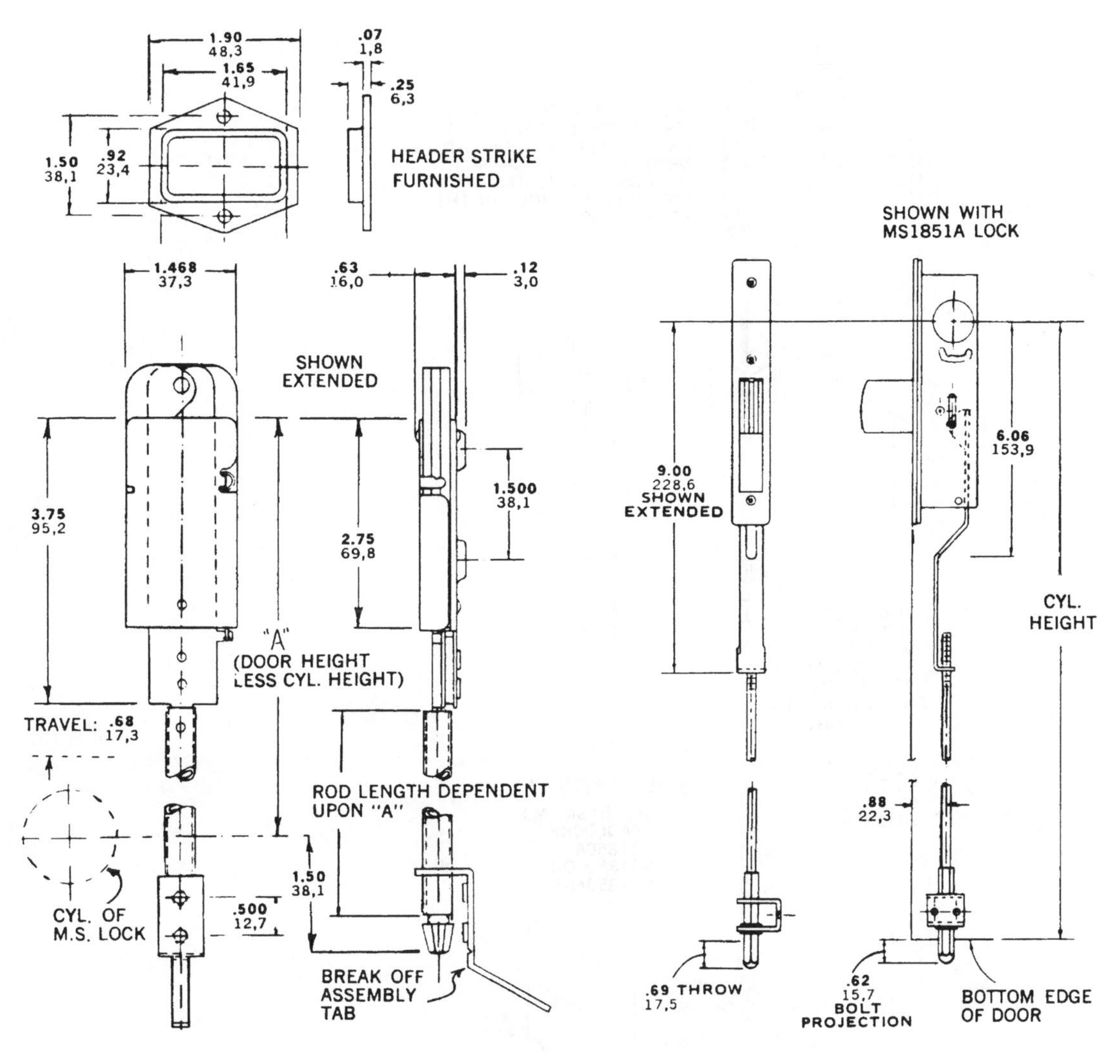

Figure 13.155 Threshold bolt and header critical dimensions (*Adams Rite Mfg. Co.*)

The *trim strike* (Fig. 13.158) is a simple strike plate that can be surface-mounted on a wood or metal jamb or mortised flush. It uses #10 flathead machine screws and/or two #10 wood screws. It is made of aluminum.

The *box strike* has a dust box added (Fig. 13.159). It is customarily used only for wood construction where the dust box prevents chips, sawdust, and other debris from entering the strike.

The *armored strike* (Fig. 13.160) has the basic trim plate, but this is backed up by a massive steel doubler designed to prevent the method of forced entry known as *jamb peeling*. It fits within aluminum or other hollow jamb sections; with a trim face flush, the steel is completely hidden.

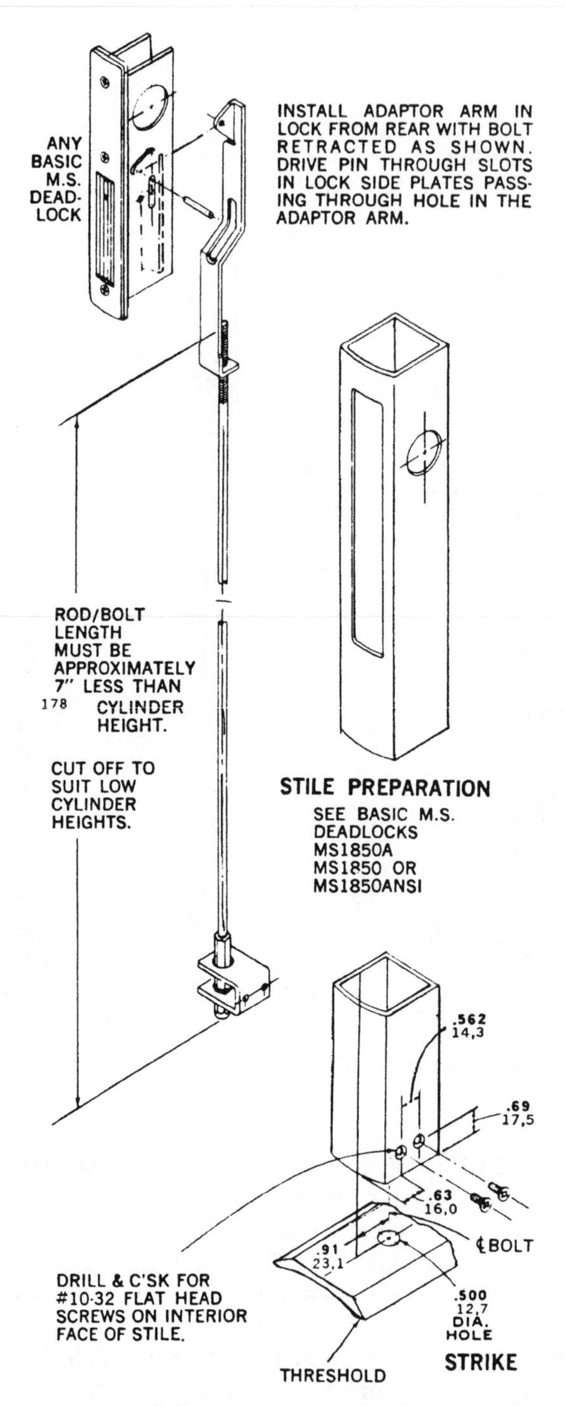
ANY
BASIC
M.S.
DEAD-
LOCK
INSTALL ADAPTOR ARM IN LOCK FROM REAR WITH BOLT RETRACTED AS SHOWN. DRIVE PIN THROUGH SLOTS IN LOCK SIDE PLATES PASSING THROUGH HOLE IN THE ADAPTOR ARM.
ROD/BOLT LENGTH MUST BE APPROXIMATELY 7" LESS THAN CYLINDER HEIGHT.
178
CUT OFF TO SUIT LOW CYLINDER HEIGHTS.
STILE PREPARATION
SEE BASIC M.S. DEADLOCKS MS1850A MS1850 OR MS1850ANSI
.562
14,3
.69
17,5
.63
16,0
.91
23,1
℄BOLT
DRILL & C'SK FOR #10-32 FLAT HEAD SCREWS ON INTERIOR FACE OF STILE.
.500
12,7
DIA.
HOLE
THRESHOLD
STRIKE

Figure 13.156 Stile preparation and installation for the threshold bolt. (*Adams Rite Mfg. Co.*)

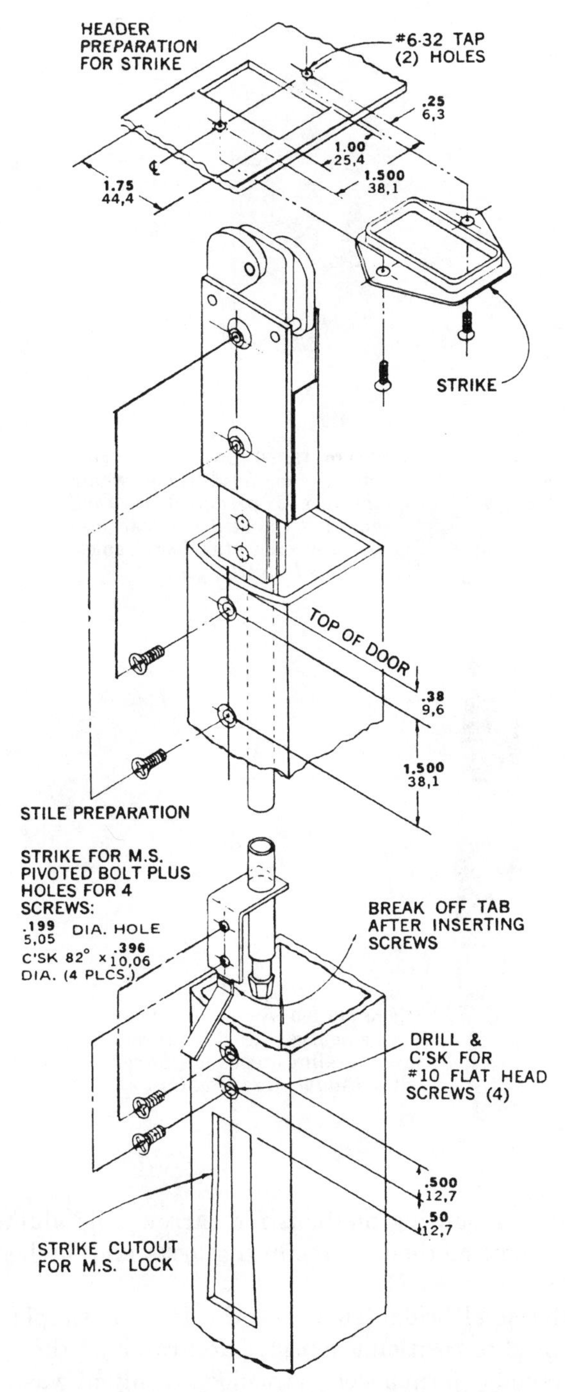

Figure 13.157 Installation details for the header bolt. (*Adams Rite Mfg. Co.*)

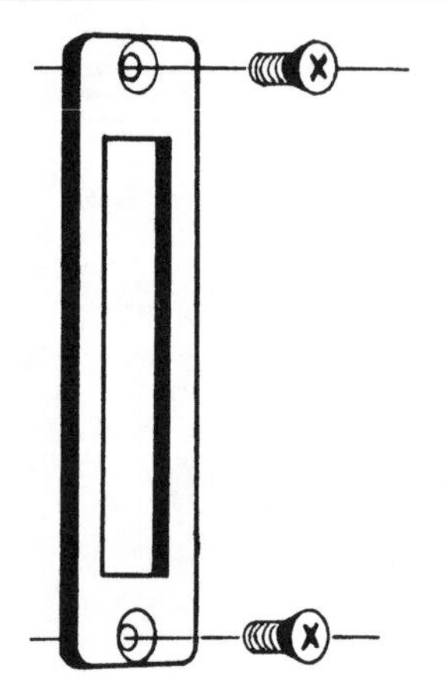

Figure 13.158 Simple trim strike plate for deadlocks in a metal stile door unit. (*Adams Rite Mfg. Co.*)

Figure 13.159 Box strike, customarily for wood frames where the dust box permits debris from entering the strike. It can also be used on metal door frames. (*Adams Rite Mfg. Co.*)

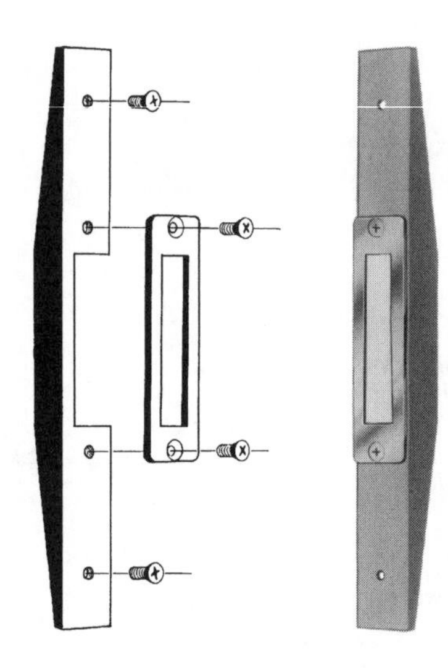

Figure 13.160 Armored strike unit, essentially used to prevent jamb peeling types of forced entry. (*Adams Rite Mfg. Co.*)

Figure 13.161 shows mounting methods for narrow stile aluminum doors. The three most prevalent methods of mounting locks and latches within the hollow tube stile of a door are easy.

Method A uses the steel bridge spanning the stile. A simple handle, the installation tool, is used to position the bridge accurately in the stile while its two screws are tightened to form a web in the door. Resilient washers allow for minor adjustment to fit the lock flush with the door.

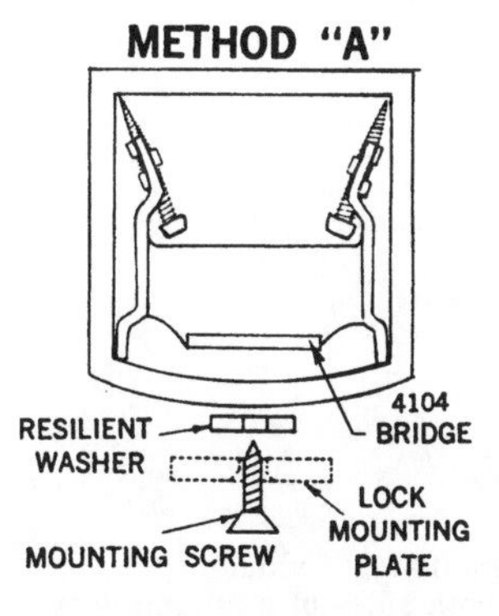

Method "A" spans the stile with a steel bridge at top and bottom of the lock. A simple handle (Adams Rite Installation Tool 4075) is used to position the bridge accurately in the stile while its two screws are tightened to form a "web" in the door. Resilient washers allow for minor adjustment to fit the lock flush in door.

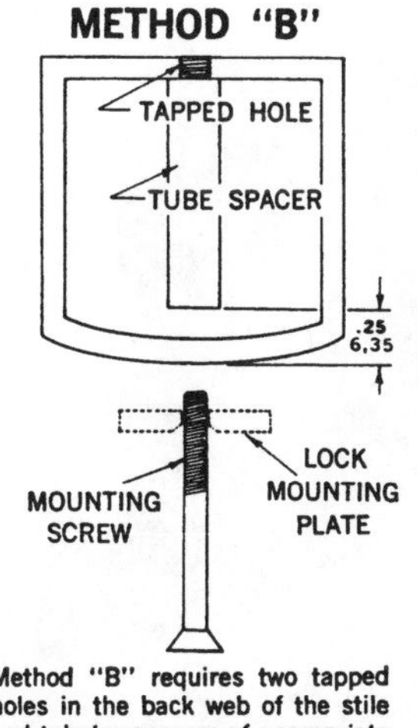

Method "B" requires two tapped holes in the back web of the stile and tubular spacers of appropriate lengths. Some door manufacturers use a heavy coil spring in place of the tube to allow for minor adjustment.

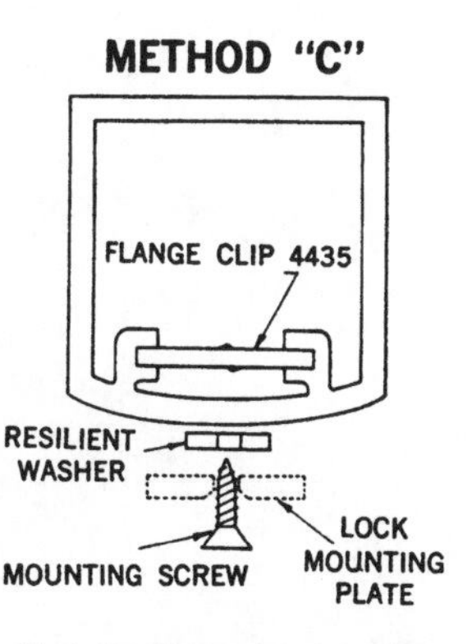

Method "C" is used with doors whose stile extrusion shape contains a special inside channel to accept flange clips on which the lock is mounted. Several door manufacturers have incorporated this channel in their newest designs. It provides a convenient lock mounting means as well as adding overall strength to the stile.

Figure 13.161 Three methods of mounting a lock in the narrow stile doors using the basic component parts of the Adams Rite installation kit. (*Adams Rite Mfg. Co.*)

Method B requires two tapped holes in the back web of the stile and tubular spacers of the appropriate lengths. Some door manufacturers use a heavy coil spring in place of the tube to allow for minor adjustments.

Method C is for doors whose stile extrusion shape contains a special inside channel to accept flange clips, on which the lock is mounted. Several door manufacturers have included this channel in their latest designs, so it is something you must be aware of and look for prior to installing the lock. It provides a convenient lock mounting means and adds overall strength to the stile. *Note:* Refer to the Adams Rite Locksmithing Installation Kit, which discusses some of the specific tools necessary for the three methods of mounting locks.

Cylinder pulls. Cylinder pulls (Fig. 13.162) are designed for certain deadlocks and latches. These surface-mounted pulls allow you to key a sliding glass door similarly to other types of doors that have key-in-knob or mortise cylinder locks installed. Dimensions for the cylinder pull are in Fig. 13.163 and installation instructions shown in Fig. 13.164.

Lock cylinder guards. The standard mortise cylinder is made of brass and is literally the soft spot in the narrow stile door's security. Using special pliers or other leverage devices, a burglar can tear the cylinder out of the door, leaving an opening through which he or she can operate the deadlock by hand. The cylinder guard by Adams Rite addresses this problem (Fig. 13.165). It offers a

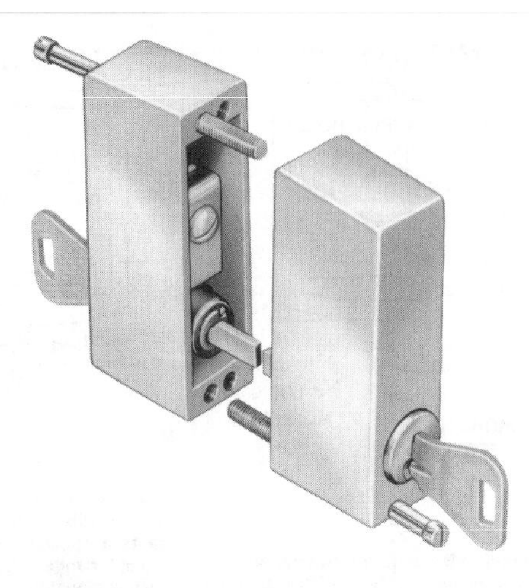

Figure 13.162 Cylinder pull set for both sides of a sliding door. (*Adams Rite Mfg. Co.*)

three-way defense. First, the outer shield ring offers a poor purchase for either prying or twisting. Second, the ring is of hardened steel, so the combination of shape and hardness makes it virtually impossible to grip, even with sharpened tools. Third, in the event that a prying tool, such as a cold chisel, is driven into the stile behind the shield ring, the would-be burglar must pull a heavy steel plate through the round hole in the metal door stile. This degree of leverage is far beyond that available with most hand tools.

The standard package includes the security ring of hardened steel (and free-swiveling when properly installed) and the steel retainer plate to permit the security ring to swivel. The latter is hardened to act as an antidrill shield. The package also contains spacers for flush fitting in thin stile walls.

Figure 13.166 shows the cylinder guard dimensions. Figure 13.167 provides installation information.

Extracting Broken Keys. If a broken key is so far down the cylinder that you can't see it, then use a screwdriver to turn the cylinder to the open position (the position the key usually enters at). Then use a broken key extractor to hook the top of the broken key, and work it back out. See Fig. 13.168. If part of the broken key is protruding from the lock, again turn it to the open position and then use a saw blade type of extractor to work the broken key back out. See Fig. 13.169.

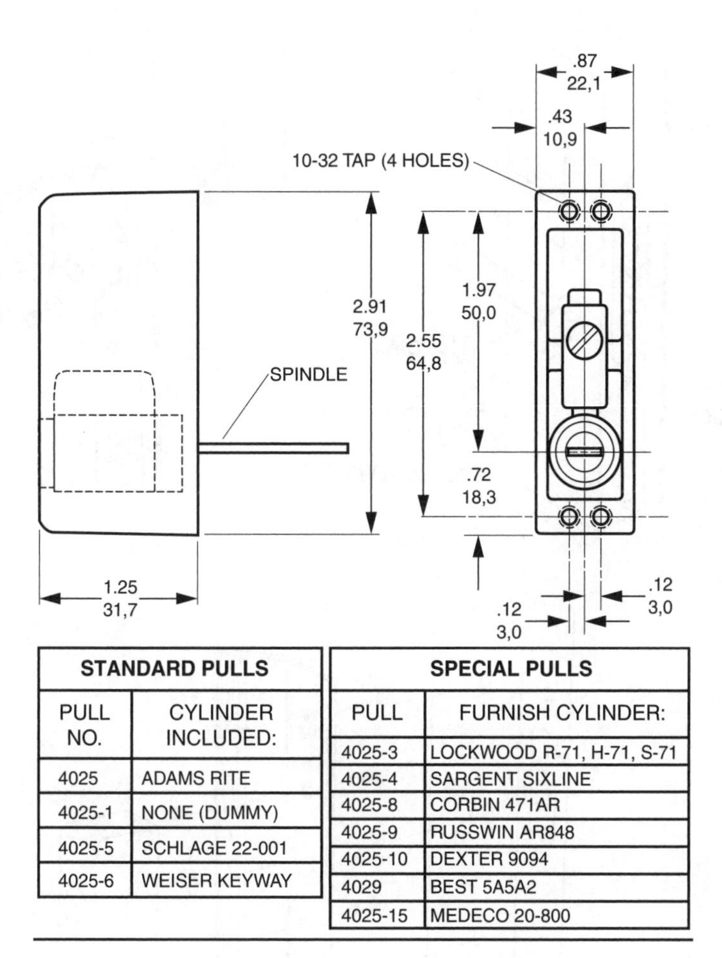

STANDARD PULLS	
PULL NO.	CYLINDER INCLUDED:
4025	ADAMS RITE
4025-1	NONE (DUMMY)
4025-5	SCHLAGE 22-001
4025-6	WEISER KEYWAY

SPECIAL PULLS	
PULL	FURNISH CYLINDER:
4025-3	LOCKWOOD R-71, H-71, S-71
4025-4	SARGENT SIXLINE
4025-8	CORBIN 471AR
4025-9	RUSSWIN AR848
4025-10	DEXTER 9094
4029	BEST 5A5A2
4025-15	MEDECO 20-800

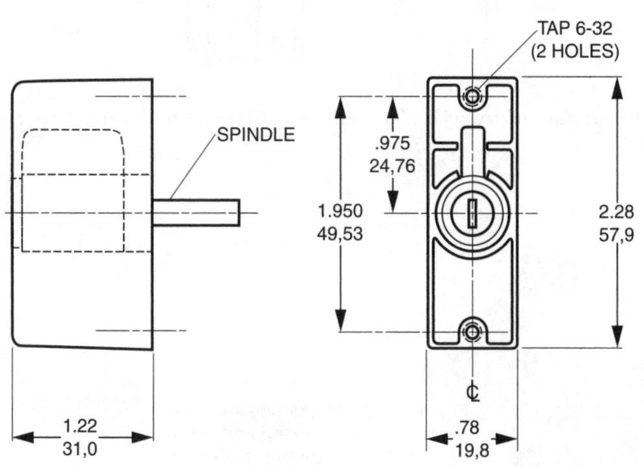

Figure 13.163 Cylinder pull dimensions that can be used with the unit. (*Adams Rite Mfg. Co.*)

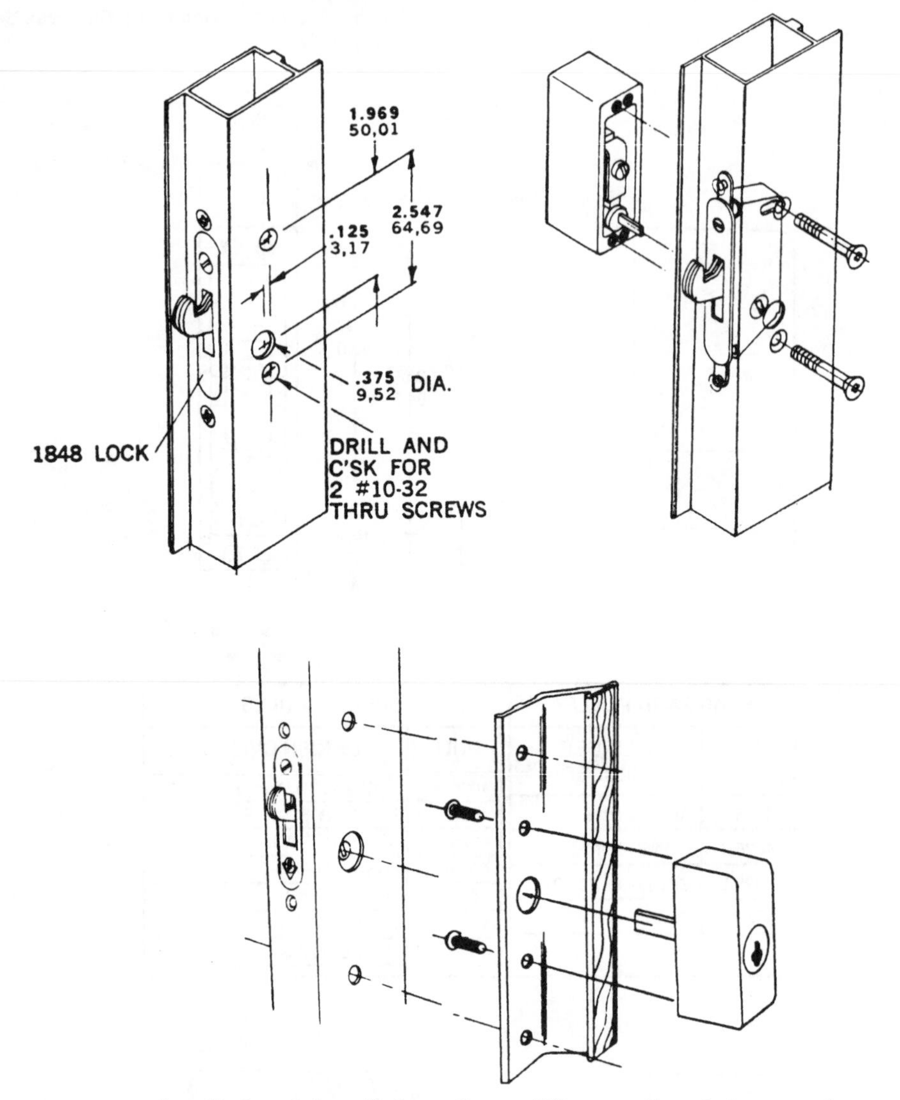

Figure 13.164 Installation of the cylinder pull; two different pull variations are shown. (*Adams Rite Mfg. Co.*)

Figure 13.165 Hardened steel free-swinging security ring and retainer plate to provide protection against direct attack on the locking unit. (*Adams Rite Mfg. Co.*)

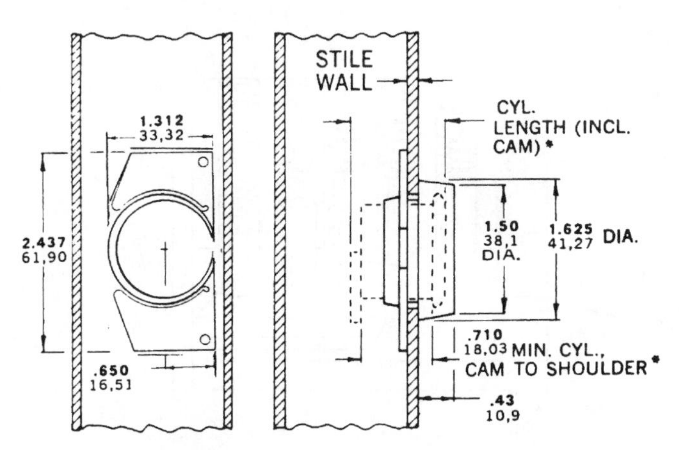

Figure 13.166 Cylinder guard dimensions and positioning when installed. (*Adams Rite Mfg. Co.*)

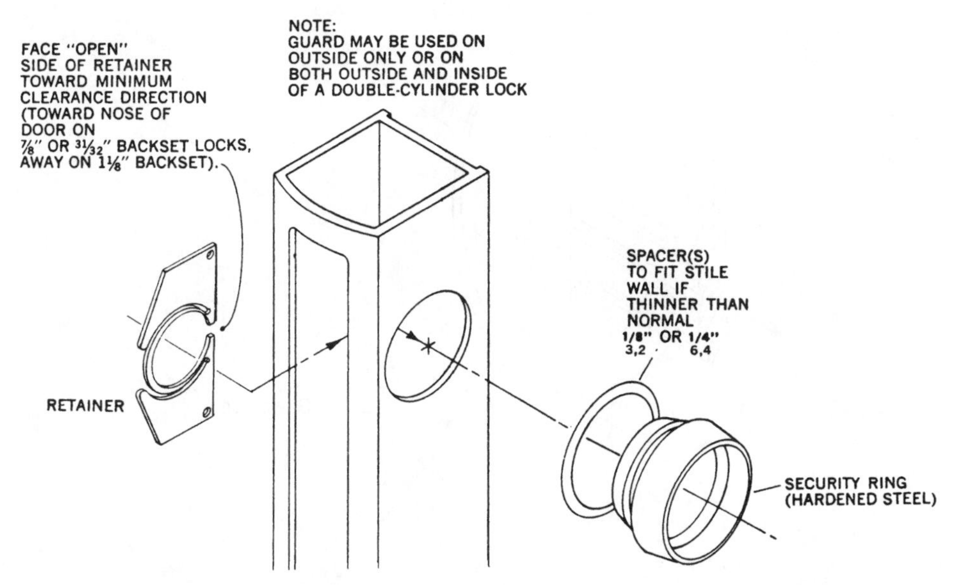

Figure 13.167 Cylinder guard installation. (*Adams Rite Mfg. Co.*)

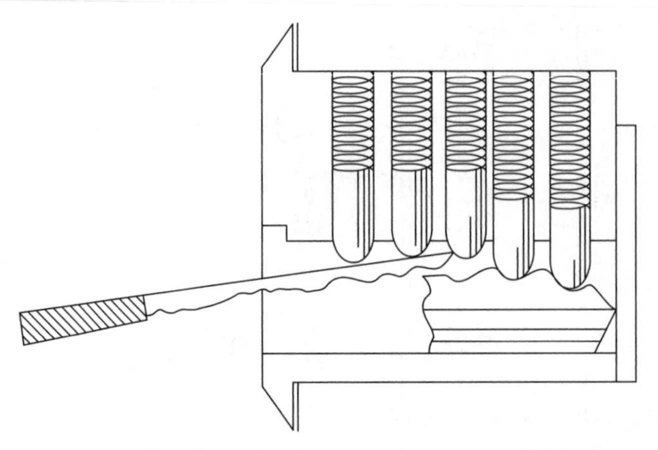

Figure 13.168 Insert a broken key extractor under the bottom pins to hook a broken key.

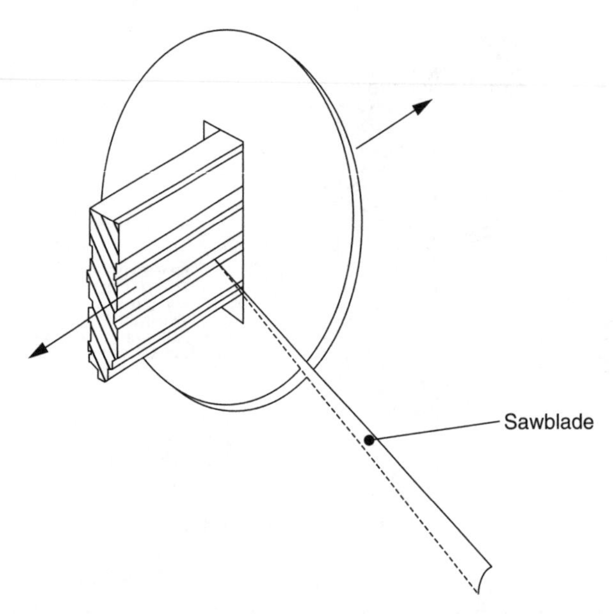

Figure 13.169 If a broken key protrudes from the cylinder, use a sawblade type of broken key extractor.

Chapter

14

Lock Picking and Impressioning

Picking open and impressioning locks quickly are fundamental abilities that every locksmith should have. This chapter gives all the information you need to begin picking and impressioning all types of locks—including high-security models.

Picking Pin Tumbler Locks

With practice you should be able to pick open most standard pin tumbler locks within a few minutes—unless there's a problem with the lock. In theory, any mechanical lock that is operated with a key can be picked. Tools and techniques can be fashioned to simulate the action of any key.

The secret to picking locks fast is to focus on what you're doing and to visualize what's happening in the lock while you're picking it. If you know how pin tumbler locks work, then it's easy to understand the theory behind picking them. If you're not sure how they work, read Chap. 7. Figure 14.1 shows an exploded view of a pin tumbler lock.

Why Pin Tumbler Locks Can Be Picked

The slight spaces and misalignments within a cylinder allow it to be picked. When locks are manufactured, there has to be room in each of the lower and upper pin chambers for the pins to move back and forth freely. The lower pin chambers are those holes along the length of the plug that hold the bottom pins. The upper pin chambers are those corresponding holes inside the cylinder housing that hold the upper pins. To create a locked condition, one or more of the upper pins must fall into the lower pin chambers, or one or more lower pins must be raised partially into the upper pin chambers. When a pin tumbler is lodged between an upper and lower pin chamber, the plug can't be turned. In other words, it's locked.

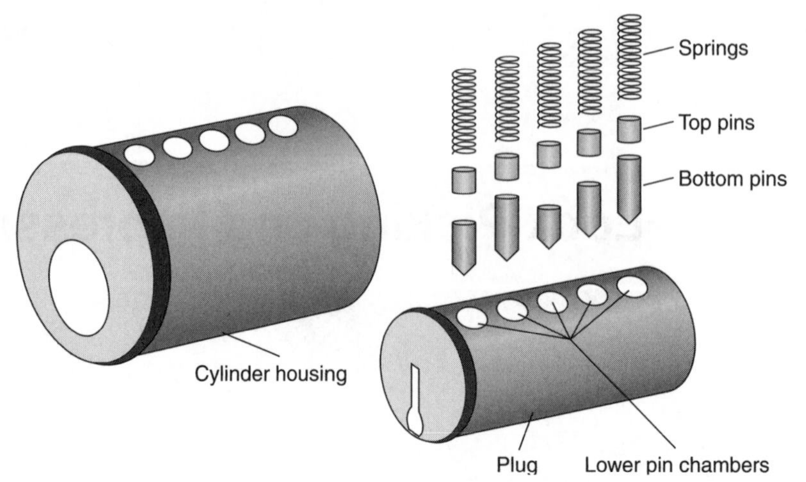

Figure 14.1 In a pin tumbler cylinder there are many parts that all need to move freely; that freedom of movement makes a lock pickable.

Pin chamber holes aren't perfectly aligned, but there has to be enough room in each of the pin chamber holes for the pins to move in and out easily. Likewise, there has to be space between the plug and cylinder housing to allow the plug to be turned to the locked and unlocked positions. (When you're turning the lock with a key, you're turning the plug.) Figure 14.2 shows the plug in an unlocked position. This is when the bottom and top pins meet at the shear line. Lockpicking takes advantage of the necessary spaces and slight misalignments in a lock.

In the locked position, one or more pins are lodged between the upper and lower pin chambers—preventing the plug from turning. The right key has cuts that are positioned to move all the pins to the shear line at once, which frees the plug to be turned. The shear line is the space between the plug and cylinder housing. If all the pins are at the shear line, the plug is free to rotate (Fig. 14.3).

The upper and lower pin chambers aren't drilled in a perfectly straight line across the plug and cylinder. If you look at them closely, you'll see more of a zigzag pattern. Figure 14.4 shows the lower pin chambers aligned along the cylinder plug. In all but the most expensive locks, there is a lot of play for the pins within the cylinder. This is because manufacturing processes aren't perfect.

When you use a pick to lift a bottom pin, you're also lifting the corresponding top pin (or top *pins* in the case of a masterkeyed lock). As that pin stack is being lifted and you're applying varied pressure to your torque wrench (Fig. 14.5), the top of the bottom pin and the bottom of the top pin both reach the shear line at about the same time.

The top pin goes into its upper pin chamber and leans against the chamber wall. The bottom pin, depending on its length, either stops at the shear line or falls back into the plug. At this point the torque wrench is able to turn the plug slightly. When it turns, a small ledge is created on the plug for the top pin to

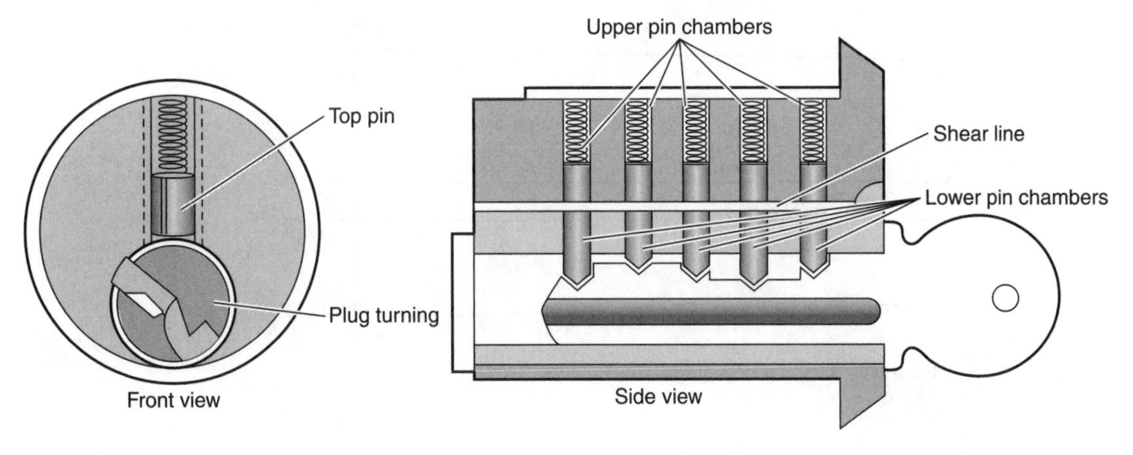

Figure 14.2 The right key lifts all the pin sets to the shear line, allowing the plug to turn.

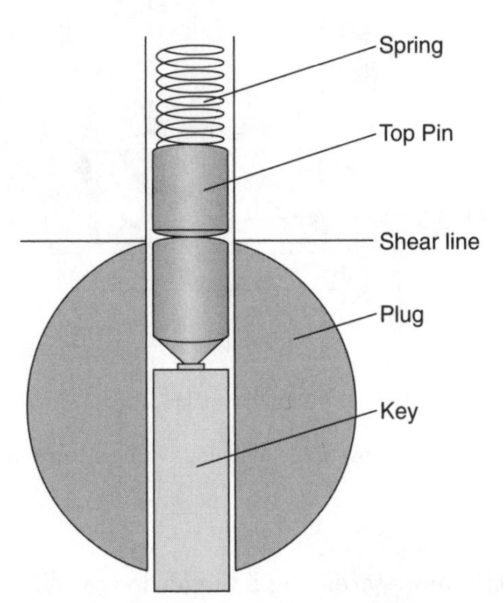

Figure 14.3 When the bottom and top pins meet at the shear line, the plug can be turned.

sit on. This ledge prevents the top pin from falling back into the lower pin chamber when the pick is removed. The top pin will stay on the plug's ledge at the shear line as long as adequate pressure is being applied to the torque wrench. The shear line is the space between the upper and lower pin chambers.

The object of lock picking is to have the top and bottom pins meet at the shear line so that none of the pins are obstructing the plug from turning. If you're picking the pins in the proper order, as each top pin is lifted and set on

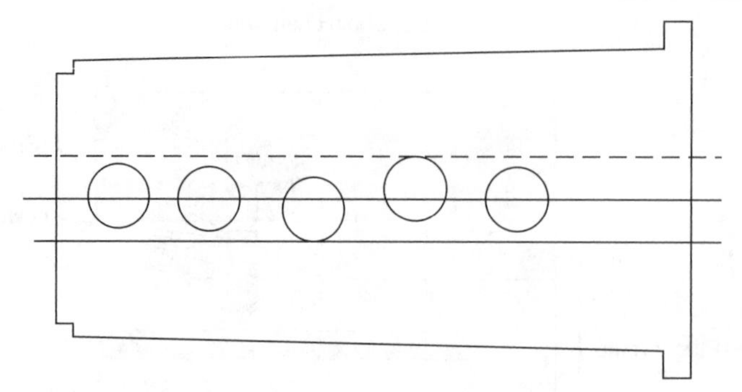

Figure 14.4 Top view of a plug showing lower pin tumbler chamber.

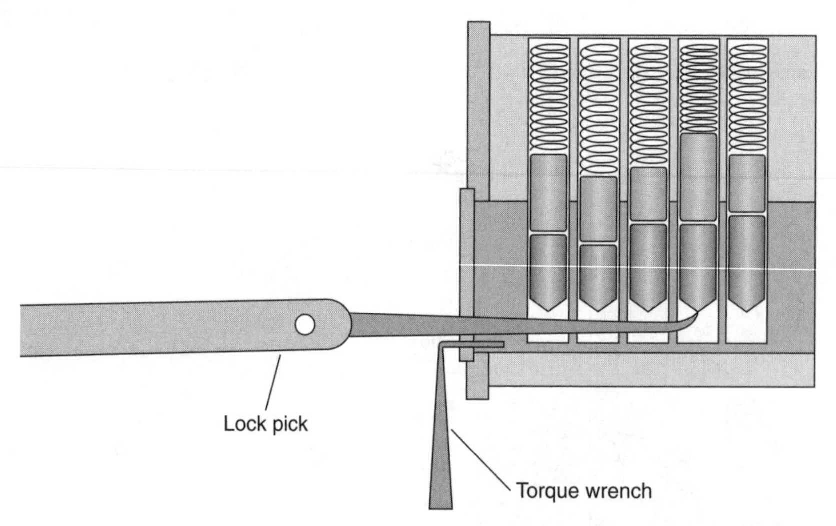

Figure 14.5 To pick a lock, lift one pin set at a time while applying torque or turning pressure with a torque wrench.

the plug, the plug turns a little more, creating a bigger ledge. When all the top pins are on the ledge, all the bottom pins will be at or below the shear line, and the plug will be free to be rotated to the unlocked position by the torque wrench.

Raking

A common method of picking locks is the rake method, or raking. To rake a lock, insert a pick (usually a half-diamond or rake) into the keyway past the last set of pin tumblers, and then quickly move the pick in and out of the keyway in a figure eight movement while varying tension on the torque wrench. The scrubbing action of the pick causes the pins to jump up to (or above) the shear line, and the varying pressure on the torque wrench helps catch and bind the top pins above the shear line. Although raking is based primarily on luck, it sometimes works well. Figure 14.6 shows how a rake pick works.

Frequently, locksmiths will rake a lock first to bind a few top pins, and then they will pick the rest of the pins one at a time.

Using a Pick Gun

A pick gun can be a great aid in lock picking. To use one, insert its blade into the keyway below the last bottom pin. Hold the pick gun straight. Then insert a torque wrench into the keyway. When you squeeze the trigger of the pick gun, the blade slaps the bottom pins, which knock the top pins into the upper pin chambers. Squeeze the trigger several times. Immediately after each squeeze, vary the pressure on the torque wrench. You likely will capture one or more upper pins in their upper pin chambers and set them on the plug's ledge. Then you can pick each of the remaining pin sets one by one.

Lock Picking Tips

Before attempting to pick a lock, make sure that it is in good condition. Turn a half-diamond pick on its back, and try to raise all the pin stacks together. Then slowly pull the pick out of the lock to see if all the pins drop or if one or more are frozen (Fig. 14.7). If the pins don't all drop, you may need to lubricate the cylinder or remove foreign matter from it. (But don't lubricate the lock if you might want to impression it.)

To give yourself maximum working space, use the narrowest pick available. Hold the pick as you would hold a pencil—with the pick's tip pointing toward the pins (Fig. 14.8). Don't use wrist action; your fingers work better in manipulating the pick in the lock.

With your other hand, place the small bent end of your torque wrench into the top or bottom of the keyway—whichever position gives you the most room to maneuver your pick properly. Make sure that the torque wrench doesn't touch any of the pins. Use your thumb or index finger of the hand that's holding the

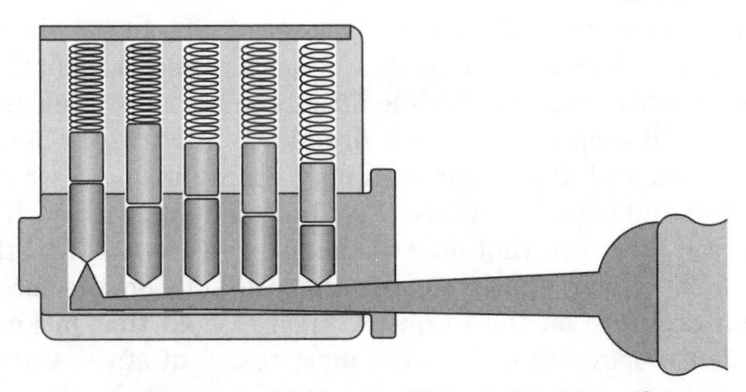

Figure 14.6 Raking a cylinder can pick one or more sets of pins at once.

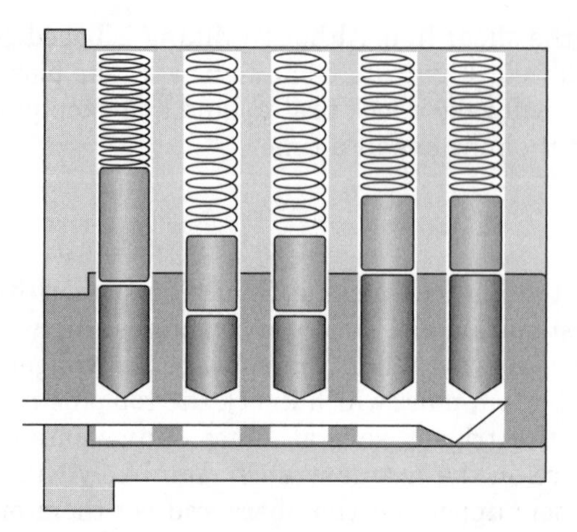

Figure 14.7 Before picking a lock, use the back of a pick to make sure that each tumbler set drops freely.

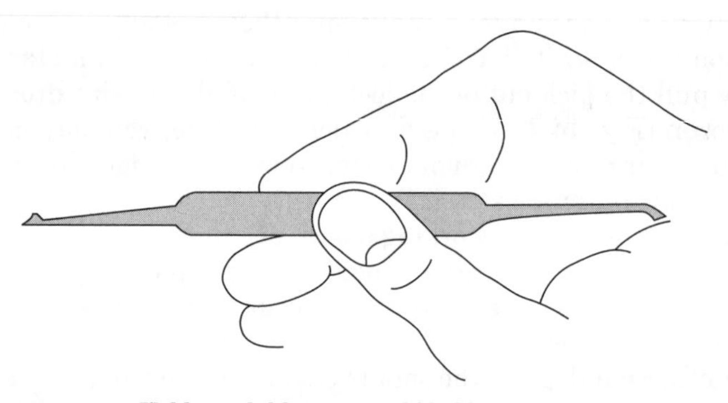

Figure 14.8 Hold a pick like you would hold a pen.

torque wrench to apply light pressure on the end of the torque wrench in the direction you want the plug to turn.

While using a pick, carefully lift the last set of pins to the shear line while applying slight pressure with the torque wrench. Make a mental note of how much resistance you encountered while lifting the pin stack. Release the torque wrench pressure, letting the pin stack drop back into place. Then move on to the next pin stack and do the same thing, keeping in mind which pin stack offered the most and least resistance (Fig. 14.9). Repeat with each pin stack.

Then go to the pin stack that offered the most resistance. Lift the top of its bottom pin to the shear line while varying pressure on the torque wrench. Apply enough pressure on the torque wrench to hold that picked top pin in place. Then gently move on to the next most resistant stack. Continue lifting each pin stack (from most resistant to least resistant) to the shear line. As you lift each pin stack into place, you will be creating a larger ledge for other top

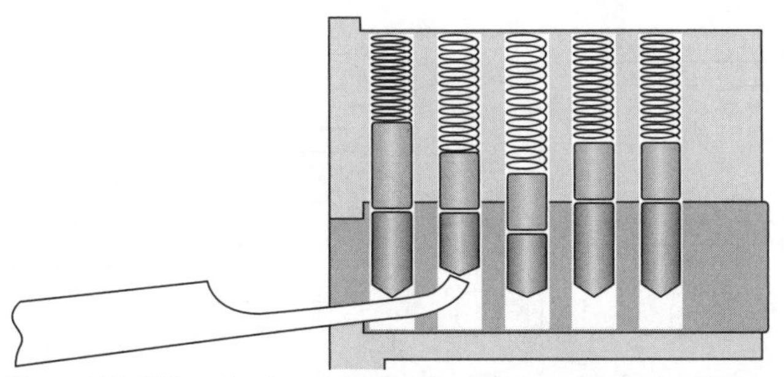

Figure 14.9 Lift each pin set to the shear line, one set at a time.

pins to rest on. When all the top pins are resting on the plug, the plug will be free to turn to the unlocked position. It's important to pick the pin stacks from most resistant to least resistant, because the less resistant stack creates a smaller ledge on the plug and will cause the already picked, more resistant stacks to fall off the ledge.

Picking warded locks

Warded locks are easy to pick. Sometimes you can even use a pair of wires—one for throwing the bolt and another for adjusting the lock mechanism to the proper height for bolt movement. Here, as with any other kind of lock picking, it is a matter of practice on a variety of locks.

Warded padlocks have only one or two obstacles, thus simplifying your choice of pick. Some warded locks that have keyways with corrugated cross sections can be opened in no time at all by having several precut blanks available (Fig. 14.10). You can buy a set of warded padlock picks for a few dollars from a locksmith supply house.

Smaller locks with correspondingly smaller keyways require that you get a second set of warded padlock picks and cut them down to size.

Warded bit-key locks are also easy to pick. (For more information on warded bit key locks, see Chap. 4.) Many can be unlocked by using an "L" shaped piece of coathanger wire about 6 inches long. The easiest way to unlock most warded bit-key locks is by buying a set of skeleton keys for a few dollars.

Picking lever tumbler locks

Figure 14.11 shows a lock with the faceplate removed, indicating the position of the tension tool in relation to the bolt. Holding the levers with the tension tool (as shown in the side view) enables you to manipulate the levers with the pick.

In learning to pick the lever lock, it is best to start with a lock with only one lever in place. At first, work on the lock with the faceplate removed so that you can get an idea of how much pressure to apply to the bolt and how much movement is required of the pick to move the lever into position for the bolt to move through the lever gating.

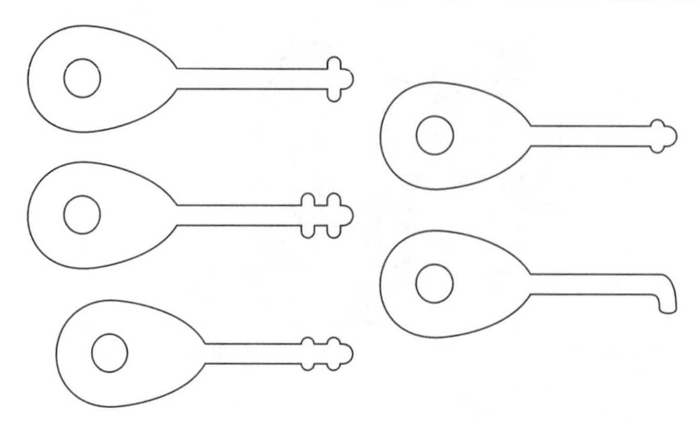

Figure 14.10 A few warded padlock picks can be used to open most warded padlocks.

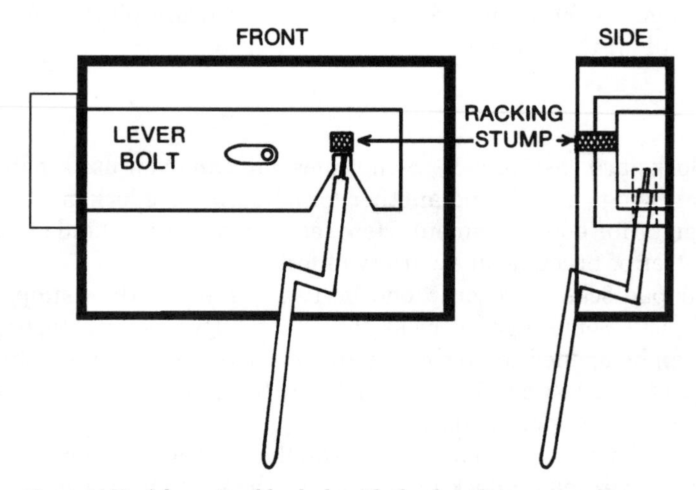

Figure 14.11 A lever tumbler lock with the faceplate removed.

As you progress to locks with several levers, keep the faceplate on. When you encounter a problem, remove the faceplate so that you can see what you're doing wrong. Always remember to insert the tension tool first. Push it to the lowest point within the keyway so that the pick will have maximum working space.

As with the pin tumbler lock, one tumbler tends to take up most of the tension, so work on this one first. As you move this tumbler slowly up to the proper position for the bolt to pass through the gate, you will feel a slight slackening in tension from the bolt as it attempts to force its way into the gating. This tension will be transmitted to you through the tension tool. Stop at this point and do the same to the next lever with the greatest amount of tension. Continue until all the levers reach this point. Then, by shifting the tension tool against the bolt, the bolt will pass through the gate and open the lock.

There must be at least some pressure on the tension tool at all times. A lack of pressure will cause any levers in position to drop back into their original locations.

Picking disc tumbler locks

Standard disc tumbler locks (those using a single-bitted key) take the same picks as pin tumbler locks. Like the pin tumbler lock, a disc tumbler lock can be picked by bouncing the tumblers to the shear line. Usually a rake is the best tool for this job, but other picks can be used too.

To develop your proficiency, try opening a disc tumbler lock with the feeler pick, working on each tumbler individually. This helps you to learn which picks you are most comfortable with for different types of locks.

When working on double-bitted disc tumbler locks, the bounce method is best. Insert the tension tool into the keyway, and apply a slight pressure to the core as you pull the pick out of the lock. These locks also can be opened with a standard pick set and tension tools, but it can take from 10 minutes to half an hour.

Picking High-Security Cylinders

All cylinders that use a key can be picked open. High-security cylinders often are harder to pick than a standard one because they have patented features and require special tools and techniques.

ASSA Cylinders

To pick an ASSA cylinder, you'll need a key blank that fits the cylinder. Cut off most of the length of the blade to give you room to pick the tumblers. Then use the blank as a turning tool while picking the cylinder.

Schlage Primus

To pick a Schlage Primus, you'll need a key blank that fits the cylinder (that has the same milling). Trim the length of the blade down to the milling to give you room to pick the tumblers. Then use the blank as a turning tool while picking the cylinder.

An alternative method for picking a Schlage Primus is to use a key blank that fits the cylinder and cut all the spaces 0.010 inch deeper than the lowest possible cut. Insert the key, pull it back out one notch, and use the key as a turning tool. While applying torque, smack the key into the lock hard. Repeat until the lock opens. This method works on the same principle as a pick gun.

Medeco Biaxial

To pick the Medeco Biaxial, use a rigid turning tool and pick the pins to the shear line. When the plug turns a couple of degrees, release a bit of pressure from the turning tool and gently rake each side of the keyway to rotate the pins to the sidebar. You also can try to rotate each pin one by one.

An alternative method for picking a Medeco Biaxial is to use a key blank that fits the cylinder, cut all the spaces 0.010 inch deeper than the lowest possible cut, and widen the cut flats about one-half their widths. Insert the key, pull it back out one notch, and use the key as a turning tool. While applying torque, smack the key into the lock hard. Repeat until the lock opens.

The Need for Practice

No amount of reading will make you good at picking locks. You need to practice often so that you develop the sense of feel. You need to learn how to feel the difference between a tumbler that has been picked and one that is in a locked position.

To practice lock picking, start with a cylinder that has only two pin stacks (or tumblers) in it. When you feel comfortable picking that, add another pin stack. Continue adding pin stacks or tumblers until you can at least pick a five-tumbler lock.

When you're practicing, don't rush. Take your time, and really focus on what you're doing. Always visualize the inside of the lock and try to picture what's happening while you're picking the lock. For best results, practice under realistic circumstances. Instead of sitting in a comfortable living room chair trying to pick a cylinder, practice on locks that are on a door or on a display mount.

With a lot of focused practice, you'll find yourself picking all kinds of locks faster than ever.

Impressioning Locks

From the outside, impressioning is inserting a key blank into a keyway in a way that leaves tumbler marks on the blank. You then file the marks, clean the blank, and reinsert the blank to make more marks. Then you file the new marks. You repeat the procedure until you no longer see marks and you have a working key. With practice, you should be able to impression most pin tumbler locks within 5 to 10 minutes.

Pin tumbler locks

To impression a pin tumbler lock, you need to choose the right key blank. It needs to fit smoothly in the lock. If the blank is too tight, you won't be able to rock it enough to mark it. The blank also needs to be long enough to lift all the pins. If you use a five-pin blank on a six-pin cylinder, you probably won't be able to impression the lock because the sixth pin won't mark the blank. To choose the right size blank, use a probe or pick to count the number of pin sets in the lock.

The material of the blank needs to be soft enough to be marked by the pins but not so soft as to break off while you're twisting or rocking the blank. Nickel-silver blanks are too hard for impressioning because they don't mark well. Aluminum blanks are soft enough to mark well, but they break off too easily. Brass blanks work best. Nickel-plated brass blanks are also good for impressioning because the nickel plating can be filed off.

Filing the blank. New key blanks have a hardened glazed surface that hinders impressioning unless you prepare them. To prepare a key blank, shave the length of the blade along the side that comes in contact with the tumblers. Shave the blank at a 45º angle without going too deep into the blank (Fig. 14.12). You want the blank's bitting edge to be sharp (a knife edge) without reducing the width of the blank because some pins may require a No. 0 cut to reach the shear line. File forward only. Don't draw the file back across the blank.

Use a round or pippin file with a Swiss pattern No. 2, 3, or 4 cut. A courser file will shave the blank quicker but will leave rougher striations on the blank. This will make the impression marks harder to see. A finer file will make the marks easy to see but will clog quickly while you're filing. This will make impressioning take a long time. You probably won't find impressioning files at a hardware store or home improvement center, but they're sold though locksmith supply houses.

Another popular way to prepare the blank is to turn the blank over and shave the other side of the blank along the length where the tumblers touch the blank. After shaving both sides of the bitting edge at 45º, you will have a double knife edge.

Other Useful Equipment and Supplies

In addition to a file, you'll need a key-holding device, such as an impressioning tool or a 4- or 5-inch pair of locking pliers (Fig. 14.13). A magnifying glass can be helpful for seeing impression marks. A head-wearing type lets you see the marks and file at the same time. Head-wearing magnifying glasses are

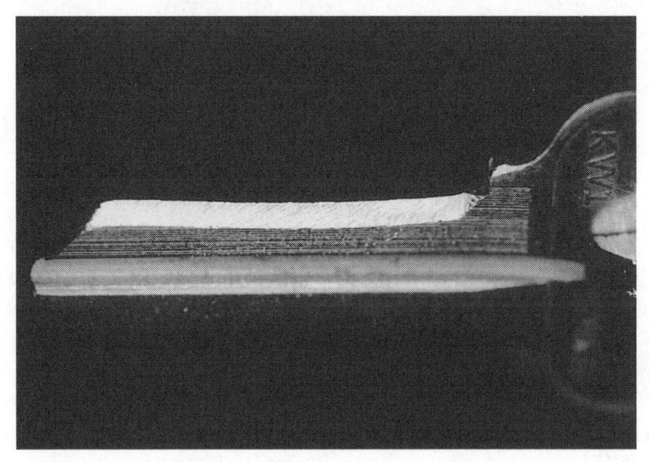

Figure 14.12 Shave the bitting edge of the blank at 45°.

shown in Fig. 14.14. Although they aren't essential, you can use depth and space charts and a caliper to file marks more precisely.

Popular Impressioning Technique

The pull method of impressioning is often used on pin tumbler locks. After preparing the blank, insert it fully in the keyway with locking pliers or an impressioning tool and turn clockwise (Fig. 14.15). This will bind the bottom pins against the chamber walls (Fig. 14.16). Maintain the turning pressure, and rock the blank up and down several times while slightly pulling the blank out, about 1⁄16 inch. The rocking and pulling will cause the pins to press into the

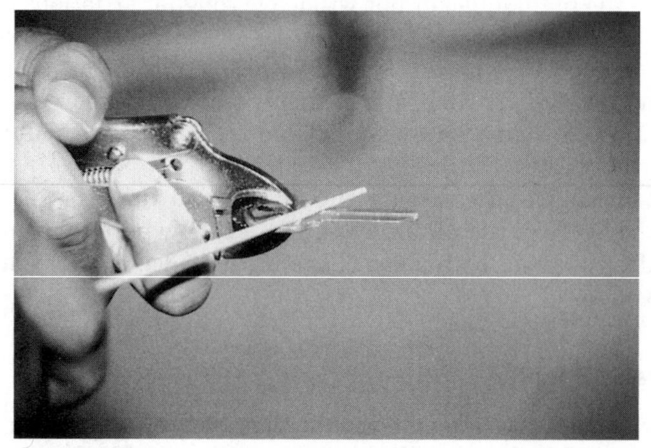

Figure 14.13 Use a pair of locking pliers to hold the blank while you're filing it.

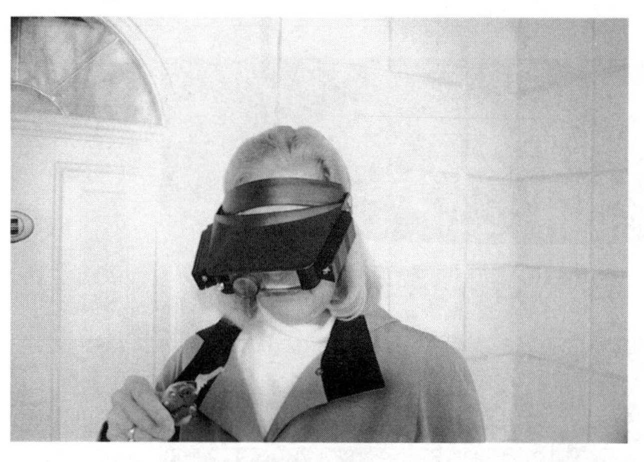

Figure 14.14 Head-mounted magnifying glasses can help you to see impression marks.

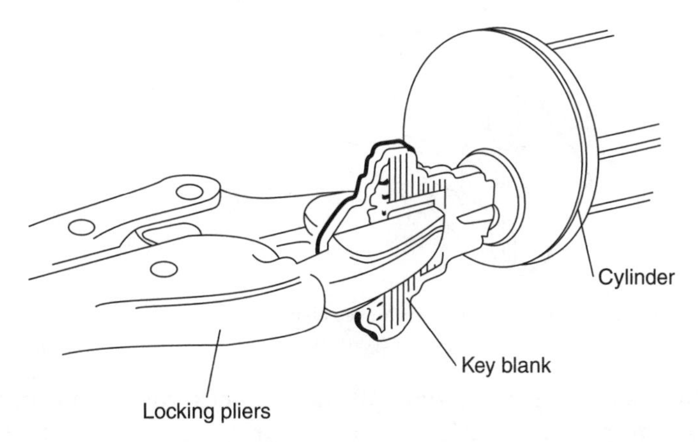

Figure 14.15 Using locking pliers or an impressioning tool to turn the blank in the cylinder.

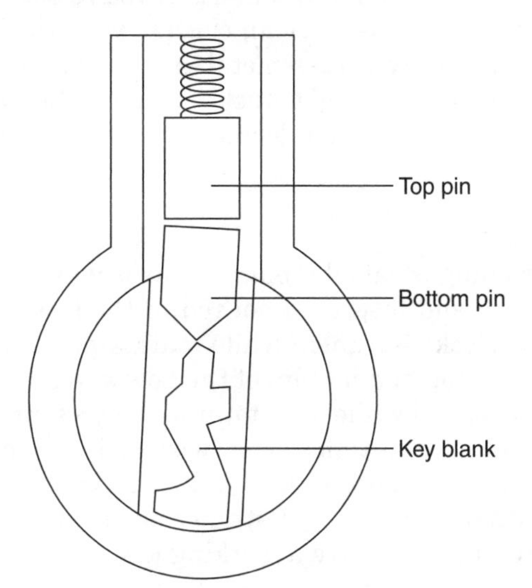

Figure 14.16 When you turn the plug that has a prepared blank, the bottom pins will bind and leave an impression on the blank.

blank, leaving small marks or "impressions." Then turn the blank counterclockwise, and rock the blank up and down several times while slightly pulling the blank out. (If the lock turns in only one direction, then twist the blank in that direction only.)

Remove the blank from the keyway, and look for impression marks. You may see one mark or several marks. File down about one depth at each mark (Fig. 14.17). You can use depth and space charts and a caliper to precisely file the blank.

Don't file where you don't see a mark. The marks will be subtle; they aren't dark imprints like a felt tip pin. They are a change in the texture of the key blank filing. When the light is properly reflected, the markings look like small, smooth spots. You may need to view the blank at different angles to see the marks.

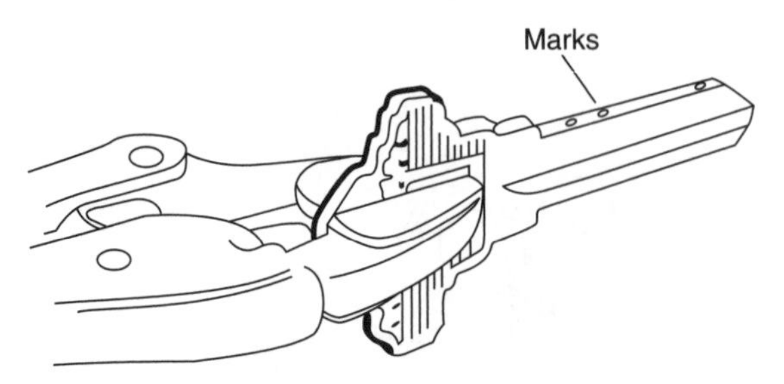

Figure 14.17 Only file the marks that you can see clearly.

Never file more than one key depth for the lock you're working on. If you don't have a caliper and depth and space charts, just file two or three strokes at a time. (File forward only.) It's better to file too little away than to file too much. Sometime dirt or debris may seem like a mark. If you're not sure if what you're seeing is an impression mark, use a clean cloth to wipe the blank.

After filing the marks, clean away excess metal on the bitting side of the blade by again filing it a little at a 45º angle. Just a couple of forward strokes on each prepared side will be enough. Don't decrease the width of the blade.

Alternative Impressioning Method

Another method of impressioning is called *tapping*. This is where you insert a prepared blank in the keyway and insert a steel rod in the bow. Use the rod for leverage to turn the bow clockwise; then, while maintaining the pressure, use a small hammer to tap the top and bottom of the bow several times. Then twist the bow counterclockwise, and while maintaining the pressure again, tap the top and bottom of the bow several times. The twisting is to bind the pins, and the tapping is to force the pins to mark the blank. File down at each mark you see. Clean the blank with one or two light strokes across the length of the blade. Repeat the procedure until you have a working key.

Problems with Impressioning

One common problem that happens during impressioning is that the blank cracks or breaks off near the bow. This is more likely to happen if you use aluminum blanks or locking pliers that are too large. If your brass blanks are breaking off, you're turning too hard. You don't need to turn very hard. It's the rocking that marks the blanks, not the turning.

If your blank cracks or breaks, duplicate it on a key machine or by hand. Then use the new blank to finish impressioning the lock.

Anything that makes it harder for the pins to mark the blank will hinder your impressioning attempts. Problems can include: missing or worn tumbler springs, worn tumblers, a long pin next to a short one, extremely close tolerances within the lock, and oil, dirt, or debris in the cylinder.

There isn't much you can do about it if the lock is too old and worn. But if you have a hard time making marks, use a nonoil, wax-free spray cleaner for electronic parts. Shoot a couple of squirts into the keyway and let it dry. That should clean the lock.

Impressioning Practice

No amount of reading will make you skilled at impressioning because during impressioning you have to make a lot of judgment calls—which will get better only with experience. You have to judge how hard to twist and rock the blank, how hard to file, and how deep and how wide to make a cut based on the type of mark you see.

If you've never impressioned a lock, practice by using a lock with only two sets of pins. Prepare the blank to a knife point. Insert the blank into the keyway and twist the key clockwise, and while maintaining the pressure, rock the blank up and down and pull it out slightly. Then twist the blank counterclockwise, and while maintaining the pressure, rock it up and down and pull it out slightly. Remove the blank from the lock, and observe the marks made by the pins. File at the strongest mark only. Cut just one mark at a time before cleaning the blank and reinserting it in the keyway to make new marks. Repeat the turning, rocking, pulling, and filing procedure until you have a working key.

When you're able to impression a two-pin lock, add another pin stack, and practice impressioning it. Continue adding pins until you can quickly impression a five-pin tumbler cylinder.

Warded Bit-Key Locks

Impressioning a warded bit-key lock relies on the ability to decipher small marks made on a smoked key blank that has been inserted into a lock and turned. Interpretation of the marks tells the locksmith what cuts to make, where to make them, and how deep they must be. The advantage of impressioning is that you need not disassemble a lock or remove it from a door to make a key.

The first cut allows that blank to enter the keyhole. Since you do not have the original key, smoke the end edge of the blank, and insert it into the keyhole so that the edge comes into contact with the case ward. Scribe mark the top and bottom of the ward on the blank. Remove the key; the candle black that was removed indicates the depth of the cut, and the scribed marks show the position of the cut on the blank. Transfer the scribe marks to the near end of the bit, and draw lines connecting the two pairs of marks. Use a small piece of metal to make a depth gauge so that you will not file too deeply. Mount the blank in the vise, and cut the ward slot. When you're finished, the key should pass the case ward.

Next, prepare the blank for impressioning the internal wards. Recall that the key must pass certain side wards. When the lock is assembled, how can you be sure exactly where to cut the key so that it will pass these wards? You can't. You will have to prepare the key blank using one of two methods. The first is to coat the key with a thin layer of wax. This is unprofessional and can harm the lock. The wax may clog the mechanism, forcing you to disassemble and clean the mechanism at your expense—you can't charge a customer for your mistake. A more professional method is to smoke the key. The smoking must be thick enough to form a stable marking surface. If, for example, the blank is thicker than it should be, enough blackening must be present to give true readings. A thin coat will speckle as you turn the key and send you on wild goose chases. The technique for impressioning the key is as follows:

1. Insert the key into the lock and turn it with authority. Remove the key. You will notice one or more bright marks where the blackening has been removed. These marks indicate obstructions. File the appropriate cut for the key to pass.
2. Blacken, insert, turn, and remove the key. If a mark is present below the point you filed, the cut is too shallow. Carefully file it deeper. When the cut is deep enough, use emery paper to clean off the fine burrs left by the file.
3. Reinsert the key, and turn it again. It should pass the wards. If it tends to stick slightly, quickly and lightly pass over the cut with the warding file to alleviate the problem. *Note:* The cut should be square on all sides. Think of each ward cut as a miniature square, perfect and even on all sides.
4. Smoke the key, insert it, and turn it. The edges of the cuts should be shiny. The brighter the spot, the greater is the pressure at that point. As the individual cuts are filed deeper, the bright spots grow dimmer. When this happens, you are close to the point where you should stop filing.

You must be more cautious each time you insert the key to test the depth of the cut. You are working "in the blind"; a small overcut means that the depth is permanently wrong on the blank, and you have to start over. This is the reason for making a few light but firm strokes as you near the completion of each cut. Take your time and make one perfect key instead of rushing the job and making many incorrectly cut keys.

The cuts on the key should be as deep and as wide as necessary but no more. Overly large cuts interfere with the action of the lock and may force you to cut another blank.

Once the key is complete, it must be dressed out. Use emery paper to remove burrs before they break off and fall into the lock, and then polish the entire key with emery paper. For appearance, you also may give the key a light buffing on your wheel. It does you no good to give the customer a dirty and smudged key. Show pride in your work!

In practice, however, there isn't much reason to impression a warded lock. For only a few dollars, you can buy a set of four to five skeleton keys that will open most bit-key locks or a set of warded padlock picks that will unlock most warded padlocks.

Chapter

15

Automobile Lock Servicing

From the outside, most automobile locks look the same, but the internal constructions often differ. The construction of an automotive lock depends on the purpose of the lock (ignition, door, trunk, etc.) and the manufacturer, model, and year of the vehicle for which the lock was made. This section explains how to service the locks on a wide variety of vehicles.

Vehicle Identification Numbers

Every automobile in the United States is supposed to have a vehicle identification number (VIN) on a plate that's near the vehicle's windshield on the left side. The plate should be visible from outside the vehicle. You can use a VIN to determine several things about the vehicle, including its model and model year. (In many cases, you need that information to choose the right lock- or car-opening tool.)

Some VIN positions vary depending on the manufacturer and year of manufacture. However, the letters and numbers in positions 1, 2, and 10 (reading from left to right) are codes for the vehicle's country of origin and make and model year, respectively. Table 15.1 shows what the various numbers and letters mean.

Basics of Automobile Lock Servicing

Servicing vehicle locks is a specialty field of locksmithing that can be very profitable. To be proficient in this field, you must stay informed about the constant changes that are made to automobile locks and locking systems. Most of the changes are made to increase vehicle security and convenience.

Locksmiths often are called on to service automobiles that have lock-related problems. Such problems include a foreign object (such as a broken piece of key) that has become lodged in the cylinder and keys that won't operate the lock.

TABLE 15.1 Reading VIN Codes

Position	Meaning	Code option
1	Country of origin	1 = U.S. 2 = Canada 3 = Mexico J = Japan
2	Make	B = Dodge F = Ford 1 = Chevrolet 2 = Pontiac 3 = Oldsmobile 4 = Buick 5 = Cadillac
10	Model year	T = 1996 V = 1997 W = 1998 X = 1999 Y = 2000 1 = 2001 2 = 2002 3 = 2003 4 = 2004 5 = 2005 6 = 2006 7 = 2007 8 = 2008 9 = 2009

A broken piece of key can be removed from an automobile lock in the same way it can be removed from other locks. A key that is stuck in a door lock usually can be removed by first turning the broken piece to the upright position (the position the key is in when it's first inserted into the lock) and then using a broken-key extractor to pull the broken piece out of the lock. For more information on broken-key extraction, see the last part of Chap. 13.

When a key turns hard in a cylinder, it's often the result of an improperly bent rod between the walls of the car door. This can happen when someone improperly uses car-opening tools. To solve the problem, remove the trim panel from the vehicle, and locate and straighten the problem rod (Fig. 15.1).

When the key turns smoothly but won't operate the door, the problem could be with the cylinder. You may be able to solve the problem by repairing or replacing the cylinder. More often, however, the rod within the door that should be connected to the lock will have become disconnected. If this happened, you'll need to remove the inside trim panel, locate the disconnected rod, and reconnect it to the lock. You'll probably need a new retainer clip to fasten the rod to the lock.

The procedure for removing an inside trim panel differs for different vehicles. In general, however, you need to do the following:

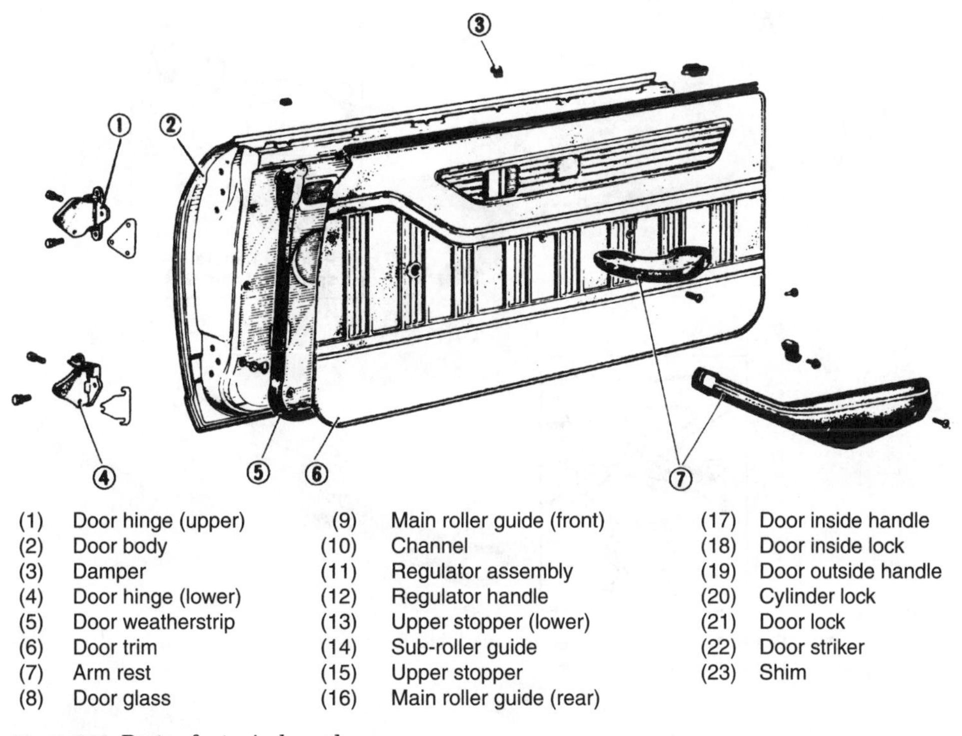

Figure 15.1 Parts of a typical car door.

1. Wind the door window all the way up.
2. Remove the door handle. Sometimes it's held on by a screw; other times, by a clip.
3. Remove the arm rest from the door. It's usually secured by screws and clips.
4. Remove all the screws located around the edge of the trim panel, and carefully pull the panel from the door. You can use a screwdriver, but if you use a clip-removal tool, you'll be less likely to damage the clips. In any case, it's a good idea to have extra door clips for the vehicle you're working on.

Door Locks

Modern door locks are disassembled from inside the door. Remove the arm rest and inside handles. Most handles are secured by a spring clip that is accessible with the tools shown in Fig. 15.2. Note the positions of the handles on their spindles for assembly.

Remove the Phillips screws holding the door trim to the door body. Gently pry the trim away from the body and set it aside. The door lock mechanism is visible through access holes in the inner door body panel.

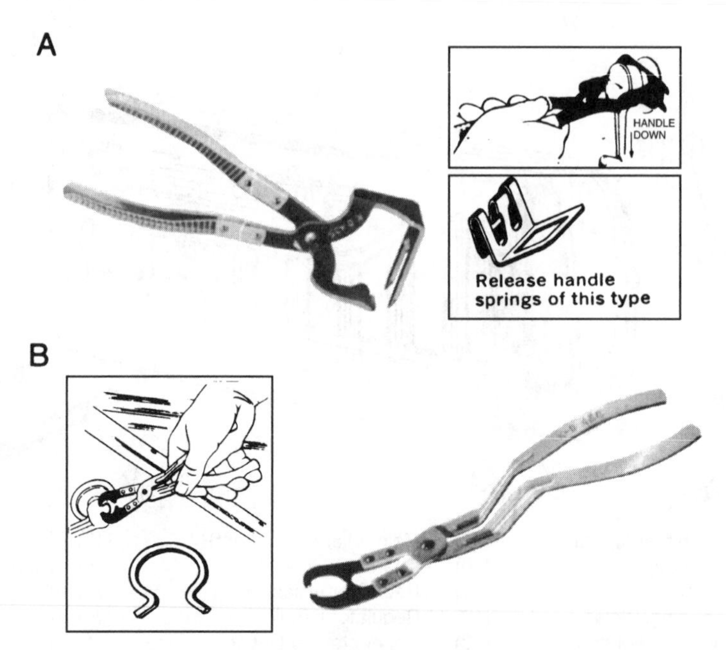

Figure 15.2 Special tools are available from auto parts stores to help remove automobile clips.

Most problems involve the lock mechanism and not the cylinder. Check for broken springs, missing retainer clips, and rust on working parts. The cylinder is secured by a U-shaped retainer clip. Cylinders can be disassembled, although modern practice is to replace the cylinder as a unit rather than attempt repairs.

Move the striker to align the door with the body panels and compensate for door sag. Loosen the mounting screws a quarter turn or less, and using a soft mallet, tap the striker into position (Fig. 15.3). Shims are available from the dealer to move the striker laterally out from the door post.

While this is out of the province of locksmithing, as such, it is useful to know that the door can be adjusted at the hinge on many automobiles (Fig. 15.4). Loosen the pillar side bolts slightly and manipulate the door.

Ignition Locks

Early production ignition locks are secured to the dash by bezel nuts or retainer clips and feature a *poke hole* on the lock face or in the back of the cylinder that gives access to the retaining ring. In a few cases, screws hold the lock from the underside. Late-model automobiles (those with MATS, PATS and assorted transponders) require specialized knowledge. Those and other vehicle security technologies are constantly changing. In addition to having current information, you'll also need special tools that can be very expensive (and that may be obsolete in a

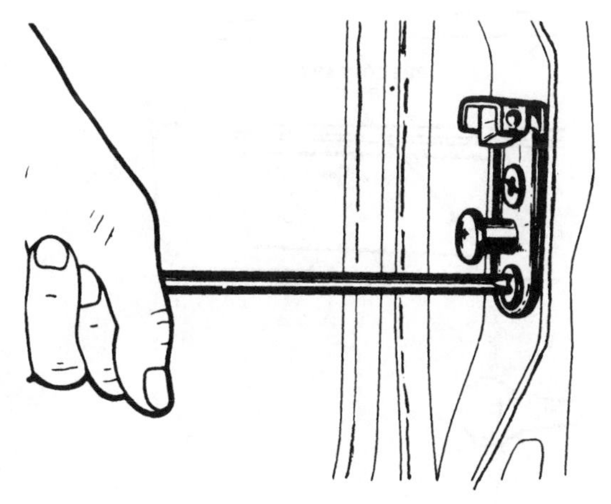

Figure 15.3 Door strikes are adjustable. (*Chrysler Corp.*)

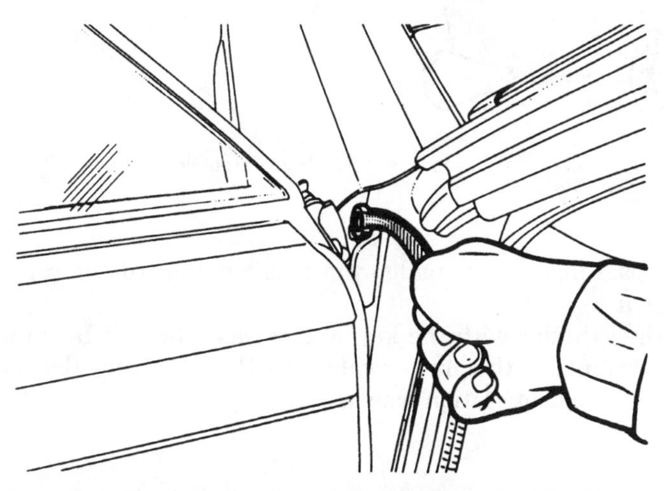

Figure 15.4 The door adjustment is at the pillar side of the hinges for a lot of vehicles. (*Chrysler Corp.*)

short period of time). If you want to service high-tech vehicle locks, join organizations that will help keep you up to date on the ever-changing technologies—such as the National Locksmith Automobile Association. Also, attend automobile lock servicing classes that are given by locksmithing trade associations.

Glove-Compartment Locks

Modern glove-compartment locks typically are secured by a retainer accessible from the front of the lock. Figure 15.5 shows how this is done. Make a 14-inch hook in the end of a piece of Bowden (hood-latch) cable. Insert the hook into the keyway, and work the retainer free. Most are installed with the open ends

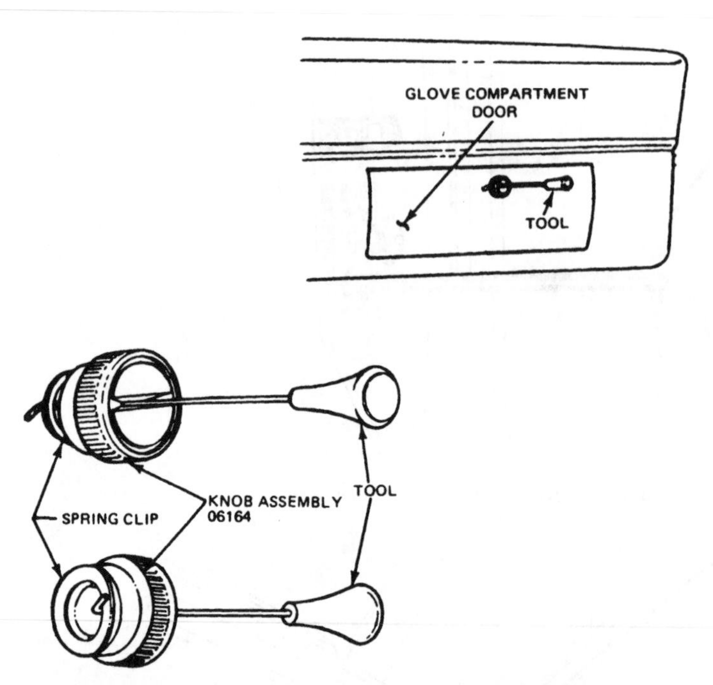

Figure 15.5 The glove-box cylinder can be removed from the outside with the tool shown. (*Ford Motor Co.*)

toward the passenger door; nudge the retainer toward the center of the car to disengage it.

Extract the cylinder with the key, and release the bolt by hand. The cylinder may be stamped with a numerical code—the same code that is used for the trunk or, in the case of station wagons, the tailgate.

Trunk Locks

Figure 15.6 shows a luggage-compartment latch typical of those found on middle-priced automobiles. More expensive cars use a more complex lock mechanism that may be solenoid-tripped from inside the vehicle. The basic parts and their relationships are the same. The cylinder is secured to the trunk lid by a retaining clip and works the lock assembly through a tailpiece or extension. (Some cars mount the cylinder on the lower body panel and the striker on the lid.) The striker is located on the aft body panel.

Many lock assemblies are adjusted only at the striker plate. The plate is mounted on elongated bolt holes and can be moved up, down, left, and right. Other types have an additional adjustment at the latch. The idea is to position the parts so that the lid closes without interference from the striker and tightly enough to make a waterproof seal between its lower edge and the weather stripping.

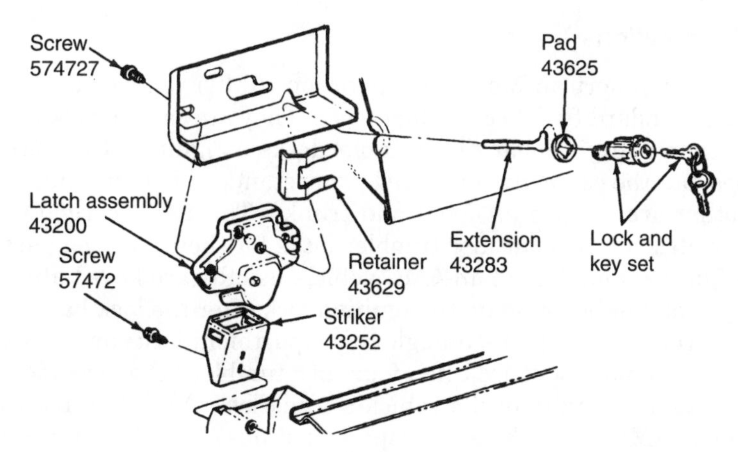

Figure 15.6 The trunk lock assembly used on Mustang and Cougar automobiles. (*Ford Motor Co.*)

Usually you can get to a trunk latch easily by going through the back seat. The back seat of many Toyota Corollas is held in place with two bolts. After you remove them and take out the back portion of the seat, you'll have an area that's big enough to crawl through. You can then easily use a screwdriver to trip the latch to open the trunk. Before tripping the latch, disconnect the trunk lock linkage from its connector. Sometimes the trunk lock will break if you don't do this.

In some cases, you also can unscrew the sidelight that's near the taillight and use an across-the-car tool with a flashlight to release the latch. You can open some older BMW trunks by taking off the BMW cover plate that sits on the rear of the trunk and using a probe wire.

Transponders

The introduction of transponder-equipped vehicles has shaken up automotive locksmithing. Locksmiths are desperately trying to keep up with the new technology, but it keeps changing. There are a lot of vehicles today that use transponders. There are both original and aftermarket transponders, and many nontransponder keys look like transponder keys. Making transponder keys is expensive, requiring equipment that costs several thousand dollars. You need special equipment to identify the type of transponder key and to program it, and in many cases (especially with Japanese models) you'll need to buy a special high-security key machine.

It can be tricky figuring out which keys are and which aren't enhanced electronically. However, even if you decide not to make transponder keys, you still may want to buy a device that tells you which keys are transponder keys. A small handheld transponder detector can be used to identify transponder keys.

American Motors Corporation

All locks for American Motors Corporation (AMC) vehicles up to model year 1967 are standard five-disc tumbler locks. They are easy to pick open. In 1968, AMC began using sidebar disc tumbler locks as ignition locks on all its new vehicles, but the company continued using standard five-disc tumbler locks for most other areas, such as doors and trunks. The 1975 to 1980 Gremlin and Pacer models use sidebar disc tumbler locks for their rear compartments.

The ignition and door of an AMC vehicle usually are keyed alike. The original key code can be found on the ignition lock after the lock has been removed from the vehicle. The trunk and glove compartment locks are also keyed alike. The glove-compartment lock has four disc tumblers that are identical to the last four tumblers used in the vehicle's trunk lock. A sidebar disc tumbler lock used on an AMC vehicle's rear compartment has the code number stamped on the tailpiece.

It's usually easy to impression a key to one of the door locks. The blank should be twisted very lightly because the dust shutter can break. When replacing ignition locks for AMC models manufactured after 1986, use General Motors locks and keyways.

Fitting a key

Usually, impressioning the lock is the fastest way to fit a key for an AMC vehicle. Another method is to remove the door lock and lock pawl and read the bitting numbers stamped on the back of the tumblers. You also can remove the ignition lock by disassembling the steering wheel column, and then you can cut a key by the code on the cylinder.

To fit a secondary key, cut a key for the glove compartment, and cut one depth at a time in space number 1 (the space closest to the bow) until you've cut that space to the depth that opens the trunk. After cutting to each depth, insert the key into the trunk lock to see of it operates the lock.

You can use depth and space charts to learn the proper spacing and depth increments for the lock.

Removing an AMC ignition lock

Ignition locks for AMC vehicles through the 1978 model year can be pulled out of the steering wheel column without first disassembling the column. Locksmithing supply houses offer tools for this purpose. Most are modified versions of a dent puller. Beginning in 1979, the ignition locks for new models were bolted in; these shouldn't be removed without first disassembling the steering wheel column.

Because of the potential danger that can result from reassembling a steering wheel column improperly, don't attempt to disassemble one until you've seen the procedure done several times on the vehicle model you're planning to service. To be on the safe side, have an experienced locksmith help you disassemble and reassemble a few steering wheel columns before you do it alone.

Audi

Most Audi vehicles made after 1971 use either eight- or ten-disc tumbler locks. Usually all the locks in such a vehicle are keyed alike; in some cases, a primary key is used to operate the doors and ignition.

To fit a primary key for an Audi, use the code number found on the door lock after the lock has been removed from the vehicle. For a secondary key, use the code number on the glove-compartment lock.

BMW

Most locks on BMWs have ten-disc tumblers and are hard to pick open. All the locks on a BMW are keyed alike, but sometimes a secondary key is used. Many BMWs are operated with dimple keys. Some late models have a high-security deadlocking system that allows the owner to lock the car only by using a key; this makes it hard to get locked out of the car.

To fit a key on a BMW, cut the key by code from the number on the door lock or disassemble the door lock and use the tumblers to visually fit a key. The door handle usually is held on by a Phillips head screw under the weather stripping or by a screw and bolt.

Chrysler

In most cases, Chrysler vehicles use pin tumbler locks on the ignition, doors, and trunk. The glove-compartment locks have either three- or four-disc tumblers. The tilt/telescoping steering wheel columns found on some vehicles use a sidebar disc tumbler lock.

The doors and ignition locks of a Chrysler are keyed alike and are operated by the primary key. The code number for these locks can be found on a door lock.

The secondary key operates the trunk and glove-compartment locks. When the glove-compartment lock has three-disc tumblers, they correspond to the three middle tumblers of the vehicle's trunk lock. When the glove-compartment lock has four-disc tumblers, they correspond to the last four tumblers of the trunk lock. Many times you can find the code number for a secondary key on a trunk lock.

Beginning in 1989, some Chrysler vehicles, such as the Plymouth Acclaim and the Dodge Spirit, began using disc tumbler locks that are operated with double-sided convenience keys. These locks are designed to resist picking attempts.

Fitting a key

To fit a primary key for a Chrysler vehicle that uses pin tumbler locks, remove and disassemble a door lock and measure the bottom pins with a caliper. The measurements for Chrysler bottom pins are No. 1, 0.158; No. 2, 0.168; No. 3, 0.188; No. 4, 0.208; No. 5, 0.228; and No. 6, 0.248. If a pin is worn, the pin measurements will be slightly different. In that case, replace the pin with one that is the original pin size. The new pin will be a little longer than the one it is replacing.

To fit a secondary key, fit the key to the glove-compartment lock. If the key has four cuts, find the first cut of the trunk by cutting the first space one depth at a time until the key operates the trunk. If the glove-compartment key has three cuts, you can impression the first and last cuts. When impressioning a key for a Chrysler pin tumbler lock, remember that the locks shouldn't have a No. 5 or 6 pin in the first lower pin chamber and shouldn't have a No. 6 pin in the second lower pin chamber. When a No. 5 pin is found in the first or second lower pin chamber or a No. 6 pin is found in the second lower pin chamber, the lock probably was rekeyed improperly.

During the 1991 model year, pin tumbler locks were discontinued and wafer locks were used.

Datsun

Most locks on Datsuns have six-disc tumblers and are easy to pick open. All the locks on a Datsun are keyed alike. They are operated by a double-sided convenience key. Both sides of the key are cut, but either side can operate a lock because Datsun locks have only one set of tumblers.

A key can be fitted for Datsun locks by cutting a key by the code that's written on a piece of paper glued to the glove-compartment lid. The code also can be found on a door lock. A door lock also can be used to make a key by impression.

Ford

Up to the 1984 model year, pin tumbler locks with five sets of tumblers were used for all Ford ignitions, doors, and trunks. Glove-compartment locks were either pin tumbler or disc tumbler. (Since 1981, only four-disc tumbler locks have been used on glove compartments.) In late 1984, many Ford vehicles began using only disc tumbler locks.

On pre-1982 Ford models, the ignition and doors are keyed alike and are operated by the primary key; the glove compartment and trunk are keyed alike and are operated by the secondary key.

On pre-1977 models, the code for the primary key can be found on one of the door locks, usually on a passenger-side door. On pre-1980 models, the secondary key code can be found on the glove-compartment lock latch housing. Beginning in 1980, codes were no longer stamped on Ford glove-compartment locks.

The primary key for a post-1980 Ford fits only the ignition lock; the secondary key operates all the other locks. Many of those locks don't have key codes stamped on them, but the tumblers have bitting numbers stamped on them. The bitting numbers can be seen when a lock is disassembled.

Fitting keys to pre-1984 Fords

You can fit a primary key to a pre-1981 Ford lock by removing the door lock and measuring the bottom pins. The measurements for such Ford bottom pins are as follows: No. 1, 0.155; No. 2, 0.165; No. 3, 0.185; No. 4, 0.205; and No. 5, 0.225. To fit a primary key for a 1981 to 1984 Ford, either impression the ignition lock or remove the lock and read the bitting numbers on the tumblers.

Fit a secondary key to a pre-1981 Ford lock by cutting according to the code number on the glove-compartment lock. To fit a secondary key for a 1981 to 1984 model, remove the disc tumbler lock and read the bitting numbers on the tumblers.

Fitting keys to post-1984 Fords without PATS

Excluding the PATS system (more about PATS later in this chapter), there are two basic locking systems for post-1984 Fords. The newer system uses only disc tumbler locks (a sidebar disc tumbler is used for the ignition). The older system uses pin tumbler locks.

During midyear production of the 1984 Mercury Cougar and the Ford T-Bird, Ford began using a sidebar disc tumbler ignition lock (similar to the one used by General Motors) and disc tumbler door locks. A convenience key is used to operate the locks—either side of the key blade can be used. With the newer system, one convenience key is used to operate a vehicle's ignition and door locks.

The double-sided key (Ilco blank 1184FD), sometimes called the *10-cut key,* has 10 cuts on each side. The first six cuts (starting from the bow) are for the doors; the last six cuts (spaces 5 through 10) operate the ignition lock. Spaces 5 and 6 on the key correspond to the tumbler depths that the ignition and doors have in common. When the key is inserted into a door, the first six cuts are aligned with the tumblers; when the key is inserted into an ignition, the last six cuts are aligned with the tumblers.

The older locking system uses two keys to operate all the locks of a vehicle; each double-sided convenienece key has only five cuts on each side. (The primary blank is an Ilco 1167FD; the secondary blank is an Ilco S1167FD.) In the older system, all the locks except the ignition lock are keyed alike and use the secondary key; the primary key operates only the ignition lock.

Post-1984 Ford disk tumbler locks (non-PATS)

Neither the ignition nor door locks of post-1984 Fords with disc tumbler locks have key codes stamped on them. The easiest way to fit a key to those vehicles is to cut a key by code if the code is available. The codes are stamped on key tags given to the purchaser of the vehicle. If no code is available, you can impression a key at one of the doors. If you need a key for the ignition, use the cuts of the key you made for the door to help you impression the key.

Remember, the last two tumblers of the door lock have the same depth as the first two tumblers of the ignition lock. Another option is to remove and disassemble the lock.

The ignition lock must be rotated about 30º to the right (to the On position) before it can be removed. This allows the retaining pin to be depressed and the pilot shaft to bypass an obstruction in the column so that the lock can slide out.

You can rotate the cylinder by using a drilling jig (available from locksmith supply houses) to drill out the sidebar. The jig is held in place by a setscrew; insert a key blank into the lock to align the jig, and the jig aligns your drill bit. The jig allows you to drill easily through the roller bearings on either side of the lock's keyway. After drilling through the lock, rotate and remove it.

Now you can rekey an uncoded ignition lock and use that as a replacement lock. Or you can use the service kit supplied to Ford dealerships to replace the lock.

PATS ignition locks

Ford's Passive Anti-Theft System (PATS) relies on low-frequency radio transmissions to identify the correct key for starting the vehicle. The system was first available in Europe on 1993 Ford vehicles. It was introduced in the United States on the 1996 Ford Taurus SHO and LS models and the Mercury Sable LX.

The system uses a transponder (a glass vial about the size of a car fuse) sealed in the head of each ignition key. The transponder has 1 trillion possible electronic codes. When the properly coded key is inserted in the ignition of a PATS-equipped vehicle, the proper code is transmitted to the vehicle's control module, allowing the vehicle to be started. If a key with no code or the wrong code is used, the engine will be disabled.

Until 1998, the PATS system was easy for locksmiths to service. In 1998, Ford modified the system. The new versions aren't locksmith-friendly and are commonly referred to by locksmiths as *PATS II* and *PATS III*.

Vehicles equipped with the original PATS (or PATS I) include the 1998 Ford Contour, 1997–1998 Ford Expedition, 1996–1997 Ford Mustang, 1996–1997 Ford Taurus, 1998 Lincoln Navigator, 1996–1997 Mercury Sable, and 1997–1998 Mercury Mystique.

Models equipped with PATS II in 1998 include Ford Crown Victoria, Ford Explorer, Ford Mustang, Ford Taurus, Lincoln Mark VIII, Lincoln Town Car, Lincoln Continental, Mercury Sable, Mountaineer, and Mercury Grand Marquis. The 1999 Mercury Cougar is also equipped with PATS II.

General Motors

A General Motors (GM) vehicle ordinarily uses sidebar disc tumbler locks for the ignition, door, and trunk and standard disc tumbler locks for the glove compartment and utility compartments. Some top-of-the-line GM vehicles have sidebar disc tumbler locks on the glove compartments.

A pre-1974 GM has its ignition and door locks keyed alike, and all those locks are operated by the primary key. The code for the primary key can be found on the ignition lock.

Pre-1970 models have the code for the primary lock stamped on the door locks.

A post-1973 GM vehicle uses a primary key that fits only the ignition lock; its secondary key operates the other locks. Until the late 1970s, the secondary key code number could be found on the vehicle's glove-compartment lock. The glove-compartment lock has four tumblers that correspond to the last four tumblers of the trunk lock (the trunk lock has six tumblers).

Fitting a key

Fit a primary key for a pre-1970 GM vehicle by removing the door lock and using the code number stamped on it. To fit a primary key for a 1970 to 1973 model, remove, disassemble, and decode the door lock. Lock decoders are available from locksmith supply houses for this purpose. For a 1974 to 1978 GM vehicle, save time by just pulling the ignition lock and installing a new one. Otherwise, you must disassemble the steering wheel column. A primary key can be fitted to a post-1978 GM model (not including vehicles with VATS or PASSKey) by disassembling the steering wheel column, removing the lock, and using the code stamped on the lock.

To fit a key for a secondary GM lock, remove the glove-compartment plug and cut the key by code. Some GM glove-compartment locks are tricky to remove without using a bezel nut wrench. The plug can be removed as follows: Pick the lock open if it's locked, open the door, and then pick the lock back to the locked position. Insert an ice pick or similar instrument in the small poke hole, and depress the retaining pin. You should be able to remove the plug easily. If no code is available, fit a key to the glove-compartment lock, and use the following GM progression method to find the remaining two cuts.

The GM progression method

Since 1967, GM has adhered to the following three rules for making a factory-original key:

- The sum total of the cut depths must equal an even number.
- There cannot be a more than two cut-depth difference between any adjacent cuts.
- There can never be more than two of the same cut depths in a row.

The GM progression method is based on logically using these rules to progressively decrease the possible key-cut combinations until the proper combination is determined. The first rule refers to the fact that whenever the cut-depth numbers of a GM key are added together, the sum should be an even number. A GM key may have the cut-depth numbers 3-2-4-3-2 (which equals 14), for example, but not 3-2-4-3-3 because the sum of the latter is an odd number.

Suppose that you obtained four cut depths from a glove-compartment lock and need to find the remaining two cuts for operating the trunk lock. If the sum of the four cut depths is an odd number, the two unknown depths also must equal an odd number because any odd number plus any other odd number always equals an even number. If the sum of the four cut depths is an even number, the remaining two must equal an even number because any even number plus any other even number always equals an even number. Likewise, any odd number plus any even number equals an odd number.

The second of the three rules refers to the fact that cuts directly next to each other should never differ by more than two cut depths. For example, a No. 1 cut depth should not directly precede or follow a No. 4 cut depth.

The third of the three rules means that no GM key should have three consecutive cuts of the same depth. A key bitting of 2-2-2-3-1, for example, is forbidden.

Using these three rules, it's easy to use the four cut depths found on a glove-compartment lock to figure out the remaining two cuts for the trunk lock. First, make a key with those four cuts; the first and second spaces on the key should be left uncut. There are only 25 possibilities for the first two cuts (five possible depths for two spaces equals 5^2, or 25). The 25 possible depths for the two remaining spaces are 1-1, 1-2, 1-3, 1-4, 1-5, 2-1, 2-2, 2-3, 2-4, 2-5, 3-1, 3-2, 3-3, 3-4, 3-5, 4-1, 4-2, 4-3, 4-4, 4-5, 5-1, 5-2, 5-3, 5-4, and 5-5.

Based on the second GM keying rule, 6 of those 25 possible depths can be ignored because they have more than two depth increments between them. They are 1-4, 1-5, 2-5, 4-1, 5-1, and 5-2. This leaves only 19 possible depth combinations in any instance where the first two depth cuts must be found for a GM vehicle.

Based on the first GM keying rule, you can eliminate about half those 19 possibilities immediately. You would eliminate either all the odd pairs or all the even pairs depending on whether you need a pair that equals an even number or a pair that equals an odd number. If you need a pair that equals an even number, you would have only the following 11 choices: 1-1, 2-2, 3-3, 4-4, 5-5, 3-1, 4-2, 5-3, 1-3, 2-4, and 3-5. If you need a pair that equals an odd number, you would have only the following 7 choices: 2-1, 1-2, 3-2, 2-3, 4-3, 5-4, and 4-5.

The second GM rule then would allow you to eliminate several more of those pairs. If your four glove-compartment lock cut depths are 1-1-2-3, for example, then the second rule would be violated by preceding those cuts with cuts 3-4, 5-4, or 4-5. If a No. 4 or No. 5 depth cut were next to a No. 1 cut on a key, the key would have adjacent cuts with more than 2 cut-depth differences. This means that only five possible cuts would be available: 1-2, 2-3, 2-1, 3-2, and 4-3.

Using the same glove-compartment lock cut depths as the example, you then would take the key and cut a No. 2 depth in the first space and a No. 1 depth in the second space; then try the key in the lock.

If the key doesn't work, you would then progress to cutting a No. 3 depth in the first space and a No. 2 depth in the second space (both spaces would be cut a little deeper). After cutting three of the five possibilities, you would then need to use another key with the four cuts from the glove compartment on it to cut the remaining two pairs of depths on that key, beginning with the shallowest pair. One of them will operate the lock.

By using the GM progression method, you should never have to waste more than one key blank if you're searching for an odd combination for the two cuts. You should never have to waste more than two blanks if you're searching for an even number for the two cuts. You can use Table 15.2 as a simple way to follow the GM progression method.

There are only two reasons this method can fail: The factory made an error in keying the lock, or someone rekeyed the lock without adhering to GM's rules.

Servicing General Motors Vehicles with VATS

The GM Vehicle Anti-Theft System (VATS), also called the *Personalized Automotive Security System* (PASSKey), has been used in select GM models since 1986. The system has proven helpful in preventing automobile thefts.

TABLE 15.2 General Motors Progression Chart

The following chart can be used for originating a key for the doors and trunk locks of GM vehicles when four cuts are known. Usually, when the vehicle has a keyed glove-box lock, cuts 3 through 6 can be based on that lock. Add those four known cuts together. If the result is an odd number, refer to the odd-number column below. If the result is an even number, refer to the even-column below. By using the listed cuts from top to bottom, starting with the left column, you should be able to progressively cut blanks without ever having to cut more than three to find the correct bitting.

	EVEN CHART (used when the sum of cuts 3 through 6 equals an even number)			ODD CHART (used when the sum of cuts 3 through 6 equals an odd number)	
If third cut is a	First key	Second key	Third key	First key	Second key
1	1-1	2-2	1-3	2-1	1-2
	3-1	4-2		2-3	3-1
	3-3			4-3	
	5-3				
2	1-1	2-2	1-3	1-2	2-1
	3-1	4-2	2-4	3-2	2-3
	3-3	4-4		3-4	4-3
	5-3			5-4	
	5-5				
3	1-1	2-2	1-3	1-2	2-1
	3-1	4-2	2-4	3-2	2-3
	3-3	4-4	3-5	3-4	4-3
	5-3			5-4	4-5
	5-5				
4	1-3	2-2	2-4	1-2	2-3
	3-3	4-2	3-5	3-2	4-3
	5-3	4-4		3-4	4-5
	5-5			5-4	
5	1-3	2-4	3-5	2-3	3-4
	3-3	4-4		4-3	5-4
	5-3			4-5	
	5-5				

VATS is an electromechanical system that consists of the following basic components: a computer module, keys that each have a resistor pellet embedded in them, an ignition cylinder, and a wire harness that connects the ignition cylinder to the computer module.

When a properly cut VATS key is inserted into the ignition cylinder, the cylinder will turn. The resistor pellet in the key will neither hinder nor aid the mechanical action of the ignition cylinder.

If the key is embedded with an incorrect resistor pellet or has no resistor pellet, the computer module shuts down the vehicle's electric fuel pump, starter, and power train management system for about four minutes.

This happens because vehicles with VATS are designed to operate only when one of 15 levels of resistance is present. The 15 levels are represented in 15 different resistor pellets. A VATS key is a standard GM key with one of 15 resistor pellets embedded in its bow.

When a VATS key is inserted into a VATS ignition cylinder, contacts within the cylinder touch the resistor pellet in the key, and the resistor pellet's resistance value is transmitted to the VATS computer module by the wire harness connecting the cylinder to the computer module. Only if the resistance value is the right level can the vehicle be started.

It's important to remember that the turning of the ignition cylinder is a mechanical process that is independent of the system's electronics. Any properly cut key that fits the ignition cylinder can be used to turn the cylinder to the start position. However, unless the VATS control module also receives the correct resistor information, the vehicle won't start.

VATS keys look similar to other late-model GM keys but come with a black rubber bow and contain a resistor pellet. They use standard GM depths and spacings and fit into an A keyway. All VATS key blanks are cut the same way other GM blanks are cut.

When cutting a key for a VATS vehicle, however, it is first necessary to determine which of 15 VATS blanks to use. This can be determined by measuring the resistor value of the pattern key (the one the customer wants duplicated) with an ohmmeter or multimeter and comparing the reading with the VATS pellet's resistor values.

When VATS was used in the 1986 Corvette, each module had a predesignated resistor pellet value. This system was used through 1988.

Starting with the 1988 Pontiac Trans Am GTA, GM began using a modified VATS. The new VATS was called the *Personalized Automotive Security System.* All GM vehicles with VATS manufactured after 1989 use this new system.

For locksmiths, there are two major differences between the two systems. First, the keys used with the new system are 3 mm longer than the keys used with the old system. The old key blanks have to be modified before they can be used to operate the new VATS ignition locks.

The other big difference between the two systems is that the older system had a sticker on the VATS module showing the module's resistor number; the new system doesn't have such stickers.

Making a VATS first key

When no pattern VATS key is available for you to duplicate, you can make a VATS key by first determining the proper bitting in the same way that you would determine the bitting for non-VATS late-model GM vehicles. Then determine the VATS key blank to which to transfer the cuts.

The most expensive way to determine the correct blank is to cut and try a different VATS key blank until you find one that starts the vehicle. The high costs of VATS blanks make this method impractical. A less costly method involves using an ohmmeter or multimeter, an extra VATS ignition cylinder, and 15 different VATS blanks.

Disconnect the wire harness connecting the control module to the cylinder, and connect that wire to the extra VATS cylinder. Insert one of the VATS key blanks into the extra cylinder, and attempt to start the vehicle by using the correctly cut mechanical key to turn the vehicle's ignition cylinder to the start position. If the vehicle shuts down, you'll need to wait four minutes and then repeat the procedure with another VATS key blank in the extra ignition cylinder until the vehicle starts. After the vehicle starts, transfer the cuts from the correctly cut mechanical key to the VATS key blank that allowed you to start the vehicle. Then disconnect the wire harness from your extra ignition cylinder and reconnect it to the vehicle's ignition cylinder.

Using a VATS decoder

A VATS decoder can be very helpful for servicing vehicles with VATS. Several companies manufacture this device. A typical model can be used to perform four functions: identify the correct VATS key blank from the customer's original, decode the correct VATS blank from the vehicle, diagnose steering column connection problems, and diagnose VATS computer problems.

The resistor value of a VATS key can be determined simply by inserting the key into a slot in the decoder. This feature can be useful for quickly finding the right VATS key blank to use to duplicate a VATS key.

The decoder also can help to determine which VATS blank to use when no VATS key is available.

First, you need to cut a correct mechanical key that will turn the vehicle's ignition cylinder to the start position. Then connect the decoder's two tester connectors to the mating VATS connectors at the base of the steering column under the dash. After turning the decoder's key code switch to 1, try to start the engine with your properly cut mechanical key. If the engine doesn't start, press the four-minute timer on the decoder. After four minutes, the timer light will go out.

That lets you know that the vehicle should be ready to try another key code number. Then turn the decoder's key code switch to the next number and try the mechanical key again. Follow this procedure until the vehicle starts. When the vehicle starts, transfer the cuts from your mechanically correct key to the VATS blank that corresponds to the number of the key code shown on the VATS decoder.

Many locksmiths don't appreciate VATS. They don't like having to stock a lot of different expensive key blanks or having to take so much time making a first key for a vehicle. However, it's likely that VATS will be used in GM's vehicles for many years. Anyone who wants to service automotive locks should be prepared to handle vehicles with VATS. This means getting the proper tools and staying informed about changes in VATS.

Honda

Most locks used on Hondas have six-disc tumblers. These are operated with a double-sided convenience key and are easy to pick open. All the six-disc tumbler locks on Hondas are keyed alike. A lot of pre-1976 Hondas use locks that have eight-disc tumblers.

Impressioning is usually the fastest way to fit a key for a post-1976 Honda. Another way is to use the code number found on the door lock.

A key for a Honda made in 1989 or later is 4 mm (0.175 inch) longer from the shoulder to the bow than is a key for an older-model Honda. For pre-1989 Honda Accords, Preludes (from 1982), and Civics, use key blank HD83. For Honda Accords and Preludes made in or after 1989, use key blank HD90. Key blank HD91 can be used for Honda Civics made in or after 1989. If you use the HD83 blank to cut a key for a Honda model made in or after 1989, you might have a hard time removing the key from the ignition lock.

Chapter

16

Opening Locked Cars

While the automobile has been with us since the beginning of the twentieth century, the lock was adopted slowly. However, by the late 1920s nearly every auto had an ignition lock, and closed cars had door locks as well. Current models can be secured with half a dozen locks. This chapter explains how to open and service all kinds of vehicles.

Opening Locked Cars

Car opening can be a lucrative part of any locksmithing business. For some, it's the biggest source of income. To offer car-opening services, you only need a few inexpensive tools and some technical knowledge. In this chapter I show you how to buy and make the tools you need and give you detailed instructions on how to open most cars. I also tell you about the business matters you need to know.

In the interest of self-disclosure, I should point out that several years ago I was hired by a major automotive lock manufacturer to prepare and edit its car entry manual, which included creating new entry techniques and designing tools. At the time, it was one of the most comprehensive and best-selling publications of its kind. Although the manual is out of date, copies are still being sold. I no longer work on that publication. The suggestions and tool designs I give here are original and aren't meant to promote any company's products.

Tools you need

Car-opening tool sets sold through locksmithing supply houses may include 40 or more tools. Toolmakers point out that the variety is necessary (or at least helpful) because of the constant lock-related changes made to new cars. Some of the uniquely shaped tools are designed for one specific make, model, and/or year of car. Whether or not all the new specialized tools are worth the money is debatable. But a continuous supply of new tools means recurring revenue for the toolmakers.

You can open most cars with only five simple tools, all of which you can make yourself. In some cases, not only will it be cheaper, but the tool will work better if you make it yourself. Later, I'll tell you how to make the tools. The most important car opening tools are a slim jim, a hooked horizontal linkage tool, an L tool, a J tool, and an across the car tool (or long reach tool). They have different ways of reaching and manipulating a car's lock assembly and lock buttons.

The slim jim is a flat piece of steel with cutouts near the bottom on both sides (Fig. 16.1). The cutouts let you hook and bind a linkage rod from either side of the tool. The tool can also be used to push down on a lock pawl. Slim jims come in different widths. It's good to have a wide one and a thin one. You can buy them at most automobile supply and hardware stores. You can get better models from a locksmith supply house. They're often sturdier, have more notches, have a handle, and just generally look more professional. To make your own, you'll need a 24-inch piece of flat steel or aluminum, from 1 to 2 inches wide. You can use a ruler or other item that's the right size made from the proper material. Just draw the slim jim shape onto the metal and then grind away the excess material.

Hooked horizontal linkage tools go by many names and come in all kinds of sizes and configurations. The small hook on the end of the tool lets you catch and bind a horizontal rod and slide it to unlock the door. Some hook down onto

Figure 16.1 A slim jim shown outside of a door.

the rod (Fig. 16.2); others hook from the bottom of the rod (Fig. 16.3). It's good to have both of them. Two other kinds of horizontal linkage tools include the three "fingers" type that spreads to clamp onto the rod (Fig. 16.4), and the tooth-edged type that bites into the rod (Fig. 16.5). I don't like either of those two, because when using them, you have to be especially careful to avoid bending the linkage rods.

The J tool is one of the easiest to use (Fig. 16.6). It goes within the door, between the window and weather stripping, and then under the window and beneath the lock button to push the button up to the unlocked position.

An L tool is used to push or pull on bell cranks and lock pawls (Fig. 16.7). You can also use it to access the lock rod by going under the lock handle.

For versatility, buy or make a tool that is an L tool on one end and a J tool on the other. The part of the tool that enters between the door needs to be a specific shape; the rest of the tool is the handle. It's useful to make or buy tools that have a different or a different-size tool on each end.

The across the car tool is a 6-foot (or longer) piece of 3/16-inch round stock bar with a small hook on one end. Its name comes from the fact that the tool can be used to enter a window and reach across the car to get to a lock or window button. But sometimes you use it on the same side of the car on which you inserted it. Most of the ones you buy come in three pieces, and you screw them together before each use. They often bend and break at the joints. If you buy one in three pieces, you should braze the pieces together. But it's best to make your own out of one piece of steel.

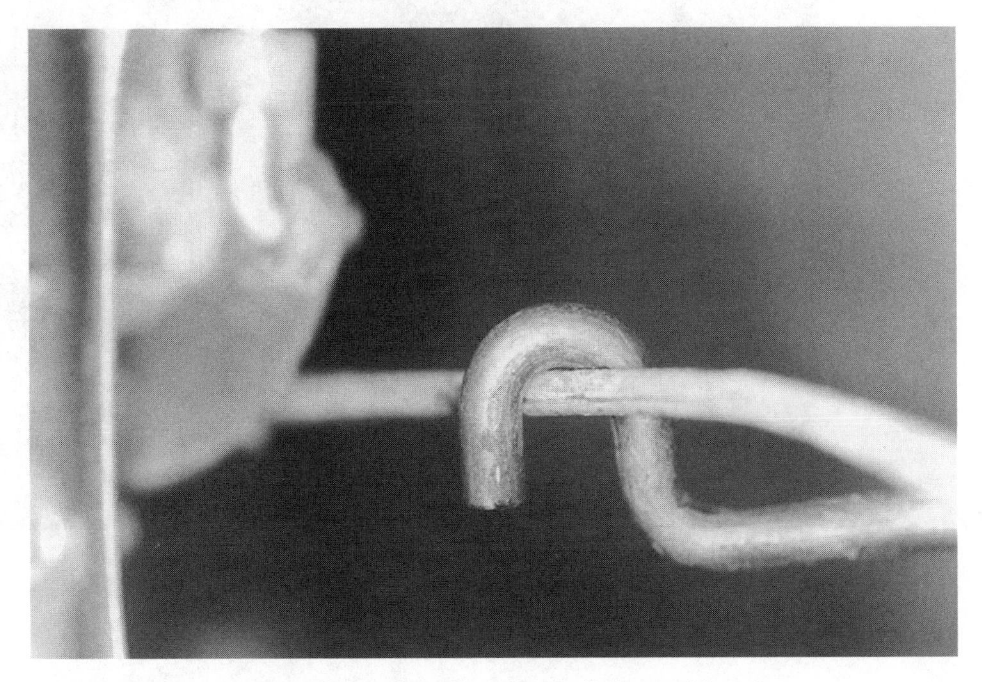

Figure 16.2 Some car-opening hook tools catch and bind a linkage rod from the top.

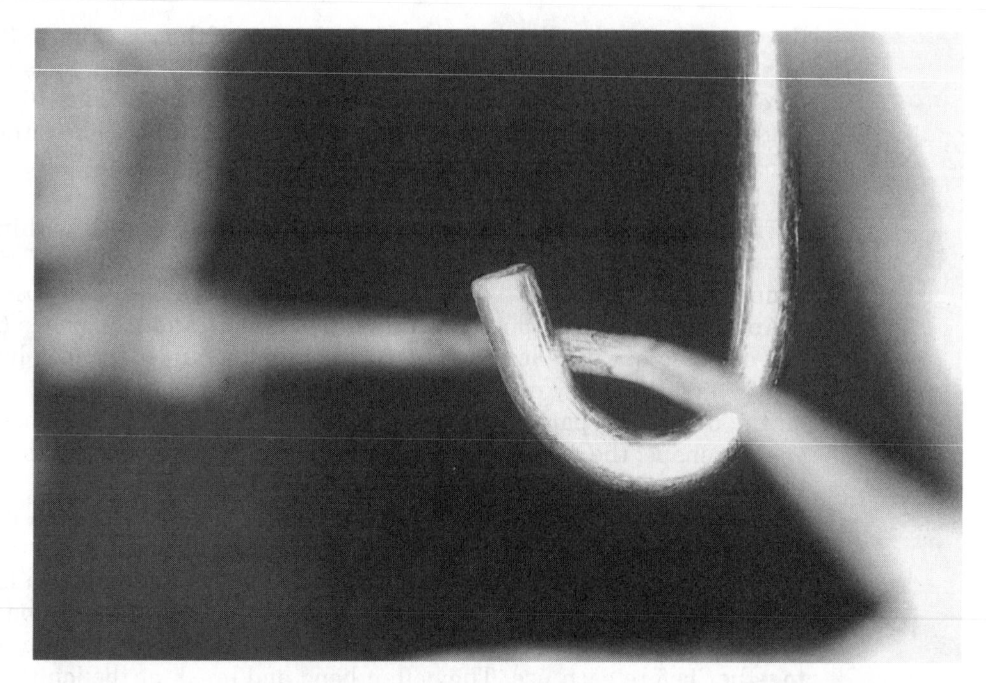

Figure 16.3 Some car-opening hook tools catch and bind a linkage rod from the bottom.

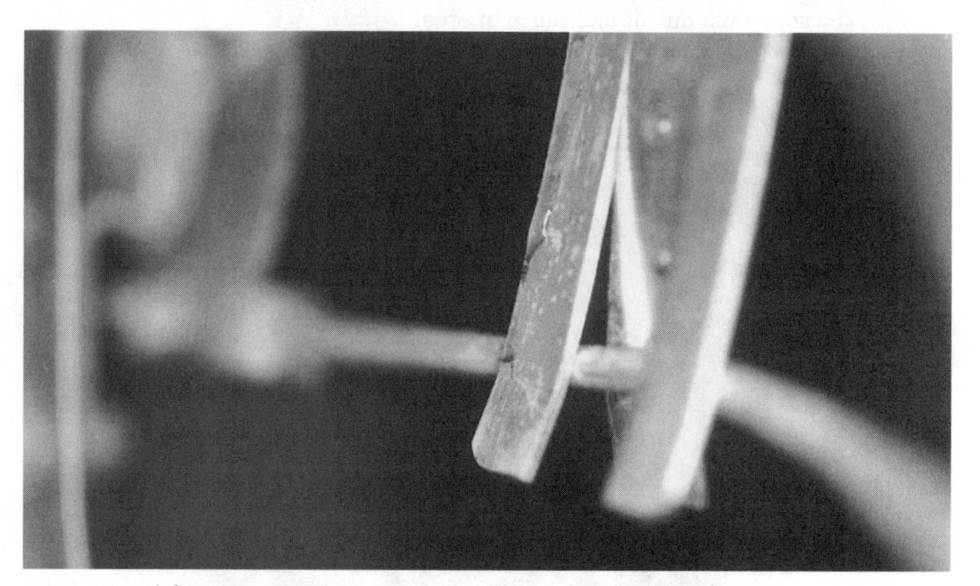

Figure 16.4 A finger-type car-opening tool spreads to grip the linkage rod.

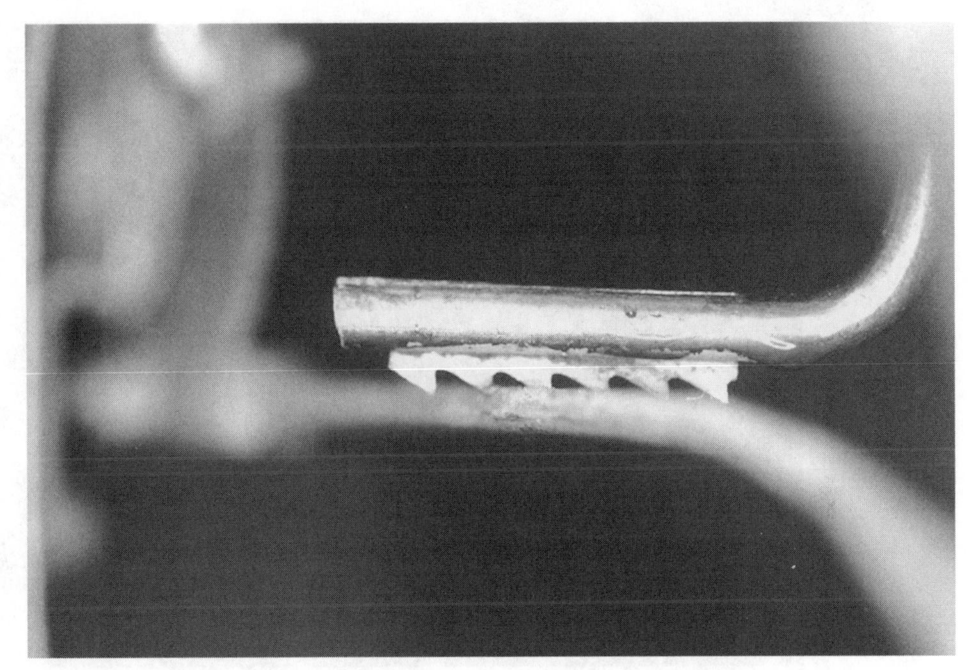

Figure 16.5 A car-opening tool with teeth grips a rod by biting into the metal.

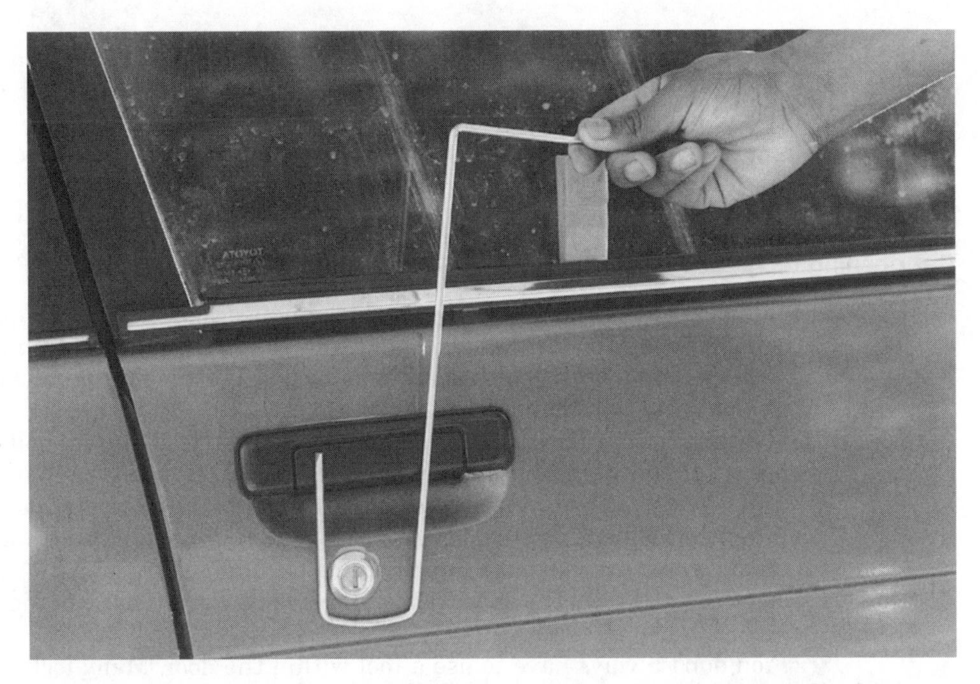

Figure 16.6 A J tool shown outside the car. The tool goes beneath a lock button and pushes it up to unlock a vehicle.

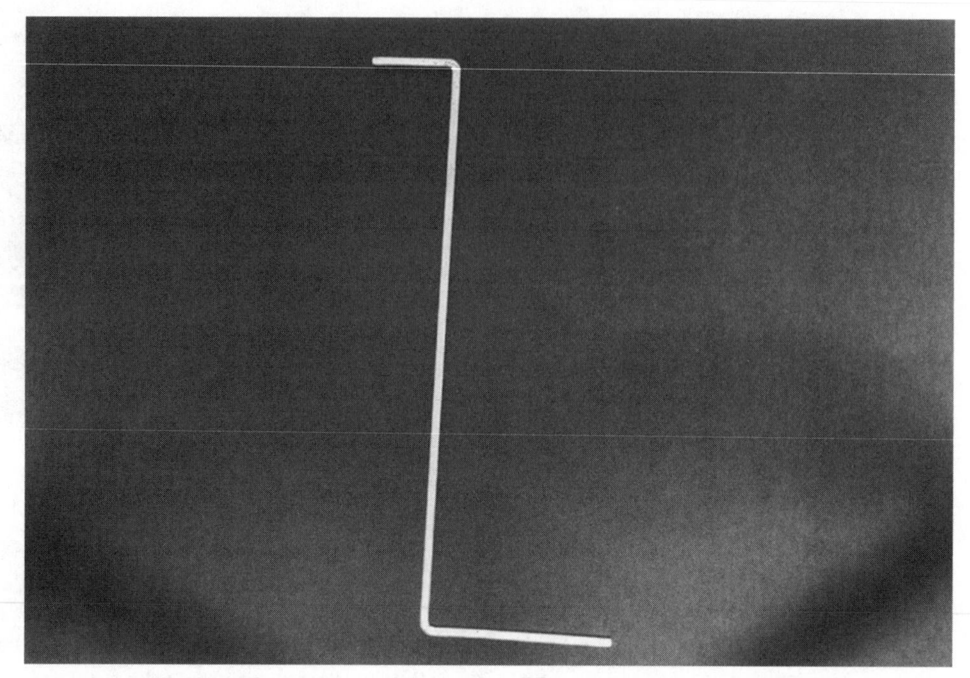

Figure 16.7 An L tool of the right size can reach and manipulate a lock pawl or bell crank.

To use those car-opening tools, you'll also need a flexible light and a couple of wedges. The wedges should be made of plastic, rubber, or wood. The wedges pry the door from the window to allow you to insert the light and tool. The light lets you see the linkage assembly so you can decide what tool to use and where to place the tool.

Car opening techniques

With most cars, there are many good techniques that will let you quickly open them. A locksmith who opens a lot of cars will tend to favor certain techniques. There's nothing wrong with that. Whatever way gets you in quickly and professionally without damaging the vehicle is fine.

Parts of the car to reach for opening include the lock button, bell crank, and horizontal and vertical linkage rods. The bell crank is a lever that connects to a linkage rod that's connected to the latch or another linkage rod. One popular style of bell crank is semicircular (Fig. 16.8); another style is L-shaped. A horizontal rod, as the name implies, runs parallel to the ground. A vertical rod runs vertically from the lower part of the door toward the top of the door (often to a lock button).

You don't always have to use a tool within the door. Many locks are easy to impression or pick open. Standard torque wrenches used for deadbolt and key-

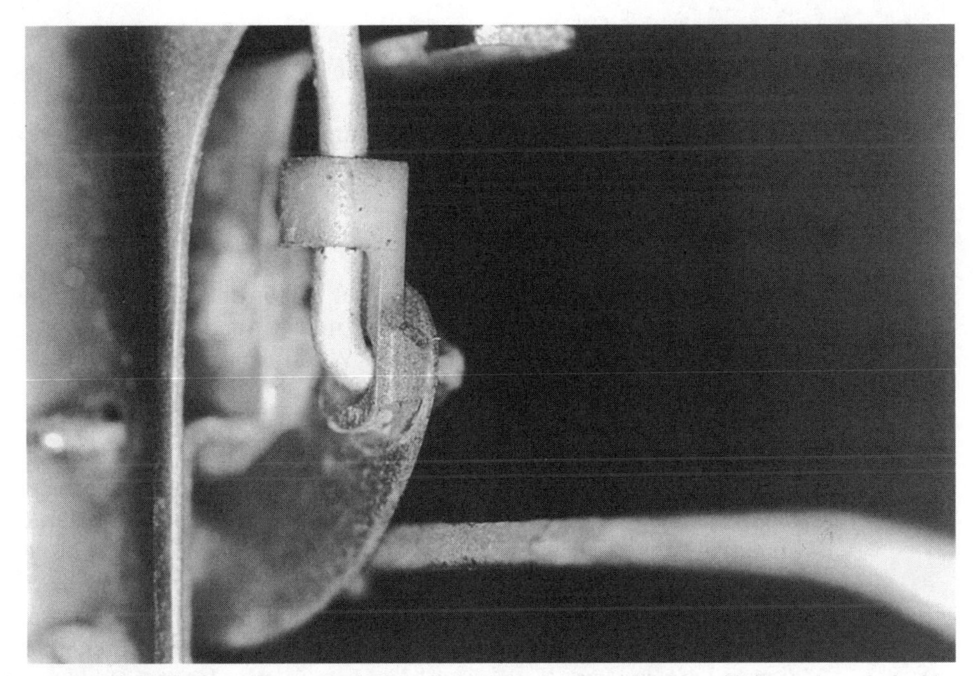

Figure 16.8 Linkage rods are attached to the bell crank. Many times, a vehicle can be unlocked by pushing or pulling on the bell crank.

in-knob locks don't work as well when picking a car lock. To make a better torque wrench for cars, grind the small end of a hex wrench.

When approaching an unfamiliar car model, walk around it, looking through the windows. As you walk around the car, consider the following:

1. Does it have wind wings (vent windows)?
2. Is there a lock button at the top of the door?
3. Are there any gaps around the doors and trunk where you may be able to insert an opening tool?
4. What type of linkage is used?
5. Can you gain access to the vehicle by removing the rear view mirror?
6. Can you manipulate the lock assembly through a hole under the outside door handle?
7. What type of pawl is used? As a rule, pre-1980 locks have free-floating pawls, and later models have rigid pawls.

Using a J tool

If the vehicle has a lock button on top of the door, you may be able to open it with a J tool. First, insert a wedge between the door's weather stripping and window, to give you some space for the tool (Fig. 16.9). Insert the J tool into the door until it passes below the window. Then turn the tool so its tip is under the lock button (Fig. 16.10). Lift the lock button to the open position (Fig. 16.11). Carefully twist the tool back into the position in which you had inserted it, and remove the tool, without jerking on it, before removing the wedge.

The Long Reach Tool

If you learn to use it, the long reach tool will be one of the most useful car opening tools you have. You can quickly unlock about 90% of vehicles with it—including many of the latest models. When you use the tool it's like you have a very long and very skinny arm. The tool lets you reach inside a crack of a car door to push, pull, press and rotate knobs and buttons. You can even use it to pick up a set of keys.

To use the long reach tool, first you place an air wedge near the top of a door to pry the door open enough to insert the tool. (Sometimes you may need to use an extra wedge.) Use a protective sleeve at the opening, and slide the tool into the sleeve. The protective sleeve is to prevent the tool from scratching the car. (You could also use cardboard or the plastic label off of a bottle of soda pop.)

Most of the long reach tools you can buy are about 56 inches long; that isn't always long enough. If you purchase one, get the longest you can find. You can make your own with a 6-foot-long, ¼-inch-diameter stainless steel rod. On one side

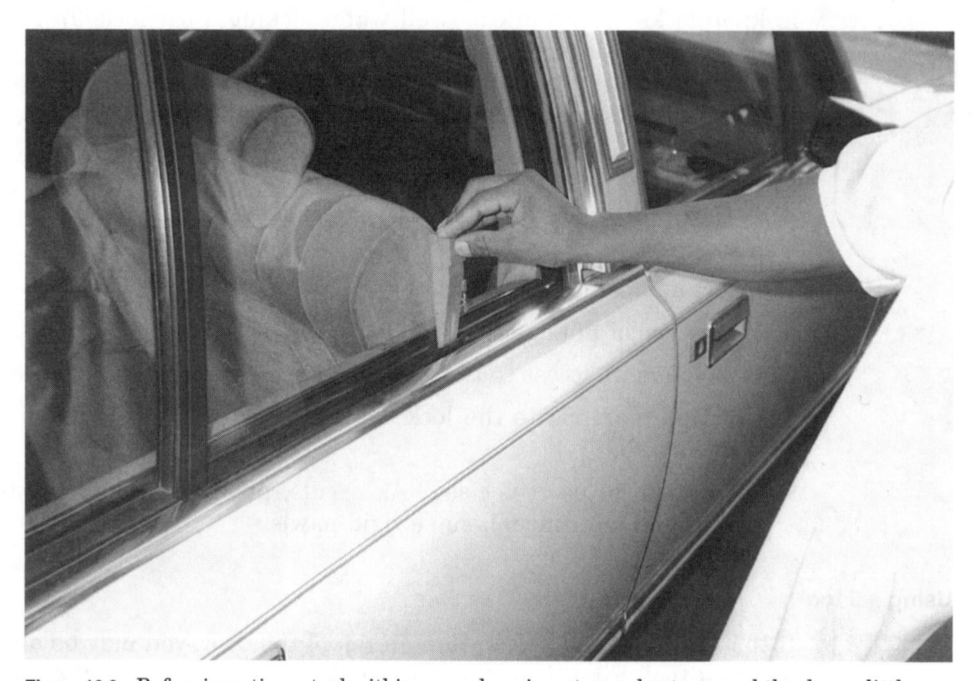

Figure 16.9 Before inserting a tool within a car door, insert a wedge to spread the door a little.

Figure 16.10 The J tool goes within the door near the lock button.

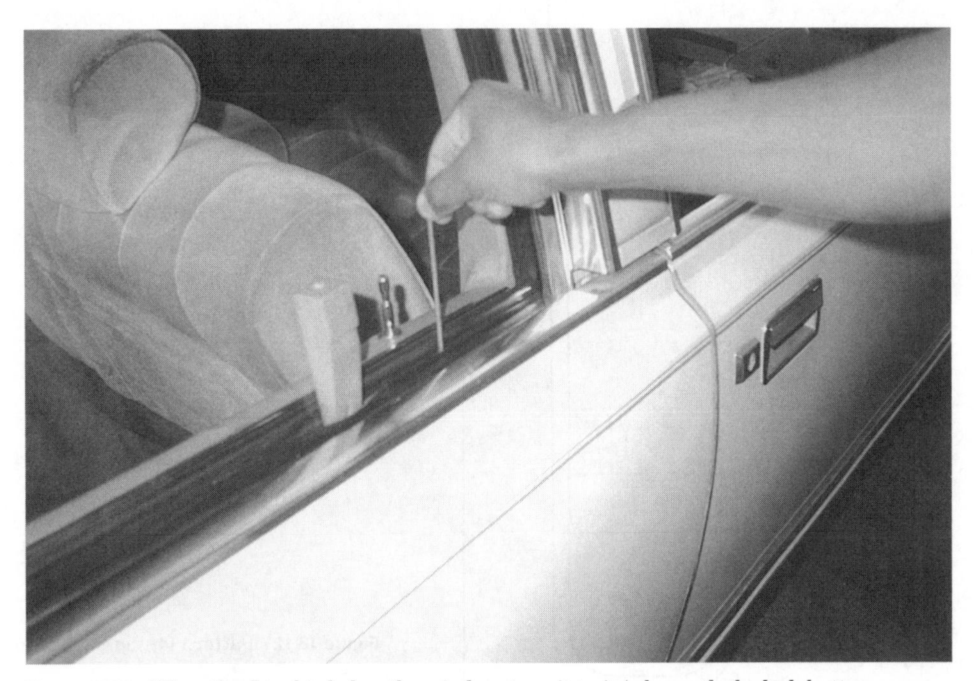

Figure 16.11 When the J tool is below the window, turn it so it is beneath the lock button.

of the rod make a 1-inch bend at a 90-degree angle. Dip that 1-inch bend into plasti-dip or some other rubber-like coating (to give it a non-scratch coating).

Making other tools

You can find supplies at many hardware and home improvement stores to make your own car-opening tools. You'll need flexible flat stock and bar stock of different sizes. See Figs. 16.12 to 16.14 for patterns for making some useful tools.

Another tool that I like a lot can be made from the plastic strapping tape that's used for shipping large boxes. It's hard to find it for sale in consumer outlets. I get mine for free from department stores before they throw it away. Take about 2 feet of strapping tape, fold it in half, and glue a small piece of fine sand-

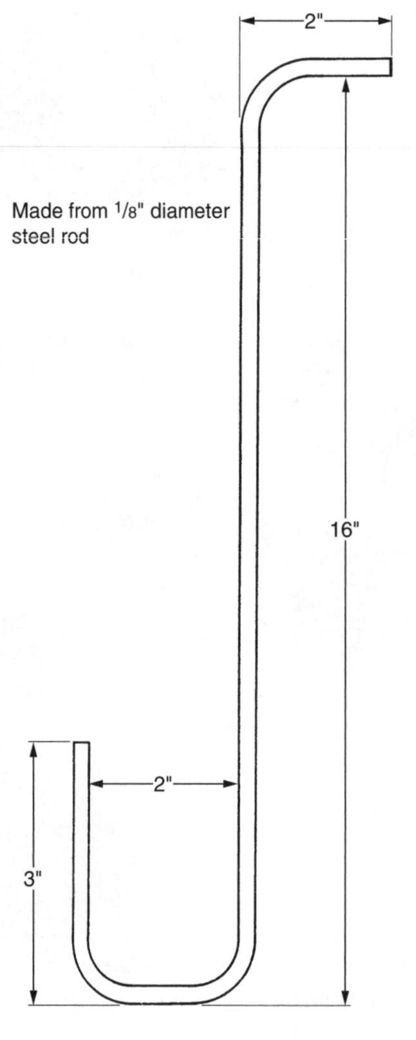

Figure 16.12 Pattern for making a J and L tool.

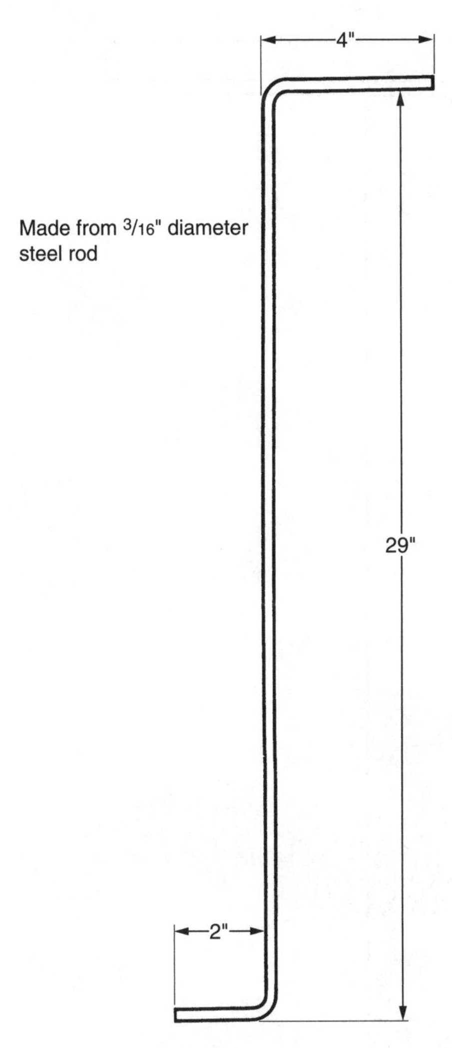

Figure 16.13 Pattern for making a double L tool.

paper to the center (Figs. 16.15 to 16.17). When it dries, you have a nice stiff tool that can easily slide between car doors to loop around a lock button to lift it up. It works like the J tool but from the top of the button instead of the bottom. The sandpaper isn't critical, but it helps the tool grab more easily.

Business considerations

Often a locked-out person will call several locksmiths and give the job to the one that gets there first. Or he or she may get the door open before the locksmith gets there. Either way, you may not be able to collect a fee, unless you made it clear when you received the call that you have a minimum service charge for going on all car-opening service calls.

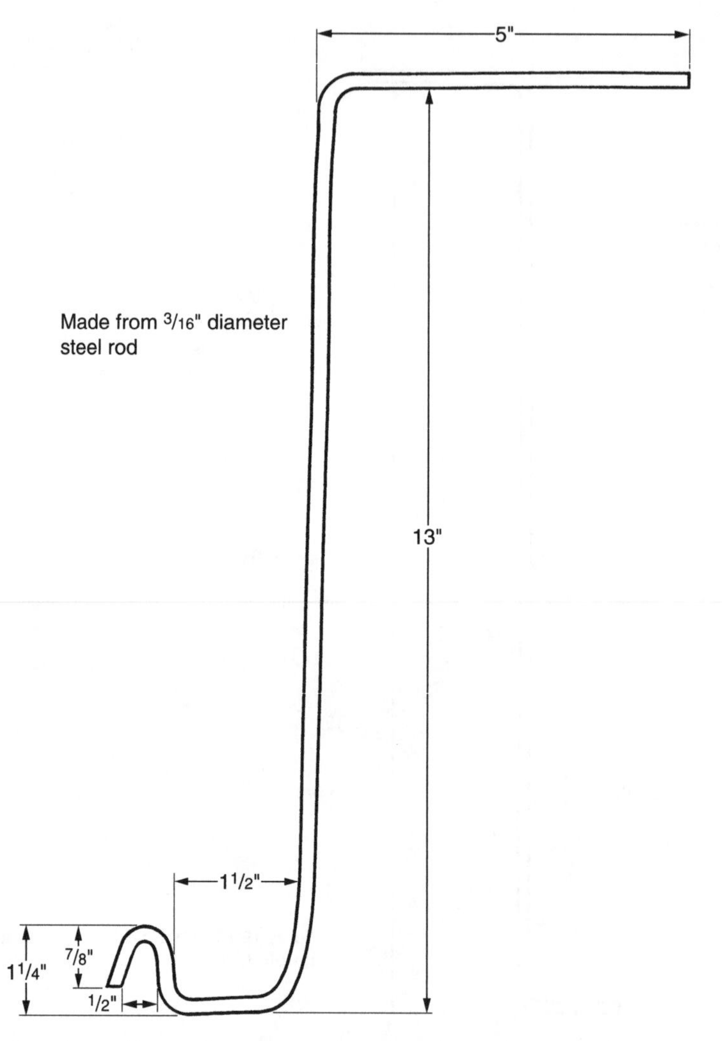

Figure 16.14 Pattern for making a down-hooking horizontal rod/L tool.

Before working on a door, ask what attempts have been made to open the door. To improve your chance of getting an honest answer, ask in a manner that sounds like you're just gathering technical information to help you work. If you learn that someone has been fooling with a door, don't work on that one. You don't want to be held responsible for any damage someone else may have caused.

To open a lock with vertical linkage rods, you can often use a slim jim to pull up the rod to the unlocked position, or use an under the window tool to lift up the lock button. Before using an under the window tool at a tinted window, lubricate the tool with dishwashing liquid. That will reduce the risk of scratching the tint off. You may also be able to use an L tool to pull up the bell crank, which is attached to the vertical linking rod.

Figure 16.15 Fold the strapping material in half.

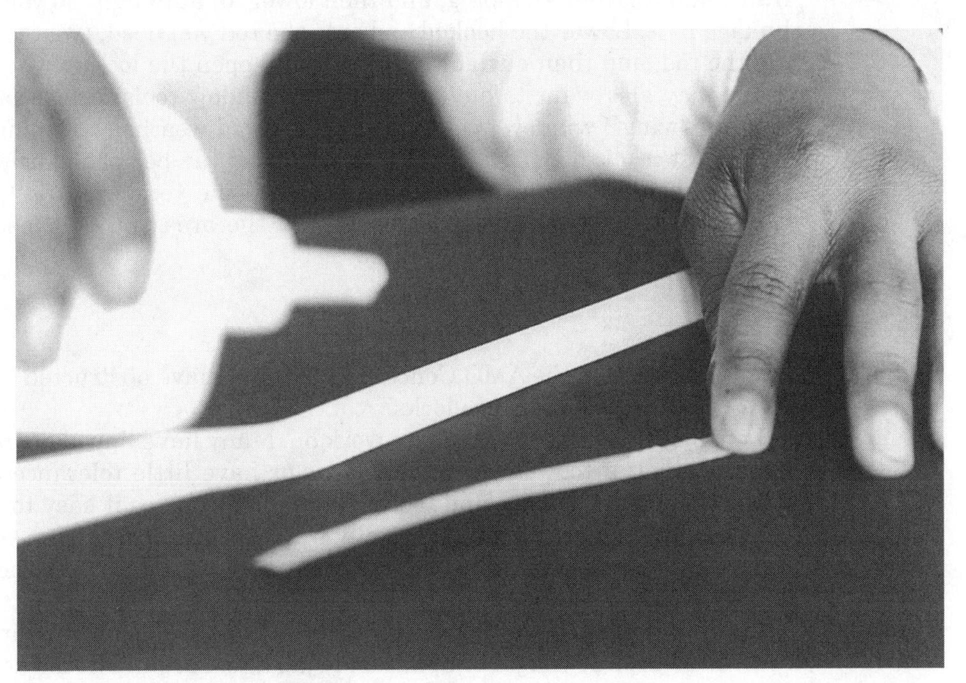

Figure 16.16 Glue a piece of sandpaper to the center.

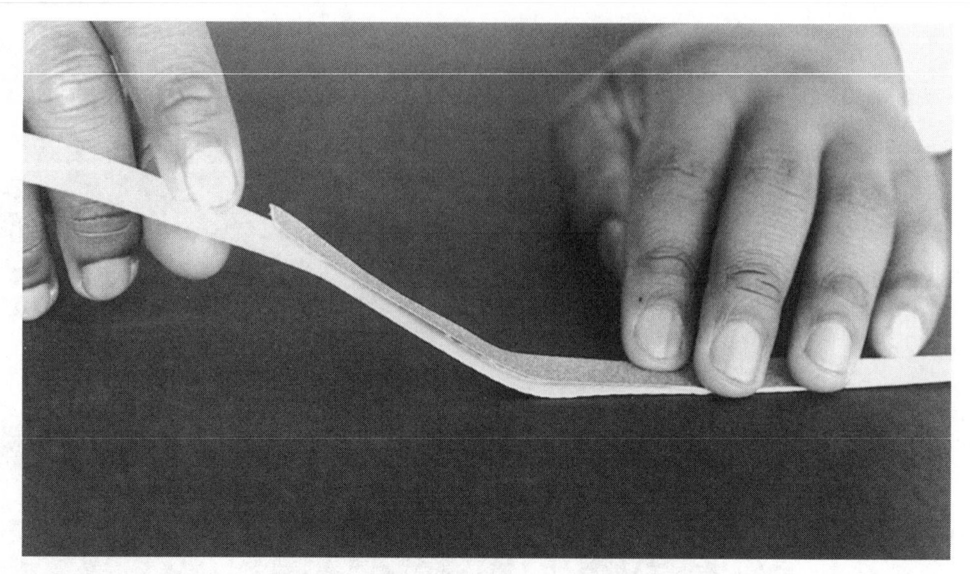

Figure 16.17 Affix the sandpaper, let dry, and refold.

To use a hooked horizontal rod tool, first insert a wedge between the door frame and weather stripping, and then lower an auto light so you can see the linkage rods. Lower the hooked end onto the rod you need, twist the tool, binding the rod, and then push or pull the rod to open the lock.

You may also want to buy a set of vent window tools for special occasions. Vent (or "wing") window opening is easy, but old weather stripping tears easily. In most cases, if it has a vent window, the car can be opened using basic vertical linkage techniques. If you decide to use the vent window, lubricate the weather stripping with soap and water at the area you will insert the tool. Then take your time, and be gentle.

Special considerations

Some models, like the AMC Concord and Spirit, have obstructed linkage rods. It may be best to pick those locks.

Late model cars can be tricky to work on. Many have lipped doors that make it hard to get a tool down into the door or have little tolerance at the gaps where you insert wedges and tools. The tight fit makes it easy to damage the car. Also, the owners may be especially watchful of any scratches you make. To reduce the risk of scratching the car, use a tool guard to cover the tool at the point it contacts the car.

Be careful when opening cars that have airbags. They have wires and sensors in the door. If you just haphazardly jab a tool around in the door, you may damage the system. Use a wedge and flexible light and make sure you can see what you're doing.

Why people call you to open their car

A lot of people know about using a slim jim. They're sold in many hardware and auto supply stores. And many people know about pushing a wire hanger between the door and window to catch the lock button. People typically try those and other things before calling a locksmith. They call a locksmith because it's freezing cold, late at night, raining, or all three, and they grow tired of trying to unlock it themselves. Newer model cars are harder to get into using old slim jim and wire hanger techniques.

People seldom break their windows on purpose, even in emergencies. Replacing a car window is expensive and inconvenient, and there's a psychological barrier to smashing your own car window. I've been called to unlock cars that have young children in them on hot days. That isn't an uncommon situation.

Car-Opening Dispatch Procedure

Having a good dispatch protocol will help you stay out of legal trouble, get the information you need to unlock the vehicle quickly, and make sure you get paid. Modify this protocol to fit your needs:

1. Speak directly to the owner or driver of the vehicle and not to a middle-person. If the owner or driver can't come to the telephone, don't go to the job.
2. Have the person verbally confirm that he or she wants you to do the job and is authorized to hire you.
3. Always quote an estimated price (or the complete price) and a minimum service call fee. Explain that the service call fee is for the trip and will be charged even if no other services are performed.
4. Ask how the charges will be paid (credit card, cash, or check). Explain that all charges must be paid in full and are due upon your arrival.
5. Get the make, model, year, and color of the vehicle and its license plate number.
6. Get the exact location of the vehicle. If the customer isn't sure, ask to speak to someone who is.
7. Get a phone number to call back on, even if it's a pay phone. Tell the person someone will call back in a moment to confirm the order.
8. Call the phone number to confirm that someone is really there. If no one answers, don't go to the job.
9. When you get to the job, ask to see identification, and make sure keys are in the car. Also, have the person sign an authorization form.
10. When you open the car, grab the keys and keep them until you've been paid. Any hassles about payment, toss them back in the car and close the locked door.

Chapter

17

Emergency and Forced-Entry Procedures

Frequently, locksmiths gain entry into locked places by impressioning or picking open a lock. A good locksmith also can gain entry through special techniques called *Emergency Entry Procedures* (EEPs). It might not seem like good locksmithing practice to force a lock open, but economics, time, and other requirements may make it necessary. In an emergency, or when a lock resists picking and impressioning, the EEPs are used. Even though you may rarely use EEPs, you need to know them to help make buildings more burglary resistant.

As a professional locksmith, you should have the proper equipment to make emergency entries as quickly and neatly as possible. Some entry tools are sold through locksmith supply houses; some tools you can make yourself. Experience will show you which tools and techniques work best for you.

Several preliminaries should be undertaken when preparing to make an entry using EEPs. Try to get as much information as possible. You can then best decide what methods will work. Ask questions about the type of lock, the key numbers, nearby open windows, a possible extra key held by someone else, the condition of the lock, alarms and other devices attached to the door, and so on. All this information will aid your preparations for the entry. You don't want to damage your customer's property when it isn't necessary.

Drilling Pin Tumbler Locks

Drill a lock only as a last resort. Two drilling methods can be used with pin tumbler locks. The first method involves drilling of the cylinder plug. This has the advantage of saving the cylinder itself. The inner core can be replaced. This method destroys the lower set of pins just below the shear line, allowing the plug to be turned. Follow these procedures to drill a cylinder plug:

1. Drill the plug below the shear line.
2. Insert a key blank and a wire through the drill hole to keep the upper pins above the shear line and the destroyed pins below it.
3. Turn the cylinder core and open the lock.

Once the door is open, dismantle the lock; remove and replace the core, and fit new pins to the lock. If you used the plug follower when removing the core, you will only have to fit new lower pins to the lock to match a key.

The second method involves drilling just above the shear line into the upper pins.

1. Insert a key blank into the lock to push all the upper pins to the upper pin chambers.
2. Drill about ⅛ inch above the shear line or shoulder of the plug and directly above the top of the keyway (Fig. 17.1). A drilling jig can be useful to ensure that you drill at the proper point on the lock face. Use a ⅛- or ³⁄₃₂-inch drill bit.
3. Poke a thin wire or needle into the hole, and withdraw the key to just inside the keyway. Then use the tip of the key to turn the core. Removing the key partway allows the bottom pins to drop below the shear line while the wire or needle keeps the upper pins above the shear line.

Once the door is open, dismantle the lock, remove the cylinder, and replace it with a new one.

Cylinder Removal

Sometimes the cylinder must be removed to open a lock. This means shearing off the screws that hold the cylinder in place. Several different rim cylinder

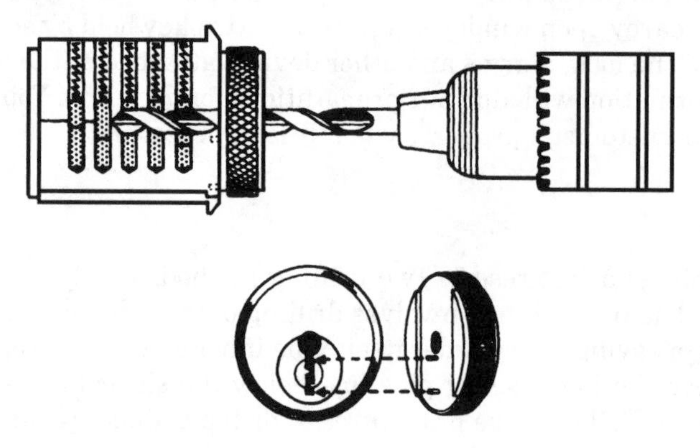

Figure 17.1 A drilling jig can help you drill straight. (*Desert Publications*)

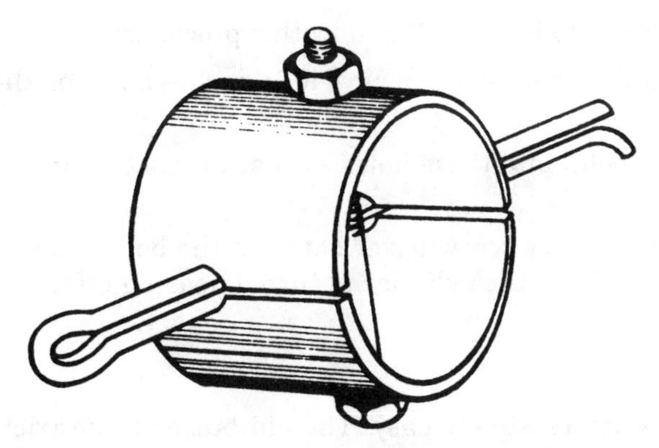

Figure 17.2 A cylinder removal clamp. (*Desert Publications*)

removal tools are available. The one in Fig. 17.2 can be made in your own shop. It's called a *cylinder removal clamp*. To make one, follow these steps:

1. Use a piece of steel tubing about ⅛ inch larger in diameter than the cylinder. The tubing should be no more than 2½ inches long.
2. Cut into the tubing about 3⁄16 inch.
3. Cut a center hole, ¼ to ⅓ inch in diameter at the end of both cuts, and insert a steel rod about 10 inches long. The rod becomes your handle for turning the cylinder.
4. Drill two holes in the tubing so that you can insert a bolt perpendicular to the handle. Ensure that the bolt will be ½ inch in front of the handle. The bolt threads must extend on both sides of the tubing.

To remove the cylinder, follow this procedure:

1. After first removing the cylinder rim collar, set the removal clamp over the edge of the cylinder. Tighten down the tension bolt; this provides the gripping pressure on the cylinder.
2. Twist the cylinder clamp by the handles, and force the cylinder to rotate, shearing off the retaining screw ends.
3. Remove the cylinder from the lock. When removed, reach in and open the door by reversing the bolt.

You also can use a standard Stillson wrench to remove the cylinder.

Cylinders also can be pulled out of the lock unit, but this method ruins the cylinder threads, requiring you to replace the entire unit. The tool used is called a *nutcracker*. The sharp pincer points are pushed in behind the front of the cylinder face. Clamp down and pull the cylinder out. Many times an unskilled locksmith can ruin the entire lock and a portion of the door by not knowing how to use this tool properly.

Some cylinders have to be drilled. Follow this procedure:

1. Drill two 3⁄16-inch-diameter holes about 7⁄8 to 1⅛ inches apart on the face of the cylinder.
2. Drive two heavy bolts into these holes so that at least 1½ inches are sticking out.
3. Place a pry bar or heavy screwdriver between the bolts, and use it as you would a wrench to force the cylinder to turn, shearing off the long screws.

Window Entrances

Window entrances are relatively easy. The old butter knife trick is usually successful. Since most window latches are located between the upper and lower windows, sliding a knife up between the windows allows you to work open the latch.

Should the area be too narrow for a knife, shim, or other device, drill a 1⁄16-inch hole at an angle through the wood molding to the base of the catch. Insert a stiff wire, and push back the latch.

Office Locks

Most office equipment can be opened using a few basic methods and a handful of tools.

Filing cabinets

Although most filing cabinets have locks in essentially the same position, the locking-bar arrangement for the drawers will vary. These variations have to be considered in working on the locks.

One method is to work directly on the lock itself. Slide a thin strip of spring steel ⅛ inch thick into the keyway, and pull the bolt downward. This will unlock the lock (Fig. 17.3).

If the lock has a piece of metal or a pin blocking access to the locking bolt, use a piece of stiff wire with one end turned 90°. Insert the wire between the drawer and the cabinet face, and force the bolt down with the wire. This will allow you to open the drawers.

If there is not enough room for you to work with the wire, use a piece of thin steel or a small screwdriver to pry back the drawer from the cabinet face to allow you to see and work with the wire.

You also can open the drawers individually if necessary.

1. Use a thin piece of spring steel or a wedge to spread the drawer slightly away from the edge.
2. As your opening tool, use another strip of steel about 18 inches long, ½ to 1 inch wide, and 0.020 inch thick.

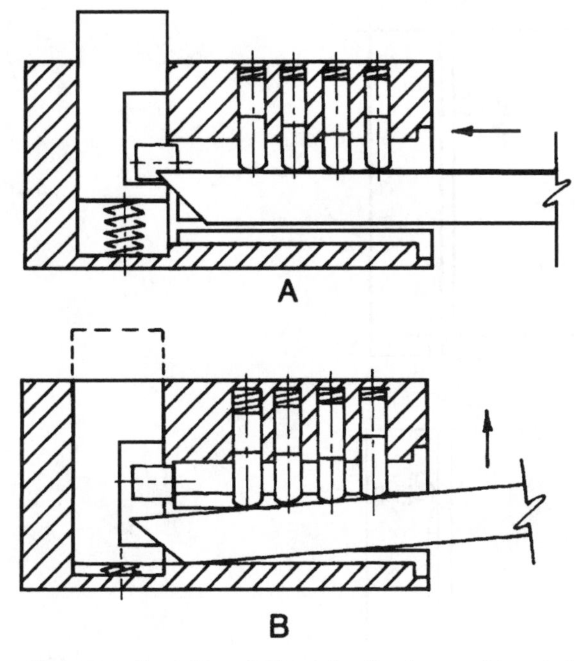

Figure 17.3 Sometimes a thin strip of spring steel inserted into the keyway of a filing cabinet lock (*A*) can be used to pull the bolt down (*B*).

3. Insert the opening tool between the drawer catch and the bolt mechanism (Fig. 17.4).
4. With a healthy yank, pull the drawer open. The opening tool creates a bridge for the drawer catch to ride on and pass the bolt.

Other filing cabinets can be inverted to release gravity-type vertical engaging bolts. When the lock mechanism itself is fouled up, the best way to proceed is to drill out the cylinder and replace the entire assembly.

Desks

Desks with locking drawers controlled from the center drawer can be opened in a couple of ways besides picking and drilling.

Look at the desk from underneath to see what the locking mechanisms for the various drawers look like. Notice that the locking bar engages the desk by an upward or downward pressure depending on the bolt style. Closing the desk drawer all the way pushes the bolt into the locked position. Herein lies the weakness in the desk's security. The bolt usually needs to be pushed up from under the desk by hand to open most of the drawers. The center drawer has its own lock.

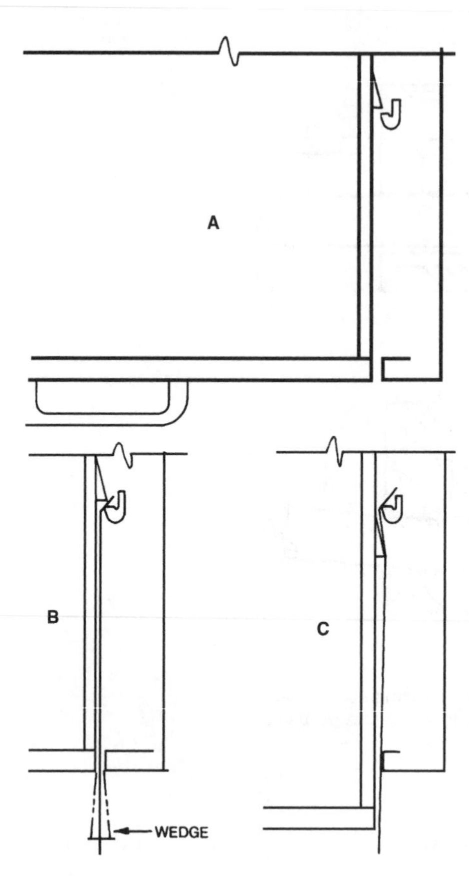

Figure 17.4 Sometimes the drawer of a filing cabinet (*A*) can be opened by wedging the drawer away from the cabinet frame (*B*) between the drawer catch and the bolt mechanism (*C*). (*Desert Publications*)

In other desks, you might have to use a little force and pull outward on the center drawer to push the bolting mechanism downward slightly to open the various drawers.

To open the center lock, follow this procedure:

1. Use two screwdrivers and some tape or cardboard. Put the cardboard or tape between the drawer and the underside of the desktop so that you don't mar the desk.
2. Insert a screwdriver, and pry the drawer away from the desktop.
3. With the other screwdriver, pull the drawer outward to open it. With practice, this can be done with only one screwdriver.

If you drill a small hole in the drawer near the lock, you can insert a piece of stiff wire, such as a paper clip, to push down the plug retainer ring. In doing so, you pull the plug free of the lock, causing the bolt to drop down into the open position.

Doors

Possibly the simplest way to open most doors is with a pry bar and a linoleum knife. Insert the linoleum knife between the door and the jamb with the point tipped upward. Insert the pry bar as shown in Fig. 17.5. Exerting a downward motion on the pry bar spreads the door slightly and allows you to disengage the locking safety latch.

When this is done, bring the linoleum knife forward, pushing the latchbolt into the locking assembly and opening the door. If there is no safety catch, the knife alone can be used to move the bolt inward. It's also possible to use a standard shove knife or even a kitchen knife.

Sometimes the deadlatch plunger is in the lock, but there isn't room to insert a pry bar. What do you do? Use wooden wedges. Insert one of them on each side of the bolt, about 4 to 6 inches from the bolt assembly. Spread the door away from the jamb. Then use a linoleum knife to work the bolt back.

Some doors and frames have such close clearances that you cannot insert a wooden wedge or pry bar. Instead, use a stainless steel door shim. Force it into the very narrow crevice between the door and the frame, and work back the bolt.

Often a door lock can be opened with a Z-wire. This tool is made from a wire at least 0.062 inch thick and 10 to 12 inches long (Fig. 17.6). Insert the Z-wire between the door and the jamb. When the short end is all the way in, rotate it toward you at the top. As you do this, the opposite end will rotate between the door and the jamb. It contacts the bolt and retracts it. If the bolt binds, exert pressure on the knob to force the door in the direction required.

Sometimes you may be required to open locked chain latches. You can, of course, force the door and break the chain, but there are better ways. The rubber band technique works most of the time.

1. Reach inside and stick a tack in the door behind the chain assembly (Fig. 17.7).
2. Attach one end of a rubber band to the tack; attach the other end to the end of the chain.

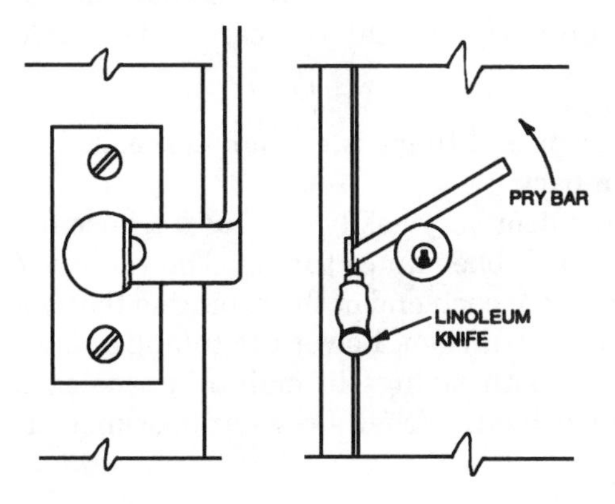

Figure 17.5 Opening a door using a pry bar and linoleum knife. (*Desert Publications*)

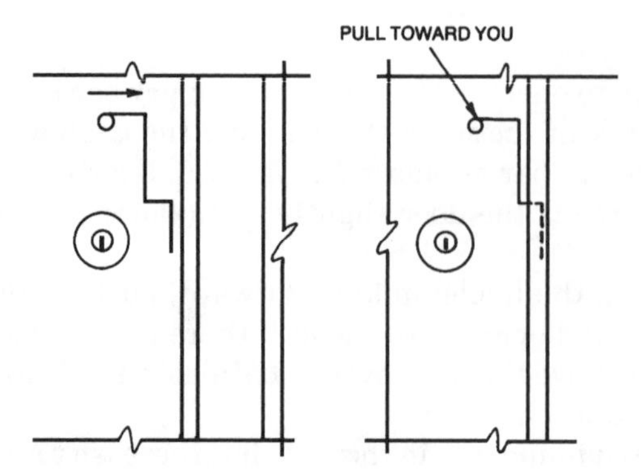

Figure 17.6 Opening a door with a Z-wire. (*Desert Publications*)

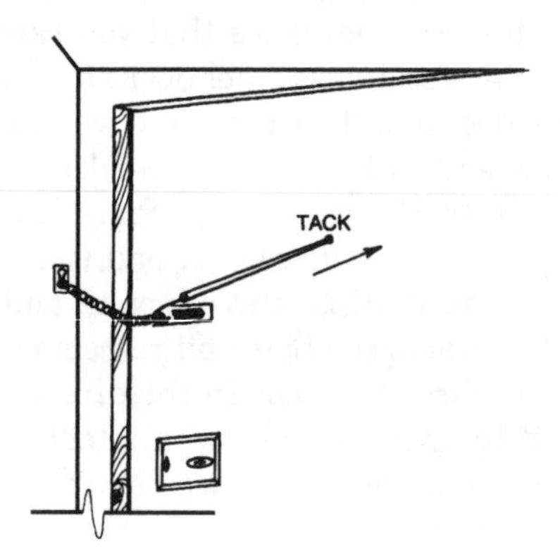

Figure 17.7 The rubber band technique is usually the easiest way to unlock most chain locks.

3. Close the door. The rubber band will pull the chain back. If it doesn't pull the chain off the slide, shake the door a little.
4. If the door surface will not receive a tack, use a bent coat hanger to stretch the rubber band (Fig. 17.8). Make sure that the coat hanger is long enough and bent properly so that you can close the door as far as possible.

If the door and jamb are even and there is enough space, a thin wire can be inserted to move the chain back.

Sometimes you can open a door very easily if it has a transom. Use two long pieces of string and a strip of rubber inner tubing. The tubing should be 8 to 10 inches long. Attach a string to each end of the tubing so that you can manipulate the tubing from the open transom. Lower the tubing and wrap it around the knob. Pull up firmly on both strings to maintain tension and turn the knob. This method can be used with either a regular doorknob or an auxiliary latch unit.

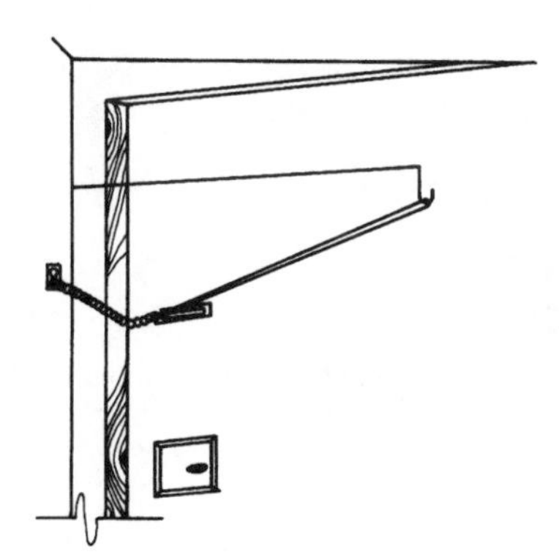

Figure 17.8 A variation of the rubber band technique.

Unlocking Thumbcuffs, Handcuffs, and Legcuffs

Standard thumbcuffs, handcuffs, and legcuffs all work the same way and have the same basic parts (Figs. 17.9 and 17.10). The information here can be used for unlocking all of them.

Typically, when you're asked to unlock cuffs, someone will be in them. Don't automatically assume that something illegal is going on. Cuffs are sold in army surplus stores, martial arts stores, and magic and novelty shops. People may call you because they've lost the key or the key isn't working. Before opening cuffs, assure yourself that you're not helping a fleeing criminal. Feel free to ask what happened, and consider what the person is wearing (or not wearing) and the surroundings. A kid locked in handcuffs while wearing a top hat and black cape just didn't learn to do the trick correctly. The scantily dressed woman cuffed to brass bedposts, with adult novelty items around the room, probably isn't a fleeing felon either.

Also consider how the cuffs are on. When properly handcuffed by law enforcement, the person's arms will be behind his or her back with his or her hands facing out, both thumbs up, and the cuff's keyholes facing out. And if the cuffs have a double-lock feature, they should be double-locked. If you feel uncomfortable about the situation, call the police station and ask if anyone recently escaped from custody. You also can just tell the caller that you don't do handcuff work.

How cuffs work

You don't have to know how cuffs work to be able to unlock them. By knowing how they work, however, you'll be able to open them faster and come up with your own techniques when you face unique situations or don't have the tools I mention here.

Two cuffs are connected by chain or hinges. Each cuff looks and works like the other. The main parts of a cuff are the keyway, pawl, and ratchet. The keyway has a post for the barrel key to slide onto (Fig. 17.11). The ratchet has a

Figure 17.9 A standard pair of handcuffs and key.

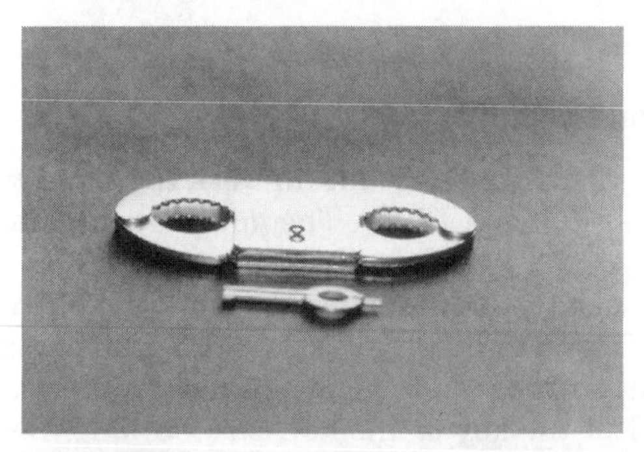

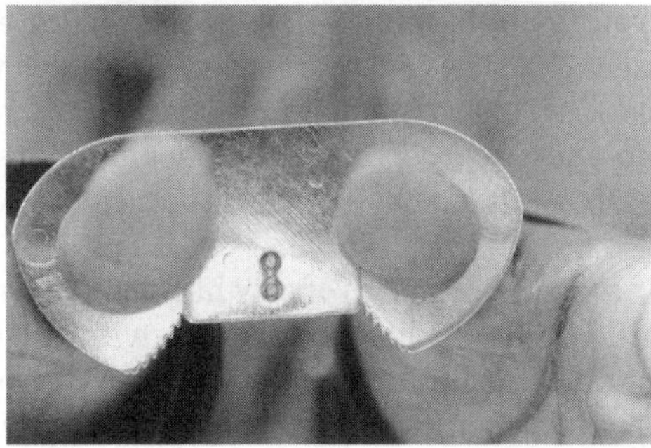

Figure 17.10 Thumbcuffs lock thumbs much as handcuffs lock wrists.

long row of teeth and slides against the pawl in the direction that tightens the cuff's grip (Fig. 17.12). The pawl is a spring-loaded bar with a small row of teeth that are angled opposite to the ratchet's teeth (Fig. 17.13). The opposing angles of the pawl and ratchet's teeth allow the ratchet to slide across the pawl in one direction but don't allow the ratchet to back out until the pawl's teeth are lifted off the ratchet. The bit of the key can lift the pawl out of the way, freeing the ratchet, of course. Figure 17.14 shows how all the parts fit together.

You also can unlock a cuff by using a bent paper clip to pick the lock. An easier way is to shim the cuff. Insert a 3-inch-long strip of spring steel between the ratchet and pawl until it covers the pawl's teeth (Fig. 17.15). A standard L-shaped torque wrench works well because you can use the bent part as a handle. The ratchet then should be easy to pull out, unless the cuff was double-locked.

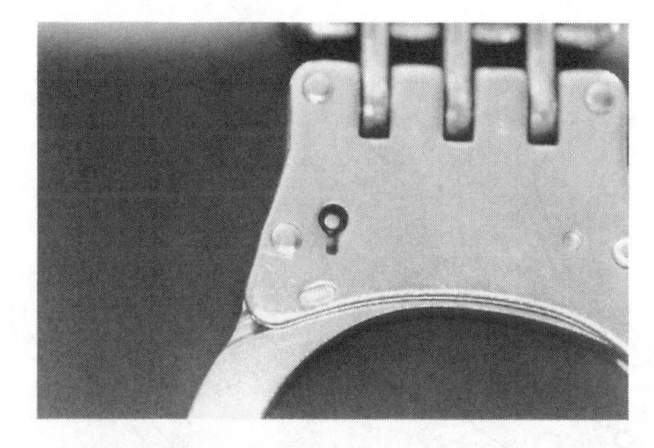

Figure 17.11 The keyway of most cuffs is for a small-barrel key.

Figure 17.12 The ratchet of a cuff has a row of teeth that are angled in a direction to allow the cuff to close freely but not to open without a key.

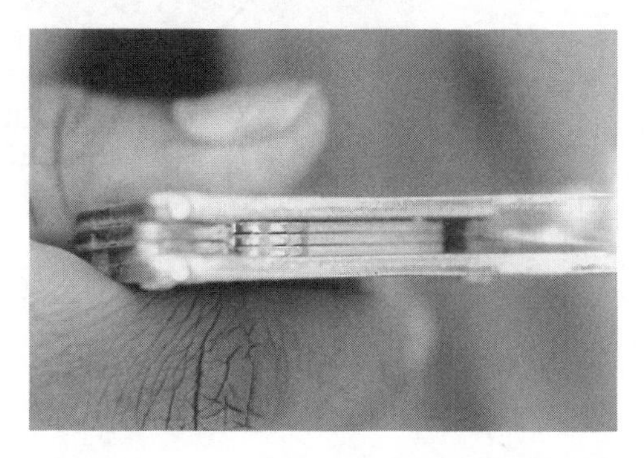

Figure 17.13 The paw of a cuff is a spring-loaded bar with teeth angled opposite of the teeth of the ratchet, which lets the ratchet close but not back out without a key.

Better cuffs have a double-lock feature that allows someone to use the back of the key to push a pin into the cuff to block the pawl. When a cuff is double-locked, the key has to be turned in two directions to unlock it. One direction is to free the double-lock pin (Fig. 17.16); the other direction is to lift the pawl. It isn't hard to pick open a double-locked cuff, but you need to pick it both ways, as with the key.

The easiest way to unlock any cuff is to use a key. Most of their differences are ornamental and apply to the shape and design of their bows or grips. One design lets the key double as a screwdriver. Another is removable to make the key easier to hide. The handles have no effect on which cuffs the key can unlock, however. Only the part that enters the keyway, the bit, does the unlocking. If two

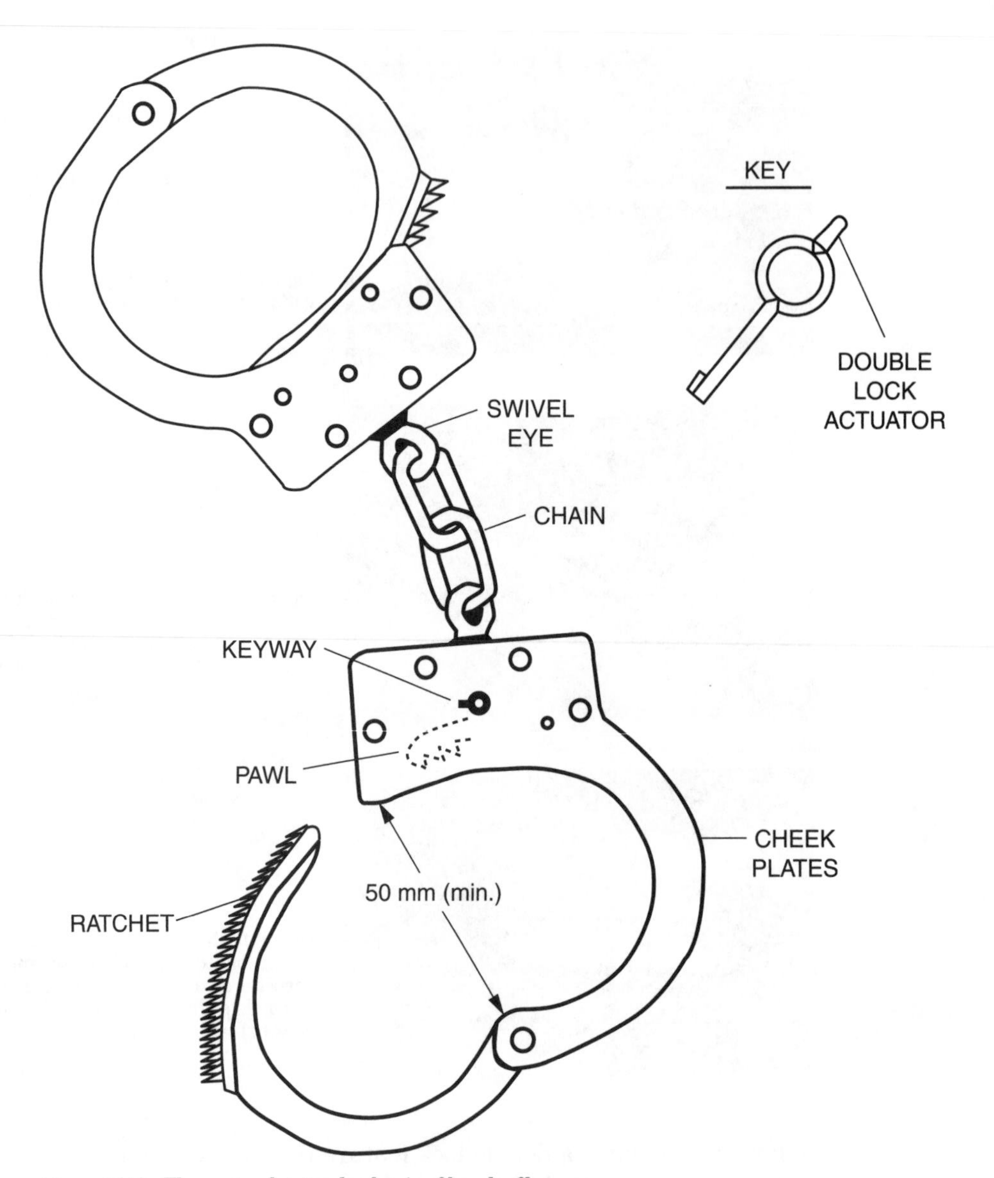

Figure 17.14 The parts of a standard pair of handcuffs.

keys have bits of about the same size and design, the keys should be able to unlock the same cuffs. Among standard cuffs, there are few planned bit variations. Most variations are the result of poor machining and show up in low-cost off-brand models.

It's useful to have two or three different standard cuff keys from different companies. This should let you open almost any standard cuff quickly. One way to get the keys is to buy a couple of pairs of handcuffs and thumbcuffs from differ-

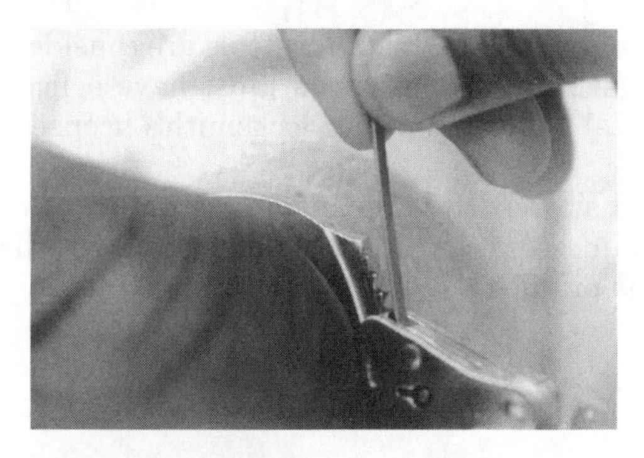

Figure 17.15 Place the shim between the ratchet and pawl to open cuffs.

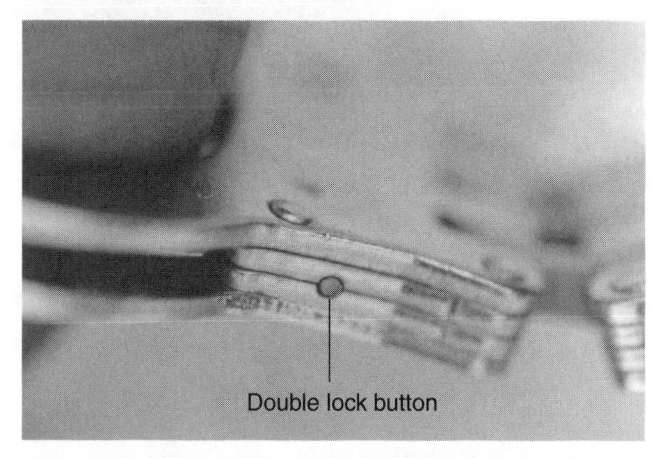

Figure 17.16 When you depress the double-lock button, the pawl is blocked from moving, and the cuff can't be shimmed.

ent companies (at least one brand-name and one generic model). You also can buy keys for a few dollars from locksmith or law enforcement supply houses.

Forced Entry

Forced entry in place of proper professional techniques is never recommended except in emergencies or when authorization is given by the owner. There are some simple rules about forced entry that you should consider:

- Attempt a forced entry only as a last resort. Try other techniques first.
- If you must jimmy the door, do it carefully. Today, antique doors and frames can be a valuable part of a home. Don't risk a costly lesson by being too hasty.
- If you must break a window to gain access to a lock, break a small one. Replacing broken windows can be expensive.
- Don't saw the locking bolt. This is very unprofessional.

Remember that lock picks and many other entry tools are considered burglar's tools in some jurisdictions. In many cases you must have a license to carry them. Check with your local police, keep your locksmith's license current, and destroy worn-out tools.

Be aware of the trust that your city and customers put in you by allowing you to be a locksmith; you possess certain tools and knowledge that others do not have. Don't abuse this privilege.

Chapter

18

Combination Locks

Like other locks, combination locks are available in a variety of types and styles. In this chapter we will consider a number of representative types. Some of these locks are more secure than others, but each has its role. For example, it would be inappropriate to purchase a high-quality combination padlock for a woodshed; by the same token, it would be foolish to protect family heirlooms with a cheap lock. Even the best lock is only as good as the total security of the system. A thief always looks for the easiest entry point.

Parts

Though combination locks have some internal differences, all operate on the same principle—rotation of the combination dial rotates the internal wheel pack. The pack consists of three (sometimes four) wheels. Each wheel is programmed to align its gate with the bolt-release mechanism after so many degrees of rotation. Programming may be determined at the factory or it may be subject to change in the field. As the wheels are rotated in order (normally three turns, then reverse direction for two turns, then reverse direction again for one turn), the gates are aligned by stops, one for each wheel and one on the wheel-pack mounting plate. When all the gates are aligned, the bolt is free to release. Padlocks are built so the shackle disengages automatically or manually by pulling down on the lock body. The captive side of the shackle, the part known as the *heel*, has a stop so the shackle will not come free of the lock body.

One weakness of low-priced locks is their ease of manipulation. With practice you can discriminate between the clicks as the wheels rotate. Once you learn to do this, manipulation of the lock is child's play. Fortunately, manufacturers of better locks make false gates in the wheels that make manipulation much more difficult. A real expert will not be thwarted, but experts in this arcane art are few and far between. It takes training and intensive practice to distinguish between three or more false gates on each wheel and the true gate.

Manipulation

Manipulation of combination locks is a high skill—almost an art. A deep understanding of combination-lock mechanisms and many, many hours of practice are needed to master this skill. But the rewards—both in terms of the business this skill will bring to your shop and the personal satisfaction gained—are great.

No book can teach you to manipulate a lock any more than a book can teach you to swim, box, or do anything else that is essentially an exercise in manual dexterity. But a book can teach the rudiments—the skeleton, as it were, that you can flesh out by practice and personal instruction from an expert.

Manipulation is a matter of touch and hearing. It also involves an understanding of how the lock works. Electronic amplification devices are available to assist the locksmith. These devices are useful, especially as a training aid.

Padlocks usually do not have false gates and so are ideal for beginners (Fig. 18.1). Pull out the shackle as you rotate the wheels. This tends to give better definition to the clicks.

Work only for one number at a time. Stop when the bolt hesitates and touches the edge of the gate. The bolt has touched the far side of the gate, so move back a number on the dial and note it. Turn the dial in the reverse direction, going past the first number at least twice. Pull on the shackle and slowly continue to turn the dial. When you sense that the bolt has touched the gate, note the number that came up just before the bolt responded. This

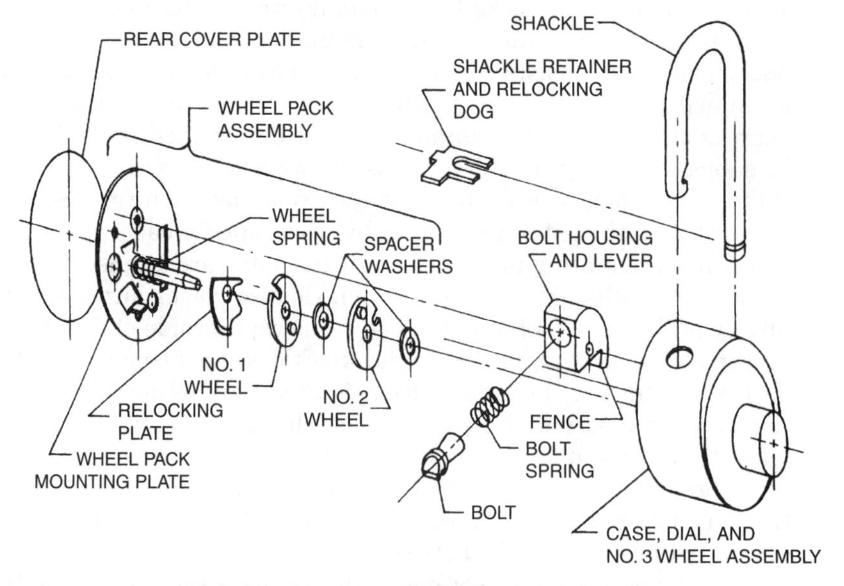

Figure 18.1 A typical dial combination padlock in exploded view. (*Desert Publications*)

is the second number of the combination. The third number comes easier than either the first or second one. The dial may stop at either the first or second number. Since many of the earlier combination locks did not have the accuracy of modern locks, the bolt catches at any one of the three numbers at any time.

You have the three numbers, but perhaps not in their correct order. Vary the sequence of the numbers until you hit the right one.

Because of imperfections in older and inexpensive modern locks, the bolt may stop at points other than the gates. Only through practice can you learn to distinguish between true and phantom gates.

For practice, obtain two or three locks of the same model. Disassemble one to observe the wheel, gate, and bolt relationship and response. These insights will help you manipulate the other two locks.

Drilling

As a last resort, all combination locks can be opened by drilling. To drill these locks follow this procedure:

1. Drill two ⅛-inch holes in the back of the lock.
2. Turn the dial and determine that the gate of one wheel is aligned with one of the holes. You might be able to see the gate. If that fails, locate the gate with a piece of piano wire inserted through one of the holes.
3. When the gate and the hole are aligned, note the number on the dial face. Determine the distance, as expressed in divisions on the dial, between the hole and bolt.
4. Subtract this distance from the reading when the gate is aligned with the hole. The result is the combination number for that wheel.
5. Reverse dial rotation and find the number for the second wheel. Do the same for the third.

Changing Combinations

Many locks are designed for combination changes in the field. Three of them are described here.

Sargent and Greenleaf

S & G padlock combinations are changed by key.

1. Turn the numbers of the original combination to the change-key mark located 10 digits to the left of the zero mark on the dial face.
2. Raise the knob on the back of the lock to reveal the keyway. Insert the key and turn 90°.

3. Repeat step 1, but use the new combination and the change-key mark as zero.
4. Remove the change key and test the new combination.

Simplex

The Simplex is a unique combination lock, employing a vertical row of pushbuttons rather than the more usual dial (Fig. 18.2).

1. Turn the control knob left to activate the buttons.
2. Release the knob and push the existing combination (Fig. 18.3).
3. Push down the combination change slide on the back of the lock (Fig. 18.4).
4. Turn the control knob left to clear the existing combination (Fig. 18.5).
5. Push the buttons for the new combination firmly and in sequence (Fig. 18.6).
6. Turn the control arm right to set the new sequence (Fig. 18.7).

Figure 18.2 The Simplex pushbutton combination lock. (*Simplex Access Controls Corp.*)

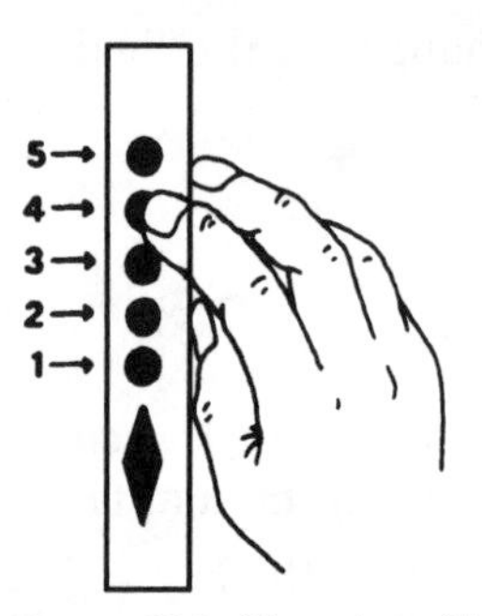

Figure 18.3 Changing Simplex combination, Step 2.

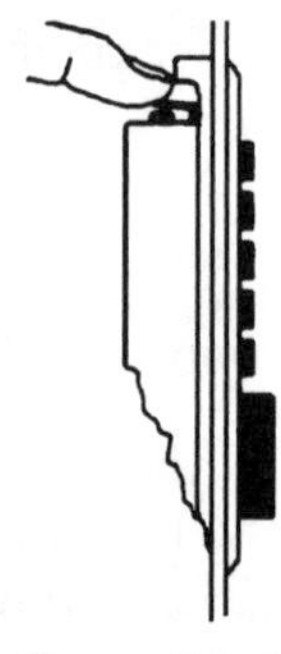

Figure 18.4 Changing Simplex combination, Step 3.

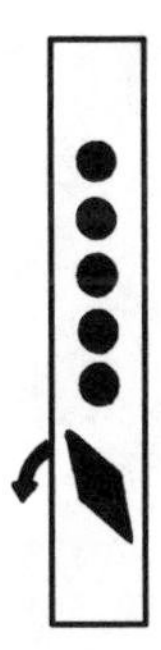

Figure 18.5 Changing Simplex combination, Step 4.

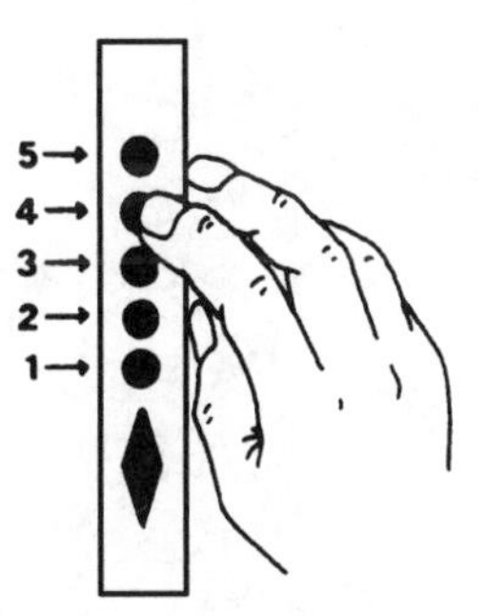

Figure 18.6 Changing Simplex combination, Step 5.

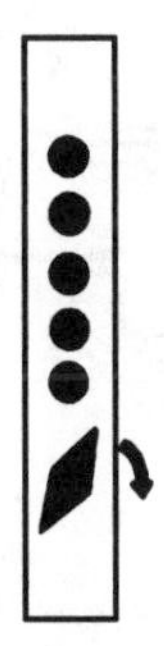

Figure 18.7 Changing Simplex combination, Step 6.

Dialoc

Partial disassembly is required to change the combination. Refer to Fig. 18.8 for the parts numbers in the following steps.

To disassemble:

1. Remove the Dialoc from the door by withdrawing the two mounting screws 487-1, holding the inside plate 821 against the door.
2. Place the outside plate 820 face down on a smooth work surface so that housing 414 is facing up.
3. Remove the three nuts 484-5 holding the housing cover 745.
4. After removing the nuts, gently lift the housing cover up and away from the housing. Take care that control-bar spring 744 is not dislodged and lost in the process. Lay it where it won't get lost.
5. Remove the secondary arm 402.
6. Grasp the end of the control bar 726 and lift out the four ratchet assemblies. Observe how they are arranged before dismantling.

Now you can rearrange these four ratchet assemblies to get a new combination. (Should you want to use numbers that are not in the present combination,

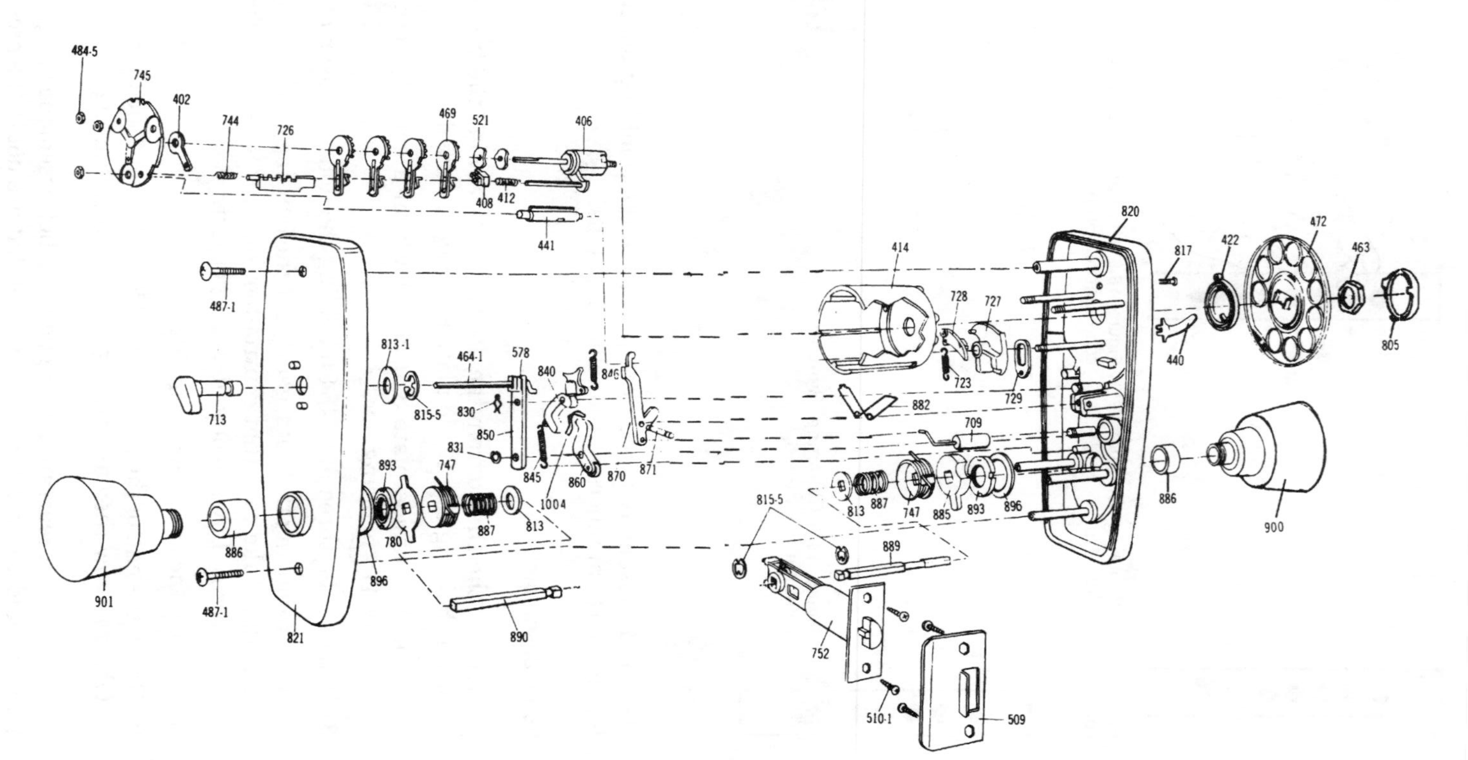

Figure 18.8 The very sophisticated Dialoc 1400. (*Dialoc Corporation of America*)

your distributor or the Dialoc Corporation can readily supply these.) Remove the ratchet assemblies 469 from the control bar and change the sequence as you desire. Put the ratchet assemblies back on the control bar. When replacing the ratchet assemblies on the control bar, make sure they go on with the ratchet part of each assembly down and that the first digit of the combination is put on first, then the remainder of the combination in order from the first ratchet assembly outward.

To assemble:

1. Place the four ratchet assemblies 469 on the control bar 726 and, grasping the assembled control bar and the ratchet assemblies, slip them over the center spindle of the primary arm 406. The ratchet assemblies can be guided over the lower mounting pin protruding from the outside plate 820.
2. Take care that the control bar 726 passes through the elongated hole in the bottom of the housing 414.
3. Replace the secondary arm 402 on the primary arm 406. Make sure that the finger 408 and the finger follower spring 412 are in position.
4. If the cam pawl trip 441 was dislodged during the combination change, replace it by inserting it into the housing with the small protruding end down. The blade on this part must be against the ratchet pawls of the ratchet assemblies.
5. Make sure that the secondary arm 402 is flush with the top of the square bar. This allows a spacing of approximately 0.020 inch between the secondary arm and the top ratchet assembly.
6. After the secondary arm is properly placed, return control spring 744 to the end of the control bar 726.
7. Replace the housing cover 745. Make sure that the cam pawl trip 441 and the control bar 726 are in their respective holes in the housing cover. A straight pin is helpful in guiding these parts into holes. Be sure reset arm 870 is inside the housing 414.
8. Replace the cover-retaining nuts 484-5. Turn these nuts snugly with your fingers, plus approximately one-half turn with a wrench.
9. Operate the lock before you replace the door to ensure its proper operation on the new code settings.

Master 1500 padlock

The Master 1500 combination padlock is covered with heavy-gauge sheet steel, rolled and pressed at the edges. The net result of this somewhat unorthodox construction is a tight-fitting lock and one that is not easily pried open. Disassembly is recommended only as a training exercise. It is not practical to open this lock for repair.

Figure 18.9 The Master 1525 masterkeyed dial combination padlock. (*Master Lock Company*)

Three wheels are employed, each with a factory-determined number. If you receive the lock with the shackle open, you can determine the combination by looking through the shackle hole. Note the dial reading as you align each wheel. If you read the numbers properly and if the wheel gates are in alignment, you should only have to add 11 to the dial readings to get the true combination. If you are slightly off, compensate by adding 10 or 12 to the original readings.

The Master 1500 series can be masterkeyed (Fig. 18.9). This lock is popular in schools. While the combination can be obtained by manipulation, picking the keyway is faster. Once the lock is open, determine the combination by the method described above.

Other Keyless Locks

Combination locks are those devices that do not require a key to open. Although many people think of such things as safes and vaults or perhaps a bicycle padlock requiring a three-digit combination as the only types of combination locks, this is not true. Any device that does not require a key but instead relies on a single or series of digits or letters in a certain order to open the mechanism is classified as a combination locking device.

The following sections discuss, illustrate, and provide specific details concerning several of the best and most popular keyless locking devices available today. Remember that the products discussed here are not the only types, nor are they representative of the only manufacturers of such products; they are the products that have proven themselves to a wide variety of customers in varied and numerous types of installations. As such, they are recommended for your consideration as some of the highest-quality, professional, and most secure keyless combination locking devices on the market today.

Simplex keyless locks

Keyless entry locks provide exceptional security in a variety of modes. Motels, hotels, various private and government institutions are using these locks with increasing success (Fig. 18.10). The main manufacturer and proponent of key-

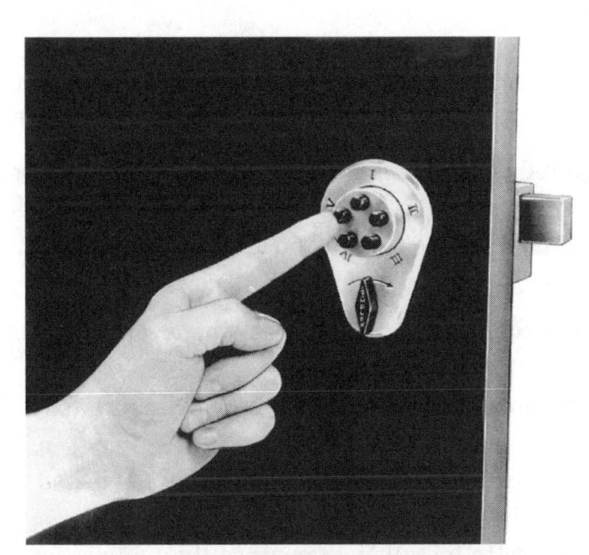

Figure 18.10 Simplex 5 pushbutton deadbolt lock. (*Simplex Access Controls Corp.*)

less locking devices is Simplex Security systems. Their locks are commonly used in a variety of security-related situations where more than one entrance control device is required. As an example, consider the case of a company research and development facility. For added security, a keyless lock is added to each door. At night, then, two locks secure the door. During the day, the main lock is in the unlocked position, but the pushbutton lock is readily available for the use of the research and development employees because they are the only individuals who know the particular combination to gain access.

In motels and hotels, the use of keyless locks has increased security, cut costs, and saved time and manpower. How? By using the Simplex lock, in one example, one motel stopped losing more than 50 keys a month and saved the costs of continually having duplicate keys made and locks rekeyed. Best of all, thefts were almost entirely eliminated.

Time and manpower were also saved with the installation of the Simplex lock. The desk clerk no longer had to hunt a duplicate key or go to a room and unlock it because the key was locked inside. The clerk simply gives the specific room combination, just as she did when the guest checked into the motel. Once the guest checks out, it requires only 3 minutes to change the combination; this keeps room security integrity very high.

The Simplex lock is extremely easy to operate and maintain, despite its very high security. The lock has five pushbuttons arranged in a circle and numbered from 1 to 5. The combinations can be set at random; to operate the lock, the various buttons are pushed in sequence, in unison, or a combination of both.

Installation of the lock is also relatively easy. The only tools required are a screwdriver, an electric drill, and a hole-saw set.

Advantages. Before we get into the various installation procedures, let's look at the specific advantage of a Simplex lock so you will understand, as a locksmith, why many businesses and homeowners are installing these locks.

- No keys are required. For businesses, this means none are issued or controlled and there are none to recover when the employee leaves.
- They are burglarproof because there is no keyhole to pick.
- There are thousands of possible combinations.
- You can change the combination in seconds whenever there is a change in personnel or tenants or for other reasons of security.
- Money is saved because there are no keys or cylinders to be changed or replaced.
- The locking units are extremely durable. They are wear-tested for the equivalent of 30 years of intensive use.
- It is an extremely easy-to-install lock, more so than any other existing lock or knob type lockset. It is completely mechanical and requires no electrical wiring or specialized accompanying equipment.

Varieties. The pushbutton combination lock (Fig. 18.11) comes in several models: the deadlock with a full 1-inch deadbolt, the automatic spring latch, the key bypass, and the special security model (for use with Department of Defense or industrial complexes requiring the security necessary for the protection of closed or restricted areas).

The *deadlock*, model DL, requires only the combination pressed in the right order and the turn knob moved to open the lock from the outside. The lock remains in the open position, with the bolt retracted, until either the inside lever or outside knob is turned to throw the bolt into the locked position.

The *automatic spring latch*, model NL, locks automatically. It opens with a finger lever from the inside. A thumb slide on the inside holds the retracted latchbolt in the unlocked position, if desired (Fig. 18.12). A variation, model NL-A, lacks the latch holdback. A deadlocking latch is standard for these units; this prevents prying the latch back on the closed door. The latchbolt is reversible without affecting the deadlatching function.

The *key bypass* models (Fig. 18.13), DL-M and NL-M, are the DL and NL models with an optional key cylinder. They provide the flexibility of a master key system and the security advantages of a combination lock. Employees, tenants, and others use the combination. The key ensures entry regardless of the combination setting used but is given only to designated management personnel. This eliminates the need to maintain a central listing of combinations for reference use by top management personnel. It is completely safe because there is no widespread proliferation of keys and key distribution can be severely restricted.

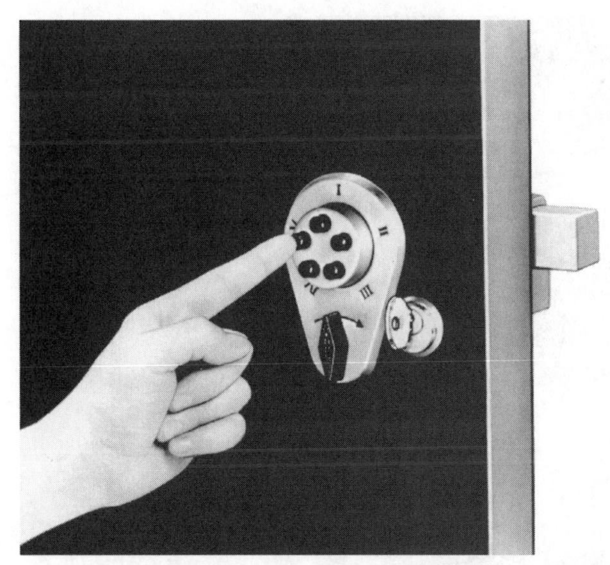

Figure 18.11 Deadbolt with a key bypass. (*Simplex Access Controls Corp.*)

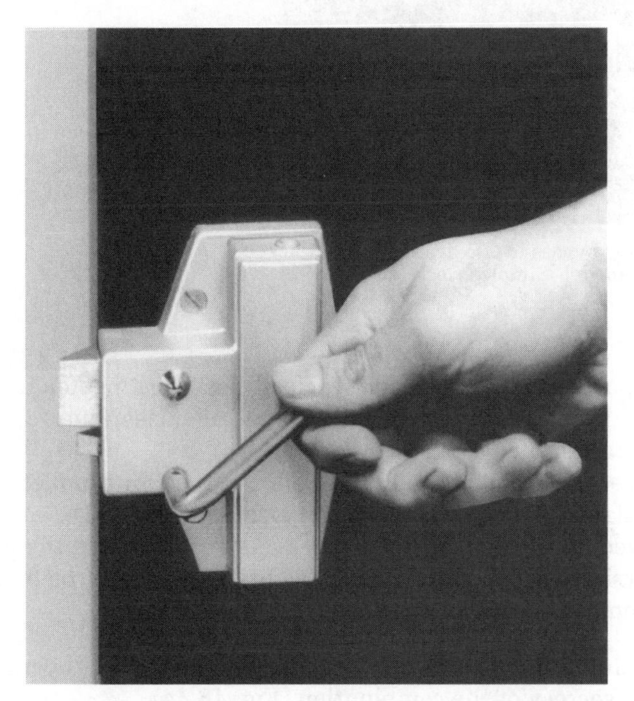

Figure 18.12 Finger lever and thumb slide for the Simplex Mode DL lock; inside view of door and back of lock. (*Simplex Access Controls Corp.*)

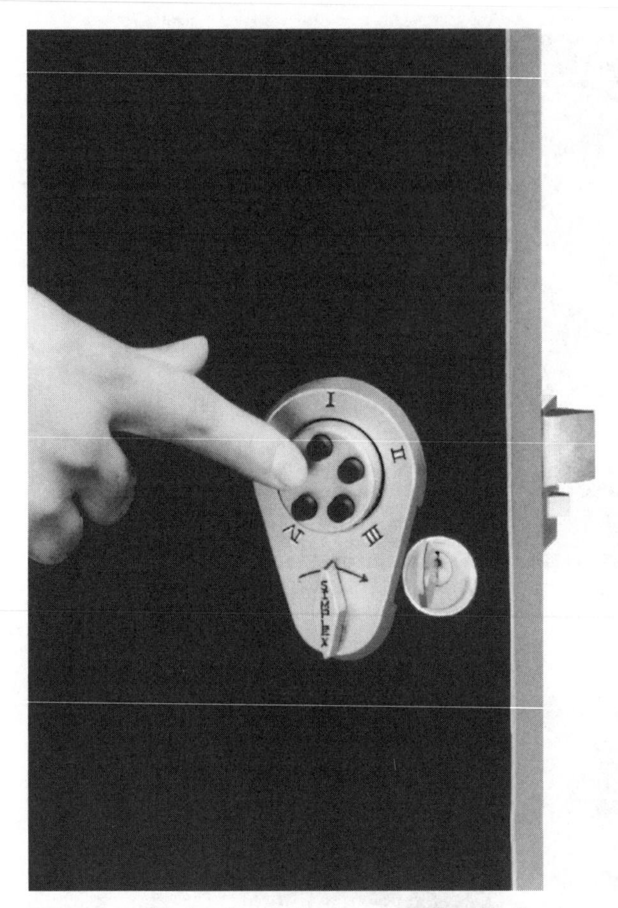

Figure 18.13 The key bypass is also available with the standard latch bolt model Simplex lock. (*Simplex Access Controls Corp.*)

The key bypass model is popular in facilities working for and with the Department of Defense and other government agencies, due to its excellent security advantages.

The special security model, NL-A-200-S, is designed for complete compliance with the Industrial Security Manual of the Department of Defense. The model is an approved substitute for the expensive guard forces in the protection of closed and restricted areas in contractor facilities. It has all the features of the model NL automatic spring latch, plus

- A faceplate shield to prevent observations of pushbutton operation and to maintain the secrecy of the combination (Fig. 18.14).
- The combination change access can be padlocked to ensure authorized combination changes only.

Figure 18.14 Faceplate shields prevent observation of the combination code by unauthorized individuals. (*Simplex Access Controls Corp.*)

- The latch holdback feature is eliminated so the lock can never be kept open. It is in locked condition at all times.

Installation. All of the Simplex models have the same basic three-piece self-aligning assembly and install easily and quickly. *Note:* All Simplex door locks require a minimum flat surface of 3½ inches on the inside of the door to mount.

Simplified, the procedures for installation are as follows.

Wood door and frame installation. Drill two holes, ¾ and 1⅝ inches in diameter, a distance of 2⅜ inches from the door edge. Key bypass models will require an additional ⅞-inch key cylinder hole. All parts are provided for a complete wood door and frame installation. Included are two styles of strikes (standard) to suit any wood door frame (*A* in Fig. 18.15).

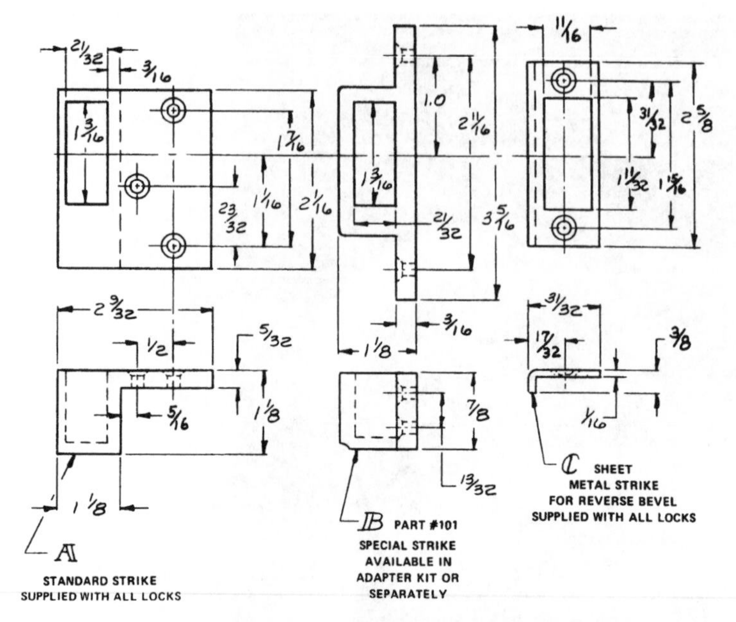

Figure 18.15 Standard and optional strikes. (*Simplex Access Controls Corp.*)

Metal door and frame installation. An adapter kit containing a special surface-mounted strike (*B* in Fig. 18.15) is available. The special strike is fastened to the inside surface of the metal frame and eliminates the need to cut or mortise the frame for the standard strike. The adjustable riser plate raises the lock housing to ensure the proper alignment of the latch or deadbolt with the special strike.

Models DL and NL installation

1. Using the template provided with the lock (Fig. 18.16), cut the template from the sheet and fold on the printed line as shown by the arrow. With the door closed, tape the template to the inside of the door with the fold line aligned with the edge of the door (or door stop if one is present). The same template applies to both inswing and outswing doors (Fig. 18.17). Using a center punch, mark the centers of the two large and three small holes.
2. Using a ⅛-inch-diameter drill, drill the centers of the ¾- and 1⅝-inch-diameter holes all the way through to the front surface of the door. Drill the three mounting screw holes approximately ½ inch deep.
3. With a 1⅝-inch-diameter hole saw, open up the larger hole by drilling from both the outside and the inside surfaces, meeting in the middle. Follow the same procedure with a ¾-inch-diameter drill for the smaller hole. *Note:* With Model NL, the latch is reversible by removing the two large

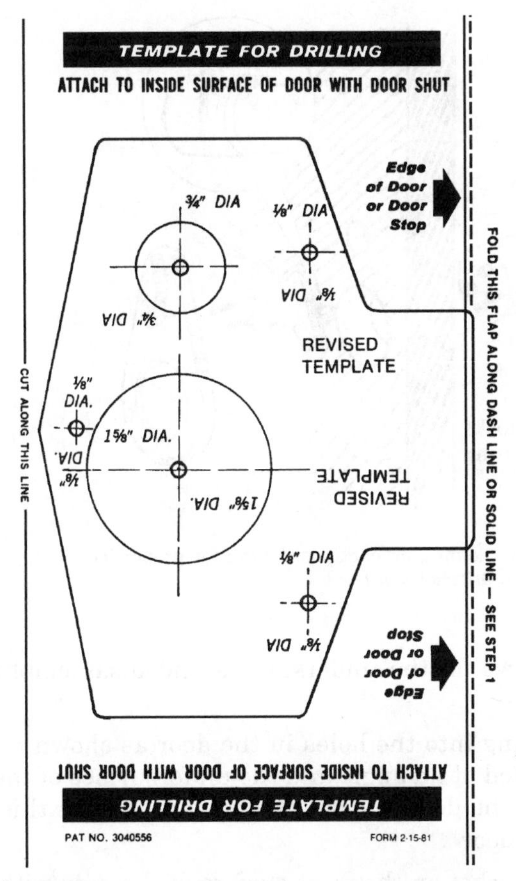

Figure 18.16 Drilling template for the 100 and 200 series DL/NL model locks. (*Simplex Access Controls Corp.*)

Fold template on Dash Line for additional clearance required on outswing doors or when using Part #100 Riser Plate on inswing doors.

Fold template on Solid Line on inswing doors when not using Riser Plate.

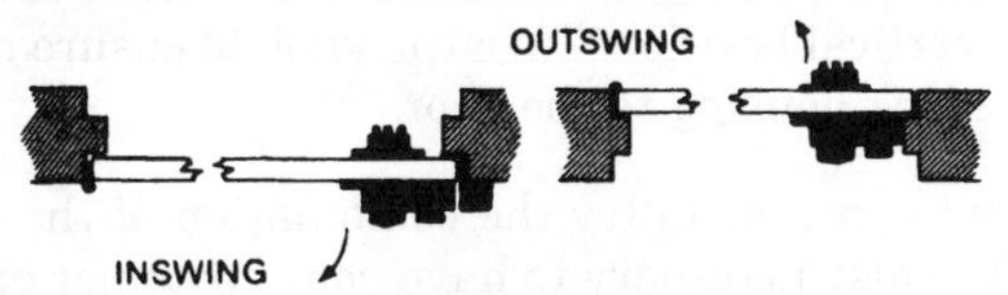

Figure 18.17 This visual provides for the proper determination of which template line will be used for positioning the lock properly to the door. (*Simplex Access Controls Corp.*)

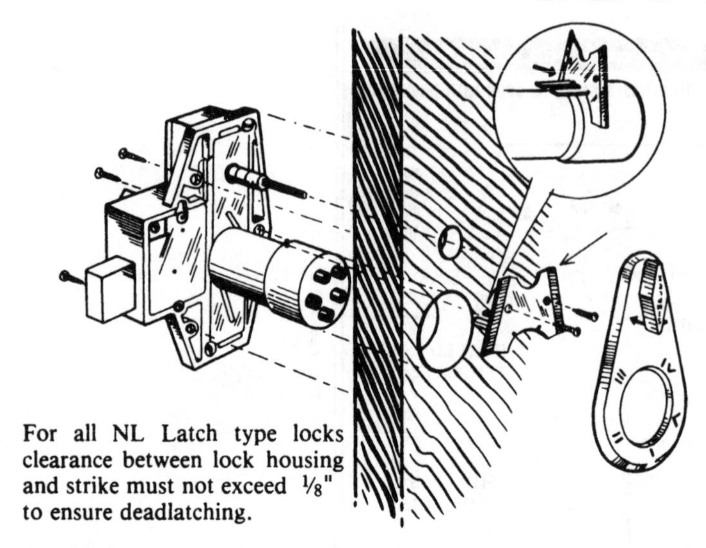

For all NL Latch type locks clearance between lock housing and strike must not exceed ⅛" to ensure deadlatching.

Figure 18.18 Lock housing and holding bracket positioning for proper lock installation. (*Simplex Access Controls Corp.*)

and two small screws on the undersurface and disassembling the lock housing.

4. Slide the lock housing into the holes in the door as shown in Fig. 18.18. If the holes are drilled at an angle and the undersurface of the lock is not in full contact with the door, file out the holes to correct this. Screw the lock housing to the door.

5. Place the holding bracket on the front surface of the door with the slotted legs engaging the aligning pin on the barrel. With the large radius of the holding bracket mating with the radius of the barrel, fasten with the screws provided.

6. Remove the lock housing from the door.

7. Slide the holding edges of the faceplate behind the formed up edges of the holding bracket. Make sure the faceplate is securely held on both sides.

8. Replace the lock housing while holding the control knob, marked "SIMPLEX," in the vertical position. Turn the knob to ensure proper engagement and fasten the lock housing to the door.

Note: It is very important to try the combination of the unit several times before locking. It is also necessary to have your customer open the lock several times before you leave. If the lock does not fit exactly to the door, there is an adjustable riser plate available from the factory. With minimum maintenance, this unit will now provide a very long and extremely useful service to your customer.

Key bypass installation instructions. Follow the above instructions for installing the standard DL and NL locks, with these exceptions:

- Use the template (Fig. 18.19) for positioning a ⅞-inch-diameter hole for the key lock cylinder.
- With two pairs of pliers, break the tailpiece of the key cylinder at the proper location as indicated (Fig. 18.20): (1) for the Simplex 100 or 200 series locks when the riser plate is not used, (2) for 100 series when either ¼- or ⅜-inch side of the riser plate is used, or with the 200 series locks when ¼-inch side of the riser plate is used, (3) for 200 series locks with ⅜-inch side of riser plate, use full length of tailpiece.
- Assemble the key lock cylinder in the ⅞-inch-diameter hole. *Note:* The hole in the flange of the cup must be positioned over the ⅛-inch-diameter

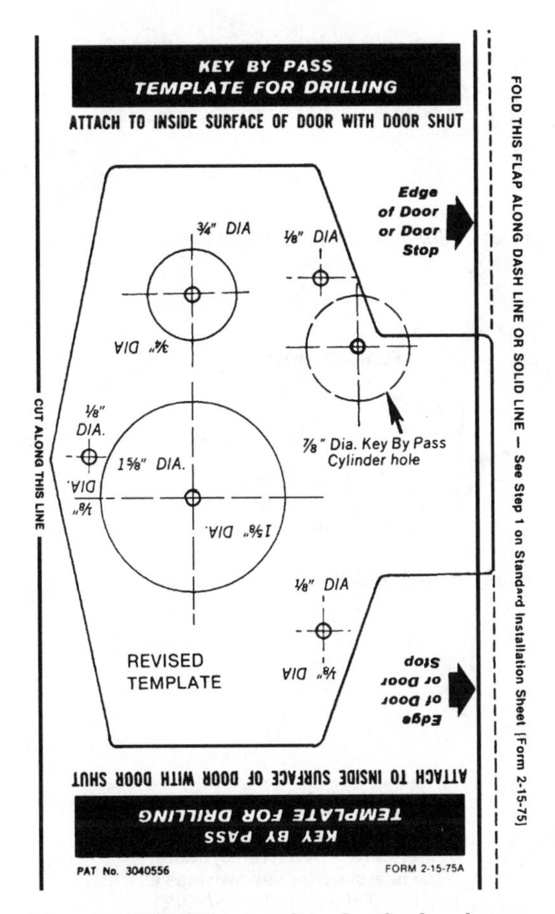

Figure 18.19 Drilling template for the key bypass feature. (*Simplex Access Controls Corp.*)

mounting screw hole used to secure the lock housing to the door (Fig. 18.21).

- Proceed with the last step of the basic instructions.

You may wish to substitute a standard rim cylinder for the ⅞-inch-diameter cylinder supplied with the Simplex key bypass models DL-M and NL-M. Due to space limitations, the Simplex face plate with Roman numerals must cover a portion of the standard key cylinder (Fig. 18.22).

When a standard key cylinder is used, you must let the factory know this; in your order you must specify the Simplex model (DL-M or NL-M, 200 series, and whatever finish is desired) plus the very important notation ". . . for use with standard rim cylinder . . ." placed prominently on your order blank.

Make sure you get the *rim cylinder*, not the *mortise*. In the proper keying of the unit, your local or regional distributor or other supply source will probably

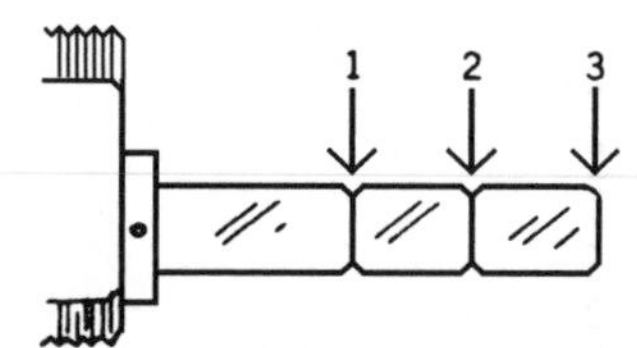

Figure 18.20 Tailpiece break points. (*Simplex Access Controls Corp.*)

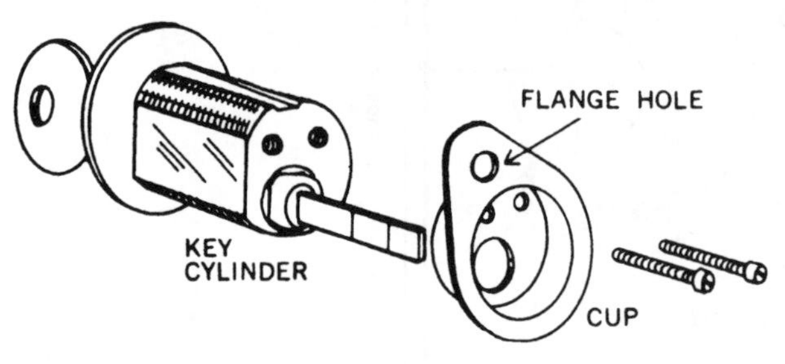

Figure 18.21 Bypass lock assembly procedure. (*Simplex Access Controls Corp.*)

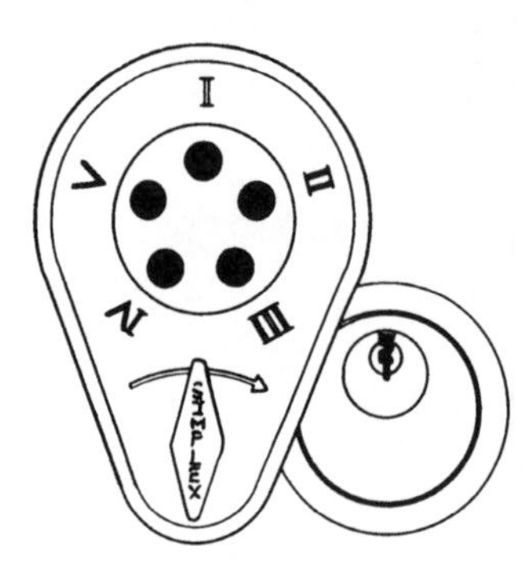

Figure 18.22 Simplex lock using a standard sized rim cylinder. Note how the Simplex overlaps a portion of the cylinder. (*Simplex Access Controls Corp.*)

be contacted also, unless you have a wide variety of various lock parts in stock. You must furnish the cylinder. An escutcheon (or bezel) must be used that allows the nose of the cylinder to be mounted flush with the front surface of the door (Fig. 18.23). Cut the length of the tailpiece to suit. Cut or grind the width of the tailpiece to 3⁄16 inch to fit the tubular coupler on the Simplex lock.

Use the Simplex key bypass template (Simplex Form 2-15-65A) for drilling. Use the center of the 7⁄8-inch-diameter hole for drilling the appropriate size hole for the standard rim cylinder and escutcheon. Assemble the cylinder, escutcheon, and back plate to the door *before* mounting the Simplex lock.

Install the Simplex by following the standard installation instruction procedures. The Simplex faceplate (with Roman numerals) and holding bracket will cover part of the cylinder. It might be necessary to notch the edge of the escutcheon to clear one of the two screws used to fasten the holding bracket to the front surface of the door. One of the two lock housing screws nearest the door edge must be fastened to the back plate of the rim cylinder (Fig. 18.24). Using the hole in the lock housing as a guide, drill a 0.161-inch-diameter hole (#20 size drill) in the back plate for the #10 sheet metal screw.

Servicing Simplex locks. After you install the Simplex lock, you might have a service and maintenance contract for the numerous Simplex locks in the building. During the course of your locksmithing career, you will be called on many times to carry out your portions of the service contract with the building man-

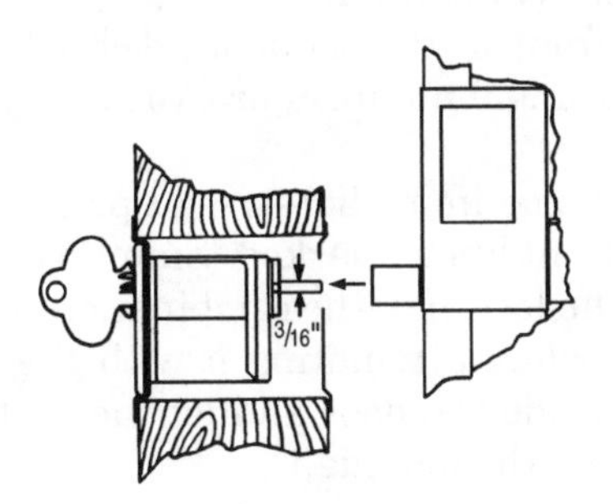

Figure 18.23 Standard lock cylinder positions with the door front surface. (*Simplex Access Controls Corp.*)

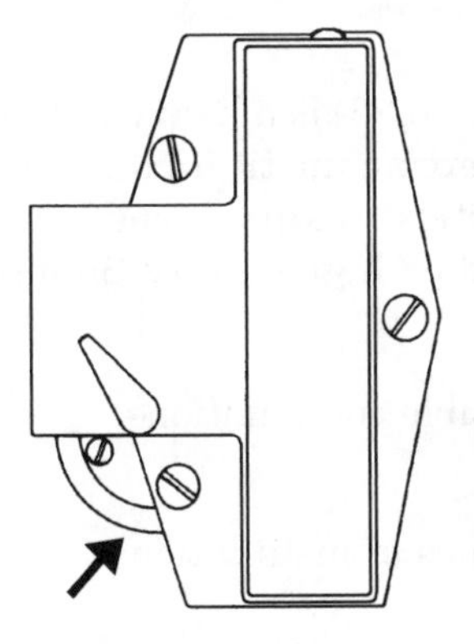

Figure 18.24 Arrow indicates lock housing screw that must be attached to the cylinder back plate. (*Simplex Access Controls Corp.*)

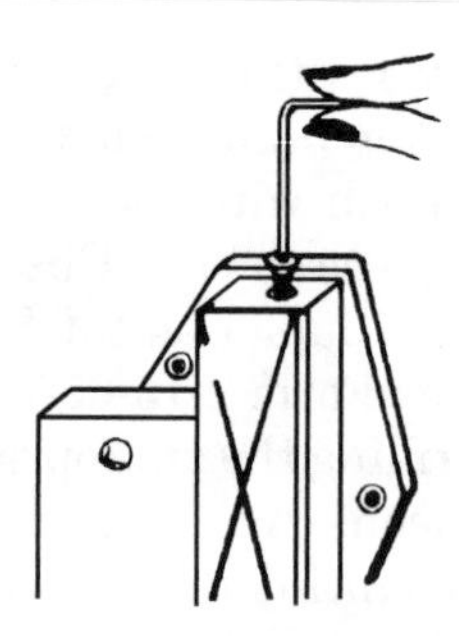

Figure 18.25 Use the Allen wrench to remove the access screw to change the combination of the lock. (*Simplex Access Controls Corp.*)

ager. Among the several procedures most common for the Simplex lock, you may change the combination for one or many of the locks or disassemble the lock for one reason or another. You will also be expected to find an unknown combination. The procedures for these various parts of the locksmith's job are detailed below.

Combination changing. With the door in the open position and the lock in the locked position, push the existing combination. Remove the hexagon screw at the top of the lock housing (Fig. 18.25) with an Allen wrench. After removing the screws, insert the wrench into the hole and depress the red button inside. Remove the wrench.

Turn the front control knob (below the pushbuttons and marked "Simplex") to the left. This removes the existing combination.

Depress the pushbuttons on the front in the sequence desired for the new combination. (This can be up to five single numbers or two at a time simultaneously.)

Write down the combination sequence immediately and present it to your customer for safekeeping. Turn the front knob (marked "Simplex") to the right. The new combination is now set in the lock. Try the combination several times with the door in the open position before attempting it with the door in the closed position and locked. (Stand inside the door and let the customer operate the combination at least once from the outside.)

The Simplex Unican 1000 series lock

The Simplex Unican 1000 series lock (Fig. 18.26) is a heavy-duty lockset ideal for securing high-traffic areas in commercial, institutional, and industrial buildings, apartment lobbies, hotels, motels, luxury homes, and restricted areas in commerce and industry. Like other high-quality Simplex products, the Unican 1000 has the following features:

- Built-in security with thousands of possible combinations
- No keyway to pick
- Weatherproofing which allows for all types of installations

Figure 18.26 Simplex Unican 1000 Series Heavy-Duty Lockset. (*Simplex Access Controls Corp.*)

- A ¾-inch deadlocking latch with a 2¾-inch backset
- Completely mechanical
- No electrical wiring
- One-hand operation

The unit is 7¾ inches high, 3 inches wide, and $3\frac{13}{16}$ inches deep. Latch front is 1⅛ × 2¼ inches with a backset of 2¾ inches. The Unican lock will fit doors with thicknesses ranging from 1⅜ to 2¼ inches.

A key bypass option is available, mainly for masterkeying purposes, using the Best or Falcon removable core cylinders in the outside knob (Fig. 18.27). The removable core cylinders are not available with the lock when ordered; they must be obtained separately. The Russwin/Corbin removable core capability can be obtained on a special order basis.

The second optional feature is the passage set function. When activated by an inside thumb turn or key cylinder, the outside knob can be set to open the lock without the need for using the pushbutton combination (Fig. 18.28). This is designed for use in offices, etc., when the door is to be left unlocked for an extended period of time.

Figure 18.27 Unican 1000 with key bypass feature. (*Simplex Access Controls Corp.*)

Figure 18.28 Key bypass or thumb turn can be employed with the unit. (*Simplex Access Controls Corp.*)

Both 9- and 14-inch-high "mag"-type filler plates are available to cover up previous lock installation holes on 1¾-inch doors. These three-sided steel plates wrap around the door extending 5½ inches on both the outside and inside surfaces.

Installation. The Unican 1000 uses an ASA 161 cutout (2⅛ inches in diameter through the hole with a 2¾-inch backset) like most other heavy-duty cylindrical locksets (Fig. 18.29). In addition, the lock requires two ¼-inch-diameter holes to bolt the top of the lock case to the back side of the door and a 1-inch-diameter hole for the key-operated combination change mechanism to operate from the back side as well. The Unican can be installed in wood or metal doors from 1⅜ to 2¼ inches thick.

In addition to the special tool provided with each lock, you, the installer, also need the following tools:

Figure 18.29 Exploded view of Unican 1000 positioning for a wood door. (*Simplex Access Controls Corp.*)

- Drill (½-inch-diameter electrical drill preferred)
- 2⅛-inch-diameter hole saw (for metal doors) or spur bit (for metal doors)
- 1-inch-diameter hole saw or spur bit
- ¼-inch-diameter drill
- Two wood chisels (wood doors), ¼ and 1 inch wide
- Phillips and standard screwdrivers
- Center punch and hammer
- Standard 12-inch ruler

The template. Before starting the actual installation procedures, let's look at a portion of the Unican 1000 three-part template that will be used. (Figure 18.30 shows the template and proper dimensions; Fig. 18.31 shows the door edge portion of the template.) Figure 18.32 provides the information and specifications for the latch cutout required. *Caution:* When using this type of template, always apply the template and drill from the outside, being sure to compensate for the door bevel, if any (Fig. 18.33).

Basic installation. Initially, make all necessary corrections to ensure the door is properly hung in the frame. Apply the template to the high edge of the door bevel and tape it in place. If a metal door frame prepositions the strike location, position the template to center the 1-inch-diameter latchbolt hole in the strike cutout in the frame. With a center punch and hammer, mark the centers of all required holes.

With the ¼-inch drill, drill the centers of the 2⅛- and 1-inch holes all the way through the door. If you want to center mark the hole on the other side of the

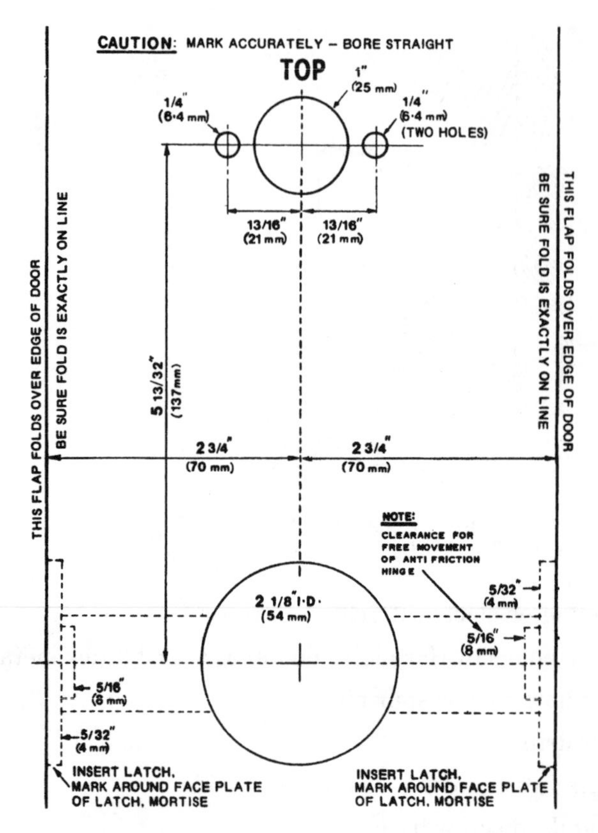

Figure 18.30 Unican 1000 door template. (*Simplex Access Controls Corp.*)

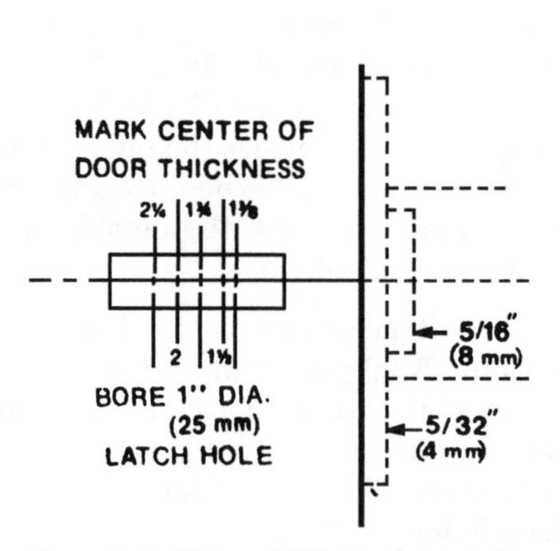

Figure 18.31 Unican 1000 template portion for marking a door thickness. (*Simplex Access Controls Corp.*)

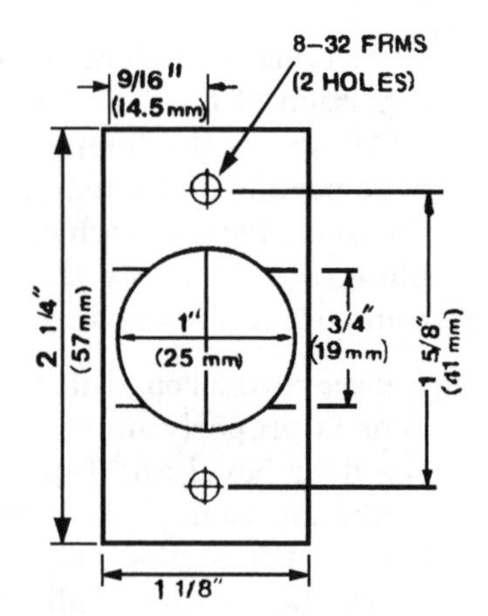

Figure 18.32 Unican 1000 latch cutout specifications. (*Simplex Access Controls Corp.*)

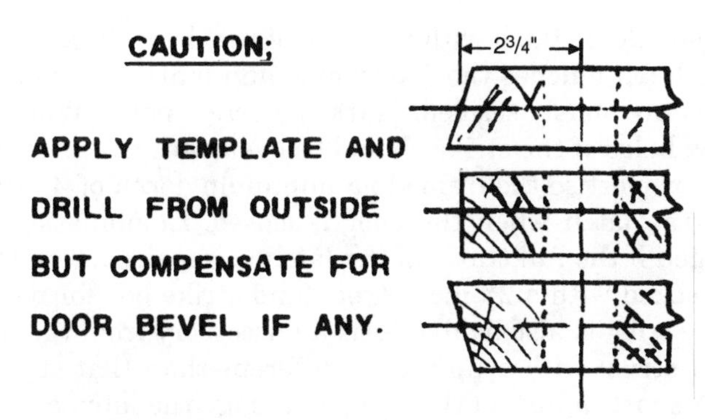

CAUTION;
2 3/4"
APPLY TEMPLATE AND
DRILL FROM OUTSIDE
BUT COMPENSATE FOR
DOOR BEVEL IF ANY.

Figure 18.33 Details of the bevel compensation for the Unican 1000. (*Simplex Access Controls Corp.*)

door (the low side of the bevel) using the template, be sure to move the "fold" line to shorten the backset to compensate for the bevel. Lay a carpenter's square on the high edge of the bevel to see how much compensation to make.

It is a good practice to bore the 2⅛- and 1-inch-diameter holes halfway in from both the front and back surfaces of the door.

If the door was originally prepared for a 2⅜-inch backset, file out the 2⅛-inch hole by an additional ⅜ inch; the lock case and inside rose are wide enough to cover.

Be sure to drill the 1-inch latchbolt hole on the door edge straight and level to prevent binding of the latch assembly. Carefully drill the two ¼-inch mounting holes through the door. *Note:* A drill jig is of great assistance in keeping your mounting hole straight and true.

Take care to allow clearance for the hinge that operates the antifriction device on the Unican latch. The latchbolt hole must be relieved adjacent to the antifriction trigger to permit the hinge to rotate freely and prevent the binding of the latch.

For a proper Unican 1000 installation, it is imperative to have a freely operating latch properly positioned to the strike. Excessive binding or dragging of the latch can cause the front knob clutch to *freewheel*, preventing latch withdrawal. Also, the deadlatching plunger must be held depressed by the strike plate to be operable. The gap between the door and the frame must never be more than ¼ inch to ensure proper deadlatching. If door silencers are present, allowances must be made in positioning the strike.

Position the strike. To properly position the strike, install the latch assembly into the edge of the door. On wooden doors, mark around the faceplate of the latch and mortise in with a chisel. Remove 5/16 inch, minimum, for the antifriction trigger hinge and 5/32 inch for the faceplate assembly. When inserted, the latch assembly should be flush with the nose of the door. Close the door to the point where the latch touches the door frame and mark the frame at the top and bottom lines previously scribed. Mark the screw holes. When the door is closed, the screw holes of the strike should be even with the screw holes of the latch plate. Cut or mortise the frame to a minimum depth of ¾ inch to ensure a full throw of the latch when the door is closed. Combinations cannot be cleared or changed if the full travel of the latch is impaired. The Unican 1000 comes factory packed with a 2¾-inch strike (and strike box for metal frames). An ASA 4⅞-inch strike is also available as an accessory from the manufacturer. Do not use a strike with a lip radius different than that supplied by the manufacturer. A short radius on the lip can damage the latch assembly.

Changing the hand. Unless otherwise specified in your order, all Unican 1000 series locks are factory assembled for left-hand operation. To change it to right-hand operation, remove the backplate containing the cylindrical unit from the lock case by removing the Phillips head screws. Remove the four Phillips head screws holding the cylindrical unit at the backplate (Fig. 18.34) and rotate the cylindrical unit so that the cutout for the latch faces in the opposite direction. Reattach the cylindrical unit using the four screws (with two lock washers for each screw), making sure the unit is centrally positioned to the hole in the backplate and rotates freely when the spindle is turned. To prevent internal jamming, two lockwashers must be used with each screw. Remount the backplate assembly to the lock case.

Adjusting for different door widths. All locks are factory set for a 1¾-inch-thick door. To adjust for a 1⅜- to 1½-inch door, remove the backplate and cylindrical unit assembly from the lock case and remove the spacer that is positioned between the cylindrical unit and the backplate (Fig. 18.35). Reassemble the cylindrical unit with the shorter (⁶⁄₃₂- × ¼-inch) screws provided with the unit. Insert the roll pin in the front knob shaft one hole closer to the front knob

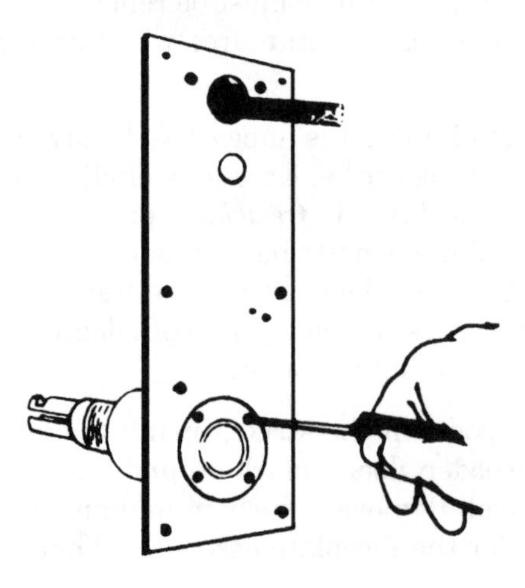

Figure 18.34 The four screws must be removed so the cylinder can be rotated, if necessary. (*Simplex Access Controls Corp.*)

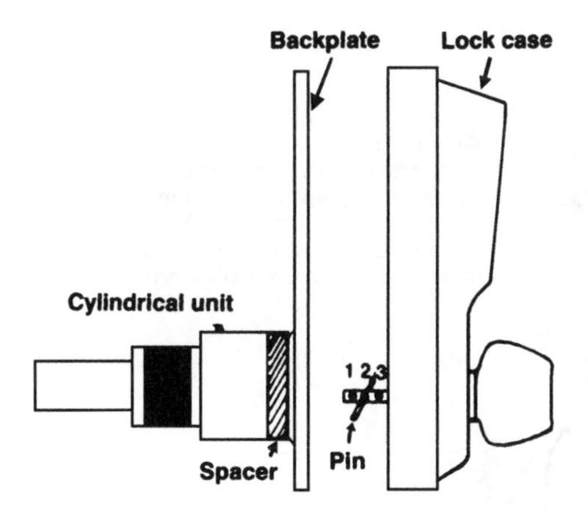

Figure 18.35 Backplate assembly and cylinder unit must be removed to adjust the lock for door thickness variations. (*Simplex Access Controls Corp.*)

(from position 2 to 3 in Fig. 18.35). Again, be sure that there are two lockwashers under each screwhead joining the cylindrical unit to the backplate.

To adjust for 2- to 2¼-inch doors, follow the previous instructions but add the extra spacer provided (lip edge to lip edge) to the one already assembled to the lock. Use the longer (⁸⁄₃₂- × ¾-inch) screws provided, and insert the pin in the shaft in the hole farthest from the front knob (from position 2 to 1) in the shaft in the hole farthest from the front knob (from position 2 to 1 in Fig. 18.35). When remounting the cylindrical unit, be sure it is centered in the hole in the backplate and turns freely. Ensure that the two lockwashers are under each screwhead.

Hold the latch assembly in the edge of the door and slide the lock case in from the front. Depress the latch slightly to engage the "T" (*A* in Fig. 18.36) in the lock retracting shoe (*B*). Viewing the unit from the back side of the 2⅛-inch hole, be sure the curved lips of the latch housing slide within the opening of the cylindrical case (*C*) in Fig. 18.36. Hold the lock case to the front of the door with one hand and insert the two top mounting screws from the back. *Finger-tighten* both screws to ensure the holes were properly positioned. If you are not able to start the screws into the back of the lock case with your fingers, the mounting holes were not positioned correctly. Forcing can strip the threads. To correct this condition, remove the lock case and file out the holes to ensure proper alignment of each to the nuts on the backplate.

Screw on the inside rose and tighten it with the spanner wrench. Slide the knob onto the inside sleeve, depressing the knob catch spring by pushing the pointed tip of the wrench through the hole in the side of the knob collar. When the knob is pushed into the stop, remove the tool and make sure the knob is securely fastened.

Mount the trim plate horizontally over the 1-inch-diameter combination change hole and the two mounting screws (Fig. 18.37). With the center punch, indent the door slightly for the two protrusions on the plate used to prevent rotation. Cut the lock screw to length as required. Insert the control key and keep

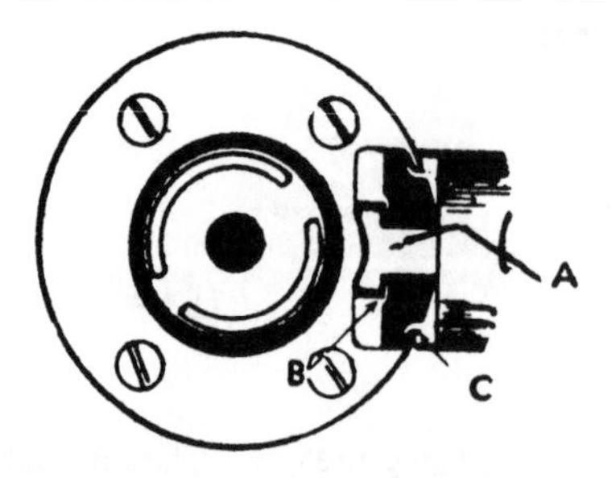

Figure 18.36 Latch assembly with critical portions indicated; these three points (A-B-C) must be worked in exact order to ensure proper lock installation. (*Simplex Access Controls Corp.*)

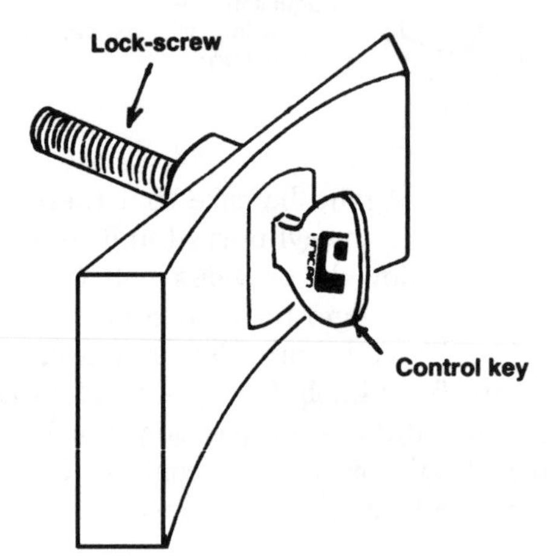

Figure 18.37 The trim plate must be positioned horizontally over the hole in the door. (*Simplex Access Controls Corp.*)

turning clockwise to thread the lock screw into the combination change sleeve assembly on the back plate of the lock case. Tighten the screw until the plate is snug to the door, but do not overtighten. The key can only be removed in a vertical or horizontal position. Overtightening may jam the lock screw in the change sleeve, requiring the removal of the screw with your needle-nosed pliers.

Changing the combination. All locks are factory set for the same combination: buttons 2 and 4 are pushed together, then 3 individually. For security reasons, you should change the combination immediately, using between one button and five buttons. When the buttons are set for pushing at the same time with other buttons, it is a different combination than if the buttons are set to be pushed individually (i.e., 1, 2, 3 individually pushed buttons is a different combination than if 1, 2, 3 are pushed at the same time). The same button cannot be used more than once in any given combination. If your Unican has a passage set function, be sure it is not in use when the combination code is changed.

To set the combination, the door must be open. Remove the inside trim plate with the control key. Turn the front knob fully clockwise, then release it. Push the existing combination, then release the buttons. The buttons should never be held depressed during lock opening. Insert the flat blade end of the special tool into the slots of the combination change sleeve and rotate clockwise to stop, but do not force. Then, rotate counterclockwise to stop and remove the tool. The combination change sleeve will rotate easily when the correct combination is pushed. It will not rotate with an incorrect combination, and forcing can damage the mechanism.

Turn the front knob fully clockwise to stop and release. Turn only once at this stage of the combination change. Depress the buttons firmly and deliberately in the sequence desired for the new combination. Record the combination at once so it will not be forgotten. Turn the front knob fully clockwise to stop to set in the new combination. Try the new combination before closing the door and be sure you have properly recorded it. *Note:* If the Unican will open without pushing a code, you are in the "0" combination mode because the front knob was turned more than once or turned out of sequence. In such a case, repeat all the steps in the change procedure *except* pressing the existing (old) combination. It is very important that the front knob be turned fully *clockwise to stop* when called for in the combination change procedure to ensure that the old code is completely cleared out and the new code is completely set in. Full rotation will be impaired by any of the following:

- Using a latchbolt assembly not supplied by the manufacturer (especially one with a shorter backset or shorter throw)
- A shallow strike box that prevents the full ¾-inch throw of the latch when the door is shut
- Any binding of the latch assembly in the door or in the strike caused by improper door or frame preparation or by door sag and/or buckling

Further installation tips:

- MAG-type filler plates (Fig. 18.38) in 9- and 14-inch lengths are available for covering up existing mortise lock preparation or badly butchered doors.

Figure 18.38 Filler plate. (*M.A.G. Engineering and Mfg., Inc.*)

These can be obtained for 1¾-inch doors only. They are handy to use for an installation template as well.

- Do not allow either the inside or the outside knob to hit against the wall. A floor mounted door stop should be used to prevent damage to the factory-adjusted clutch mechanism behind the knob. The manufacturer's one-year warranties are voided if the knobs are allowed to hit. If there is a door closer, be sure that it is properly adjusted for the Unican.
- Due to its unique design, clockwise rotation of the front knob performs two functions. It clears out any incorrect combination attempts and reactivates any buttons previously pushed, and it withdraws the latchbolt if the combination has been pushed properly. When operating the lock, *never* hold the buttons depressed while turning the front knob; damage to the unit can result. Just depress the buttons until they stop and then release them.
- Proper installation and operation of the Unican will result in many years of satisfied customer usage. Just remember that it starts with a good installation.

Servicing. You might be required to service the Unican 1000 series locks. In the majority of instances, servicing is the result of customer mishandling and abuse of the unit. Once in a very great while the unit itself may malfunction for some unknown reason. At this time, you will require some basic information concerning the various parts of the Unican 1000 unit and illustrative material for the proper assembly or disassembly. Figure 18.39 is an exploded view of the Unican 1000. Figure 18.40 is a parts item description and the assembly number and parts.

All Simplex auxiliary locks are shipped with the following combinations: 2 and 4 pushed at the same time, then 3 pushed singly.

Some people forget how to operate this lock, even though the procedure is very simple:

1. Turn the control knob to the left to activate the buttons.
2. Press the correct buttons in the proper order.
3. Release the last button(s) before turning the control knob. Turn the control knob to the right to open.
4. To relock, turn the control knob to the left. Model NL locks automatically.

Some customers will look at the lock they contemplate purchasing and wonder about the control knob. Since it operates the bolt mechanism, can the lock be opened if it is removed? The answer is a resounding No. The front control knob cannot be forced to open the lock since it is connected to the lock housing by a friction clutch. If the knob has been forced, it will be at an angle and can

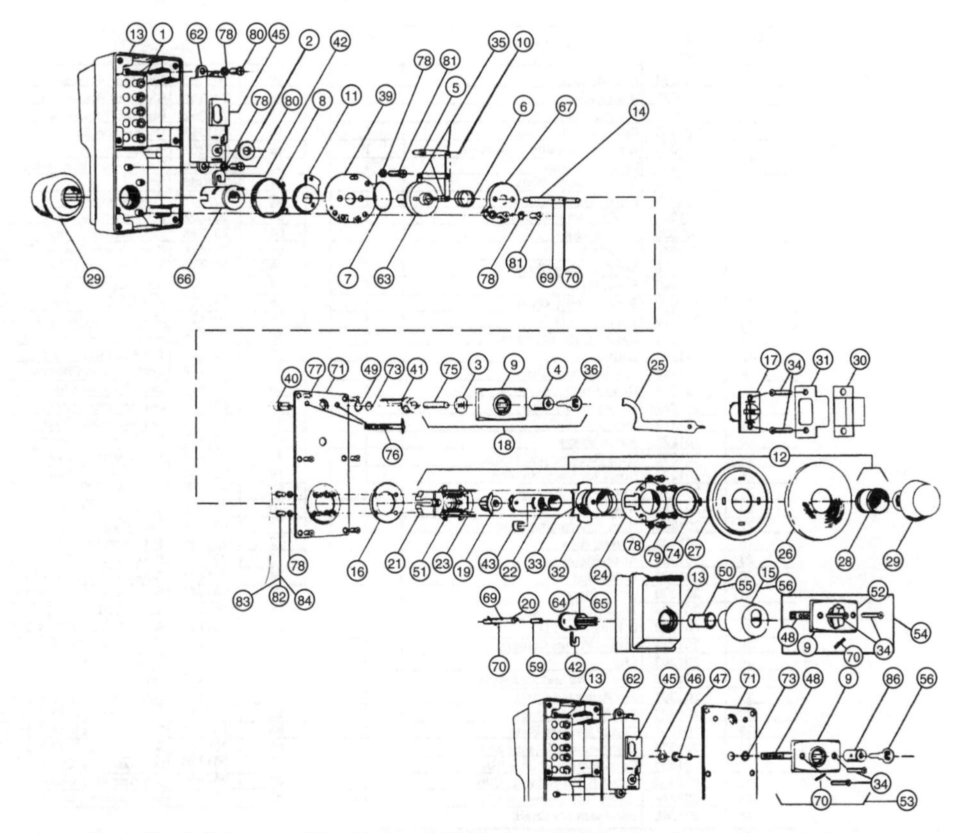

Figure 18.39 Exploded view of the Unican 1000 lock unit. (*Simplex Access Controls Corp.*)

be turned back to the vertical position by hand or with a pair of pliers without damaging the lock.

Preso-Matic

The Preso-Matic keyless pushbutton combination lock comes in either a deadbolt version (Fig. 18.41) or a deadlatch version (Fig. 18.42). This is another completely mechanically operated lock; no electricity is required. From the outside, the customer presses the selected four-digit combination. The deadbolt or deadlatch automatically retracts, opening the door. With the deadlatch model, the lock automatically locks each time the door is closed.

Several models are available, so customers can have a version with the appropriate options to meet their special needs. All models come with an instant-exit feature that requires only the depressing of an unlock button on the door inside (Fig. 18.43). Also available is a nightlatch button, which provides an added measure of protection for the home or apartment owner or the businessperson work-

ITEM NO.	PART NUMBER	DESCRIPTION	PER UNIT
1	200024	Push Buttons	5
2	200026	Bushing,Shaft	1
3	200079	Spring, Nut	1
4	200080	See Assembly Chart	1
5	201006	Link, Combination Chamber	1
6	201016	Spring, Balance	1
7	201018	Spring,Double Wave	1
8	201019	Knob Return Spring	1
9	201037	Inside Trim Plate Only	1
10	201038	Pin, Linkage	1
11	201040	Stop Plate	1
12	201048	See Assembly Chart	1
13	201049	Front Plate	1
14	201050	See Assembly Chart	1
15	201093	See Assembly Chart	1
16	201102	Spacer Door Thickness	2
17	201144	Latch	1
18	201155	See Assembly Chart	1
19	201218	Drive Insert	1
20	201267	See Assembly Chart	1
21	201272	Shoe	1
22	201275	Inside Sleeve	1
23	201276	Shoe Retainer and Bridge	1
24	201277	Cover for Shoe Housing	1
25	201278	Installation Wrench	1
26	201280	Rose	1
27	201281	Rose Reinforcing Plate	1
28	201282	Thread Ring	1
29	201284	Knob Standard	2
30	201287	Strike Box	1
31	201288	Strike Plate	1
32	201289	Shoe Retainer Cap	1
33	201290	O Ring 1 in. I.D.	1
34	201291	Screw, Philips Combination No. 8	4
35	201293	Clip	3
36	201298	Key DF 59 and Key Ring	2
37	201301	See Assembly Chart	1
38	201303	See Assembly Chart	1
39	201304	Clutch Backing Plate	1
40	201335	Combination Change Stud	1
41	201336	Combination Change Sleeve	1
42	201343	Front Knob Retainer Clip	1
43	201344	Inside Knob Retainer Clip	1
44	201345	See Assembly Chart	1
45	201350	Chamber Release	1
46	201351	Chamber Release Cam	1
47	201352	Bushing, Passage Set	1
48	201358	Connecting Bar	1
49	201362	Washer	1
50	201365	Knob Insert Best/Falcon	1
51	201366	Shoe Spring	2
52	201367	Turn Knob	1
53	201397	See Assembly Chart	1
54	201398	See Assembly Chart	1
55	201414	Knob Insert Russwin/Corbin	1
56	201416	See Assembly Chart	1
57	201417	See Assembly Chart	1
58	201422	Key DF 5 for Passage Control	2
59	201425	Shaft Self-Aligning Bushing 1000-2,4 & 6	1
60	201426	See Assembly Chart	1
61	201427	See Assembly Chart	1
62	201429	Combination Chamber	1
63	201430	See Assembly Chart	1
64	201431	See Assembly Chart	1
65	201432	See Assembly Chart	1
66	201433	See Assembly Chart	1
67	201434	See Assembly Chart	1
68	201435	See Assembly Chart	1
69	201436	Cross Pin .075 Dia.	1
70	201437	Cross Pin .100 Dia.	1
71	201438	Back Plate	1
72	201439	See Assembly Chart	1
73	201440	Retainer Ring 3/8 I.D.	1
74	201441	Retainer Ring 1 1/4 I.D.	1
75	201442	Screw Stud 1/4-20 X 1 5/8	1
76	201443	Screw 8-32 X 2 1/2	2
77	201444	Screw Philips 6-32 X 3/8 F.H.M.S.	6
78	201445	Split Washer No. 8	17
79	201446	Screw Philips 6-32 X 1/4 R.H.M.S.	4
80	201447	Screw Philips 6-32 X 5/16 R.H.M.S.	2
81	201448	Screw Philips 6-32 X 3/8 R.H.M.S.	3
82	201449	Screw Philips 8-32 X 1/4 R.H.M.S.	4
83	201450	Screw Philips 8-32 X 1/2 R.H.M.S.	4
84	201451	Screw Philips 8-32 X 3/4 R.H.M.S.	4
85	201490	See Assembly Chart	1
86	201496	See Assembly Chart	1
87	201497	See Assembly Chart	1

ASSEMBLY CHART

ASS'Y NO.	INCLUDING ITEMS NUMBERS	ASSEMBLY DESCRIPTION
200080	4, 36	Control Lock Assembly Combination Change (DF-59)
201048	19, 21, 22, 23, 24, 28, 32, 33, 43, 51, 74, 78, 79	Cylindrical Drive Unit Assembly
201050	14, 69, 70	Shaft Assembly Standard
201093	15, 50	Knob Assembly Key Override for Best/Falcon
201155	3, 4, 9, 36, 75	Inside Trim Plate Assembly
201267	20, 69, 70	Shaft Assembly, Key Override
201301	16, 19, 21, 22, 23, 24, 32, 33, 40, 41, 43, 49, 51, 71, 73, 74, 78, 88	Back Plate Assembly
201303	26, 27, 28	Rose Assembly
201345	22, 32, 33, 43	Inside Drive Sleeve Assembly
201397	9, 34, 48, 58, 70, 86	Key Actuator Assembly for Passage Set
201398	'9, 34, 48, 52, 70	Turn Knob Actuator Assembly for Passage Set
201416	55, 56	Knob Assembly Key Override for Corbin/Russwin
201417	20, 42, 55, 56, 59, 65	Conversion Assembly 1000-1 to 1000-2C
201426	1, 2, 8, 13, 14, 29, 42, 62, 63, 66, 67, 78, 80, 81	Front Plate Assembly 1000-1
201427	1, 2, 8, 13, 15, 20, 42, 50, 59, 62, 63, 64, 67, 78, 80, 81	Front Plate Assembly 1000-2B
201430	6, 7, 63	Clutch Sub-Assembly
201431	11, 64	Sleeve Drive Assembly Key Override for Best/Falcon
201432	11, 65	Sleeve Drive Assembly Key Override for Corbin/Russwin
201433	11, 66	Sleeve Drive Assembly Standard
201434	39, 67	Clutch Cover Assembly
201435	5, 10, 35	Link Assembly
201439	40, 41, 49, 71, 73	Back Plate Sub-Assembly
201490	15, 20, 42, 56, 59, 64	Conversion Assembly 1000-1 to 1000-2B
201496	58, 86	Control Assembly for Passage Set (DF-5)
201497	1, 2, 8, 13, 15, 20, 42, 55, 56, 59, 62, 63, 65, 67, 78, 80, 81	Front Plate Assembly 1000-2C

Figure 18.40 Parts identification and assembly chart for the Unican 1000 lock unit. (*Simplex Access Controls Corp.*)

Figure 18.41 Preso-Matic push-button combination deadbolt lock. (*Preso-Matic Lock Company*)

Figure 18.42 Deadlatch version of the Preso-Matic lock. (*Preso-Matic Lock Company*)

Figure 18.43 Instant exit feature and also locking ability with just a touch of a button. (*Preso-Matic Lock Company*)

ing late at night. It means that nobody is going to interrupt (or surprise) him or her at work or at home.

The unit can also be used for businesses that rely on electric strikes operated from remote locations. The Trine model 012 electric strike unit is given a unique configuration by the Preso-Matic factory to accept the deadlatch bolt.

The Preso-Matic lock has several combination possibilities, including four-digit combination, seven-digit combination, and master combination (for use with the four-digit combination version).

The *four-digit combination* provides a maximum of 10,000 possible combinations; the *seven-digit combination* can give up to 10 million possible combinations. The last option adds a master combination to the individual four-digit combination. All locks of a given system will open for 1 six-digit master combination, but each unit within the system can still be opened by its own four-digit combination. The master combination variations allow for security protection and flexibility that are not available with other pushbutton keyless combination locks.

The numbers of the combination for each lock unit are determined by the two *combination slides* (Fig. 18.44) that are inserted into the unit. To change the combination, you simply remove the inside cover plate, turn two spring clips clear of the slot in the lock housing, lift out the combination slides, and slip in new ones.

Installation. All the Preso-Matic locks are carefully designed and engineered to aid you in quick, easy, troublefree installation. There are only three basic steps:

1. Bore the opening for the lock and latch (Fig. 18.45) using the required template that comes with the unit (Figs. 18.46 and 18.47).
2. Mortise the door edge for the bolt face and insert the bolt housing and tighten the screws (Fig. 18.48).
3. Insert the lock and tighten only one screw, connecting the bolt housing to the lock and making the unit operable (Fig. 18.49). Attach the cover plate on the inside of the door. Installation is complete.

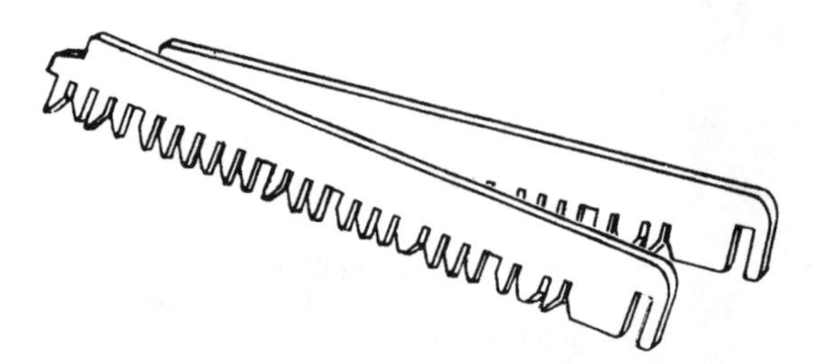

Figure 18.44 Combination slides used with the lock unit. (*Preso-Matic Lock Company*)

Figure 18.45 Preso-Matic installation template, 2⅜-inch backset.

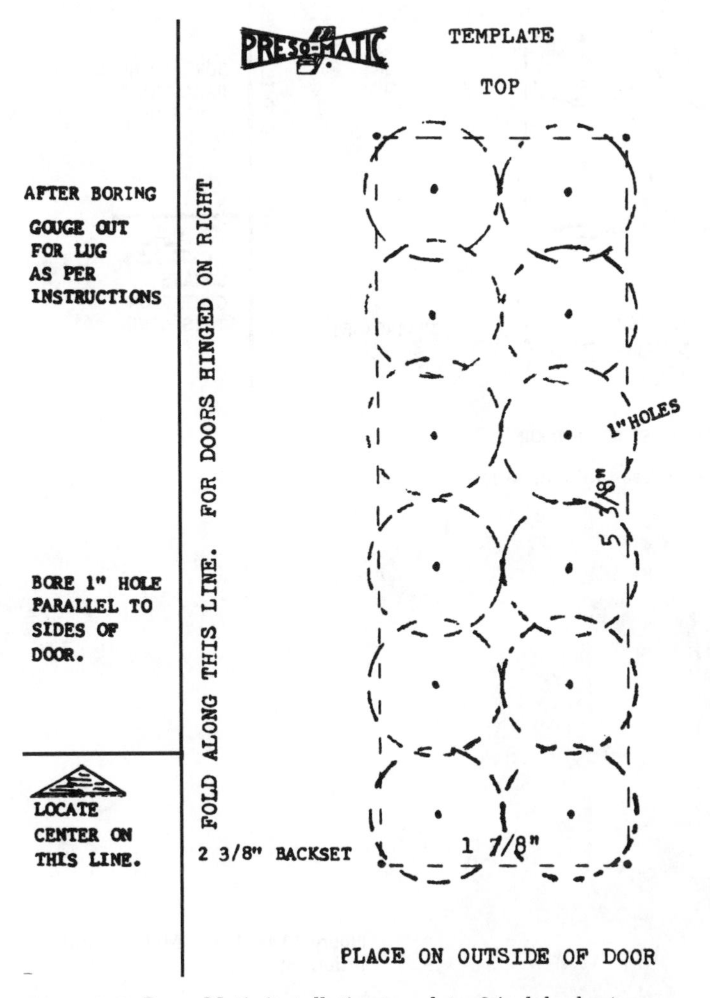

Figure 18.46 Preso-Matic installation template, 3-inch backset.

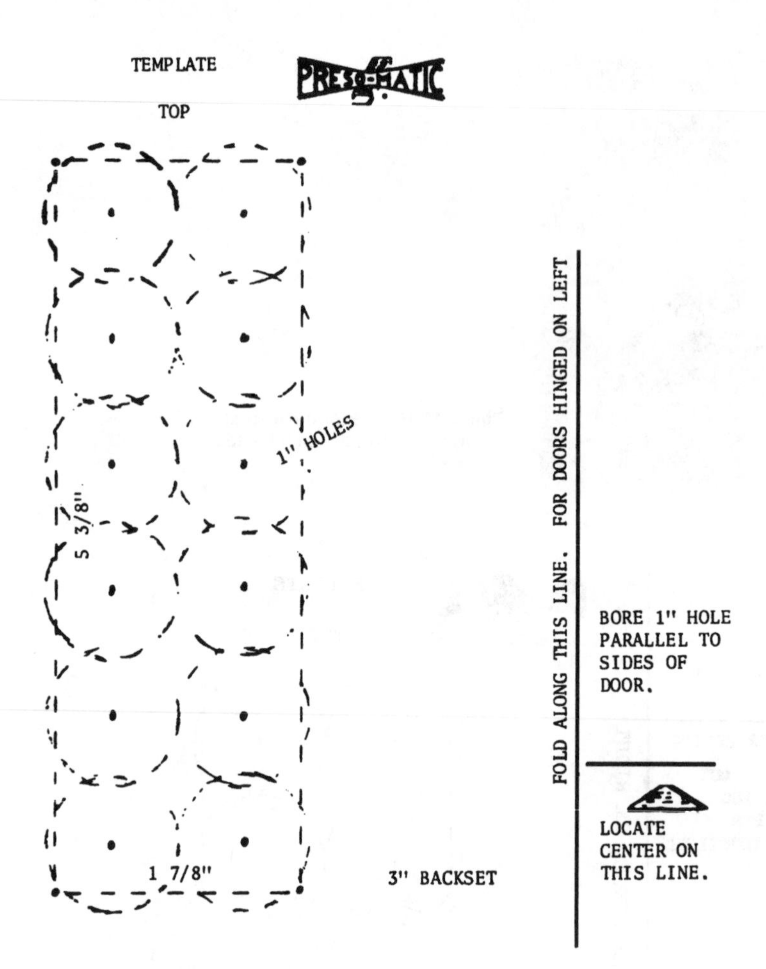

Figure 18.47 Preso-Matic installation, Step 1.

Figure 18.48 Preso-Matic installation, Step 2.

Figure 18.49 Preso-Matic installation, Step 3.

MODEL	LOCK TYPE	BOLT PROJECTION	BACKSET	NOTES
8101	Deadbolt	5/8"	2 3/8" (2 3/4 also available)	Manual locking. Built-in night latch button. Hardened steel free-turning bolts. Manually locks.
LT8102	Deadbolt	1"	3"	Same
8103	Deadlatch	1/2"	3"	Night latch button. Locks automatically. Inswinging doors only.
8103A	Deadlatch	1/2"	3"	Same, except without night latch button
8200	Deadlatch	1/2"	3"	Hardened steel latch bolt. Locks automatically. Night latch button
8200A	Deadlatch	1/2"	3"	Same, except without night latch button.

OPTIONS:
1. Change of combination slides for a 7 digit code available.
2. Lock shields.
3. Stay open function; keeps the deadlatch in an unlocked position.
4. Door stop strike plate; deadlatch models only. Case hardened; ideal for double door installation.
5. Extra wide strike plate for metal doors (2 1/2" × 2 1/2"); case hardened.
6. Heavy duty ANSI electric door opener. 24 VAC. for normally locked units.

Figure 18.50 Installation layout variations. (*Preso-Matic Lock Company*)

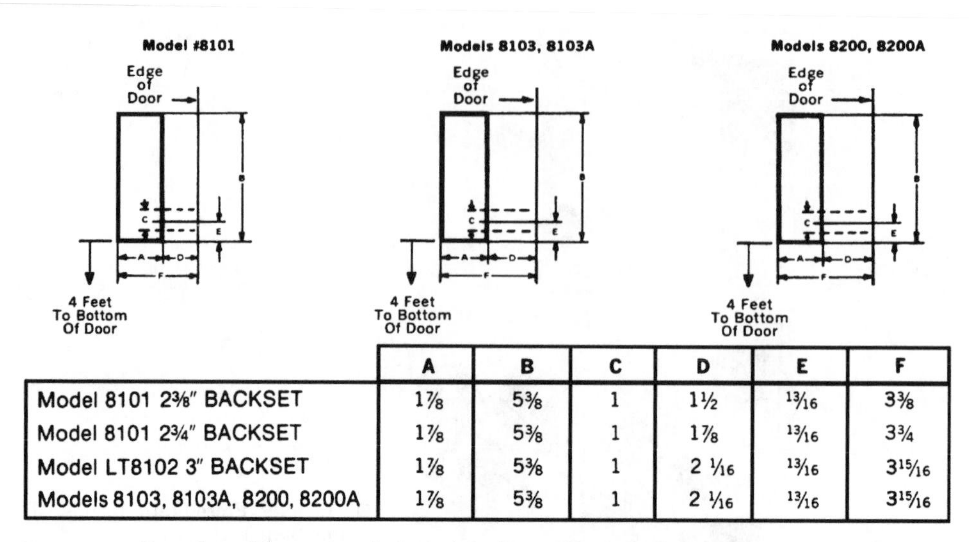

	A	B	C	D	E	F
Model 8101 2⅜″ BACKSET	1⅞	5⅜	1	1½	13/16	3⅜
Model 8101 2¾″ BACKSET	1⅞	5⅜	1	1⅞	13/16	3¾
Model LT8102 3″ BACKSET	1⅞	5⅜	1	2 1/16	13/16	3 15/16
Models 8103, 8103A, 8200, 8200A	1⅞	5⅜	1	2 1/16	13/16	3 15/16

Figure 18.51 Overview of the various lock models. (*Preso-Matic Lock Company*)

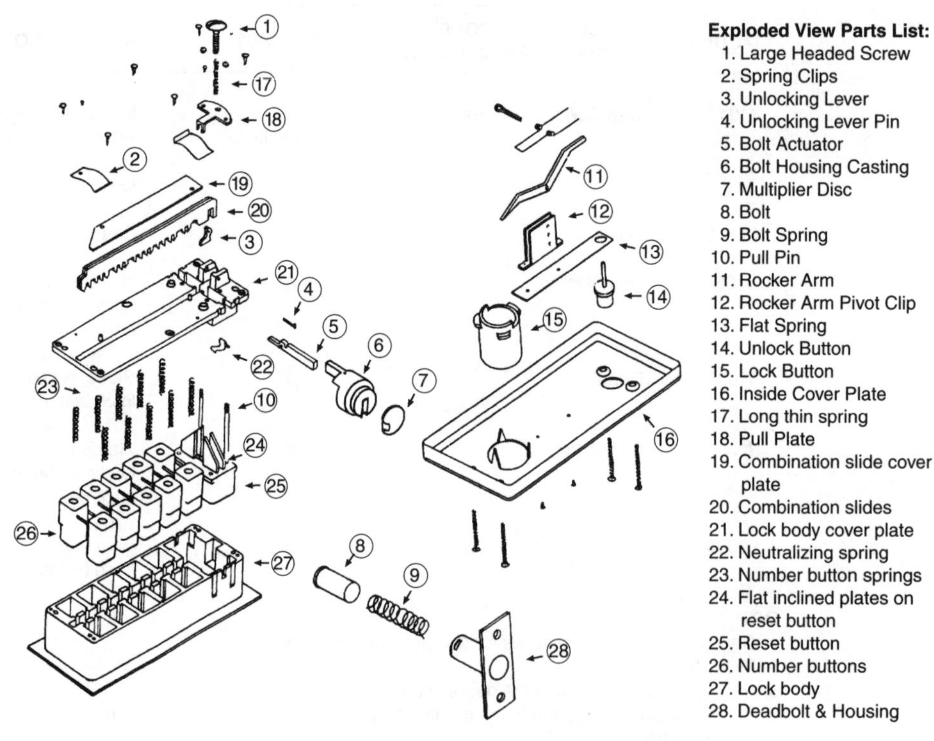

Figure 18.52 Exploded view of the Preso-Matic pushbutton combination lock and parts identification. (*Preso-Matic Lock Company*)

For volume installations in large construction projects or numerous prefabricated doors, you should use the Preso-Matic round face, drive-in bolt, which installs in seconds and saves time.

Three basic installation layouts are available, as shown in Fig. 18.50. Figure 18.51 is a quick summary of the various lock models. Figure 18.52 provides an exploded view of the lock and part identification.

Chapter

19

Electrical Access and Exit Control Systems

Electrically operated release latch strikes and locks are easily installed in place of standard units to provide remote controlled access for a door. These units are not new to business and industry and really shouldn't be that new to the locksmith—but they are. Why? Because, excepting a few instances in large cities or industrial regions, the average locksmith doesn't get involved with such units. Since it is "electrical," the job is passed to an electrician. By doing so, the locksmith loses valuable business.

This chapter discusses some electrically operated locks and release latch strikes, the circuitry behind the various types of units, where they can be installed, and the potential of such units. (For more information on electronic security see my book, *The Complete Book of Electronic Security*, published by McGraw-Hill.)

Electric Release Latch Strikes

Electrically actuated release latch strikes are more common than locks. These units normally can be reversed for either right- or left-hand doors. They operate on low voltage and the strikes fit a hollow jamb channel as shallow as 1⅝ inches (1¾ inches for mortise latches) or they can be, if necessary, installed in wooden jambs.

Electrical release strikes are gaining popularity among small and large businesses. They come in a variety of sizes (Fig. 19.1) to meet varying needs and requirements. There are many electrical strikes available on the market. While all of them meet minimum standards, prior to your purchase of them you should still consider the manufacturer and the purpose for which the units were developed.

Since this is a rapidly growing area of access control (security), it is important that you receive the latest and best information available. To obtain the

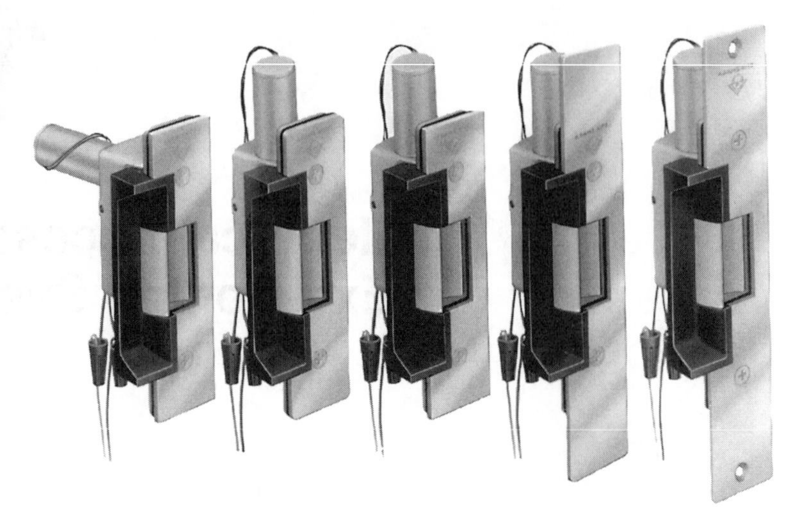

Figure 19.1 Various types and sizes of electric release latch strikes. (*Adams Rite Mfg. Co.*)

following information, I contacted Adams Rite, the manufacturer considered the nation's leader in the electrical release latch strike field. I am extremely grateful for Adams Rite's assistance.

First and foremost in the field is Adams Rite, whose electrical strikes are architecturally sound and designed to add increased security features to areas requiring a certain amount of traffic control. When properly powered and installed, these electrical release strikes will provide a very long, maintenance-free life.

The fact that electrical power is used for the operation of the strike in no way makes the unit an electrical appliance, as some people tend to think. With a few exceptions, the electrical circuitry unique to these units are designed to meet the needs of the strike, not the other way around. The hardware specifier (that's you) should select the strike and expect the electrician (which could be you) to supply the power at the point of installation. In a few cases an existing circuit can be used, so the strike is selected and installed to it. In selecting strikes, two areas of consideration must be followed: electrical and hardware.

Electrical considerations

First, determine the duty of the unit. Is operation to be intermittent or continuous? If the door is normally locked and released only momentarily from time to time, it is *intermittent*. In some cases, the strike is activated (unlocked) for long periods, so duty is *continuous*. A more rare requirement is *reverse action*, in which the strike is locked only when its current is switched on.

Intermittent duty. For a normal intermittent application, the electrical strike will have to be 24 Vac. This alternating current call-out gives enough power for almost any entrance, even one with a wind-load situation. Yet the low voltage range is below that requiring UL or building code supervision. At this voltage, good, reliable transformers are available.

Continuous duty. When a continuous duty application is required, additional components are added to the unit at the factory.

Reverse action. A long period of unlocking can also be obtained by using a reverse action strike. This might be required to provide the same service as a continuous duty strike but preserves current because it is on battery dc power. It could also be used to provide a fail-safe unlocked door in case of a power failure or building emergency (such as a fire) where quick, guaranteed unlocked doors are required. *Note:* Reverse action strikes, when ordered by the locksmith for installation, must specify the hand of the door on the order. These types are not reversible, as are some other strikes.

Monitoring. If a visual or other signal is required to tell the operator that the electric strike is doing its job properly, a monitoring strike will be required. Two sensor/switches are added: one is activated by the latchbolt's penetration of the strike, and the other by the solenoid plunger that blocks the strike's release.

Transformer. Low voltage is necessary for proper electric strike operation, and this is obtained by the use of transformers that are stepped down from the normal ac power voltage to 12, 16, 24, or other lower voltage. Whether or not you perform the electrical side of the installation (when wires must be run), three electrical specifiers are necessary:

- Input voltage (110—standard household current)
- Output voltage (12, 16, 24, etc.—from the transformer unit)
- The capacity of the transformer, called *volt amps* (output voltage times the output amperes)

Skimping on the capacity of the transformer to save a few dollars will underpower the door release and is likely to bring in complaints of "poor hardware." Intermittent duty electric strikes draw from 1½ to 3 amps; a continuous duty model draws less than 1 amp.

In any application where the current draw of the strike must be restricted to less than 1 amp, regardless of the duty, a current limiter that stores electrical energy to relieve high-use periods is available.

Wiring. The wiring must carry the electrical power from the transformer through the actuating switch to the door release. It must be large enough to minimize frictional line losses and deliver most of the input from the transformer to the door release. A small-diameter garden hose won't provide a full flow of water from the nozzle, particularly if it's a long run; neither will an undersized wire carry the full current.

Electrical troubleshooting. When insufficient electrical power is suspected in a weak door release, make a simple check. Measure the voltage at the door release while the unit is activated. If the voltage is below that specified on the hardware schedule, the problem is in the circuit—probably an undercapacity transformer, if the current length is short. A long run may mean both a transformer and a wire problem.

Hardware

Hardware is the second factor in selecting a strike. There are many considerations:

Face shape (sectional). Electrical strikes from Adams Rite have a flat face. Two exceptions have a radius-type to match the nose shape of a paired narrow stile glass door.

Face size. The basic Adams Rite size conforms to ANSI standards for strike preparation: 1¼ × 4⅞ inches. However, two other sizes are offered to fill or cover existing jamb (or opposing stile) preparation from a previous installation of another unit, such as an MS deadlock strike or a discontinued series electric strike.

Face corners. Strikes are available with two types of corners: round for installation in narrow stile aluminum where preparation is usually done by router, and square for punched hollow metal ANSI preparation or for wood mortise installation.

Jamb material. Strikes with vertically mounted solenoids are designed to slip into hollow metal stile sections as shallow as 1⅝ inches. Horizontal solenoids for wood jambs require an easily bored mortise of 3⅛ inches.

Previous strike preparation. If the jamb was previously fitted with a now-discontinued strike, such as a Series 002 electrical strike, the new strike unit to be installed—perhaps an Adams Rite 7510—will fit in with only minor alterations. In this example, you would increase the lip cutout by ¼ inch at each end. In the case where a hollow jamb was originally prepared to receive the bolt from another type deadlock, another latch such as the 4510 series would have to be substituted. In such a case, the radius or flat jamb would cover the old strike cutout.

Lip length. The standard lip on all basic Adams Rite electric strikes accommodates a 1¾-inch thick door that closes flush with the jamb. Where the door/jamb relationship is different, a long lip can be added.

Compatible latch. Certain electric strikes will mate with all latches in a given series. The key-in-knob latches and mortise latches are made to mate properly with the various electric strikes available.

Handing of the strike unit. All standard operation strikes are unhanded; thus, they can be installed for either a right- or a left-hand door. However, reverse action (locked only when the current is on) units require that you specify the hand when you place the order for the unit.

To repeat: Understand that electrical strikes are not just electrical appliances; they are hardware. In the case of Adams Rite strikes, they are made by hardware people, because they know that the hardware has to be carefully matched to its application.

Also, realize that when you order a unit for installation, you need to check with the local electrical codes to ensure that you can actually install it. Some jurisdictions may require an electrician to do the job.

Installation and procedures

The following section discusses and illustrates some of the procedures required to install a variety of strike units.

Aluminum jam installation (Series 750)

1. Prepare the door jamb per Fig. 19.2.
2. Install the mounting clips to the jamb using 8-32 × ⅜ screws and pressed metal nuts. Leave the screws slightly loose to permit easy alignment of the base assembly and clip.
3. Spacers are provided to assure flush final assembly of the faceplate and jamb. Add one or more spacers between the jamb and mounting clip when the faceplate extends beyond the jamb. When the faceplate sets inside the jamb, you must add spacers between the mounting clip and the electric strike case. See detail "A" in Fig. 19.2. To attach spacers to the mounting clip, remove the protective coating from the spacer and press to the desired mounting clip surface. Make sure the clearance hole in the spacer aligns properly with the hole in the mounting clip.
4. Using the wire nuts, connect the wires coming from the unit to the wires coming from the low-voltage side of the transformer.
5. Insert the electrical release into the jamb by tipping the solenoid coil into place behind the mounting clip, then drop the unit onto the clips.
6. Attach the unit with the #10 screws and lock washers provided.
7. Attach the faceplate using the screws furnished (8-32 × ¼ inch).
8. Secure the screws holding the mounting clips to the jamb.

Figure 19.2 also shows the dimensions of the electric door release that would be installed in the aluminum jamb.

Replacing an obsolete unit with an electrical strike (model 7510 wood jamb)

1. Prepare the jamb to the dimensions shown in Fig. 19.3. Take care to mortise the cut to clear the solenoid.

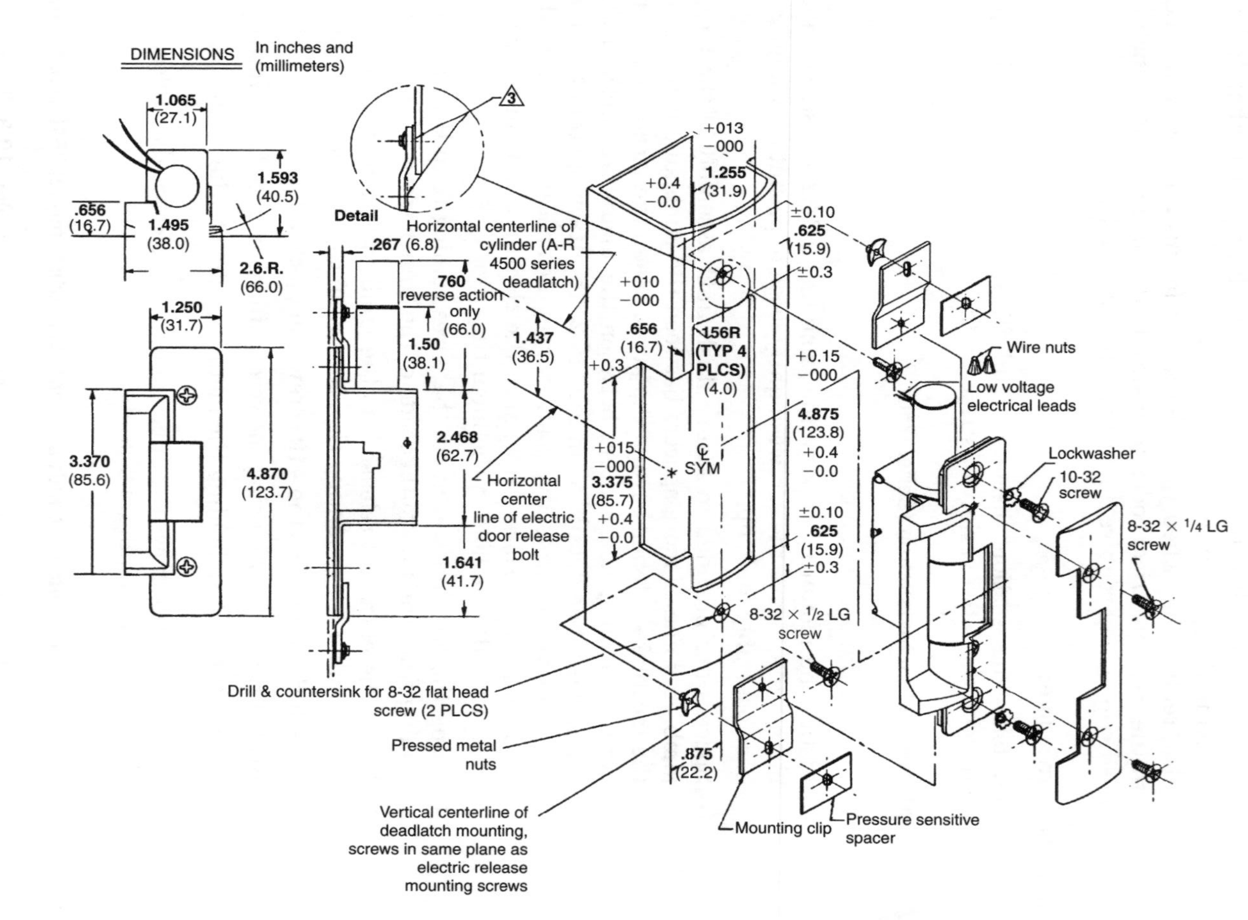

Figure 19.2 Jamb specifications and installation details for a radial (curved) aluminum door jamb. (*Adams Rite Mfg. Co.*)

2. Using the wire nuts provided, connect the wires from the 7510 unit to the wires from the low-voltage side of the transformer.
3. Insert the electric release into the jamb. Attach this with #10 flat head wood screws.

Replacing the obsolete unit. Unit 002 is used as an example.

1. Remove the 002 unit and disconnect the low-voltage wires. Do not remove the existing mounting clips.
2. Enlarge the jamb cutout by removing ¼ inch from each side.
3. Using the wire nuts, connect the 7510 wires to the low-voltage side of the transformer.
4. For a flush installation, provide a 1⁄16-inch spacer between the mounting surface and the face plate.
5. Insert the electric door release into the jamb, and attach with screws.

Figure 19.3 also shows the dimensions of the 7510 electrical door release unit.

Installing an electric strike to meet ANSI specifications. Model 7520 is used in the example. Note the molded strip down the center of the door frame in the illustration. Installation is as follows:

1. Prepare the metal jamb to the dimensions shown in Fig. 19.4.
2. Spacers are provided with the unit to assure the flush mounting of the faceplate to the jamb. To use the spacer(s), remove the protective cover and press them to the mounting bracket. Make sure the clearance hole in the spacer aligns with the tapped hole in the bracket.
3. Connect the wires from the unit to the transformer wires.
4. Insert the electric release into the jamb by tipping the solenoid coil into place behind the mounting bracket. Then drop the unit onto the mounting brackets. Attach the unit with the screws furnished.
5. Attach the faceplate.

Installing an electric strike for key-in-knob latch to ANSI specifications. The example used is an Adams Rite Model 7540 Unit.

1. Prepare the door jamb as shown in Fig. 19.5.
2. Install the mounting clips using the screws and pressed metal nuts. Leave the screws slightly loose to permit easy alignment of the case assembly and clip.
3. Use the provided spacers as necessary.
4. Connect the wires from the 7540 unit to the low-voltage wires from the transformer.

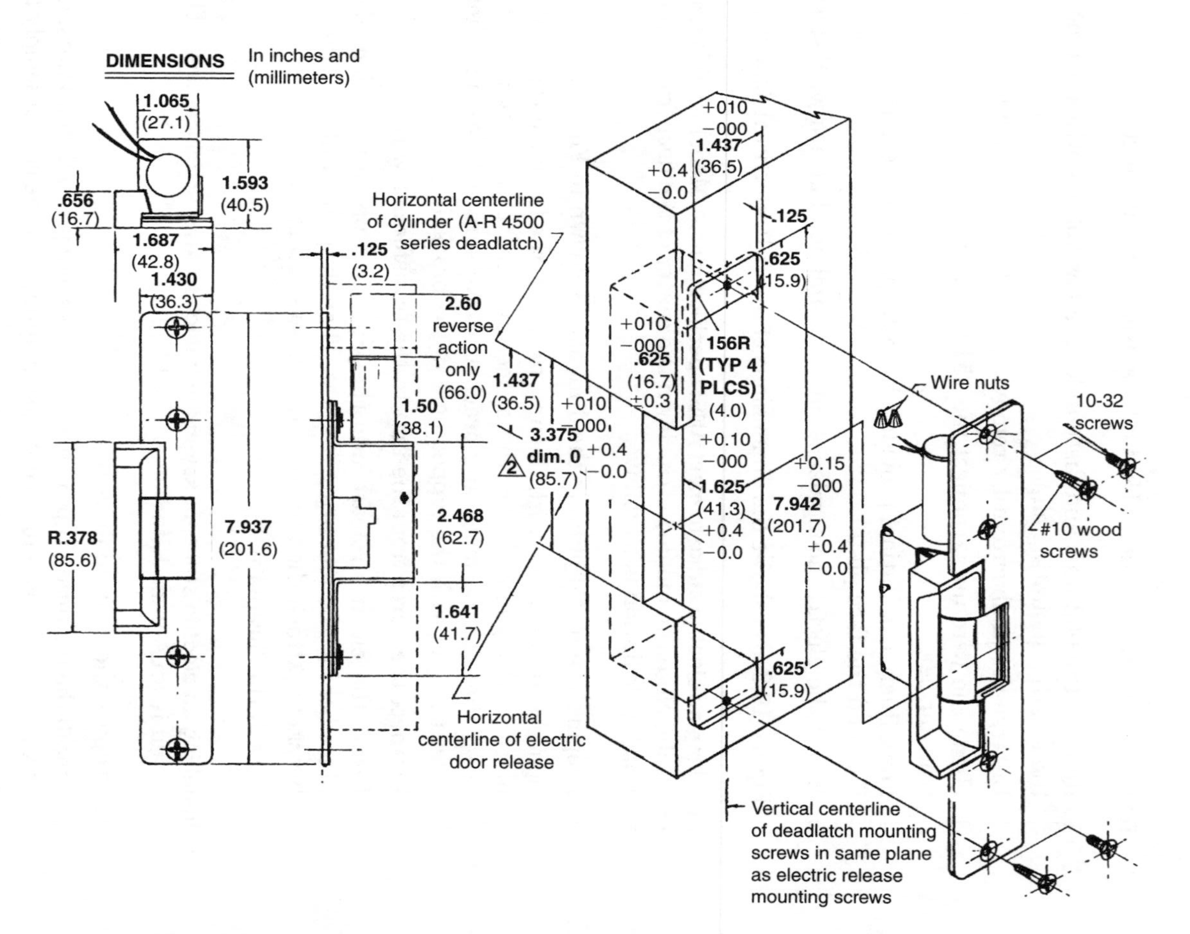

Figure 19.3 Dimensions and installation details for unit to be mounted in a wood frame. (*Adams Rite Mfg. Co.*)

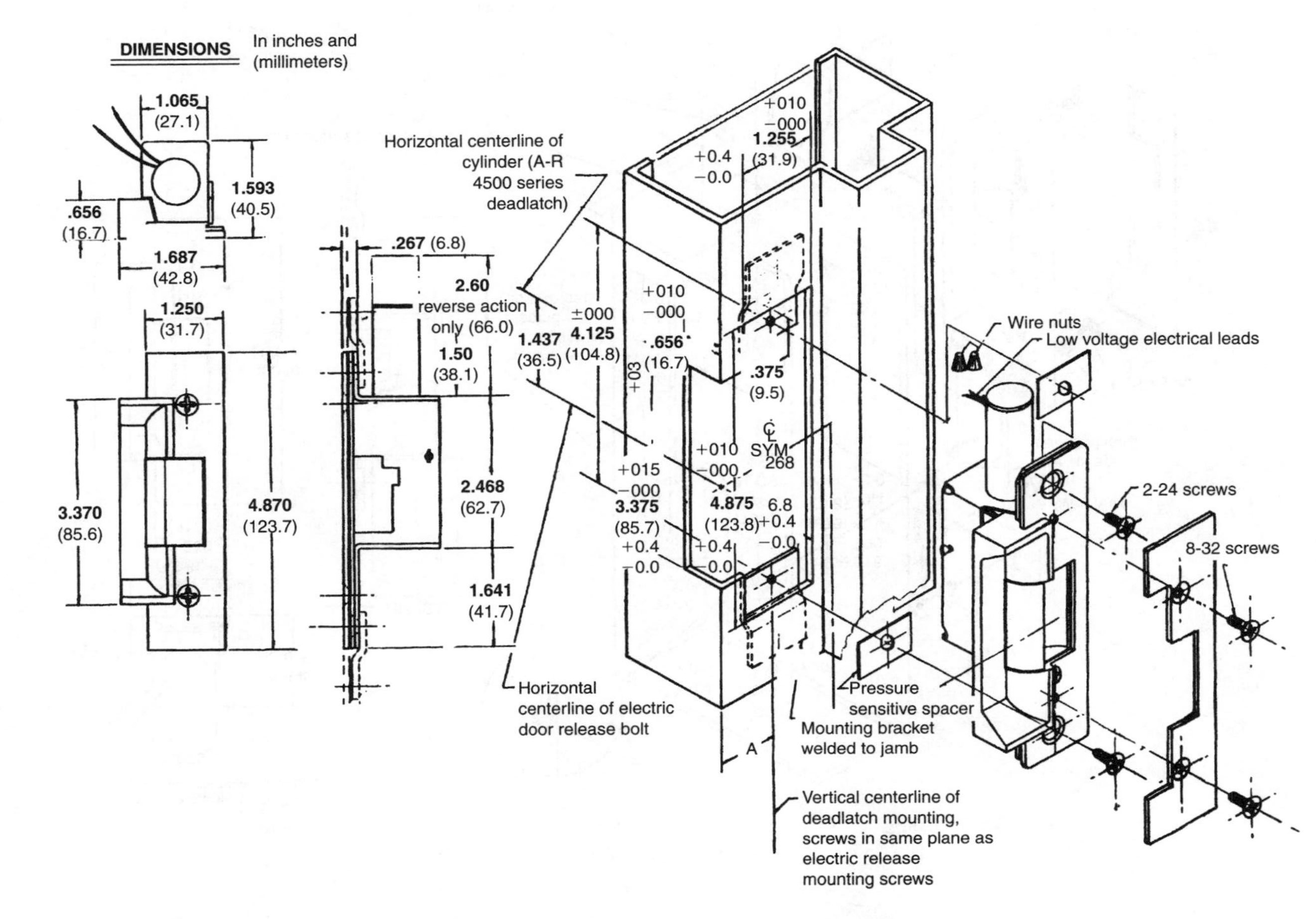

Figure 19.4 Jamb preparation and dimension data for a metal frame that has a molded strip down the frame center. (*Adams Rite Mfg. Co.*)

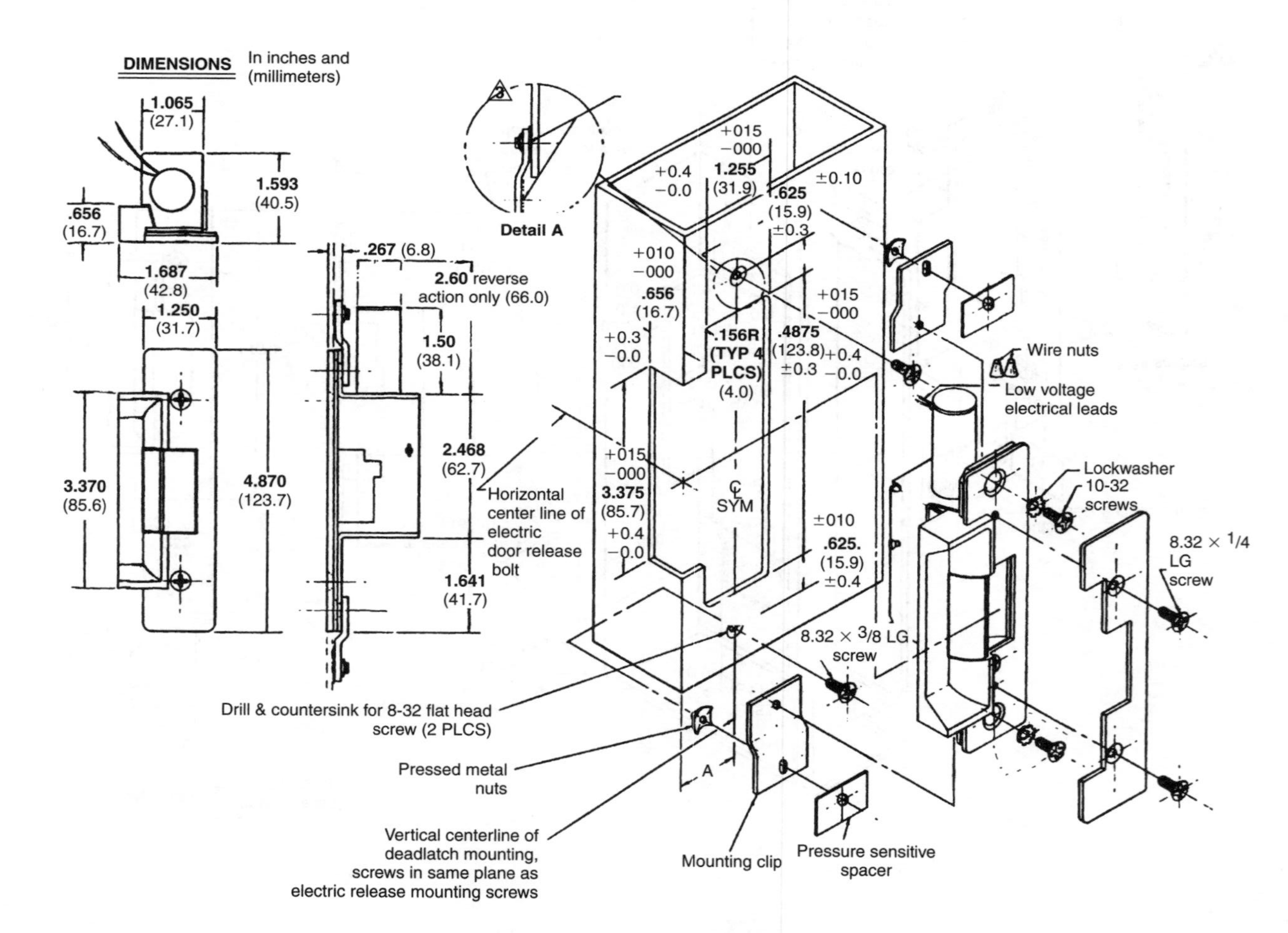

Figure 19.5 Installation details for a key-in-knob latch unit. (*Adams Rite Mfg. Co.*)

5. Insert the electric release into the jamb and attach with the #10 screws provided.
6. Attach the faceplate and secure the screws holding the mounting clip to the jamb.

Installing an electric strike in a wooden jamb

1. Prepare the wood door jamb to the dimensions indicated in Fig. 19.6.
2. Connect the wires from the unit to the transformer wires.
3. Insert the strike into the jamb and attach with the #10 wood screws.
4. Attach the faceplate.
5. Figure 19.7 provides the technical dimensions of the strike unit to be installed.

Circuitry

Figure 19.8 illustrates the typical wiring diagram for an audible (ac) operating strike unit. Figure 19.9 is the same unit with a rectifier used to provide silent (dc) operation of the strike unit.

You might have an instance where the installation requires a monitoring signal operation. Figure 19.10 provides a typical wiring diagram. Many monitoring arrangements are available. The green light indicates that the latchbolt is in the strike and that the solenoid plunger is blocking the strike. The yellow light indicates the latchbolt is in the strike but is unblocked, allowing access. The red light signals that the latchbolt is out of the strike.

Troubleshooting Adams-Rite electric strike latch units

You have seen several of the more common types of electrical strike release units. What do you do if you have trouble with a unit after you install it?

Accurately checking an electrical circuit, just as any other job, requires specific tools. You need a *good* 20,000 volts per ohm volt-ohm-milliammeter. The Simpson model 261, one of several currently available, costs about $60, but it is money well spent if you will be installing and maintaining electric strike latch units.

First, read the volt-ohm-milliammeter (VOM) instructions and practice on simple low-voltage circuits to ensure that you properly understand the operating instructions and can correctly read the results on the meter.

Checking voltage

1. Zero the pointer.
2. Turn off the power to the circuit being measured.
3. Set the function switch to the correct voltage to be measured (+dcC or ac).

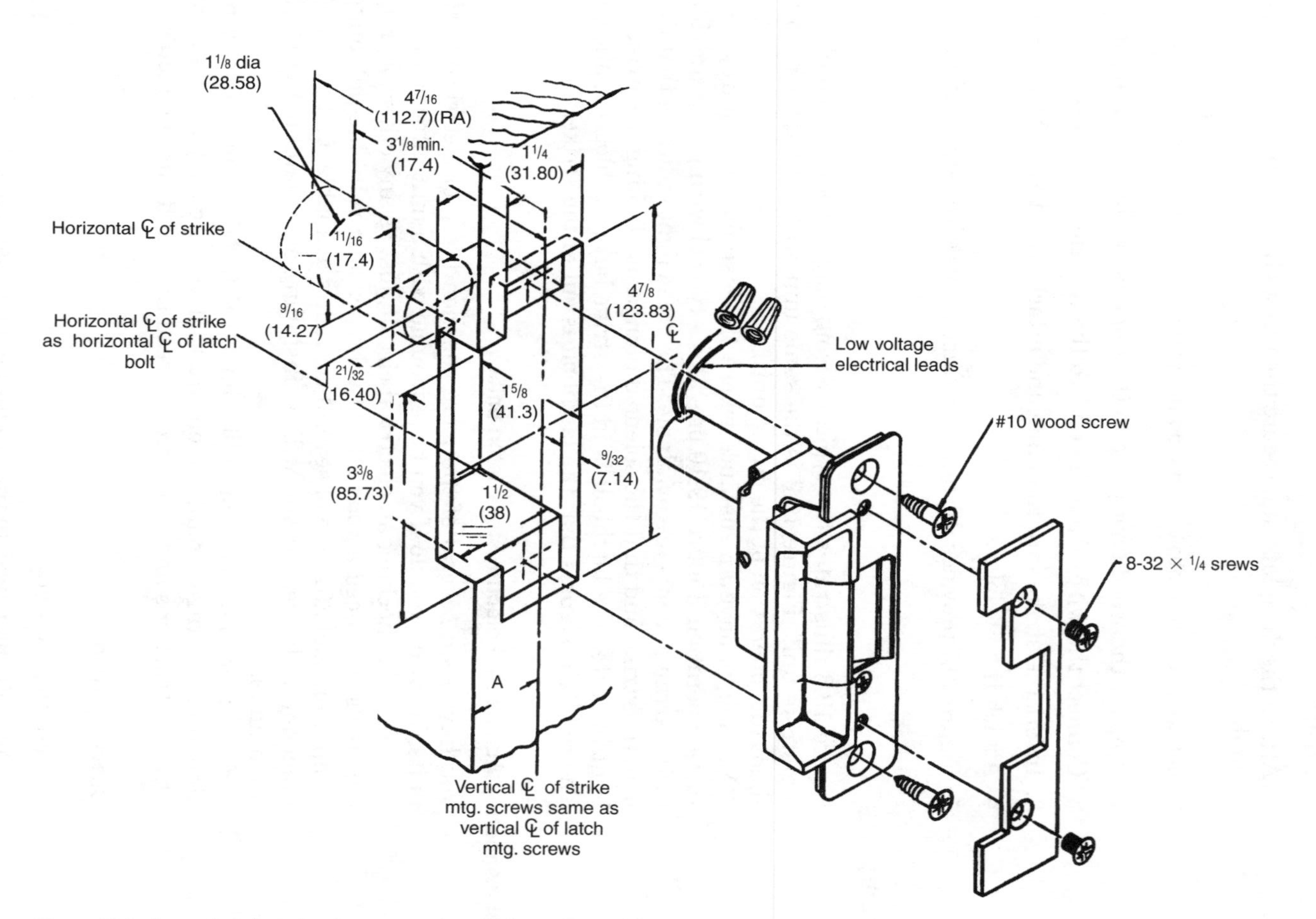

Figure 19.6 A wooden door jamb preparation. (*Adams Rite Mfg. Co.*)

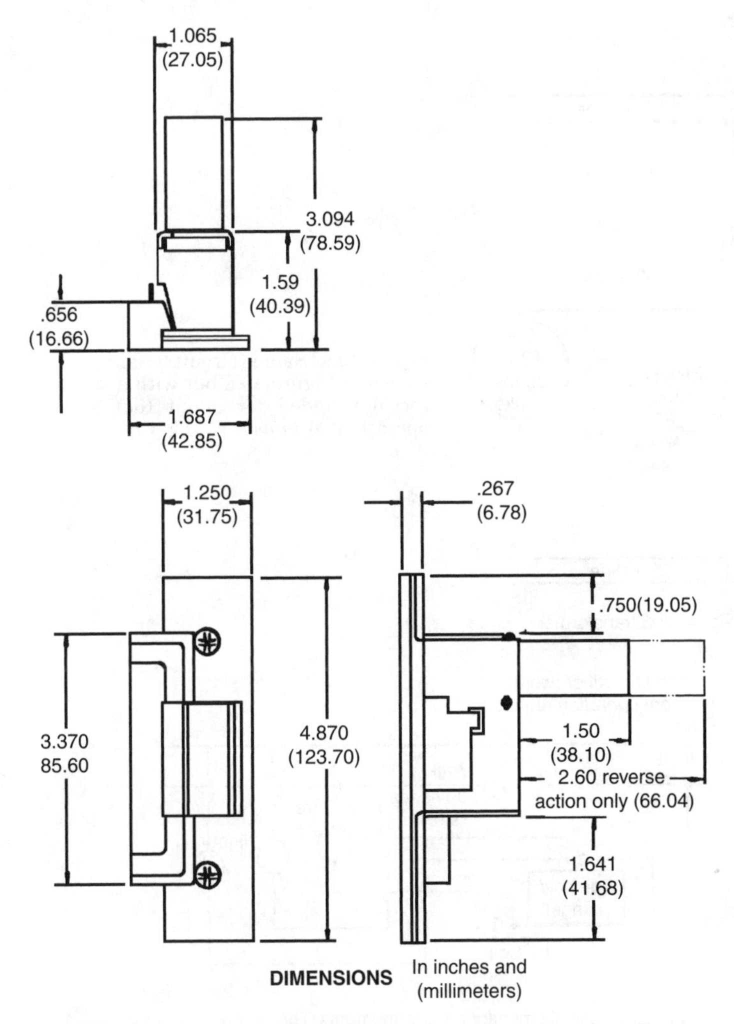

Figure 19.7 Strike unit dimensions for wood door jamb installations. (*Adams Rite Mfg. Co.*)

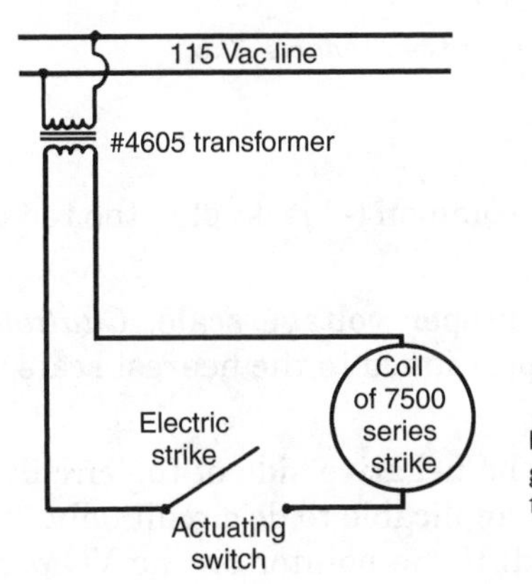

Figure 19.8 Typical wiring diagram for an audible (ac) operation. (*Adams Rite Mfg. Co.*)

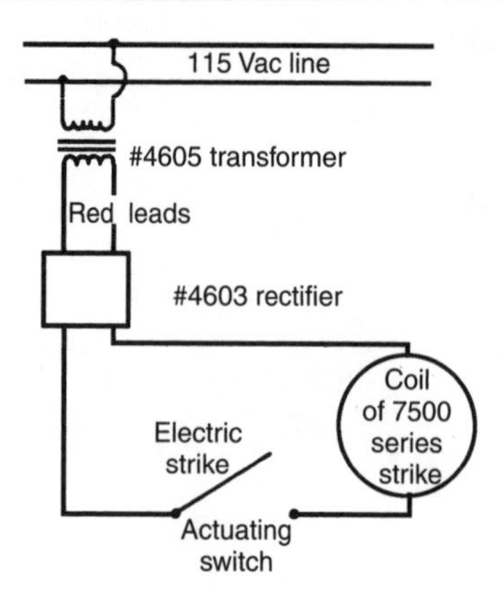

Figure 19.9 Same circuitry as shown in Figure 19.8 but with a rectifier added for silent (dc) operation. (*Adams Rite Mfg. Co.*)

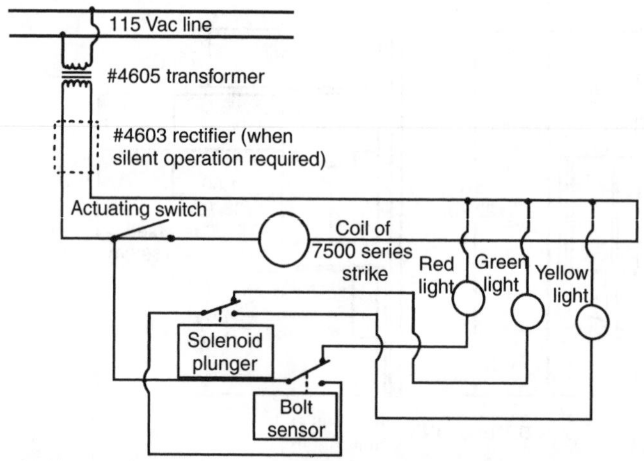

In this, one of many possible monitoring arrangements. The green light indicates that latch bolt is in strike and solenoid plunge is blocking strike. Yellow light indicates latch both is in strike but strike is unblocked, allowing access. Red light signals that latch bolt is out of strike.

Figure 19.10 Wiring for a monitor signal operation. (*Adams Rite Mfg. Co.*)

4. Plug the black test lead into the common (–) jack; plug the red test lead into the (+) jack.
5. Set the range selector to the proper voltage scale. *Caution:* It is very important that the selector be positioned to the nearest scale voltage to be measured.
6. Connect the black test lead to the negative side of the circuit and the red lead to the positive side. This is applicable to dc circuit only. Turn the power on to the circuit being tested. If the pointer on the VOM moves to the

left, the polarity is wrong. Turn the switch function to –dc and turn the power back on. The pointer should now swing to the right for the proper reading on the dc scale. For an ac circuit, the connections are the same except as noted in step 3. You don't have to worry about polarity in the ac circuit.

7. Turn the circuit off before disconnecting the VOM.

Checking current (dc)

1. Zero the pointer.
2. Turn off the power to the circuit being tested.
3. Connect the black test lead to the –10-amp jack and the red test lead to the +10-amp jack.
4. Set the range selector to 10 amps.
5. Open the circuit to be measured by disconnecting the wire that goes to one side of the solenoid. Connect the meter in series—that is, hook the black lead to one of the disconnected wires and the red lead to the other wire.
6. Turn the power on to the circuit and observe the meter. If the pointer moves to the left, reverse the leads in the jacks.
7. Turn the power back on to the circuit and read amperage on the dc scale.
8. Turn the VOM off before disconnecting.

Checking for line drop. Measure the line drop by comparing voltage readings at the source (the transformer's secondary or input side) with the reading at the electric strike connection.

Finding circuit shorts in a hurry. The VOM is the most reliable instrument for detecting a circuit short. This is accomplished by setting up the VOM to measure resistance.

1. Set the range switch at position R × 1.
2. Set the function switch at +dc.
3. Connect the black test lead to the common (–) jack and the red lead to the (+) jack.
4. Zero the pointer by shorting the test leads together.
5. Connect the other ends of the test leads across the resistance to be measured—in the case of a solenoid, one of the test leads to one coil terminal and the other test lead to the other terminal.
6. Watch the meter. If there is no movement of the pointer, the resistance being measured is open. If the pointer moves to the peg on the right side of the scale, the resistance being measured is shorted closed. If you get a reading in between these two extremes, it is very likely the solenoid is okay.

Problem solving

The following are solutions to several common problems.

The strike will not activate after installation

- Check the fuse or circuit breaker supplying power to the system.
- Check that all wiring connections are securely made. When wire nuts are used, ensure that both wires are firmly twisted together for good electrical transfer between the wires.
- Check the solenoid coil rated voltage (as shown on the coil label) to make sure that it corresponds to the output side of the transformer within ±10 percent.
- Using the VOM, check the voltage at the secondary (output) side of the transformer.
- Using the VOM, check the voltage at the solenoid. This assures you that there are no broken wires, bad rectifiers, or bad connections.
- Check the coil for a possible short.

The transformer overheats

- Make sure that the rated voltage of the transformer and the rated voltage of the coil correspond within ±10 percent.
- Make sure the VA rating is adequate. For all Adams-Rite units, it is recommended you see 40 VA; 20 VA is the absolute minimum, and you could experience transformer heating in even moderate use applications.

The rectifier overheats

- The rectifier is wired incorrectly. This means the overheating is of a temporary nature (a few milliseconds) and then it's burned out.
- There are too many solenoids being supplied by a single rectifier and you are pulling more current through it than the system diodes are rated for.

The solenoid overheats

- The coils used in an electric strike latch, when used as rated (continuously or intermittently by pulsing once per second), gave a coil temperature rise rating of 149°F (65°C) *above* ambient. To get the exact temperature, add 149°F plus 72°F (ambient), which equals 221°F. The coil insulation is rated at 266°F (130°C). Regardless of whether the coil goes to 221°F or 266°F, it is too hot to keep your fingers on it!
- With the above in mind, it is fair to say the most intermittent duty units never see that kind of use. If a coil gets extremely hot on very short pulses at 2- or 3-second intervals, you either have the wrong coil or the wrong

STRIKE VOLTAGE, CURRENT AND DUTY:	6 VDC - INT	12 VDC - INT	24 VDC - INT	48 VDC - INT	12 VDC - CONT	24 VDC - CONT	16 VAC - INT	24 VAC - INT	48 VAC - INT
COIL RESISTANCE (Ohms) Ω	1.0	3.5	15.0	61.0	15.0	61.0	7.4	3.5	15.0
PEAK INSTANTANEOUS CURRENT (Amps)*	5.8	2.7	1.42	.70	.75	.36	1.6	7.0	3.2
CONTINUOUS OR HOLD CURRENT (Amps)*	5.8	2.7	1.42	.70	.75	.36	.70	3.72	1.9
PEAK INSTANTANEOUS POWER (Watts)	34.8	32.4	34.2	33.6	9.0	8.6	25.6	168.0	154.0
CONTINUOUS OR HOLD POWER (Watts)	34.8	32.4	34.2	33.6	9.0	8.6	11.3	89.2	95.0
TURN-ON SETTLE TIME (Milliseconds)*	40	11	12	14	30	30	8	8	8

Figure 19.11 Power data for electric strike units. (*Adams Rite Mfg. Co.*)

transformer output. The same is basically true for continuous duty coils. If the coil temperature exceeds the ratings, it has to be because the coil voltage or the transformer are improperly coordinated.

- If you set up the meter as if testing for a short (see previous instructions) and obtain the exact resistance, you can compare to the specifications found in Fig. 19.11; this becomes another way of knowing you have the correct coil. Figure 19.11 also provides all the power data for the various electric strikes from Adams Rite that you might ever be required to install on various premises.

Determining the proper wire size

Some people—especially those not trained in the various phases of proper electricity—assume that a low voltage means you can use the smallest and cheapest wire size available. This is certainly not the case. The size of the wire is dependent on the use, the amount (voltage) of electricity going through the wire, the distance the wire runs, and the number of strikes to be used on the wire run.

STRIKE USED	WIRE SIZE FACTOR
6 VDC — INT	128
12 VDC — INT	50
24 VDC — INT	13
48 VDC — INT	3.5
12 VDC — CONT	14
24 VDC — CONT	3.5
16 VAC — INT	10
24 VAC — INT	34
48 VAC — INT	9

TABLE 1

CM	WIRE GAGE
1000 AND BELOW	20
1001 — 1600	18
1601 — 2550	16
2551 — 4100	14
4101 — 6500	12

TABLE 2

Figure 19.12 Tables for determining what size is required based on type of electric strike unit used. (*Adams Rite Mfg. Co.*)

To determine the correct wire size, obtain the wire size factor from Table 1 of Fig. 19.12 and multiply this number by the distance to the strike (in feet). Take the resulting number in *circular mills* (CM) to determine the wire size required from Table 2 of the figure.

Example 1. A 24-volt ac intermittent (24 Vac-int) is to be mounted 60 feet from the transformer. The wire size factor for the unit is 34. Multiply 34 times 60; this equals 2040. On Table 2, 2040 corresponds to a wire size of 16 gauge.

Example 2. What is the maximum distance that you could mount the same 24 Vac-int using a 20-gauge wire? From Table 2, the largest CM number corresponding to 20 gauge is 2000. Divide 2000 by the wire size factor for 24 Vac-int (34) to get 29. The maximum distance is 29 feet.

If two or more strikes are to be actuated simultaneously on the same circuit, multiply the wire size factor by the number of strikes. Then, multiply by the distance to the farthest strike to obtain the CM number and the wire gauge required for the installation.

Electric Door Openers

The Trine Consumer Products Division of the Square D Company is another major supplier of electric door openers. Trine units are available for standard wood and metal doors and narrow stile doors, such as those obtainable from Adams-Rite (Fig. 19.13). Fourteen different models are currently available for such diverse applications as entrance doors to apartments, banks, institutions, industrial buildings and complexes, offices, and interlocking door control.

Installed in the door jamb in place of the lock strike or in conjunction with various other locksets, these units offer the ability to lock or unlock a door from a remote control station. This variety of electric strike can be used with most standard brands of locksets in both mortise and surface mount types.

Remember that electric strike units have two systems of operation. For unlocking doors (normally in the locked position), it opens when energized. For locking doors (normally unlocked) the door is locked when the unit is energized.

For unlocking doors, the units come in standard voltages: 3 to 6 Vdc and 8 to 16 Vac. These units are electromagnetic coil types. For locking doors they come in standard voltage of 24 Vac, with a 47-ohm solenoid, and 24-Vac 119-ohm solenoid. These units are suitable for constant duty operation without overheating or burnout.

Both types of units are fail-safe; they will automatically go to the unlock position in the event of power failure, which assures an unlocked passageway in case of emergency.

Figure 19.14 provides the electrical operating characteristics for both electromagnetic and solenoid units. Figures 19.15 through 19.17 illustrate several models commonly installed by locksmiths.

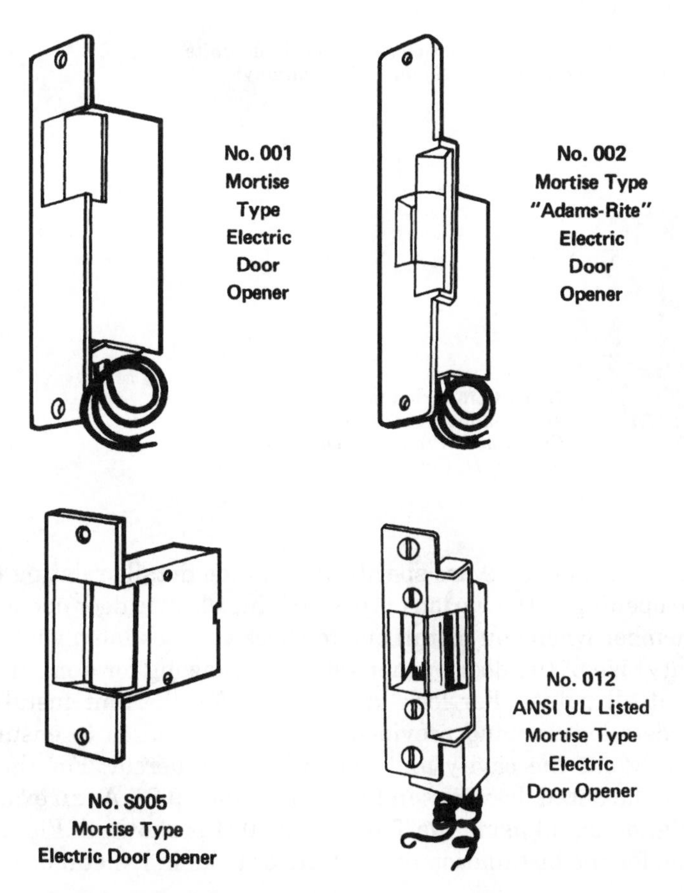

Figure 19.13 Electric door openers come in a variety of sizes and shapes to fit different installation application needs. (*Trine Consumer Products Division, Square "D" Company*)

ELECTRICAL OPERATING CHARACTERISTICS
Average Current Ratings

ELECTRO-MAGNET TYPE (normally LOCKED...UNLOCKED when energized)					
AC Voltage	*Coil-Ohms Resistance	Current Draw	DC Voltage	*Coil-Ohms Resistance	Current Draw
10	2.0	1.95	6	2.0	3.0
16	2.0	2.0			
**24	15.0	.66	**24	15.0	2.0
48	73.0	.18	48	73.0	.66

* Plus or minus 10%.
* * Suitable for CONSTANT DUTY OPERATION.

SOLENOID TYPE Available (normally LOCKED) and (normally UNLOCKED)					
AC Voltage	*Coil-Ohms Resistance	Current Draw	DC Voltage	*Coil-Ohms Resistance	Current Draw
FOR INTERMITTENT USE					
24	47.0	.42	24	119.0	.26
			48	200.0	.10
FOR CONSTANT DUTY OPERATION					
24	119.0	.26	24	119.0	.26

*Plus or minus 10%

Figure 19.14 Electrical operation characteristics for Trine units. (*Trine Consumer Products Division, Square "D" Company*)

Figure 19.15 Trine model 014 for key-in-knob locksets. (*Trine Consumer Products Division, Square "D" Company*)

Figures 19.18 and 19.19 are specification guide details relating to the actual full-size opening of the various types of Trine electric door opener latch cavities. Remember when you install one to check the lock latch width and throw to the cavity size of the door opener selected, allowing for a minimum ⅓₂-inch clearance at all points. For lock latch height for the unit installed, always check the detailed drawing provided with the lock latch to ensure that you have properly cut the cavity and not over- or undercut. For the lock latch heights, see the detail drawings in Figs. 19.20 and 19.21. As an example, a typical installation detail using the Trine Model 001 is shown in Fig. 19.22. Jamb preparation for the installation of an electric strike in a metal door frame is in Fig. 19.23.

During jamb preparation, a neat cutting job is required for good appearance, so take special care to ensure the height dimension is adequate for a loose fit

Figure 19.16 Solenoid coil unit for high-intensity use. (*Trine Consumer Products Division, Square "D" Company*)

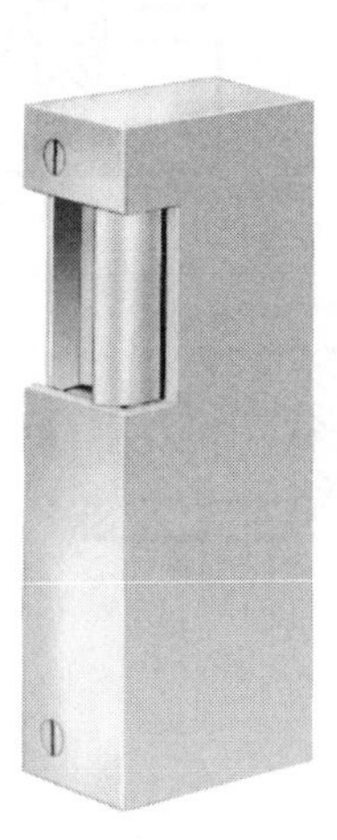

Figure 19.17 Model 003 with a magnet coil for normal use; this unit is used with rim-type locksets. (*Trine Consumer Products Division, Square "D" Company*)

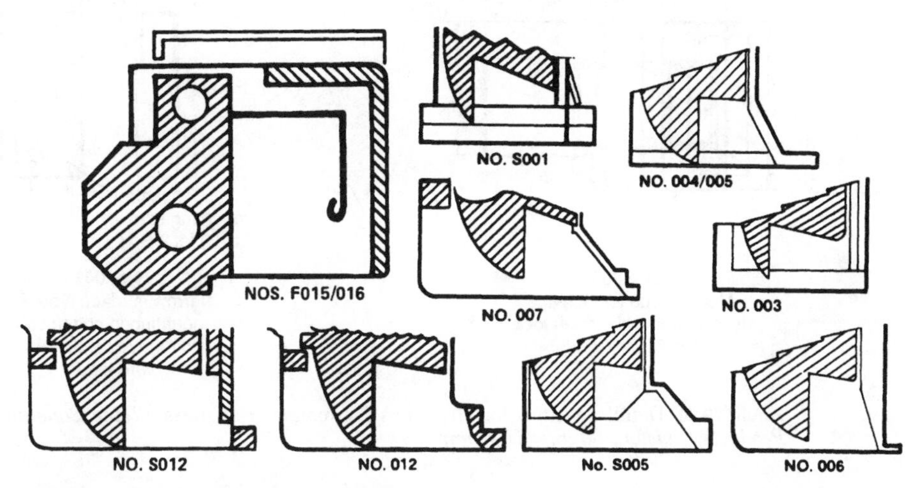

Figure 19.18 Size specifications for various latches. (*Trine Consumer Products Division, Square "D" Company*)

insertion in the cutout. A tight fit here will cause the keeper to bind in the case and result in poor performance.

The dimension indicated by an asterisk must be determined by the installer or the jamb fabricator; it is dependent on the lockset used. This dimension establishes the relationship between the keeper and the bolt. Incorrect location of the strike will prevent the door from shutting tightly or cause excessive

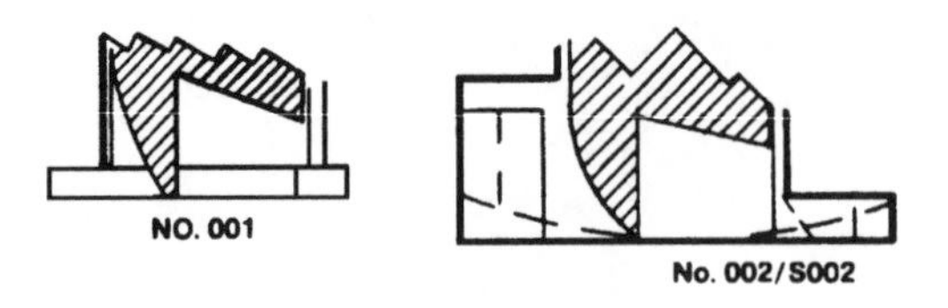

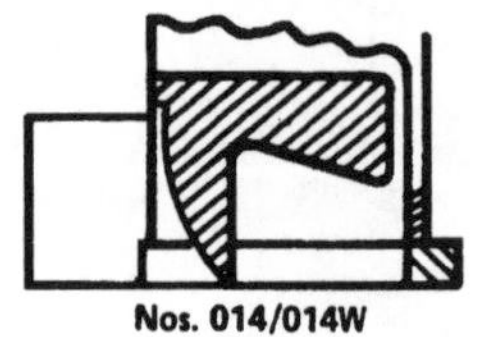

Figure 19.19 More latch cavity specifications. (*Trine Consumer Products Division, Square "D" Company*)

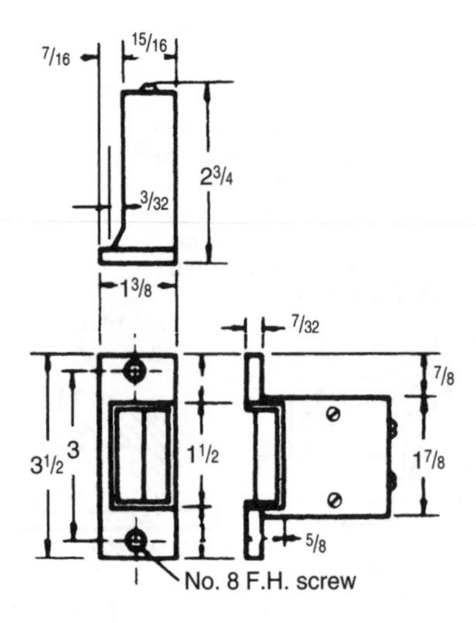

No. S005
Mortise type for Schlage type (bored cylindrical) lock sets. 6" wire leads.

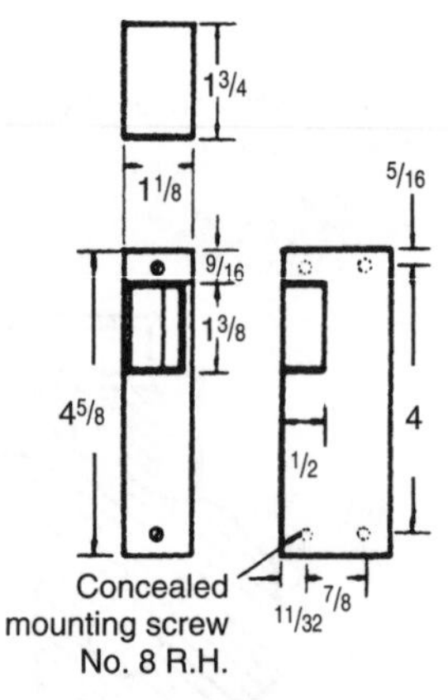

No. 003
Narrow, surface type. For mounting in conjunction with rim type lock sets. 6" wire leads.

Figure 19.20 Detail specification drawings for magnet coil units. (*Trine Consumer Products Division, Square "D" Company*)

binding between the keeper and the bolt. Leave approximately 1/32 inch of free space between the keeper and the bolt.

This is a simple installation, but the two strike wires must be attached securely by wire nuts to the supplied voltage lines. Be careful not to damage the wires when inserting the strike in the jamb cutout. This can be accomplished by hooking the wire outlet side under the jamb edge first, then pushing the other side of the strike in proper position.

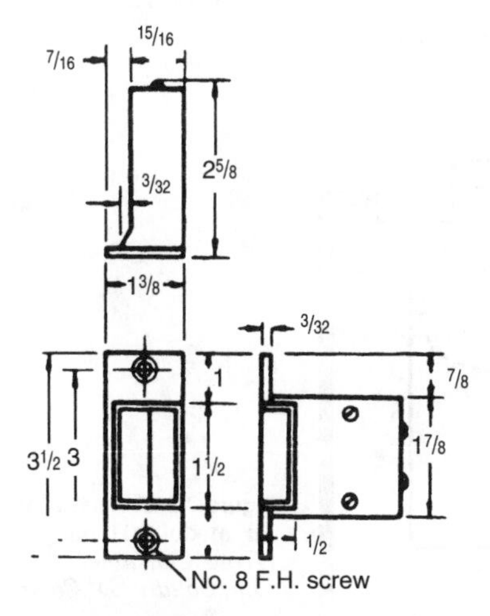

No. 005
Standard Mortise type.
6" wire leads.

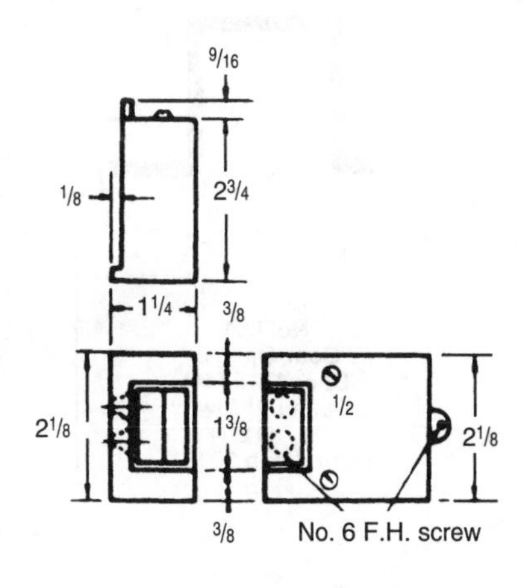

No. 006
Standard surface type. 6"
wire leads.

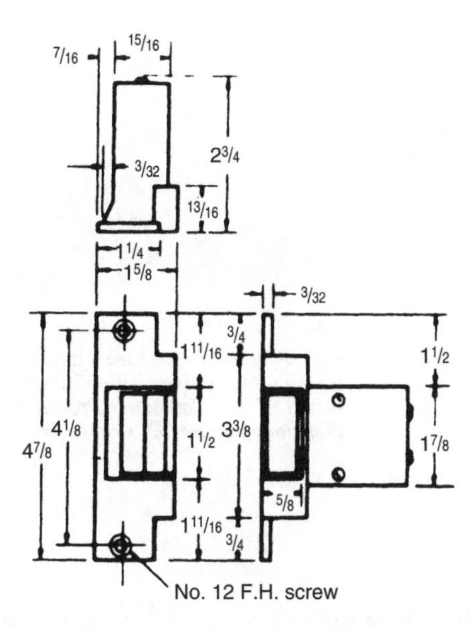

No. 007
Mortise type for A.N.S.I.
Standard strike cut-out (with
modification of jamb).
6" wire leads.

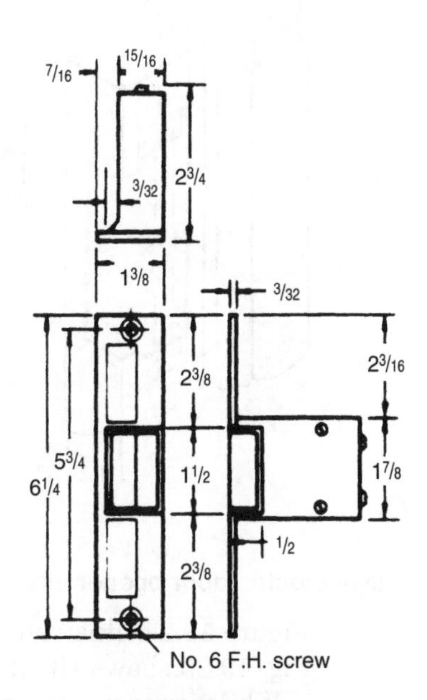

No. 004
Mortise type. For use with
locks with dead bolts.
6" wire leads. "Knock-outs"
top & bottom.

Figure 19.21 Specification details for high-intensity solenoid coil electric door opener units. (*Trine Consumer Products Division, Square "D" Company*)

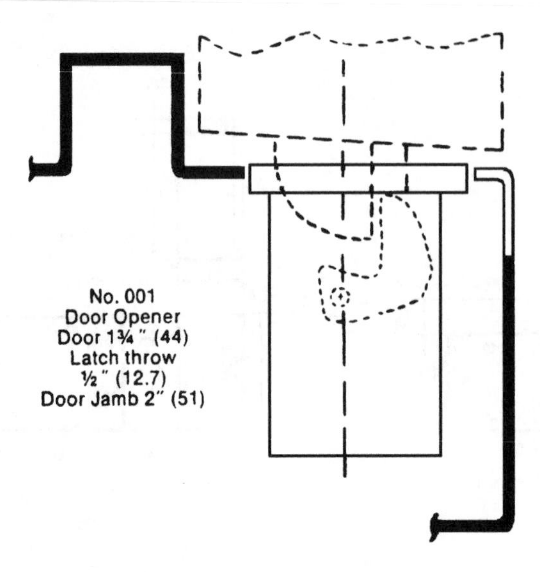

Figure 19.22 Installation detail for model 001 magnet coil unit. (*Trine Consumer Products Division, Square "D" Company*)

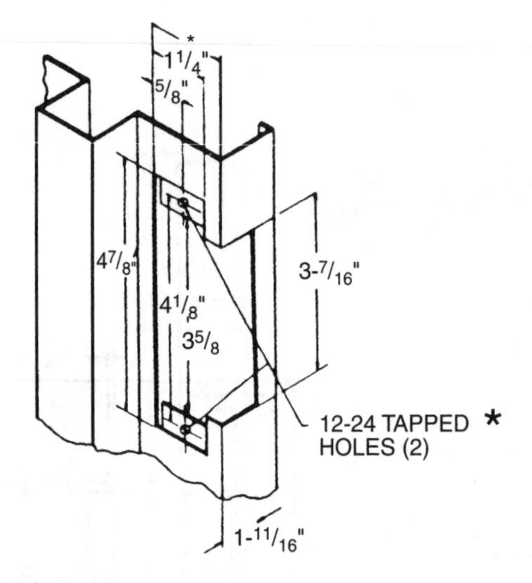

Figure 19.23 Procedural details for a mounted door frame installation. (*Trine Consumer Products Division, Square "D" Company*)

Low-voltage electric door opener circuitry

Figure 19.24 illustrates and provides specifics for audible (ac) operation, and Fig. 19.25 shows silent (dc) operation.

While it seems impracticable, some businesses even use the automatic electric interlock type of circuitry for rooms that serve more than one consumer, such as a bathroom between two offices or separate businesses located in the same building. Figure 19.26 shows how the electric interlock works in such a situation. Such interlocks can also be used for air shower rooms, money counting rooms, x-ray and photographic rooms, radiation and biological labs, and other such installations.

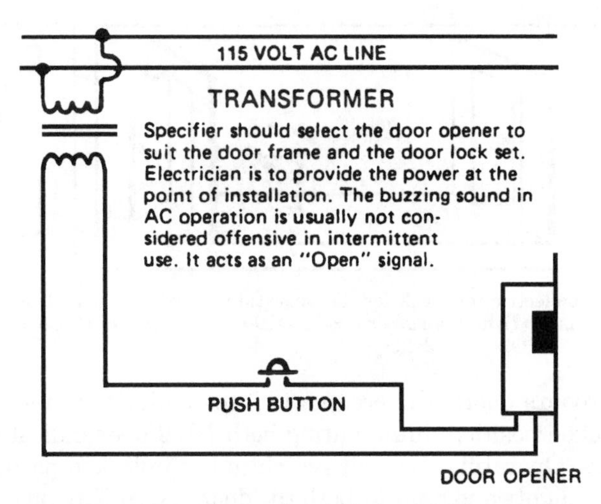

Figure 19.24 Low-voltage electrical circuit for an audible operation. (*Trine Consumer Products Division, Square "D" Company*)

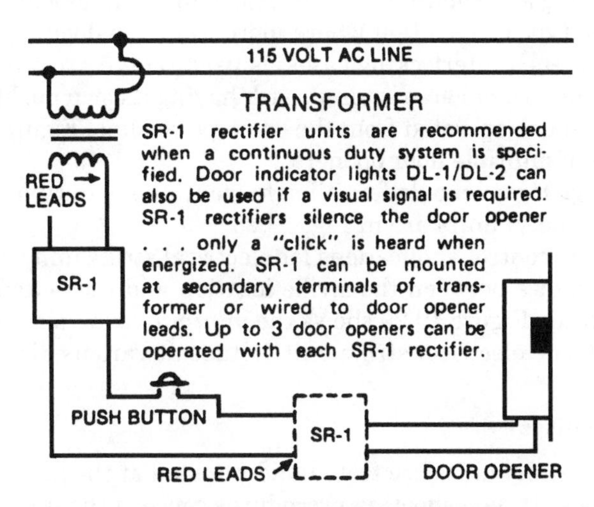

Figure 19.25 Low-voltage circuitry for silent operation of unit. (*Trine Consumer Products Division, Square "D" Company*)

In Fig. 19.26, the door opener is normally unlocked, but it becomes locked when energized. The door indicator lights (DL-1/DL-2) can be used if a visual signal is required. The opening of either door operates the door switch (a mortise-type, Trine 340, or surface-type, Trine model 316). It also activates the opposite unlocked door opener, locking the door. When the door shuts, the opposite door returns to the unlocked position.

Say, for example, that the unit is used to lock a communicating bathroom. On entering the bathroom, the occupant turns the outer knob, unlocking the lockset.

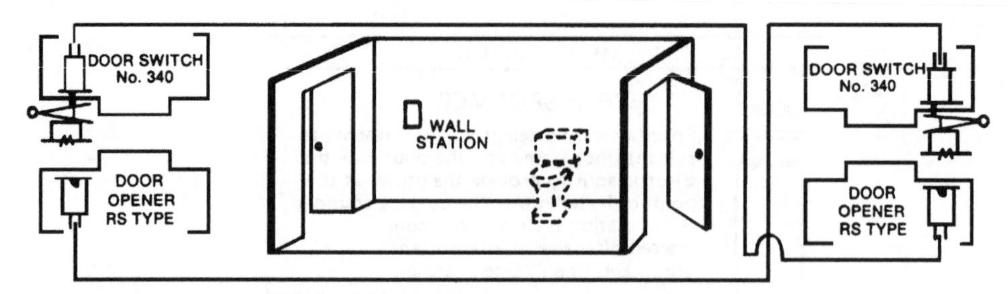

Figure 19.26 Automatic electric interlock for the operation of opposite side doors in an area servicing more than one user. (*Trine Consumer Products Division, Square "D" Company*)

To lock the bathroom's doors, the occupant closes both doors and operates the wall switch to locked position, illuminating both DL-2 door indicator lights and locking both doors. Operating the wall switch to the unlocked position turns off both door indicator lights and unlocks both the doors. After leaving the bathroom, the occupant presses the pushbutton in the lockset, preventing anyone from the adjoining room from entering the occupant's room. An emergency release is obtained by installing a pushbutton at each door, which will unlock its own door.

This is only one type of situation where more than one door opener (always an RS type unit in such interlocking cases) is used. There are other instances where multiple door openers are required, each having its own pushbutton control switch and each one operated from the same power line. Figure 19.27 illustrates the wiring of multiple door openers.

Electrical ratings for solenoids for fail-safe (continuous duty) and standard (intermittent duty only) units are in Fig. 19.28.

Due to the remote control applications for electrical strike units, there can be a considerable distance between the strike location and the actual control for that particular strike. Figure 19.29 shows the *minimum* wire sizes required for standard and fail-safe electrical strike unit voltage at various distances.

Troubleshooting Trine electrical strikes

No matter how carefully units are tested and retested at the factory, problems can still arise. These troubleshooting procedures concern the unit after installation, and they can relate to any electrical strike unit in use.

After installation, the strike hums but does not operate properly when the power is applied. *Probable cause*: ac voltage has been supplied but the rectifier has not been installed between the transformer and the strike.

Correction: Install the rectifier per the wiring diagram in Fig. 19.30.

After installation, the strike does not operate even if the power is left on for 1 minute (maximum time) and the strike is noticeably warm. *Probable cause:* (1) Static loading against the door is excessive due to weatherstripping, a warped door, wind loading on the door, or improper jamb preparation.

Corrections: (1a) Adjust the strike bolt relationship so that some slight gap exists in the locked position. (1b) Balance air conditioning pressures on opposite sides of the door.

Probable cause: (2) Insufficient voltage is reaching the strike due to an inadequate supply wire size.

Corrections: (2) Provide sufficient sized service wire (Fig. 19.29).

The keeper moves to fully open on power application but does not return to the locked position when the door closes. *Probable cause*: If the deadbolt is extended but does not pull the keeper to the fully locked position on door closure, the return spring may need adjusting.

Correction: Deform the return spring slightly by forcing it toward the keeper.

The keeper binds in the strike case on both opening and closing. *Probable cause*: The cutout in the jamb is too narrow and it is forcing the strike case walls inward toward the keeper.

Correction: Increase the vertical dimension until the keeper operates freely.

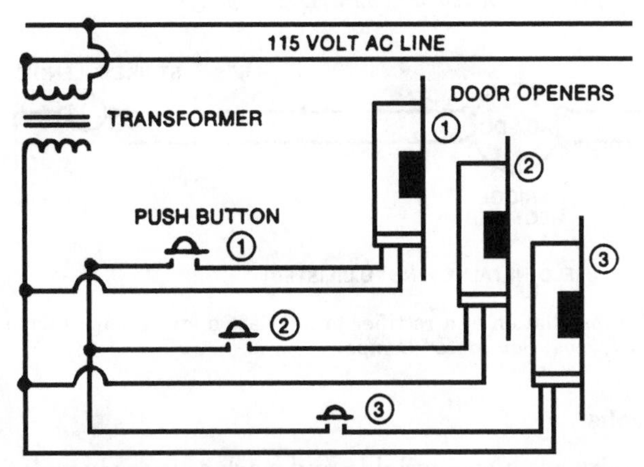

Figure 19.27 Circuitry for multiple wiring of door openers. (*Trine Consumer Products Division, Square "D" Company*)

ELECTRICAL RATINGS FOR SOLENOIDS	FAIL SAFE UNITS (CONT. DUTY)		STANDARD UNITS* (INTERMITENT DUTY ONLY)		
	24 VDC	12 VDC	24 VDC	12 VDC	48 VDC
RESISTANCE IN OHMS	115	33	16	4	90
WATTS SEATED	4.8	4.4	36	36	25
AMPS SEATED	.2	.37	1.5	3.0	.53

*Bridge rectifier may be supplied with any unit for AC operation.

Figure 19.28 Electrical ratings for intermittent use and constant operation for solenoid coil door openers. (*Trine Consumer Products Division, Square "D" Company*)

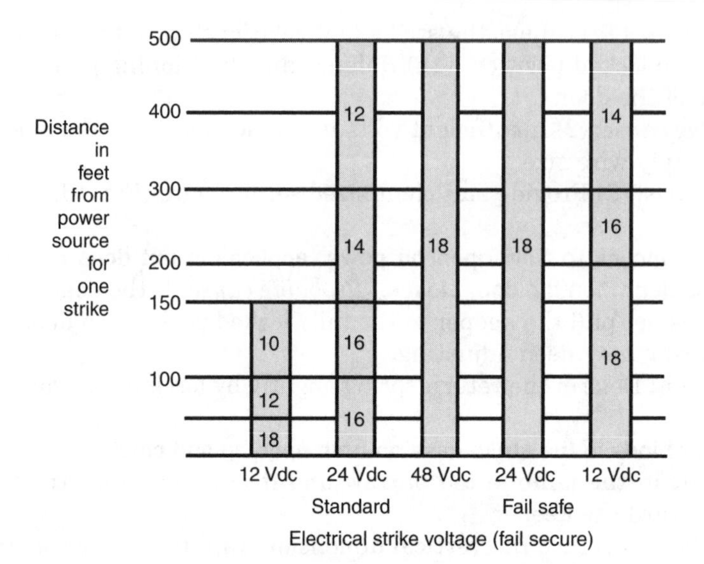

Figure 19.29 Electrical strike voltage (fail-secure mode of operation). (*Trine Consumer Products Division, Square "D" Company*)

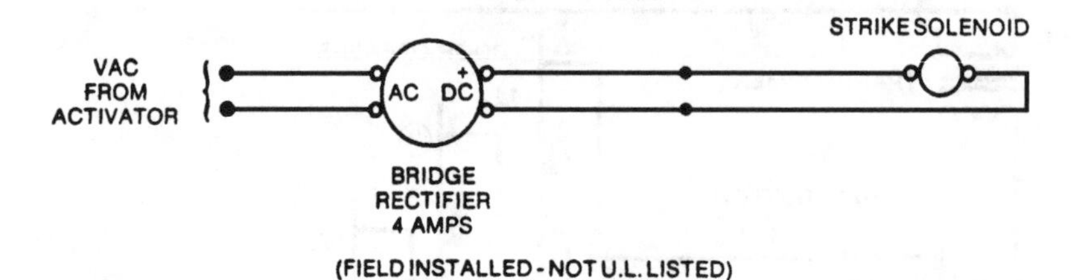

Figure 19.30 Circuitry installation of a rectifier to alleviate a strike hum. (*Trine Consumer Products Division, Square "D" Company*)

Electric door openers for deadbolts

Electric door openers are also available and work extremely well for deadbolts. They can be used with just about any manufacturer's lockset. They include the following types:

- Mortise locks with deadbolt.
- Mortise locks with deadlatch and guarded latch.
- Individual deadbolt lock.
- Panic exit devices.
- Maximum security deadbolts (¾-inch throw).
- Release up to 1-inch throw deadbolts.
- With optional adapters, they may be used with cylindrical locks, alarm locks, or with Detex bolts.

These units will also allow the strike operation of mortise locks with both latch-bolt and deadbolt release by the action of a single keeper. The keeper is held open after the door is opened if the deadbolt was in an extended position at the time of the door opening. The keeper automatically returns to the locked condition when the deadbolt reenters the strike. The door may also be closed with deadbolts extended. In strikes with options H, J, and K (Fig. 19.31), the keeper is forced open by the extending latchbolt and springs closed to the locked position to accommodate the returning latchbolt. Other options are also shown in Fig. 19.31.

Figure 19.32 illustrates two Trine unit model (015 and 016) specifications. Along with the basic unit for installation, a heavy-duty transformer (Fig. 19.33) is necessary to convert the primary voltage to the operating voltage.

Electrified Mortise Locks

The electrified mortise lock provides many additional installation sales and servicing possibilities, especially when access security controls for your customer are concerned. The electrified mortise lock (Fig. 19.34) is another excellent, durable product that has minimal installation procedures yet offers long life and reliability to your customer.

This lock unit has the standard rugged lock case construction of stainless steel, plus a normal mortise strike. The inside knob is always free, while the outside knob is rigid except when the 24-Vdc internal solenoid is energized, which releases the outside knob and allows the latchbolt to retract. Like any other standard mortise lock, it is reversible, allowing for different hands.

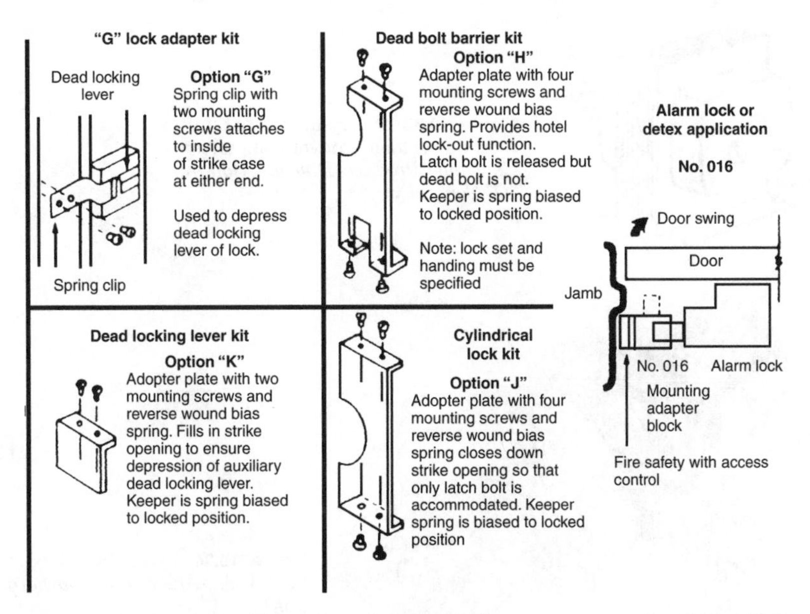

Figure 19.31 Various strike options for electric door openers used with deadbolt locks. (*Trine Consumer Products Division, Square "D" Company*)

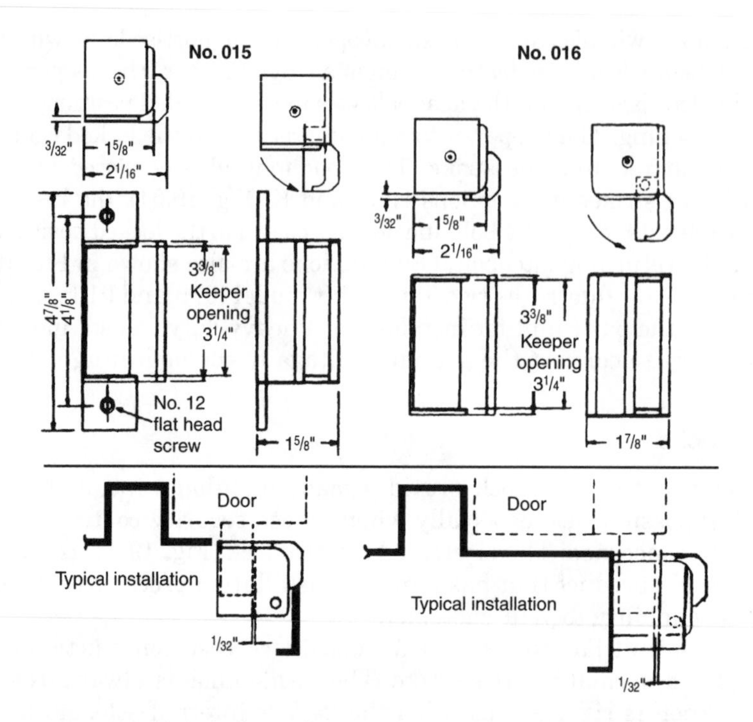

Figure 19.32 Technical specifications for two Trine door opener models that can be used with any manufacturer's lockset. (*Trine Consumer Products Division, Square "D" Company*)

Figure 19.33 Heavy-duty transformer for voltage conversion. (*Trine Consumer Products Division, Square "D" Company*)

Figure 19.34 Electrified key-in-knob lock. (*Alarm Lock Corporation*)

Installing the electrified mortise lock

Mark the door and jamb (Fig. 19.35). Mark the horizontal centerline of the lock on both sides and the edge of the door. Mark the vertical centerline of the lock at the door edge and the vertical centerline on both sides of the door as measured from the door edge.

Mortise the door for the lock case and front (Fig. 19.36). For the specific dimensions, refer to the installation template that comes with the lock unit.

Drill the holes as indicated on the template as shown in Fig. 19.37. *Caution:* Check the lock for function, hand, and bevel before drilling. Mark and drill only those holes required for your function.

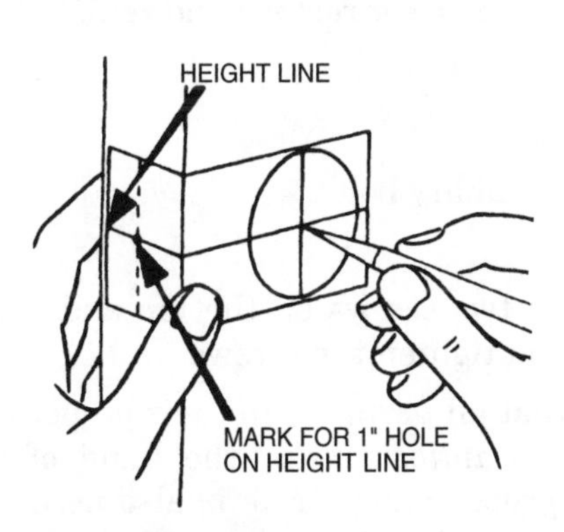

Figure 19.35 Marking the door for installation. (*Alarm Lock Corporation*)

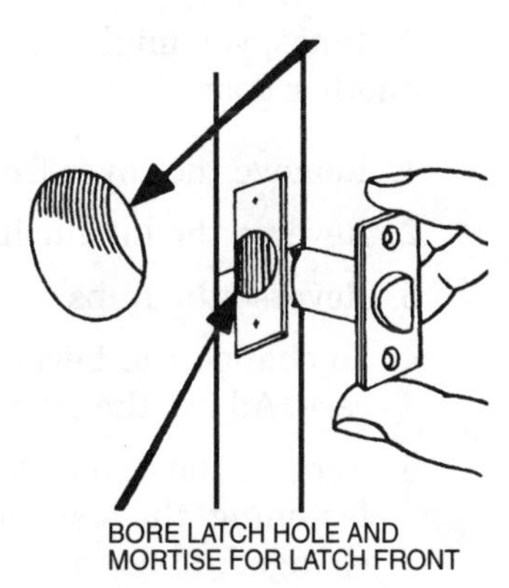

Figure 19.36 Bore and mortise for latch front. (*Alarm Lock Corporation*)

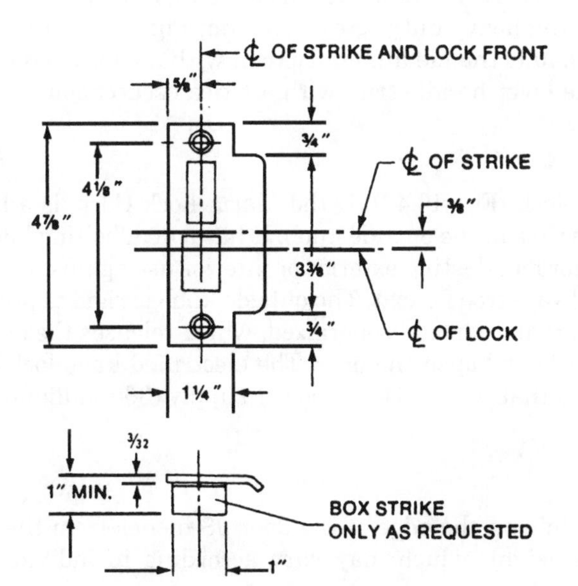

Figure 19.37 Strike installation details. (*Alarm Lock Corporation*)

Remove the front plate and mount the lock in a mortised opening. Apply the trim (four trim variations are possible; see Figs. 19.38 through 19.41 at the end of the installation instructions). Apply the cylinder (using a minimum of four turns); then, fasten with the cylinder clamp screw. Replace the front plate. *Caution:* Check the cylinder and lock for proper function before closing the door.

Install the strike (Fig. 19.37). Position the strike on the door jamb as shown. Mortise the jamb and apply the strike. When using a box strike, mortise deep enough to allow the latchbolt and deadbolt to fully extend.

Reversing the lock and bevel

At times, you might be required to replace or remove and reinstall and lock on another door.

1. Remove the cover from the lock.
2. Reverse the latchbolt and the auxiliary bolt.
3. Reverse the hubs.
4. To change the bevel, loosen the two screws on the top and bottom of the case. Adjust the proper bevel and tighten the screws.
5. Replace the cover. Be certain that all parts are in their proper place before fastening the exterior screws. *Caution:* When the hand of the lock is changed, make sure that the proper hand strike is also used. A left-hand lock requires a left-hand strike; a right-hand lock requires a right-hand strike.

Figures 19.38 through 19.41 show the various trims available for this unit. Figure 19.38 shows the heavy-duty screwless trim; Fig. 19.39 shows the lever handle trim; Fig. 19.40 is the medium duty trim with a Glenwood escutcheon; and Fig. 19.41 is the lever handle trim with a Utica escutcheon.

Electrified Knob Locks

The electrified knob lock (Fig. 19.42) by the Alarm Lock Corp. is a heavy-duty lockset with a key bypass in the outside knob. It has a deadbolting latchbolt and stainless steel trim for long-lasting exterior or interior use against the elements. The inside knob is always free for exit. The outside knob is rigid to prevent entry except when the internal solenoid is energized, which releases the outside knob to retract the latchbolt and open the door. The electrified knob lock has a continuous duty solenoid that uses a 24-Vac power supply (360 milliamps).

Installation procedures

1. Mark the height line on the edge of the door 38 inches from the floor (this is the suggested height, which may vary according to individual needs).

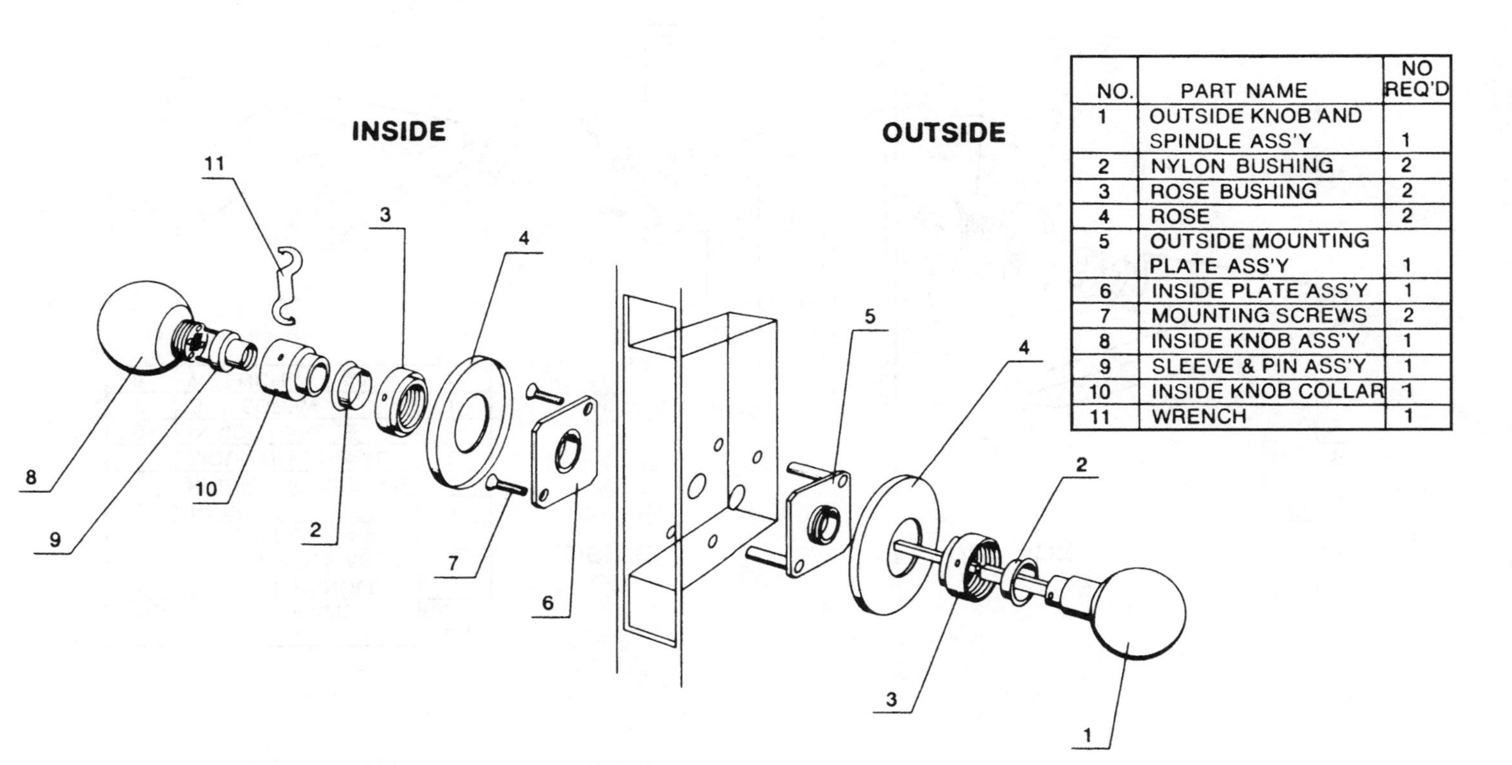

NO.	PART NAME	NO REQ'D
1	OUTSIDE KNOB AND SPINDLE ASS'Y	1
2	NYLON BUSHING	2
3	ROSE BUSHING	2
4	ROSE	2
5	OUTSIDE MOUNTING PLATE ASS'Y	1
6	INSIDE PLATE ASS'Y	1
7	MOUNTING SCREWS	2
8	INSIDE KNOB ASS'Y	1
9	SLEEVE & PIN ASS'Y	1
10	INSIDE KNOB COLLAR	1
11	WRENCH	1

Figure 19.38 Heavy-duty screwless trim. (*Alarm Lock Corporation*)

NO.	PART NAME	NO REQ'D
1	OUTSIDE HANDLE AND SPINDLE ASS'Y	1
2	NYLON BUSHING	2
3	ROSE	2
4	MOUNTING SCREWS	6
5	MOUNTING PLATE	2
6	INSIDE HANDLE	1
7	SET SCREW	1
8	WRENCH	1

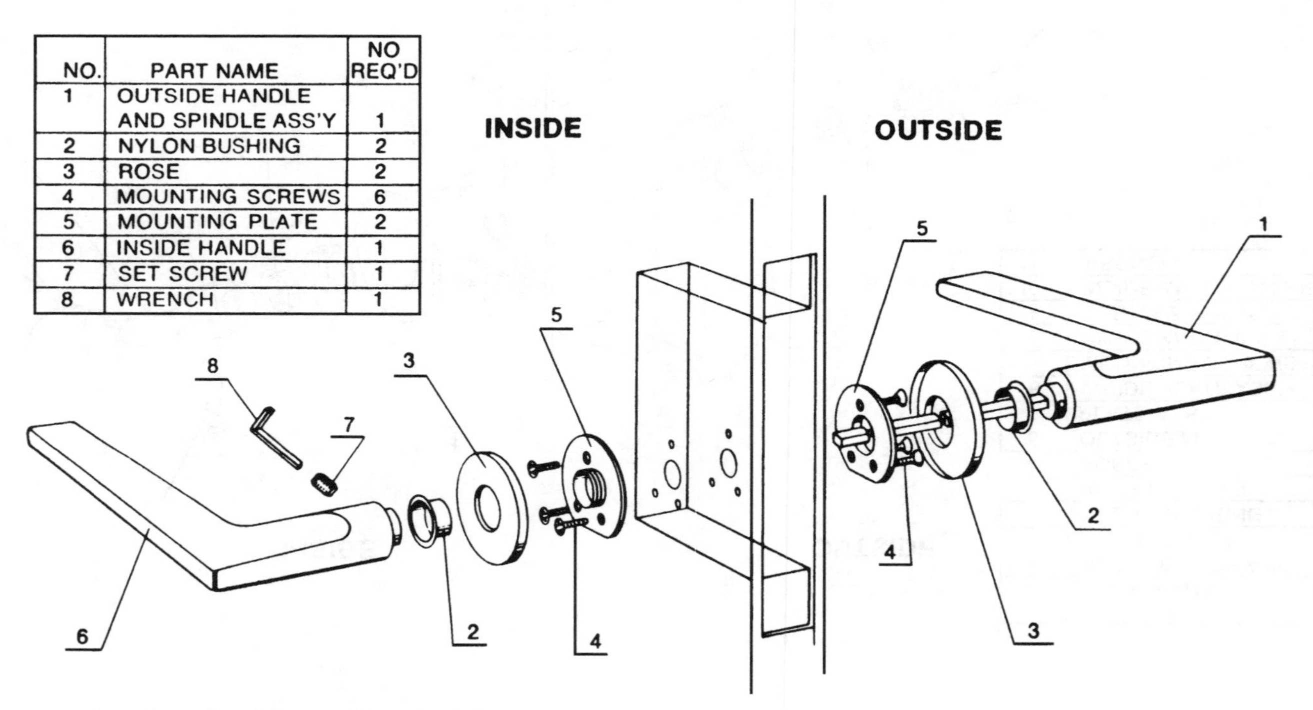

Figure 19.39 Lever handle trim. (*Alarm Lock Corporation*)

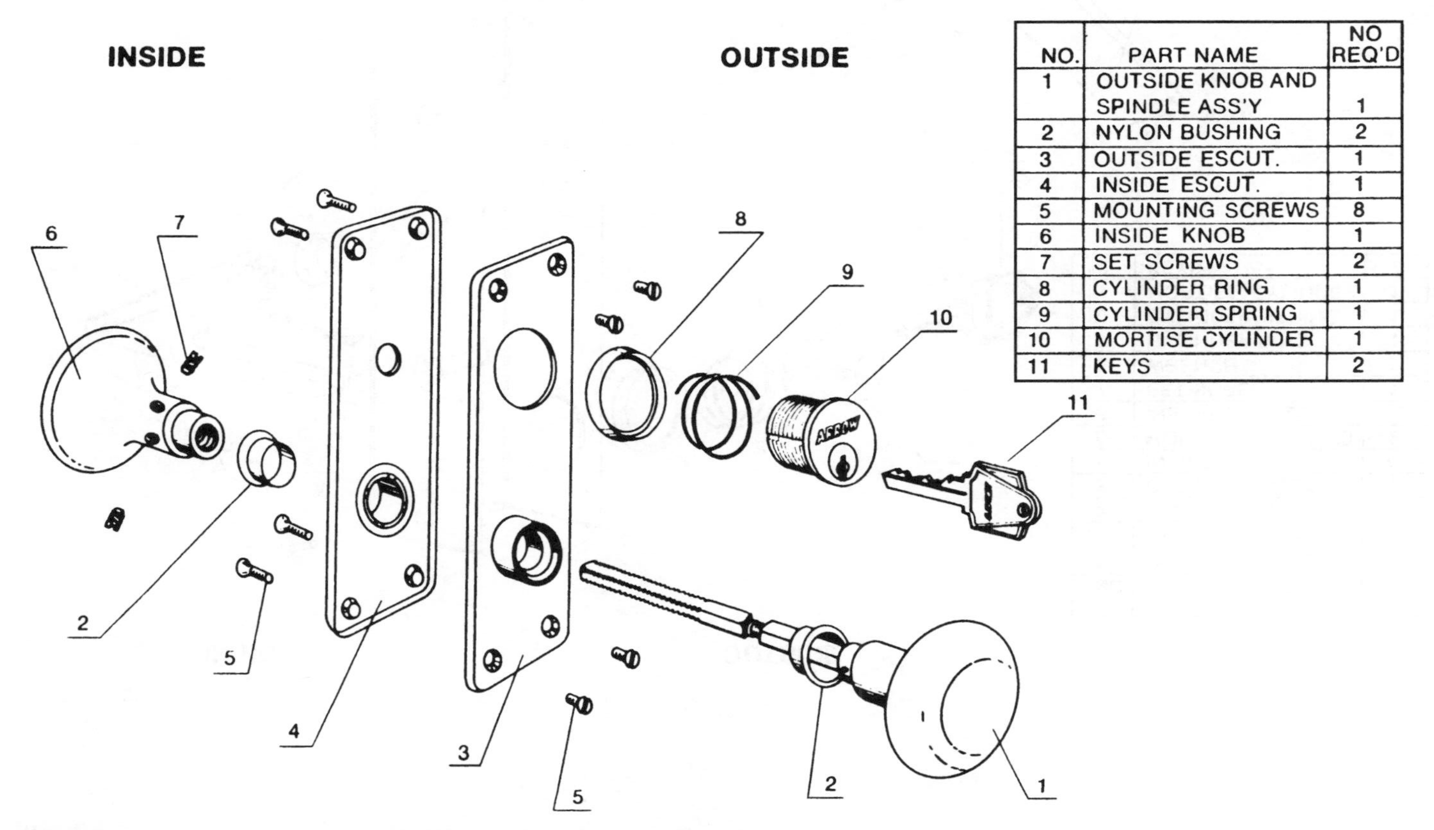

NO.	PART NAME	NO REQ'D
1	OUTSIDE KNOB AND SPINDLE ASS'Y	1
2	NYLON BUSHING	2
3	OUTSIDE ESCUT.	1
4	INSIDE ESCUT.	1
5	MOUNTING SCREWS	8
6	INSIDE KNOB	1
7	SET SCREWS	2
8	CYLINDER RING	1
9	CYLINDER SPRING	1
10	MORTISE CYLINDER	1
11	KEYS	2

Figure 19.40 Medium-duty trim with the Glenwood escutcheon. (*Alarm Lock Corporation*)

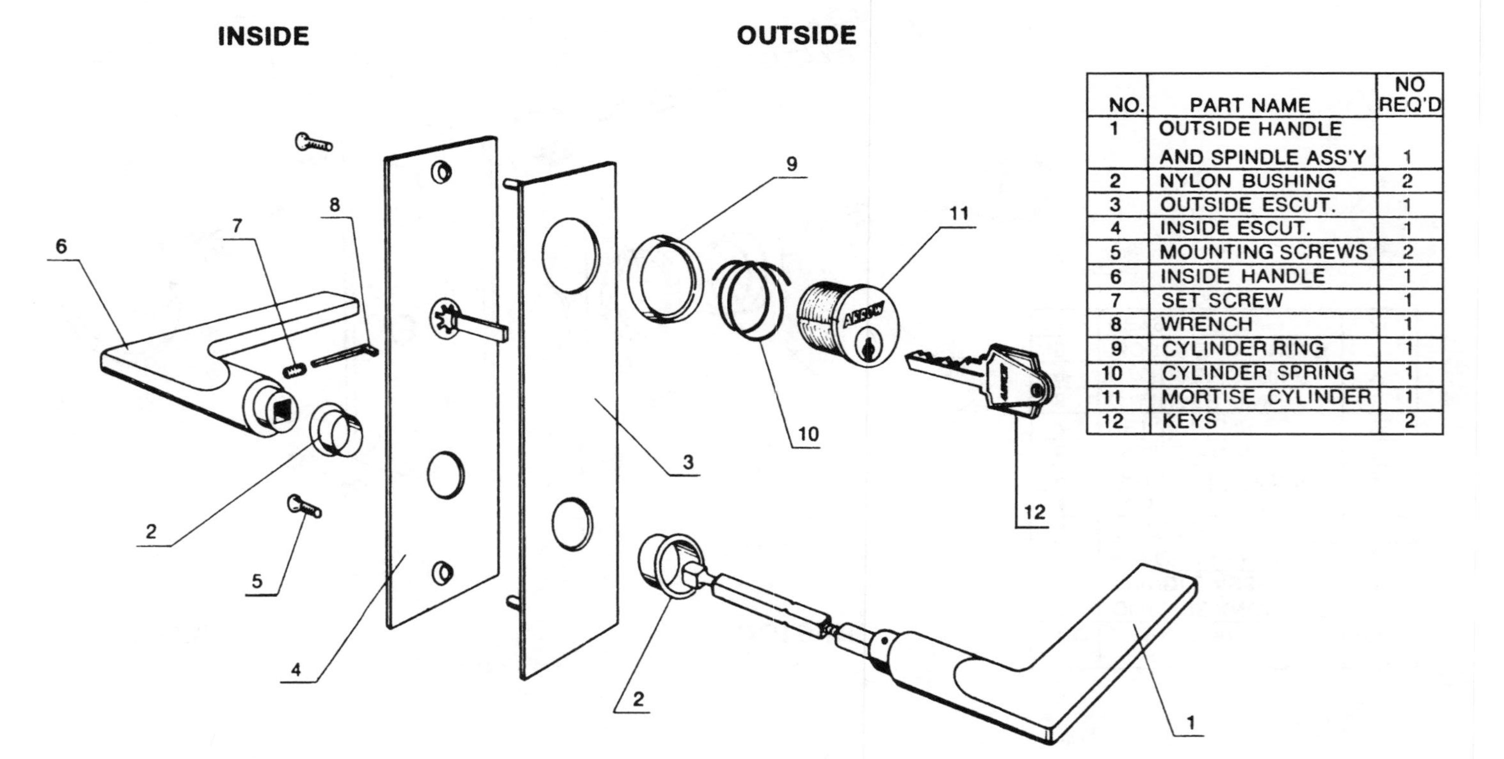

NO.	PART NAME	NO REQ'D
1	OUTSIDE HANDLE AND SPINDLE ASS'Y	1
2	NYLON BUSHING	2
3	OUTSIDE ESCUT.	1
4	INSIDE ESCUT.	1
5	MOUNTING SCREWS	2
6	INSIDE HANDLE	1
7	SET SCREW	1
8	WRENCH	1
9	CYLINDER RING	1
10	CYLINDER SPRING	1
11	MORTISE CYLINDER	1
12	KEYS	2

Figure 19.41 Lever handle trim using the popular Utica escutcheon. (*Alarm Lock Corporation*)

Figure 19.42 Model 121 electrified mortise lock. (*Alarm Lock Corporation*)

Position the centerline of the template that comes with the lock on the height line, and mark the center point of the door thickness and the center point for the 2⅛-inch hole through the door (Fig. 19.43).

2. Bore the 2⅛-inch hole at the point marked from both sides of the door. Bore the 1-inch latch unit hole straight into the edge of the door from the center point of the height line. Mortise the latch front and install the latch unit (Fig. 19. 44).
3. Remove the lock from the box and remove the inside trim by inserting the small end of a spanner wrench through the hole in the knob bearing sleeve. This hole faces the latch retractor. Press the knob catch to slide the knob off the spindle and remove the rose from the threaded sleeve (Fig. 19.45).
4. Adjust for the door thickness by moving the outside rose either in or out until the indicator on the lock shows the proper thickness for the door. Make sure the rose clicks into place (Fig. 19.46). The lock will fit on all doors 1⅜ to 2 inches thick. (If you need a lock for a thicker door, special order it from the company.)
5. Next, install the main unit (Fig. 19.47). The main housing must engage with the latch prongs and retractor with the latch tailpiece as shown in Fig. 19.48.
6. Secure the trim by replacing the rose and screwing it on to the threaded spindle with your fingers as far as it will go. Tighten the rose with the spanner wrench (Fig. 19.49).

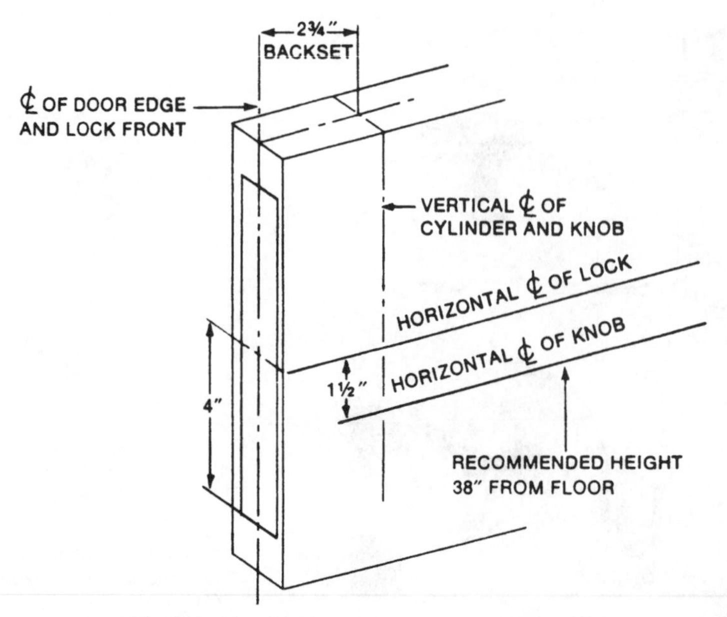

Figure 19.43 Electrified key-in-knob lock installation, Step 1.

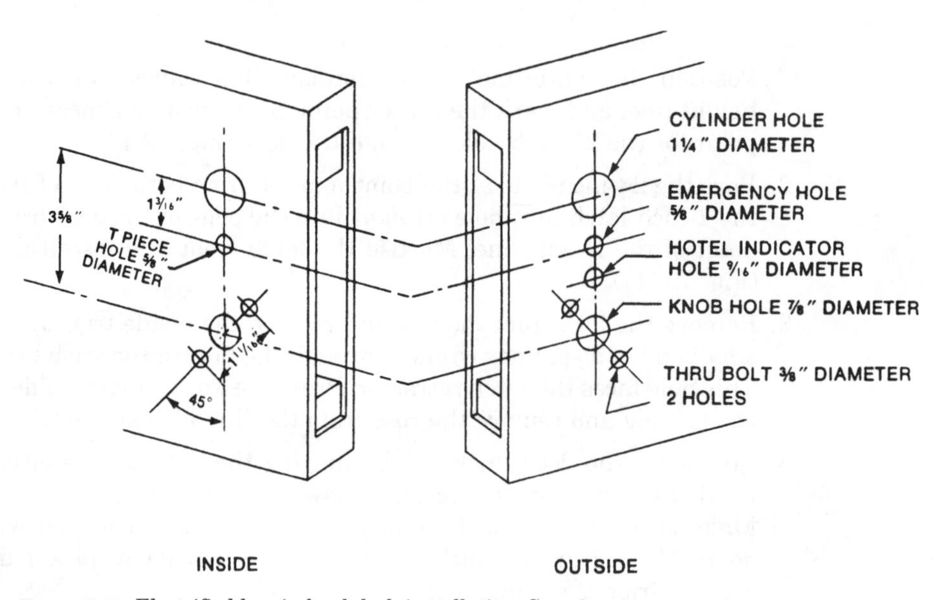

Figure 19.44 Electrified key-in-knob lock installation, Step 2.

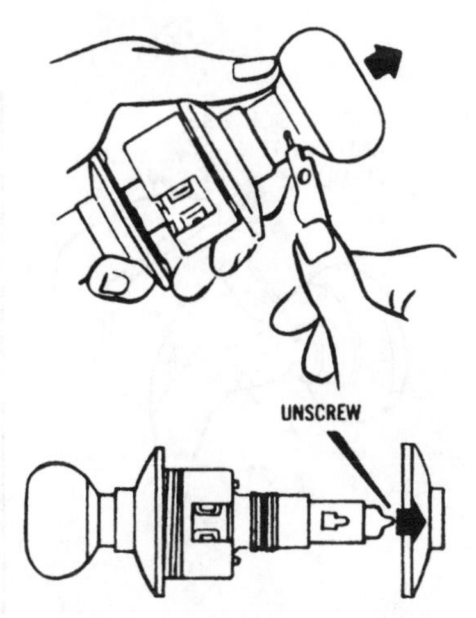
UNSCREW

Figure 19.45 Electrified key-in-knob lock installation, Step 3.

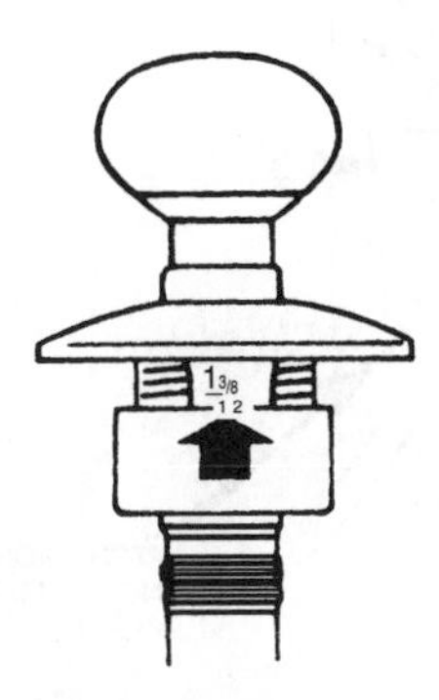
1 3/8
1 2

Figure 19.46 Electrified key-in-knob lock installation, Step 4.

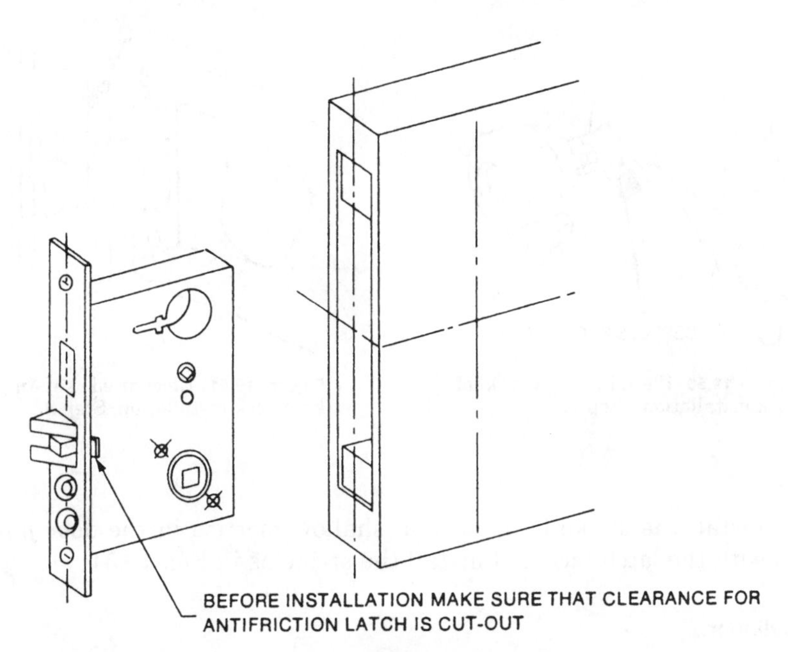
BEFORE INSTALLATION MAKE SURE THAT CLEARANCE FOR
ANTIFRICTION LATCH IS CUT-OUT

Figure 19.47 Electrified key-in-knob lock installation, Step 5.

7. Replace the knob by lining up the keyway in the knob with the slot in the spindle. Push the knob straight on until the knob hits the knob catch. Depress the knob catch and push the knob straight on until the knob catch clicks into place. Test both knobs to make sure they are securely fastened to the spindle (Fig. 19.50).

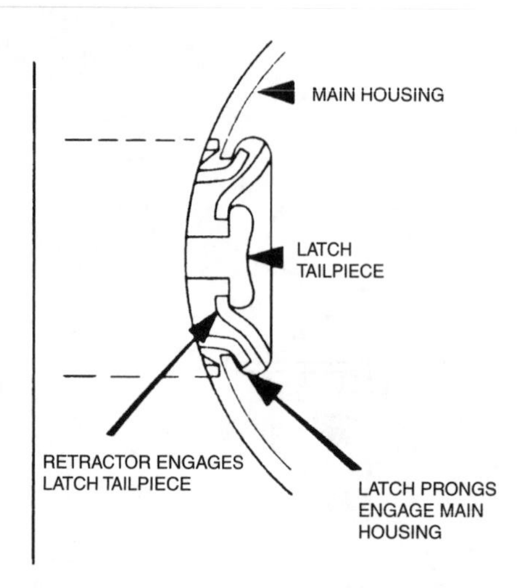

Figure 19.48 Electrified key-in-knob lock installation, Step 5.

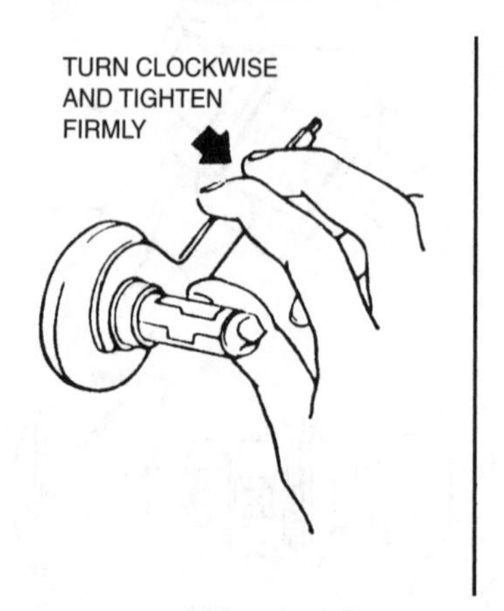

Figure 19.49 Electrified key-in-knob lock installation, Step 6.

Figure 19.50 Electrified key-in-knob lock installation, Step 7.

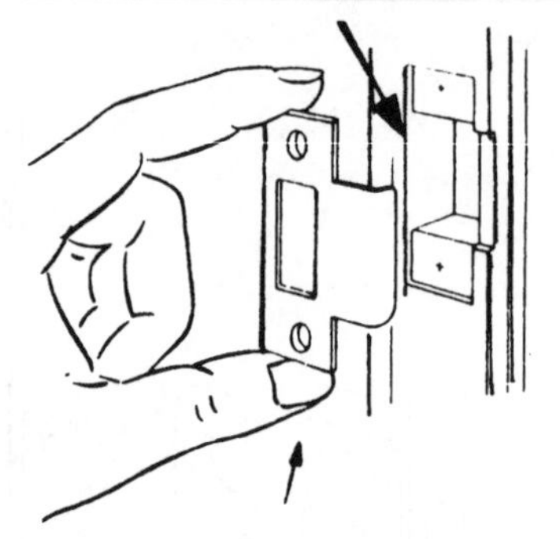

Figure 19.51 Electrified key-in-knob lock installation, Step 8.

8. Install the strike by making a shallow mortise in the door jamb to align with the latch face and install the strike as in Fig. 19.51.

Reversing the cylinder

If you must reverse the lock cylinder, follow the procedure outlined in this section. First, insert the key into the cylinder and turn the key clockwise as far as it will go. At the same time, depress the knob catch retainer. When the retainer moves down, pull the knob off.

Next, turn the knob 180°, and replace it as follows. Remove the key and push the knob on the spindle as far as it will go. Insert the key and turn it clockwise to its most extreme position while depressing the knob catch retainer. When the retainer moves down, push the knob onto the spindle, allowing the retainer to click into position.

Removing the cylinder

To remove the cylinder, first push the cylinder tailpiece sideways with your finger until the tailpiece unsnaps from the plug. Then, turn the knob over and hit the shank of the knob with a sharp blow on a flat surface. This releases the knob ferrule and permits the removal of the cylinder. Rekeying of the cylinder is conventional.

Electrical considerations

Now that you understand the specifics of installing the lock, consider its electrical components. The electrical power for the lock can either be concealed within the door by use of a current transfer unit (Fig. 19.52) or left exposed on the interior side. The use of a current transfer unit is a unique means of bringing the electrical current from the lock side of the door to the door. The face and strike must be thoroughly matched and the door closed for power to flow through. With this in place, from a remote switch activation, power will flow through, allowing the door to be opened.

The other method, leaving the wiring exposed, is not really what it seems. An armored door loop is used and it provides a much easier means for bringing electrical current from the hinge side of the door frame to the lock in the door. It consists of an 18-inch, flexible, armored cable that is attached to the frame on one end and just below the lock on the inside of the door. The lock wires can be passed through the door wall via a small hole and connected to the loop wire ends.

Electromagnetic Locks

While electric locks are available and used for many applications, there is also a need for electromagnetic locks. Such locks can be used in hospitals, convalescent homes, banks, universities, museums, libraries, factories, laboratories, penal institutions, and airports.

The electromagnetic lock, while operating at less than 12 Vdc, can exert a holding force of 1500 pounds. This is fantastic for a lock that consists of two components—the lock and its armature.

Figure 19.52 Current transfer unit. (*Alarm Lock Corporation*)

The lock mounts rigidly to the door frame and the armature mounts to the door (Fig. 19.53). When the door is closed, the two make contact; on activation of the lock, the two are attracted to each other. The door, for all practical purposes, is bonded solidly to the frame.

Ten years ago, such a lock would have been priced out of reach, but today it is possible for almost any business or institution to afford one. As a locksmith, you must be prepared for and familiar with this "unconventional" lock.

Security Engineering, the leader and major proponent of electromagnetic locks, lists these major features:

- *A high holding (1500 pounds).* Effectively seals the door from both sides, keeping intruders out and/or valuables safely in.
- *Conserves energy.* The efficient low power consumption design (2 amps) ensures a long life and economical operating cost for the lock.
- *High reliability.* Recommended for high trafficked openings with extreme usage, where conventional locks, locks with moving parts, or electromagnetic locks with a lesser holding force may not be capable of withstanding the abuse.
- *Fully fail safe.* Locking and unlocking is positive and instantaneous, with no residual magnetism or moving parts that might stick, jam, bind, wear out, and/or need replacement.
- *Extremely low maintenance.* No maintenance or adjustments are required after initial installation.
- *Exclusive mounting design.* This permits rapid efficient installation in new or existing buildings and automatically compensates for normal door wear, sag, warping, or misalignment.

Figure 19.53 Model 3900 electromagnetic lock. (*Security Engineering, Inc.*)

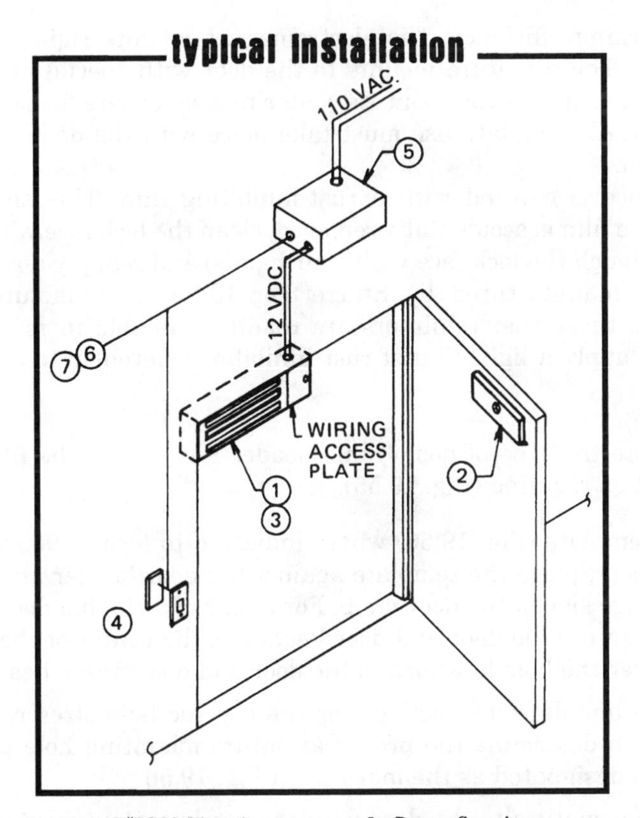

1. Model #3900 Electric Lock
2. Electric Lock Armature
3. Door Status Switch Optional*
4. Local On/Off Station Controls
5. Power Supply
6. Remote On/Off Station Controls*
7. Hazard Sensing Safety Devices*

*As System Requires

Figure 19.54 Model 3900 installation layout. (*Security Engineering, Inc.*)

- *Self-contained unit.* It is completely factory wired with all options for the unit already built in, fully concealed, and tamper-resistant. There is only one external wire connection to make.
- *Four operating voltages.* 12 Vdc, standard 12 Vac, 24 Vdc, and 24 Vac.

Figure 19.54 illustrates a typical installation and the various features (standard and optional) that can come with a unit.

Series 3900 electromagnetic lock

It is important that, prior to installing an electromagnetic lock, you handle the equipment with care. Damaging the mating surfaces of the armature may reduce its locking efficiency. The electromagnet mounts rigidly to the door frame header. The armature mounts to the door with special hardware provided that allows it to pivot about its center to compensate for door wear and misalignment. All template use must take place with the door in its normal closed position.

The lock face is covered with a rust-inhibiting film. This should not be removed. If the film is accidentally removed, clean the lock face with a dry, soft cloth (do not touch the lock face with your hands) and reapply a rust inhibitor such as MI (manufactured by Starret) or LPS3 (manufactured by LPS Laboratories); these rust inhibitors are readily available in most hardware stores. Then, apply a light film of rust inhibitor generously to the armature face.

Mounting. Note the type of door frame header and install the filler place or angle bracket as required (Fig. 19.55).

1. Fold the template (Fig. 19.56) where indicated to form a 90° angle. For a swinging door, place the template against the door header and door opposite the hinge side of the door jamb. For a pair of swinging doors, place the template against the door and door header at the center of the door opening. Transfer the hole locations to the door and door frame header.
2. Follow the template instructions for the specific hole sizes required. See Fig. 19.56 to determine the proper armature mounting hole preparation. The hole is designated as the mark ⊕ in Fig. 19.56.
3. Mount the armature to the door with the hardware provided as shown in the figures.
4. Install the mounting plate to the header with the five #10 screws.
5. Fasten the Model 3900 lock to the mounting plate with the two socket screws and compensate for any misalignment by adding or subtracting washers at the armature mounting screws.

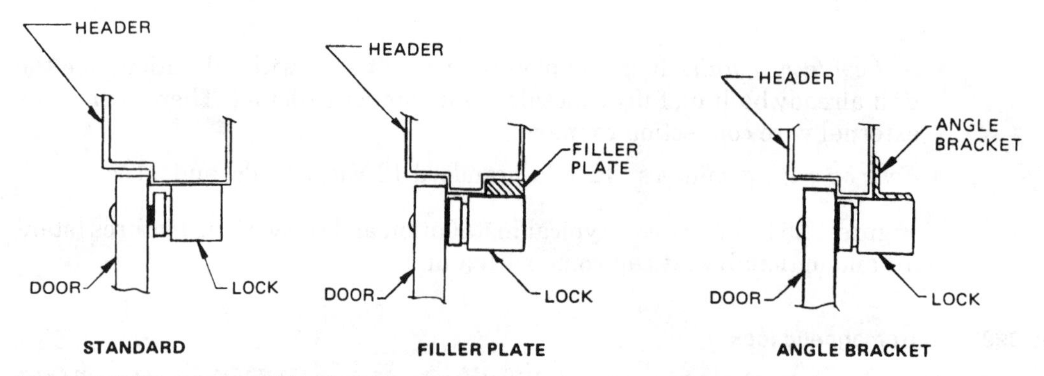

Figure 19.55 Door frame header determination tells you if a filler plate, angle bracket, or nothing will be required. (*Security Engineering, Inc.*)

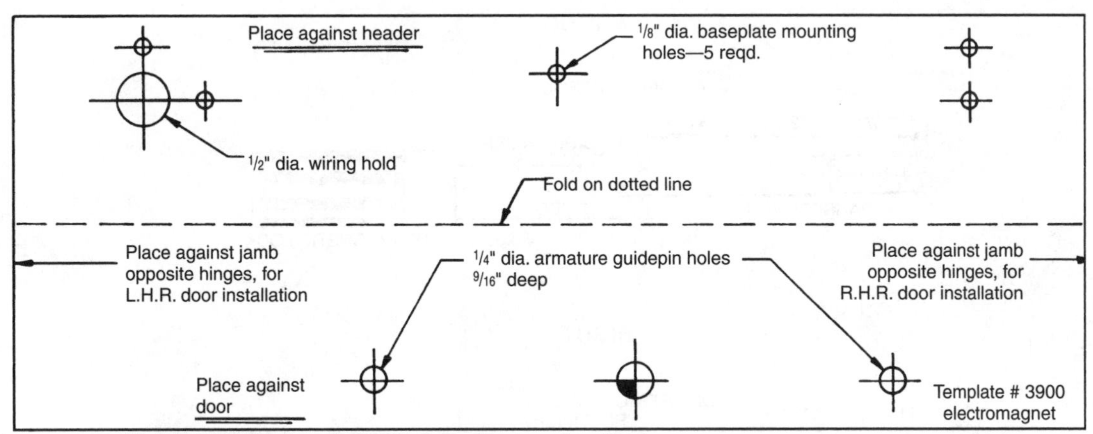

Figure 19.56 Model 3900 lock template. (*Security Engineering, Inc.*)

6. Firmly tighten all the screws. Install the antitamper plugs into the holes over each socket head mounting screw. Use a soft hammer to avoid damaging the lock case.

Wiring instructions. In some instances, you may use a prefabricated power supply for the lock unit. What if you do not? To achieve the best possible electromagnet lock installation, note the following suggestions (both right and wrong ways are illustrated for each procedure):

- Proper placement of the on-off station control in the electrical circuit will ensure instantaneous door release and avoid any delay in complete unlocking (Fig. 19.57).
- Care must be taken when adding any auxiliary controls or indicators to the electrical circuit. Improper placement may cause a slight delay in door release (Fig. 19.58).
- Solid state on-off devices must be properly protected from the back voltage generated from the collapsing electromagnetic lock coil when the unit is switched off. This may be accomplished by installing a metal oxide varistor (MOV) across the lock coil lead wires (Fig. 19.59). The MOV is available from Security Engineering; see the address in Appendix B.
- Installation of the MOV is for arc suppression in ultra-high-usage installations to prevent switch contact deterioration. In any situation, though, the installation of a diode across the leads is not recommended (see Fig. 19.58).

Series 2200 by Security Engineering

Another lock to consider is the surface-mounted electrically actuated lock, the Series 2200 from Security Engineering. This lock (Fig. 19.60) was

WRONG
ON-OFF CONTROL
110 VAC
12 VAC
RECTIFIER
TRANSFORMER
12 VDC
#3900 ELECTROMAGNETIC LOCK
RIGHT
ON-OFF CONTROL
110 VAC
12 VAC
RECTIFIER
TRANSFORMER
12 VDC
#3900 ELECTROMAGNETIC LOCK

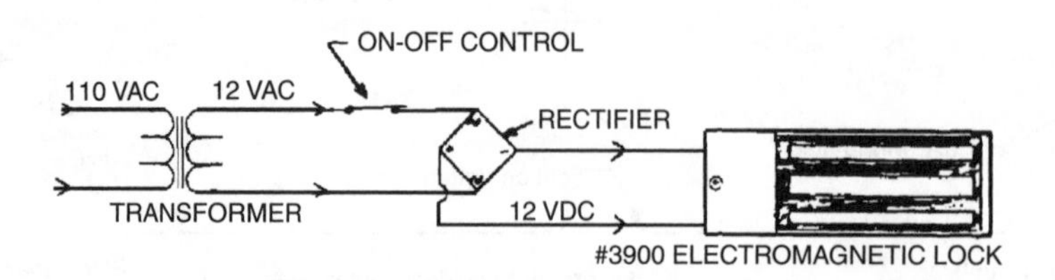

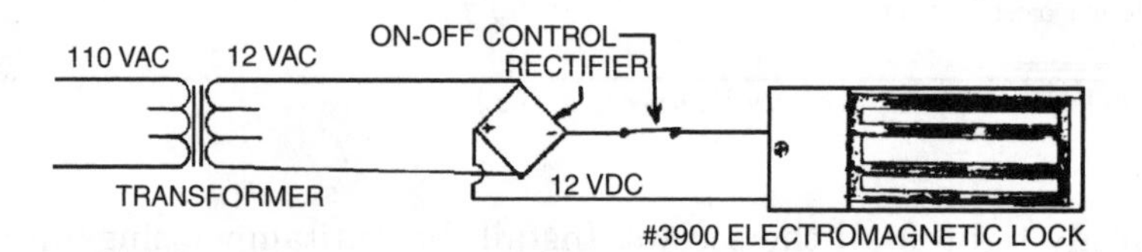

Figure 19.57 Model 3900 wiring, Step 1.

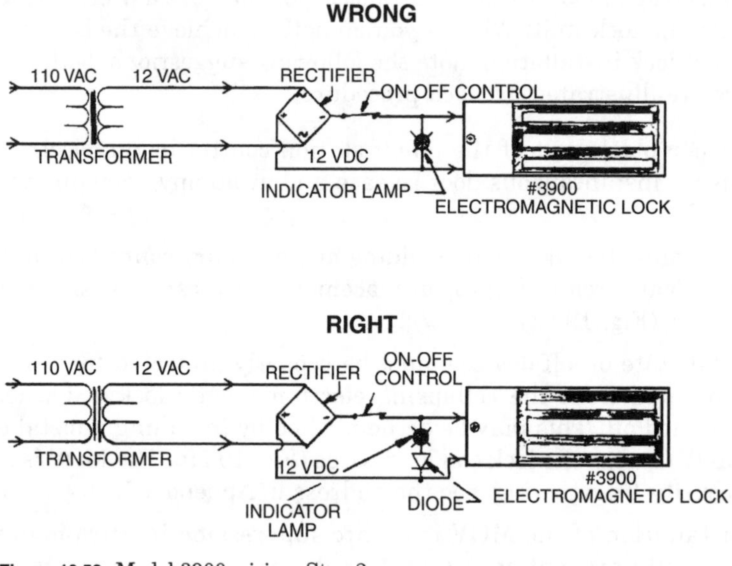

WRONG
110 VAC
12 VAC
RECTIFIER
ON-OFF CONTROL
TRANSFORMER
12 VDC
INDICATOR LAMP
#3900
ELECTROMAGNETIC LOCK
RIGHT
110 VAC
12 VAC
RECTIFIER
ON-OFF
CONTROL
TRANSFORMER
12 VDC
#3900
DIODE
ELECTROMAGNETIC LOCK
INDICATOR
LAMP

Figure 19.58 Model 3900 wiring, Step 2.

designed to offer the advantages of remote electric locking coupled with a versatility limited only by the user's imagination.

For basic operation, the lock mounts to a door frame or a door. When energized, it locks or unlocks the door when the solenoid-operated deadbolt projects from the unit and makes contact with a strike.

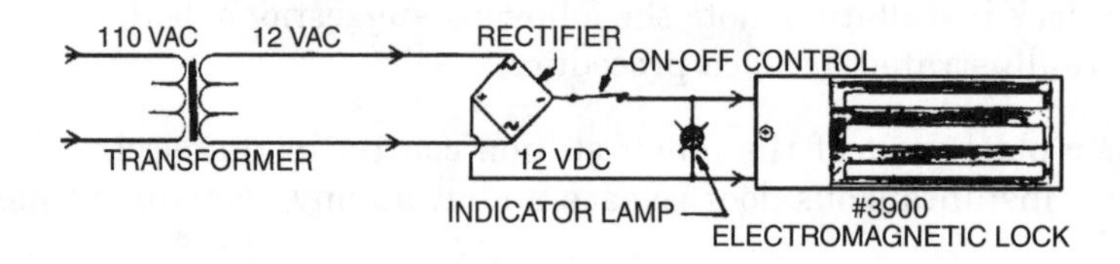

Standard features include the following:

- *Fail-safe operation*. If the power is interrupted, the unit automatically unlocks.
- *Fail-secure operation*. If the power is interrupted, the unit automatically unlocks.
- *Easy installation*. The unit is not handed and can thus be mounted on any plane.
- *Versatility*. It can be used with in-swinging, out-swinging, sliding, overhead, or roll-up doors.
- *Maintenance free*. Unit parts are ruggedly built and self-lubricating.
- *Current draw*. 0.5 amps at 24 volts.

WRONG

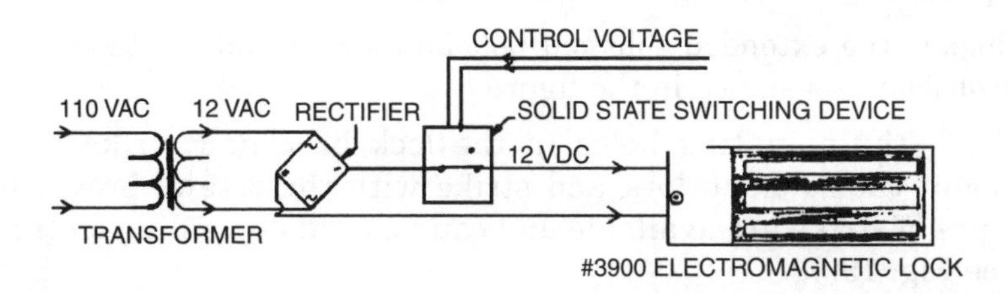

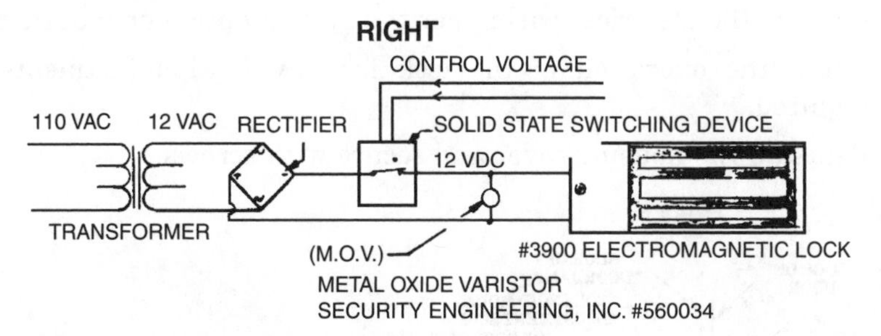

Figure 19.59 Model 3900 wiring, Step 3.

Figure 19.60 Model 2200 series surface-mounted lock. (*Security Engineering, Inc.*)

- *Low voltage*. 24 Vac standard; 24 Vdc, 12 Vac, and 12 Vdc available.
- *Long and silent operation*. The unit has a continuous duty solenoid that will function for millions of cycles.
- *Compact size*. 2 × 2 × 6¾ inches.

The unit has optional features including a choice of one or two signal lights and a built-in bolt position switch to indicate whether or not the bolt is extended or retracted for interfacing with automatic door equipment or signaling the lock status to a remote location unit. *Note:* The fail-secure locks are not recommended where life safety may be compromised or where panic crash bar hardware is the only means of emergency egress.

Mounting. Refer to Fig. 19.61.

1. Remove the four screws from the unit; remove the housing cover.
2. Manually extend the solenoid bolt by pushing the BPS pin forward (if equipped) or inserting a wire into the rear of the solenoid.
3. Engage the extended solenoid bolt and strike and locate on the door and door frame as shown in the figure.
4. Using the mounting holes in the lock base and strike as templates, mount the solenoid lock and strike with the screws provided. (Several types of strike are available and can be used depending on the situation. See Fig. 19.62.)
5. Check for the proper alignment and see that the lock is free, having no binding in operation.
6. Connect the electrical wiring per the system option or requirements.
7. Check the operation of the lock for any final adjustments; adjust as required.
8. Reinstall the housing cover and secure with screws.

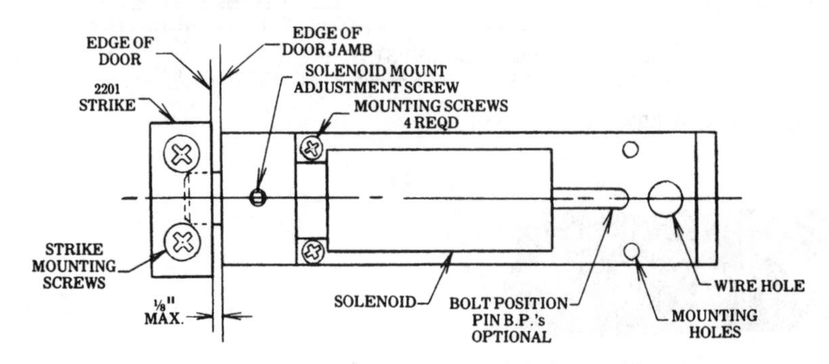

Figure 19.61 Installation procedure for 2200 series lock. (*Security Engineering, Inc.*)

Series 1300 mortise solenoid lock

A more compact but just as effective lock is the 1300 series mortise solenoid lock (Fig. 19.63). This mortise solenoid-activated deadbolt is designed primarily for installation in new construction, with emphasis on tamper resistance and unit concealment.

On activation of the solenoid, the deadbolt projects ½ inch from the faceplate or retracts ½ inch to flush with the faceplate, depending on the solenoid selected and used.

The standard features of this lock unit are as follows:

- *Fail-safe and fail-secure operation.* The unit automatically unlocks when power is interrupted.
- *Compact size.* 1⅜ inch in diameter × 4½ inches long.
- Voltage and current draw are the same as the 2200 series.

Electrical Keyless Locks

Simplex keyless security locks have also kept pace with electrically operated locks and latch units. The Simplex keyless entrance control is a wall-mounted

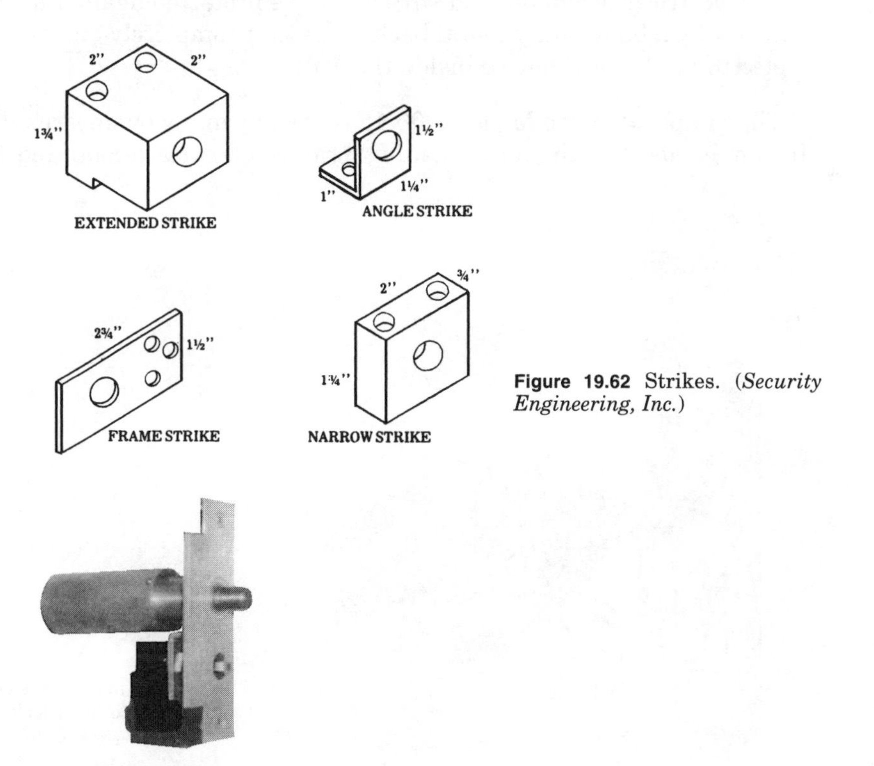

Figure 19.62 Strikes. (*Security Engineering, Inc.*)

Figure 19.63 Series 1300 mortise solenoid lock. (*Security Engineering, Inc.*)

unit control for electrical door locks (Fig. 19.64) and has a built-in, adjustable, time delay feature not found in other, currently available units. Eliminating key control problems in high-risk areas, for home or business, is a distinct and great advantage.

The Simplex entrance control is an easy yet very secure installation. The unit is secured by two rugged draw bolts that pass completely through the wall. The bolts can be cut to any desired length to permit installation in walls from 4 to 11 inches thick (Fig. 19.65). The adjustable timer delay of 15 to 20 seconds permits locating the control at any convenient position near or by the door.

Two styles of front mounting plates are included to meet varying installation needs. One is 3¼ inches wide × 9⅝ inches high; the smaller is 1¾ × 8½ inches. Both front mounting plates are made of stainless steel.

Security features for this unit include the following:

- A special visual shield for each entrance control unit to protect against unauthorized observation of the button sequence operation.
- If the control knob is forced by pliers or other tools, it will spin free without activating the electrical circuit.
- The electrical terminals and wire leads are protected against a short circuit attack by a heavy-duty metal back cover that completely encloses the entire electro-mechanical device inside the wall.

The Simplex/Unican Model 2000-15 is used by many businesses (Fig. 19.66). It can be used with many leading brands of surface-mounted panic exit

Figure 19.64 Wall-mounted electrical control unit for door lock or latch. (*Simplex Access Controls Corp.*)

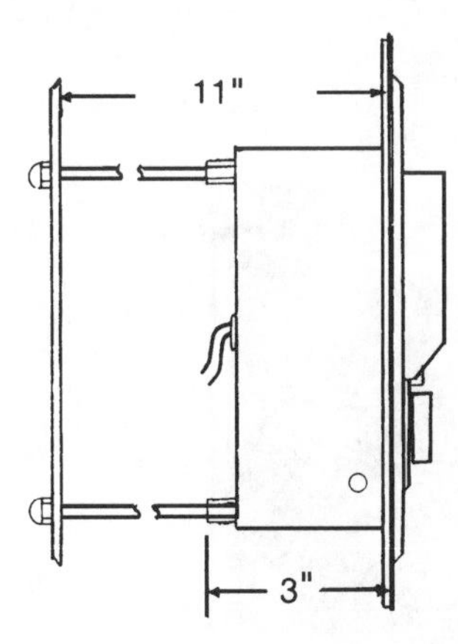

Figure 19.65 Lock unit can be mounted on any thickness wall up to 11 inches. (*Simplex Access Controls Corp.*)

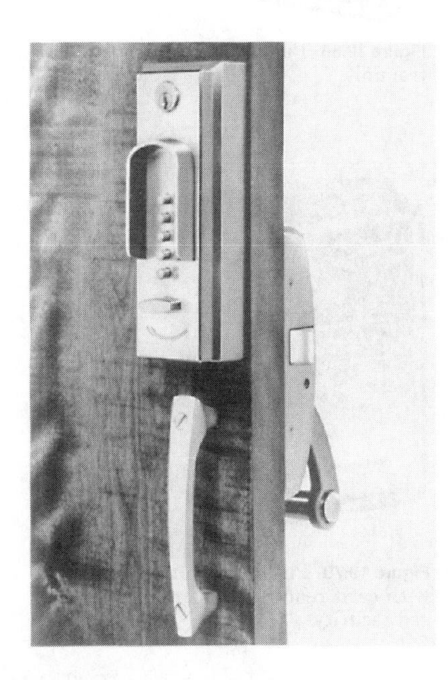

Figure 19.66 Unican 2000-15 unit used with an emergency exit panic bar. (*Simplex Access Controls Corp.*)

devices. Another popular unit is the Model TM pushbutton lock (Fig. 19.67), which can be used in both businesses and homes.

The Memorilok system

A sophisticated digital access control system that does not require a console board or monitoring station is available. The Memorilok system has two basic units: an access control keyboard (Fig. 19.68) and a program unit (Fig. 19.69). With these two units, 10,000 combinations are available to the customer. Now there is a third optional piece of equipment, the card reader (Fig. 19.70), that provides a substantially higher degree of security integrity.

Figure 19.67 Model TM lock installed in an office cabinet. (*Simplex Access Controls Corp.*)

Figure 19.68 Pushbutton control unit.

Figure 19.69 Programmer unit.

Figure 19.70 Pushbutton unit with card reader for increased security.

The Memorilok system incorporates a solid state electronic memory for the storage of the selected code combination. The entry code combination is electronically stored. Unlike other access control systems, there are no wheels, dials, or other visual indicators to betray the code combination.

Standard unit features include:

Automatic address. A preset timer is activated when the first digit of the code combination is entered. The remainder of the code entry must be completed before the expiration of this time or the entry will be canceled and the programmer will activate the penalty time.

Automatic penalty. A penalty time of approximately 5 seconds is incurred if an incorrect key is depressed during the code entry or if two or more keys are pressed simultaneously. This feature defeats random button manipulation.

Code bypass. There is a provision within the access control unit for the connection of an external (remote) switch or pushbutton to operate the lock via the programmer unit without the use of the entry keyboard. This is normally sited within the protected access area and used for giving access to personnel or visitors who do not possess the code combination but for whom access is warranted and authorized by the customer.

Memory protection. The Memorilok keeps the code combination stored and undamaged in case of a main power failure. This is due to the constantly charged nickel cadmium battery contained within the unit.

Optional features include the following:

Duress/error alarm. The duress/error alarm option is available as an integral part of the Memorilok unit and, if desired by the customer, must be ordered as original equipment. The duress alarm is activated by depressing a factory-fixed last digit (from 0 to 9) instead of the last digit of the selected code combination. If entry is made under duress, the first three digits of the programmed code combination plus the predetermined last digit—the Duress Digit—are entered on the keyboard. This automatically operates a set of Form C (Changeover) dry contacts that, in turn, triggers any desired alarm system: audible, silent, remote, or local.

The error alarm circuit will react on any number of selected unsuccessful attempts at entry. As soon as the predetermined number of incorrect attempts (from one to eight) have been made, the alarm is activated, warning that the system is being tampered with.

Flush mounting. To flush mount the programmer, keyboard, or even the battery standby, an oversized front panel is used. The panels, 8 × 9 inches, are affixed to the standard front panels to conceal the rough opening around the Memorilok unit.

The optional card reader fits into the front of the keyboard panel. It provides higher levels of security integrity because the coded card *must* be used by the individual seeking entry. It must be entered into the Memorilok before the code combination will be accepted by the electronic memory circuit.

Wiring instructions. Wiring details are in Fig. 19.71. The following notes apply, as referenced in the figure:

- These connections are not used unless the remote entry/exit is required. Use the open switch.

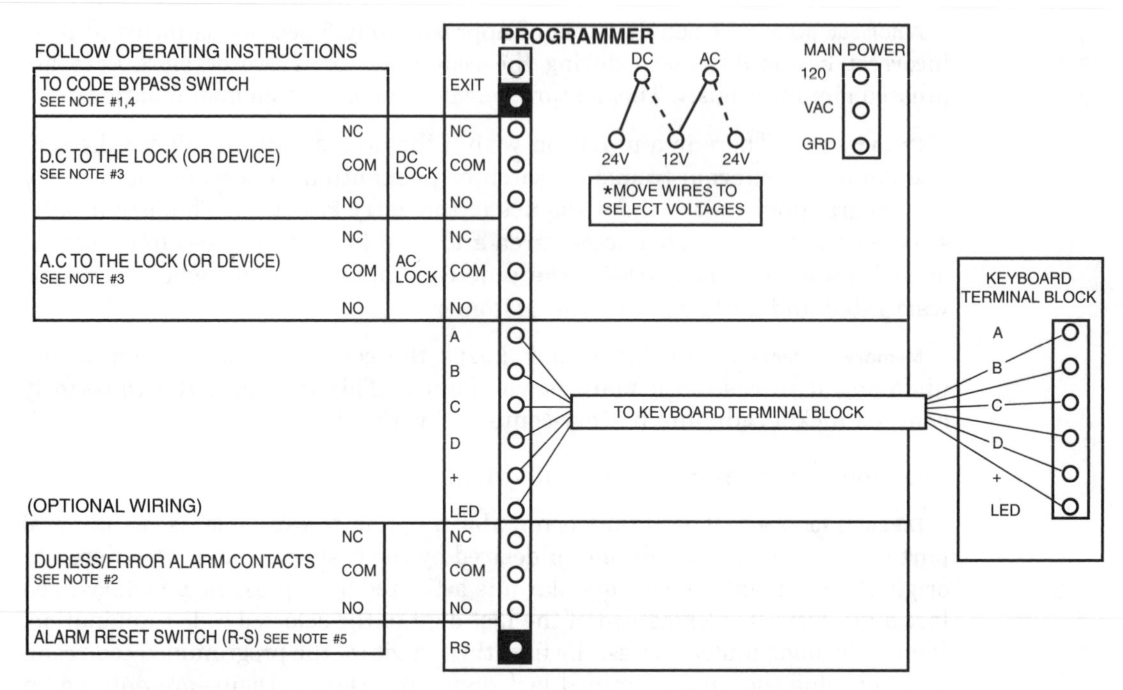

Figure 19.71 Memorilok wiring instructions (refer to text for notes). (*Alarm Lock Corporation*)

- All contacts are shown (NO and NC) without the lock being energized and without being in the alarmed state.
- On installations where there is a long wire run between the programmer and the lock or other device being operated, carefully compute the minimum requirements of the lock or device and the voltage drop that will occur in the wire.
- Code bypass switch requires a momentary switch with normally open (NO) switch contacts.
- The alarm Reset Switch (R-S) requires a momentary switch with normal switch contacts. This should be wired between terminal 2 of the exit and terminal R-S indicated on the diagram as "blacked out block" on the programmer wiring hookup. Use this only for duress/error entry.

Programming the Memorilok. To program the Memorilok, turn the programmer key to the *on* position. Enter the desired four-digit code combination on the keyboard. A green LED indicator light glows to indicate that the code is properly programmed into the memory system. Return the programmer key to the off position. Reinsert the code to test the unit for proper operation. Now the green LED light indicates acceptance of the code combination and supply of power to the lock or other device used.

The adjustable time delay for the Memorilok affects the lock open time (time delay for the unit) and it can be varied from 1 to 10 seconds by turning an adjusting screw.

Unit specifications are in Fig. 19.72. Specification installation details, template, and other specific information concerning conductor wire are provided with the unit from the factory.

Key-Actuated Switches

The full-service locksmith should have available for customers, especially those in business, a variety of key-actuated switches. Figure 19.73 shows a Micro Kaba rotary switch. This miniature switch is operated by the 12-millimeter-diameter cylinder, which has over 10,000 different possible key combinations. Four key entry variations are available, each having up to four switch elements (Fig. 19.74).

SPECIFICATIONS:

Dimensions:	
Keyboard Unit	5 1/16" (128mm) Wide 6 1/16" (153mm) High 2 5/8" (66mm) Projection
Programmer Unit	Same as Keyboard
Shipping Weight:	Complete Memorilok - 6 pounds Battery Standby Unit - 6 pounds
Main Power Input:	120 VAC or 240 VAC 50/60 HZ. Specify voltage required.
*Output to Lock:	12 or 24 VAC or VDC at 1 Amp. Voltage is user selectable.
Program Key Switch:	Contacts are closed in the "ON" position for code programming.
Wiring Information:	Includes ten feet of six conductor 22 gauge wire.

Figure 19.72 Memorilok 400 unit specifications. (*Alarm Lock Corporation*)

Figure 19.73 Micro Kaba rotary switch. (*Lori Corp.*)

KEY ENTRY	No. of Switch elements
Spring Return	1 2 3 4
One Key Entry 9h	1 2 3 4
One Key Entry 6h	1 2 3 4
Two Key Entry	1 2 3 4

Figure 19.74 Four different key positions provide an opportunity for different control features. (*Lori Corp.*)

Figure 19.75 Two-pole switch. (*Lori Corp.*)

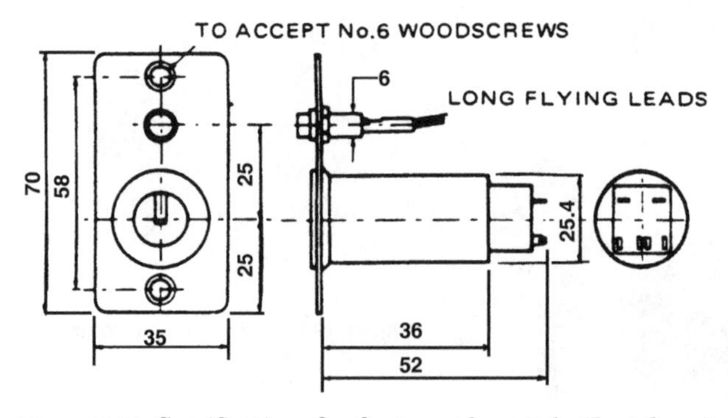

Figure 19.76 Specifications for the two-pole switch. (*Lori Corp.*)

Two-pole switches (Fig. 19.75) are also available. They provide for 90° movement with two key withdrawal positions, thus allowing the electronic unit to be held in either the on or off position. The unit can take up to 250 Vac, at 4 amps. Unit dimensions are in Fig. 19.76.

Figure 19.77 Surface-mounted remote control station with a waterproof key cover. (*Lori Corp.*)

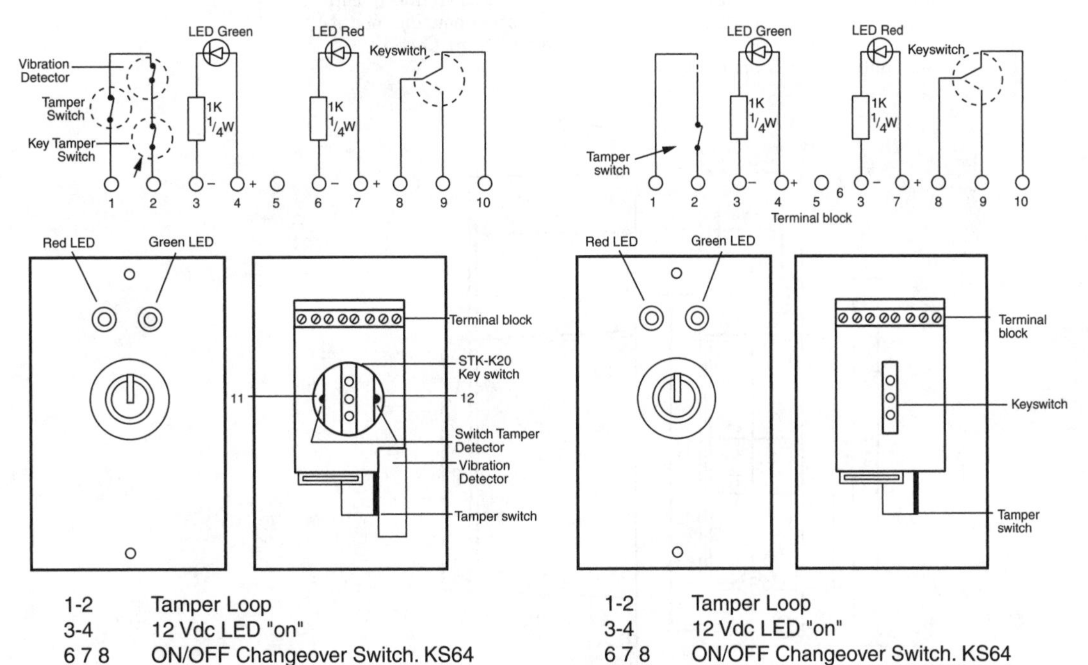

WIRING DIAGRAM FOR REMOTE CONTROL STATIONS KS - RCS/1

WIRING DIAGRAM FOR REMOTE CONTROL STATIONS KS - RCS/2

Figure 19.78 RCS/1 and RCS/2 wiring diagrams. (*Lori Corp.*)

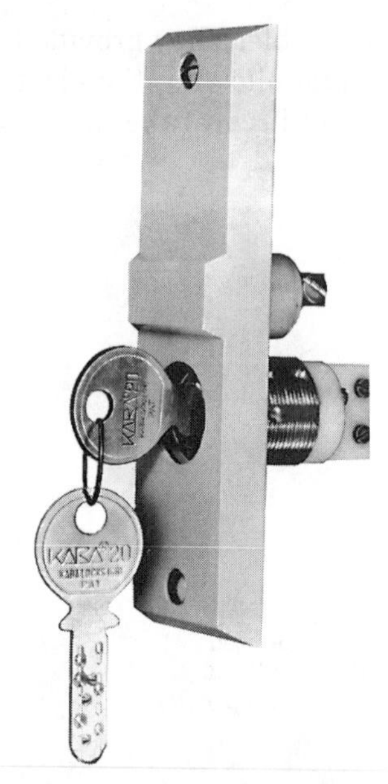

Figure 19.79 The narrow remote control switch allows for many different installation applications. (*Lori Corp.*)

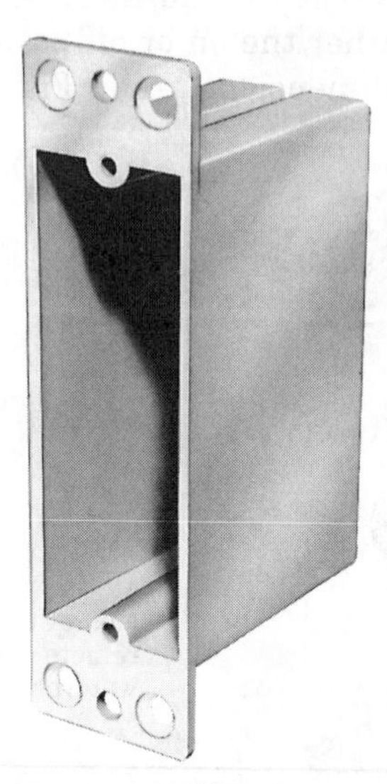

Figure 19.80 Remote control station housing mounting box. (*Lori Corp.*)

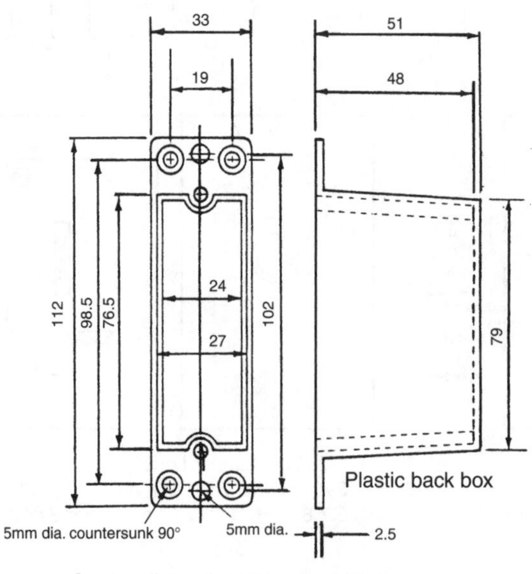

Figure 19.81 Mounting box specification data. (*Lori Corp.*)

Rated Voltage (V)	Non-Inductive Load		Inductive Load	
	Resistive (A)	Lamp (A)	Inductive (A)	Motor (A)
AC 110 to 125	5	1.5	3	2.5
220 to 250	5	1.0	2	1.5
DC 6 to 8	5	2	5	3
110 to 125	0.4	0.05	0.4	0.05
220 to 250	0.2	0.03	0.2	0.03

Figure 19.82 Electrical specifications for the remote control station. (*Lori Corp.*)

The surface-mounted remote control stations (KS RCS/1 and /2) are die-cast aluminum with a stainless steel frontplate, tempered against opening, and a red and green LED (Fig. 19.77). The waterproof version ensures unit use during all types of weather conditions. Figure 19.78 provides basic wiring information for the RCS/1 and /2 units.

Another remote control station with an antitamper switch is in Fig. 19.79. Like the other units, it uses the Mini-Kaba or the Kaba 20 key/switch. A remote control station mounting box (Fig. 19.80) is available and used for mounting the unit. Mounting box dimensions are in Fig. 19.81.

Remote control switches are UL approved. Electrical specifications are in Fig. 19.82.

Chapter

20

Working as a Locksmith

In this chapter I'll tell you how to find a job as a locksmith and how to start your own business. Every month the locksmithing trade journals have dozens of help-wanted advertisements and many businesses-for-sale advertisements. The Internet also has a lot of both.

Finding a Job

There are many money-making opportunities for a person who has locksmithing skills or who wants to learn the trade. You'll need to decide where you want to work, seek out meetings with prospective employers, meet with them, and negotiate wages and other working terms. You'll also need to make decisions about joining locksmithing associations and getting bonded, certified, and licensed.

Locksmithing associations

Belonging to a locksmithing association can help to improve your credibility among other locksmiths and in your community. Most of the associations offer trade journals, technical bulletins, certifications, classes, discounts on books and supplies, locksmith bonds, insurance, and so on. They provide membership certificates that you can display in your shop and logos that you can use in your advertisements. It also can be helpful to include association memberships on your résumé or in your Yellow Pages advertisement.

Many successful locksmiths, however, don't belong to any locksmithing association. Some don't join because they believe that the membership dues are too high. Others refuse to join because they disagree with the policies or legislative activities of the organizations. The criteria for becoming a member differ among associations, as do the policies, legislative activities, and membership dues. You'll need to decide for yourself if joining one will be beneficial for you.

Whether or not you decide to join an association, it's a good idea to subscribe to physical and electronic security trade journals and newspapers. Some are distributed free to qualified security professionals. Such publications can help you to stay abreast of new products, trends, and legislation related to the security industry. Many popular trade journals and organizations are listed in Appendix C.

Bonding

Bonding is a type of insurance that pays off if you are convicted of stealing from your customers. It's a good idea to place your bonding certificate in your shop where customers can see it. The easiest way to get bonded is through one of the major locksmithing trade journals—such as the *National Locksmith*. Subscribers may opt to be bonded for a small fee. The National Automobile Association and the National Safeman's Organization provide bonding to their members as part of their membership fees.

Certifications

A locksmith certification signifies that a person has demonstrated a level of knowledge or proficiency that meets a school or association's criteria for being certified. The significance of a certification depends on the integrity of the organization issuing it. Some of the best known certifications are granted by the Associated Locksmiths of America (ALOA), the American Society of Industrial Security (ASIS), the National Safeman's Organization, and the Safe and Vault Technician's Association (SAVTA). It's also helpful to be registered as a locksmith with the International Association of Home Safety and Security Professionals (IAHSSP). A registration test is provided in Chap. 22. By passing that test, you can qualify for the Registered Professional Locksmith designation, and after qualifying, you may access the association's website and include the initials "RPL" after your name.

Locksmiths are divided about the need for certification. Some of the most experienced and well-known locksmiths and safe technicians don't use certification initials after their names. While writing this chapter, I looked at the mastheads of the two most popular locksmithing trade journals and counted 30 regular writers between them. Of the 30, only five used certification initials.

Many locksmiths feel that they shouldn't have to meet the approval of a specific school or organization. They prefer to allow the free market to determine the competency of a locksmith. An incompetent locksmith, they argue, won't be able to compete successfully against a competent one for very long.

Others feel that locksmith certifications can benefit the locksmith industry by raising the competency level of all locksmiths and by improving the public perception of locksmiths. In any case, certifications can be helpful in finding work, especially if you don't have a lot of experience. They show prospective employers that you're serious about wanting to learn locksmithing.

Licensing

A license differs from a certification in that a license is issued by a municipality rather than by a school or an association. Most cities and states don't offer or require locksmith licenses. In those places, locksmiths are required only to abide by laws that apply to all businesses—zoning, taxes, building codes, and so on. (However, because the alarm industry is heavily regulated, a locksmith who wants to install alarms first will probably need to obtain an alarm systems installers license.)

A few places, such as New York City, Texas, Tennessee, North Carolina, and California, require a person to get a locksmith license before offering locksmithing services. Contact your city, county, and state licensing bureaus to find out if you need to have a locksmith (or other) license. Some places levy stiff fines on people who practice locksmithing without a license.

The requirement for obtaining a locksmithing license varies from place to place. Some require an applicant to take a competency test, be fingerprinted, provide photographs, submit to a background check, and pay an annual fee of $200 or more. Other places simply require applicants to register their name and address and pay a few dollars annually.

Locksmiths are divided on three major issues concerning licensing: whether or not licensing is necessary, what criteria should be required to obtain a license, and whether or not a national locksmith association should have an integral part in qualifying locksmiths for licensing.

Some locksmiths feel that licensing is needed to protect consumers from unskilled locksmiths and to improve the image of the profession. Other locksmiths feel that the free market does a good job of separating good locksmiths from the bad.

Proposed criteria for licensing—such as the fees, liability insurance, and competency testing—are also of concern to many locksmiths. The primary arguments about such matters center around whether or not a specific criterion is unnecessary, too costly, or too burdensome for locksmiths.

Some locksmithing trade associations are actively working to enact new locksmithing laws throughout the United States. But only a minority of locksmiths belongs to associations. Most locksmiths are unaware of such bills until they become law. If you're a locksmith or plan to become one, you should take an interest in the current legislation related to locksmiths in your area.

Write or call your local, county, state, and federal representatives, and ask to be kept informed of proposed legislation related to locksmiths. When licensing bills are proposed, ask your representative to mail a copy to you. In addition to letting your representatives know what you think of the bills, you can send letters to the editors of locksmithing trade journals.

Another great way to have your views heard by people who care is to place messages on locksmithing Web sites. Some of the most popular are *alt.locksmithing* (newsgroup), *www.TheNationalLocksmith.com, www.ClearStar.com,* and *www.Locksmithing.com.*

Planning your job search

Your job search will be most productive if you approach it in a professional manner. You need to plan every step of the way. When you're seeking a job, you're a salesperson—you're selling yourself. You need to convince a prospective employer to hire you, whether or not the employer is currently seeking a new employee.

Deciding where you want to work

Before contacting prospective employers, decide what type of shop you want to work in and where you want to work. Which cities are you willing to work in? Do you prefer to work in a large, midsize, or small shop? Do you prefer to work for a highly experienced shop owner or for an owner who knows little about locksmithing?

You can answer these questions by considering your reasons for seeking employment. If you're mainly looking for an immediate, short-term source of income, for example, your answers might be different than if you were more concerned about job security or gaining useful locksmithing experience.

In a small shop you might be given a lot of responsibility and be able to get a great deal of experience quickly. In a large shop you're more likely to get a lot of pressure and little respect. However, a large shop might be better able to offer you more training and a lot of opportunities for advancement.

A highly experienced locksmith can teach you a lot, but you probably will have to work a long time to learn much. Experienced locksmiths place a lot of value on their knowledge and rarely share much of it with new employees. Often, experienced locksmiths worry that if a new employee learns a lot very quickly, that employee might quit and go to a competitor's shop or start a competing company.

Some shop owners know very little about locksmithing. Working for them can be difficult because their tools, supplies, and locksmithing methods are outdated. You won't gain much useful experience working for such a person; instead, you'll probably pick up a lot of bad habits.

Locating prospective employers

You can find prospective employers by looking through the Yellow Pages of telephone directories of the cities in which you want to work. Your local library probably has lots of out-of-town directories. Also look in locksmithing trade journals, and check out locksmithing Web sites. You also might want to place "job wanted" ads in locksmithing trade journals.

When creating an ad, briefly state the cities in which you would like to work, your most significant qualifications (bondable, good driving record, have own tools, etc.), and your name, mailing address, and telephone number. Blind advertisements—those that don't include your name, address, or any direct contact information—won't get many responses. Few locksmith shop owners are that desperate for help.

Getting prospective employers to meet with you

It's usually better for you to set up a meeting with a prospective employer than to just walk into the shop unscheduled. You might come in when the owner is busy, which could give the person a negative impression of you. Also, it might be hard to interest the shop owner in hiring you without any prior knowledge about you.

One way to prompt a prospective employer to meet with you is to send a résumé and a cover letter. It isn't always necessary to have a résumé to get a locksmithing job, but it can be a highly effective selling tool. A résumé allows you to project a good image of yourself to a prospective employer, and it sets the stage for your interview.

The résumé is an informative document designed to help an employer know you better. When writing it, view your résumé as a document designed to promote you. Don't include everything there is to know about yourself; only include information that will help an employer decide to hire you. Don't include, for example, information about being fired from a previous job. It's best to wait until you're face to face with a person before you try to explain past problems.

The résumé should be neatly typed in black ink on at least 20-pound bond, 8½- by 11-inch white paper. Figures 20.1 and 20.2 show sample résumés. If you can't type, have someone else type it. Don't use graphics or fancy typestyles (remember, you're seeking work as a locksmith, not as a graphic artist). A résumé should be kept to one page; few prospective employers like wading through long résumés. There is no perfect structure for a résumé; organize it in the way that allows the employer to quickly see reasons for hiring you. List your strongest selling points first. If your education is your strongest selling point, for example, include that information first. If your work history is your strongest selling point, include that first.

Include a short cover letter with your résumé. The cover letter should be directed to an owner, manager, or supervisor by name and should prompt the person to review your résumé. If you don't know the person's name, call the shop and get it. Don't send a "To Whom It May Concern" cover letter. A sample cover letter is shown in Fig. 20.3.

After reviewing your résumé, a prospective employer probably will either call you on the telephone or write you a letter. If you're not contacted within two weeks after mailing your résumé, call the person you addressed it to and ask if your letter was received. If it was received, request a meeting. You might be told that there are no job openings there. In that case, ask if you could still arrange a meeting so that you two can get to know each other.

Preparing to meet a prospective employer

Before going to any meeting, put together a package of documents to leave with the prospective employer. Make photocopies of any relevant certifications, association membership certificates, bond certificate, locksmithing license, and the like. If the prospective employer wants you to complete a job application, attach

RESUME

John A. Smith, RPL, CJS
123 Any Street
Any Town, Any State 01234
(012)345-6789

Job Objective
Entry level position as a locksmith in a progressive shop.

Special Capabilities
- Can quickly open locked automobiles
- Can use key cutting and code machines
- Can install, rekey, and service many types of locks
- Can pick locks and impression keys
- Can install and service emergency exit door hardware
- Can operate cash registers

Certifications
Registered Professional Locksmith (RPL)
Certified Journeyman Safecracker (CJS)

Special Qualities
Dependable (missed work only 2 days over the past 5 years), good driver (no accidents), bondable (no arrests), good health, fast learner, and self-starter.

Education/Training
BA degree in Business Administration from ABC College in Quincy, CA.
Self-study with *Locksmithing,* by Bill Phillips; and the *Complete Book of Locks & Locksmithing, 4th Edition,* by Bill Phillips.

Memberships
Member of the International Association of Home Safety & Security Professionals
Member of the National Safeman's Organization
Member of the National Locksmith Automobile Association

Work Highlights
OFFICE ASSOCIATE
SEARS IN QUINCY, CA FROM JANUARY 2004 TO PRESENT Man the switchboard; handle complaints; accept money; balance money taken from the safe, registers, and other areas of the store; train other associates; and work with all of the office equipment.

ASSISTANT MANAGER
MCDONALD'S IN QUINCY, CA FROM JUNE 2003 TO JANUARY 2004.

References Available Upon Request

Figure 20.1 If you have little experience or formal training, don't emphasize those things on your resumé.

John A. Smith, RPL
123 Any Street
Any Town, Any State 01234
Phone: (012)345-6789

Job Objective
Seeking position as a locksmith/access control systems technician in a University.

Work Experience
April 2002 - Present, Manager of ABC Locksmith Shop in Quincy, CA. Supervisor of 6 locksmiths. Duties include: training apprentices; designing and maintaining masterkey systems; installing and servicing emergency exit door devices, electromagnetic locks, electric locks, electric strikes, and a variety of high-security locks; and manipulating, drilling, and recombinating safes.

January 2000-April 2002, Locksmith at DEF Locksmith Shop in Brookdale, CO. Duties included: opening locked automobiles; rekeying, master keying, impressioning, and servicing a variety of basic and high-security locks; manipulating, drilling, and recombinating safes; installing and servicing emergency exit door devices and other door hardware; servicing foreign and domestic automobile locks; installing a wide variety of electric and electronic security devices; operating the cash registers.

December 1997-January 2000, Key Cutter/Apprentice Locksmith at The Key Shop in Brookdale, CO. Cutting keys and assisting three locksmiths. Worked exclusively inside the shop during the first year, but did outside work about half the time thereafter.

June 1995-December 1997, Key Cutter/Salesperson at Building Supply Hardware in Brookdale, CO.

Licenses/Certificates
Have a current California Locksmith Permit
Certified Master Safecracker (CMS)
Member of the International Association of Home Safety and Security Professionals
Member of the National Safeman's Organization

References Available Upon Request

Figure 20.2 If you have a good, relevant work history, emphasize it.

123 Any Street
Any Town, Any State 01234
Phone: (012)345-6789

Today's Date

Ms. Patricia L. Bruce, owner
The Lock & Key Shop
321 Another Street
Another Town, Another State 43210

Dear Ms. Bruce:

I can perform most basic locksmithing tasks, and especially enjoy doing foreign and domestic automobile work. Please look over my enclosed resume. I'd like to meet with you to see if we might be able to work together.

Please call me at your earliest convenience to let me know when I can come by your shop.

Sincerely yours,

John A. Smith

Figure 20.3 A brief cover letter should be sent with a resúmé.

your documents to it. If he or she doesn't ask you to complete a job application, bring your documents to the meeting.

Before the meeting, learn all you can about the prospective employer and his or her shop. Stop by the shop to find out what products the company sells and what special services it offers. Notice the brands of locks and safes the company sells, and become familiar with them.

Call the local Better Business Bureau and Chamber of Commerce to find out how long the business has been established and how many consumer complaints have been filed against the company. Contact local, state, and national locksmithing trade associations to find out if the owner belongs to any of them.

Also, before the meeting, read the current issues of major locksmithing trade journals. Study them to find out about major news related to the trade. Then be prepared to speak confidently about such matters. This will impress the prospective employer.

Before going to the meeting, honestly assess your strengths and weaknesses. Consider which locksmithing tasks you're able to perform well. Also consider how well you can perform other tasks a prospective employer might want you to do, such as selling products, working a cash register, and so on. Then write a list of all your strengths and a list of the last four places you've worked. Include dates, salaries, and the names of supervisors.

At this point, prepare explanations for any negative questions the prospective employer might ask you during the interview. If there is a gap of more than three months between your jobs, for example, decide how to explain that gap. If you were fired from one of your jobs or if you quit one, figure out the best way to explain it to the prospective employer. Your explanations should put you in the best light possible.

However, never blame other people, such as former supervisors, for problems in past jobs; this will make you seem like a crybaby or a back stabber. If there is no good explanation for your firing, it might be in your best interest simply not to mention that job unless the employer is likely to find out about it anyway. Don't lie about problems you've had because if the prospective employer learns the truth, he or she probably won't hire you and might tell other locksmiths about you. The locksmithing industry is fairly small, and word gets around quickly.

Meeting a prospective employer

On the day of the meeting, be sure you're well rested, have a positive attitude, and are fully prepared. A positive attitude requires seeing yourself as a valuable commodity. Project this image to the person you want to work for. Do not tell a prospective employer that you "really need a job," even if the statement is true. Few people will hire you simply because you need a job. If you seem too needy, people will think that you are incompetent.

Don't go to a meeting with the feeling that the only reason you're there is so that a prospective employer can decide if he or she wants to hire you. You both have something to offer that the other needs. You have your time, personality, knowledge, and skills to offer—all of which can help the prospective employer to make more money. The employer, in turn, has knowledge, experience, and money to offer you. The purpose of the meeting is to allow both of you to get to know each other better and to decide if you want to work together. It also might lead to an employment agreement.

When you go to the interview, wear clean clothes and be well groomed. Arrive a few minutes early. When you shake hands with the prospective employer, use a firm (but not tight) grip, look into his or her eyes, and smile. The person probably will give you a tour of the shop. This will help you to assess whether or not it's a place at which you want to work. Allow the prospective employer to guide the meeting. If he or she offers you a cup of coffee, decline it unless the prospective employer also has one.

Don't ask if you may smoke; and don't smoke if invited to. In the meeting, sit in a relaxed position, and look directly into the person's eyes. If you have trouble maintaining eye contact, look at his or her nose, but try not to look away. Constantly looking away will make you appear insecure or dishonest. Listen intently while the prospective employer is talking, and don't interrupt. Smile a few times during the meeting. Be sure that you understand a question before attempting to answer it. Whenever you have the opportunity, emphasize your strengths and what you have to offer, but don't be boastful when doing so.

If you're asked a question you feel uncomfortable answering, keep your body in a relaxed position, continue looking into the person's eyes, and answer it in the way you had planned to answer it. Don't immediately begin talking about another topic after answering the tough question; if you do, it will seem like you're trying to hide something. Instead, pause after answering the question and smile at the person; he or she then will either ask you a follow-up question or move on to another topic. Don't give an audible sigh of relief when you move to a new topic.

If asked about your salary requirements, don't give a figure. Instead, say something like, "I don't have a salary in mind, but I'm sure we can agree on one if we decide to work together." Salary is usually discussed at a second meeting. However, if the prospective employer insists on discussing salary during the first meeting, let him or her state a figure first. Ask what he or she believes is a fair salary. When discussing salary, always speak about "fairness to both of us."

Making an employment agreement

The prospective employer's first salary offer almost always will be less than the amount the company is willing to pay. You might be told that the employer doesn't know about your ability to work, so he or she will start you off at a low salary that can be reviewed later.

This might sound good, but the salary you start out with has a lot to do with how much you'll earn from the company later. You want to begin working for the highest salary possible.

If the prospective employer's salary offer is too low, you can reply that you would like to devote your full attention to the job and work hard for the company but don't know if you could afford to do that with the salary offered. Explain that you simply want a fair salary. Pushing for the highest salary possible (within reason) not only will allow you to earn more money but also will give the prospective employer confidence in you. You will be proving to the person that you're a good salesperson.

If he or she refuses to agree to an acceptable salary, tell the person that you need to think about the offer. Then smile, stand up, shake hands, and leave. If the person demands an immediate answer, say you want to discuss the matter with someone else, such as your spouse. Don't get pressured into accepting an unreasonable employment agreement. It's in your best interest to give yourself time to think the matter over and to meet with other prospective employers. If you didn't attach your package of documents to a job application, leave them with the prospective employer.

Starting Your Own Locksmithing Business

When you consider starting your own locksmithing business, keep in mind the following factors:

Police clearance. Many jurisdictions require licensing and fingerprinting.

Professional qualifications. Some cities insist on tests to determine professional competence.

Basic tools and equipment. Ensure that you have all of the equipment you need to tackle most locksmithing jobs.

Supply stock. It's possible to work out special arrangements with some local supply houses.

Financing. Setting up shop involves an initial cash outlay.

Bookkeeping. You must keep records of all transactions.

Vehicles. A vehicle is necessary for out-of-shop calls.

Advertising. Sometimes it takes more than just competent locksmithing to bring in business. Good advertising usually includes signs, business cards, window displays, display racks, and display materials.

Literature. You need references and materials including locksmith supply house catalogs, manufacturers' literature, reference books, business forms, etc.

Contacts. It pays to know other locksmiths, factory representatives, etc. They can help you with locksmithing problems from time to time.

Locksmithing is a profession involving skill, competence, and public trust. It is a career that can be highly rewarding. This chapter describes the background necessary to use the basics of what you have already learned and apply it in a business.

According to the 2004–2005 edition of the U.S. Department of Labor's *Occupational Outlook Handbook,* there are about 23,000 locksmiths employed in the United States in small and large shops, as well as in-house for other businesses. It further states that "the locksmith trade itself has remained stable, with few economic fluctuations, and locksmiths with an extensive knowledge of their trade are rarely unemployed."

Fifty to 75 million keys are lost and replaced each year. Every household, automobile, and business has two to five locks. In a medium-sized city of 75,000 to 100,000 people, there are easily more than 250,000 locks. This means there are literally billions of locks and keys in this country—plenty of business for a good locksmith. Statistics show that locksmiths make good money, too.

Business Considerations

If you're thinking about launching a career in locksmithing, there are several things you should consider.

Selecting a business site

Selecting your business site is a major decision and one that will affect your business throughout its infancy.

In determining a location for your business, consider the types of businesses you will be near and what effect they will have on your customers. Shopping malls are ideal for locksmithing shops because of the walk-in business generated by people who need one or two keys duplicated. If you undertake home servicing, you will want a location that enables you to get to your customers quickly.

The type of building must be considered. It would be unwise to obtain a building that is in need of massive repairs. The building should be large enough for your needs. Thought should be given to the building's potential for future expansion.

Demographic and economic factors of the surrounding area should be considered too. Make a study of population density, population growth, projected economic growth, the tax structure, etc. Examine any potential competition with other locksmiths.

Shop layout

Window displays are a good way to attract customers. The display itself should be simple and effective, neat, and clean. Window displays cannot be cluttered; you must be discriminating and orderly in developing them. The display should tell the viewer something interesting and appealing about your products or services.

The front counter area should be carefully organized; this is where sales are made. Ensure that the key machines are near the counter—visible from the street. Spread out your keys on a keyboard; such a display stays in your customers' minds longer than a pile of keys on a table. Code books usually are kept near the key machines.

The workbench usually should be out of the customers' sight, preferably in the back room. If it is out front, it should be clean at all times.

Lighting of the work and customer service areas is most important. It should be adequate but not too bright. The lights should be of the nonglare variety. The work area can have additional small lights placed several feet above the bench.

Merchandising through advertising

Advertising takes many forms, from radio, television, and handbills to putting your name on each key duplicated. There are no hard and fast rules. Good advertising is advertising that gets results.

The selection of the proper advertising can be important for a beginning locksmith. Consider the population; does it listen to the radio, watch television, or read the paper more than in other areas, or does it look for gimmicks? What are the costs of the various advertising media? You can approach the advertising staffs of the local media to determine these answers.

Once you complete an advertising scheme, evaluate the results. Did the advertising increase business? Was the advertising worth the investment?

Ultimately, the locksmith herself is the most effective advertisement. Responsible locksmithing, a neat appearance, and a little courtesy go a long way. In spite of all the sales talk and advertising hoopla, the customer is only interested in one thing—quality services and products at a reasonable price.

Whether you work out of your home, your vehicle, or a formal locksmithing shop, you need to have some form of advertising at your shop. A sign outside that merely states "Locksmith" isn't really enough. Whenever possible, you should have and use small stickers that can be affixed to windows, vehicles, etc; they can increase your business.

Figure 20.4 shows a variety of such aids, including decals, window banners, and a sign. Whenever possible, you should use these types of aids. They promote business and tell your potential customers exactly what services you can provide. Signs also save you time in answering general inquiries from walk-ins off the street who are not sure if you can provide a specific service, such as cutting a circular (Ace) key. Having a sign indicating that you duplicate these keys can increase business.

A hanging sign (Fig. 20.5) with a slightly different slant to it (Keys Cut While You Watch) is another visual aid to customers. By using properly aligned key machines, you can ensure that the phrase "guaranteed fit" is accurate.

Decals for your office or on the service van also aid in promoting your business and security in general (Fig. 20.6).

Figure 20.4 Various sales aids used in locksmith shops. At upper left is a key sign; others are window banners. (*Dominion Lock Co.*)

Figure 20.5 Advertising sign to hang in window of locksmithing shop. (*Dominion Lock Co.*)

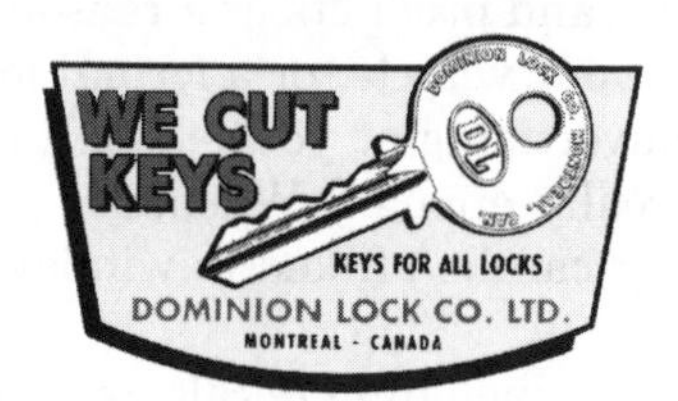

Figure 20.6 Decals for use in office or on side of van. (*Dominion Lock Co.*)

Selecting key blanks

It is important to carry a variety of key blanks to meet the continuing needs of your customers. More important, you must determine which blanks are most popular in your area. Purchasing one or two boxes of every key blank that you might use during your career in locksmithing is ridiculous and extremely costly. Purchasing such items randomly shows poor business judgment. Large companies that supply all kinds of locks and security devices will, naturally, have the widest variety of key blanks available for your purchase and use. That assortment doesn't mean that you really need such an extensive variety of key blanks. In almost every case, you do not.

To decide what you should have, examine the key determination process from the viewpoint of a key blank manufacturer. The following information comes from Herbert Stein, President of Star Key & Lock Manufacturing, Co., Inc., of New York. For many years, this company has been in the business of carefully foreseeing the key blank needs of locksmiths.

The first step is to consult a number of the largest competitors in different parts of the country for their opinions of the relative popularity of a particular blank and the anticipated annual demand. (Some blanks may be extremely popular in certain areas of the country, but very slow in other parts.) Then, consider such factors as whether the blank's lock is made by a major company or not and whether it is a big or small volume lock (for example, if the blank is for an automobile, is it for the major line of one of the top domestic or foreign automobile makers, or is it just for one of the small volume imported cars?). Naturally, if a deluge of requests for a specific type of blank comes in from across the nation, it will gain priority on the production list.

Knowing how a key blank is selected for production helps, but how do you find out what the most common blanks are in your area? Here is where your regional representative of the manufacturer, or direct contact with the factory, can help. Knowing what keys are most used in your area is part of their job. Through various national and regional trends, regional and area market studies, and previous sales to your area, the manufacturer can determine with great accuracy what the most common and popularly requested blanks will be for any given part of the country.

While manufacturers keep such records, you might feel that you are imposing on them with a "trifling request." Not so, but even then, some locksmiths just look through a catalog, especially when starting out in a business, and contact a local locksmith distributor/supply house in the area. The supply house can assist you, but many sales of key blanks don't go through the supply house; they are purchased from other area distributors or directly from the manufacturer. The supply house, while in touch with your area, isn't always as fully cognizant as the manufacturer in this regard.

At the supply house you will probably be shown one or several catalogs with a myriad of key blanks indicating which ones that you should have on your shelf. These may or may not meet your requirements. Again, Star Key & Lock has thought of this; their catalog lists the most commonly used key blanks for just about every situation. This is where it is to your advantage to deal directly with the manufacturer. By purchasing assortments of different types, you reduce your initial key blank costs, reduce your stockroom inventory, have more space for other products, and know exactly what your customers will expect. In addition, you don't have several hundred or more blanks to look through for any given type, but perhaps only 150 or so.

By purchasing an assortment, you have the advantage of getting the keys in display mounts that provide a subtle buying hint to the customers who walk into your shop. These displays hold the most popular key blanks (usually between 90 and 150 different ones) that will meet just about all of your needs. The few oddball keys that come in for duplication are also the ones that give every locksmith a headache, because it may be only once every two to five years that a key of that specific type is needed.

A little footwork at your end helps you assess the needs of your shop. Check out the local general supply houses and see what blanks are the most popular with them. Ask them about the most commonly purchased locksets in your area and get to know these locks, the different keyways possible and available for them, and also what the big sellers were within the past ten years—not just the big sellers in today's market.

From all of this information, you should have learned:

- The most popular key blanks in your area
- The volume of blanks used in the past several years for different types of locks
- The top dozen locks, by manufacturer, selling currently in your area
- The top dozen locks, again by manufacturer, that were big sellers during the past decade
- The various keyways necessary for these locksets

This knowledge (coupled with the size of your shop and projected business) gives you a very good idea of which key blanks will be the most in demand, which ones will require a medium stock, and which ones you have to consider only once every two or three years. The minimal stock blanks will last for a long time, but the more popular ones may need frequent restocking—possibly every month.

After you complete your research, you can properly select the blanks you feel are necessary. Remember that it is probably best to consider the general assortment of key blanks when first starting out. Get the catalogs that provide these blanks and consider their assortments. Later on, as your business expands, you can move on to include other key blank manufacturers for less common blanks that you see a need to stock.

After completing your key blank research, you should also know the types of locks you require, as well as exactly what types of key reference catalogs for key codes your shop should carry. You can start listing the specific cylinders, pin sizes, and dimensions required, specific repair parts kits, key machines that may be necessary, and some other specifics that will be necessary to perform as a locksmith.

"Instant" blank identification. The professional locksmith knows his or her key blanks and doesn't need to constantly refer to charts and catalogs for this information. In the beginning, every locksmith has trouble identifying one key among hundreds that may be on the shop wall.

Star Key & Lock has come up with a nifty idea that is extremely helpful (and something that I wish that all manufacturers of key blanks would do): they put an "instant" identity on the key blank. This new feature has put Star far out in front of other manufacturers. What it amounts to is stamping the make of the lock on the key blanks in addition to the comparative numbers. This achieves immediate identification of the key blank by the locksmith, saving time and trouble, and is a valuable learning tool for student and apprentice locksmiths. The make of lock stamped directly on the key also helps convince the customer that he is receiving the correct key for his lock, which aids in establishing the locksmith's credibility.

Embossing. Embossing is an added feature that you can supply to customers who require something special or an identifying logo. This means imprinting the bow of the key blank with anything from a fancy logo design to several lines of advertising copy. Imprinting can be done on any number of the brass keys available from Star. This is not a common or standard service, though some other manufacturers offer it as well.

Special embossing usually consists of three lines or less of standard type copy. This is based on a maximum of 14 figures on the first line, 12 on the second line, and 9 on the third line. The manufacturer must make back the costs of embossing; as a minimum, consider an order of 1000 embossed keys of assorted key blank numbers in increments of at least 50 pieces per number ordered.

Embossing is an excellent way to advertise your business. It costs less than average advertising and still gets your message across to the public that needs and uses your products and services (Fig. 20.7). The chief advantages of special embossing are excellent advertising, increased locksmith shop prestige, and indisputable proof that the keys were made by the name embossed thereon.

Key materials. Most keys in use today are made of brass. This is important. All-brass key blanks are best for impressioning because they have structure which shows distinct impression marks. The blanks cut very easily on key machines or by hand filing, but don't believe the myth that brass is soft; it's not true. The all-brass key retains maximum strength (77,000 p.s.i.); the special brass alloy which is used by Star Key & Locks (and other

Figure 20.7 Excellent advertising increases the locksmith's prestige and provides evidence that the keys were made by you. (*Star Key & Lock Co.*)

manufacturers) gives the best combination of desirable properties of impression-sensitive structure, free-cutting duplication, and the maximum strength necessary for a key blank.

Both as a locksmith and as an individual who requires keys in your own day-to-day movement (home, auto, office), you will see through experience that certain key blanks are better than others. Use the products that are of the best quality, not those that cost the least

Key cutting machines

Figure 20.8 shows a key machine for the office or for use on the job in remote locations. This compact, highly dependable, cylinder key cutter, the Companion, will go wherever your locksmithing work may be. It has dual voltage (110 Vac or 12 Vdc) capability with no extra wiring necessary. It comes in a sturdy carrying case and weighs only 25 pounds.

The machine has four-way jaws to grip any cylinder key; no special adapters are required. The precision alignment of the cutter ensures a correctly duplicated key every time. (Of course, check the cutter alignment periodically to ensure that the precision alignment is true. Rough handling and occasional bumps may throw the alignment out slightly. Be safe, be sure, and cut a true duplicate key every time.)

The combined key display and machine is a visual selling aid in your locksmith shop. Usually located near the counter or cash register, it promotes sales subtly by showing your customers that you have the commonly used keys in stock and can duplicate a key for them while they wait. In the case of the display in Fig. 20.9, it is important that you know exactly what is available. This particular key assortment and cutting machine is a self-standing, revolving counter rack and covers both U.S. and foreign automotive keys, in addition to keys for business and residential locks.

The jaws on this key machine can be rotated to present four different gripping surfaces to cut four different types of keys without using any adapters. The jaws simply lift up and rotate, eliminating the fumbling necessary with adapters for the key machine. The *standard* station of the cutter jaws holds

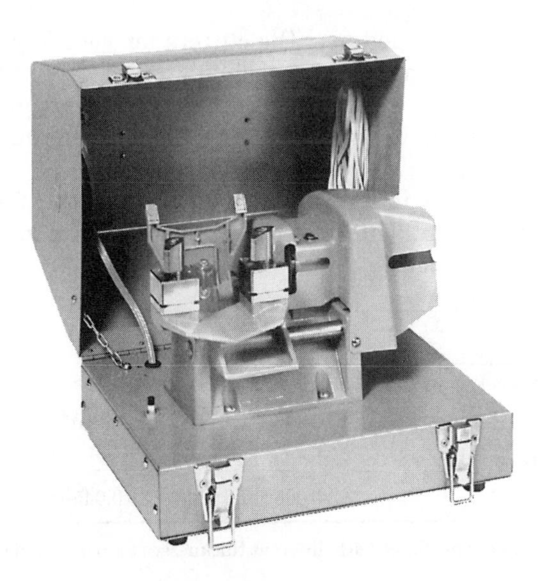

Figure 20.8 The "Companion" model portable key machine operates off 110 Vac or 12 Vdc current. (*Dominion Lock Co.*)

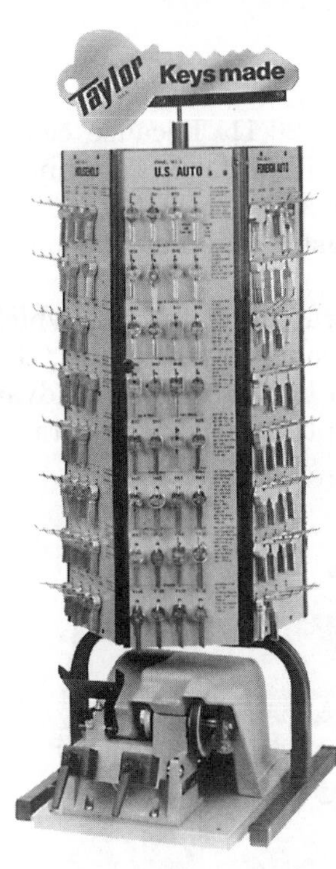

Figure 20.9 Key machine with common key blanks for quick duplication. (*Taylor Lock Company*)

regular cylinder keys, while the *wide* jaws grip Ford double-sided keys. The *A* station jaws hold the Schlage wafer key (SC 6) and the *W* station jaws are for the Schlage wafer key (SC22). Other keys, such as Chicago and double-sided Datsun keys, can be cut with the *A* station jaws.

Key machine cutters. After continued use, the cutters for your key machine(s) will become dull. This, in turn, means that each key you cut will not have the precision cuts required. Any such key may operate in the lock roughly or not at all. To alleviate this, you should have at least one replacement cutter of each type you normally use on hand.

The basic cutters are shown in Fig. 20.10 and include the file, slotting, milling, and side milling slotter. The cutter specifications and key type uses are as follows:

Material	Use	Cutter Type
File	High-carbon steel	Cylinder keys
Slotting	High-speed steel	Flat keys with 0.045-inch cut*
Milling	Tungsten-chrome alloy steel	Cylinder keys
Side Milling	High-speed steel	Cylinder and flat keys (0.045-inch slotter cut)

*Two different slotting cutters are available with different thicknesses for 0.030- and 0.045-inch cuts on flat steel keys.

Displays

The color key carousel is shown in Fig. 20.11. Because colored keys are becoming more popular, some customers will prefer to have another color key when having their key duplicated. Color keys assist the customer in rapid identification of specific keys. In some instances, they will not accept a key that is not colored.

A key comparison board allows you to rapidly determine which blanks, by manufacturer, are the same in their component makeup (i.e., the specifics of the key identification). A key comparison board is a great advantage to the locksmith and even more so for the apprentice or locksmith trainee.

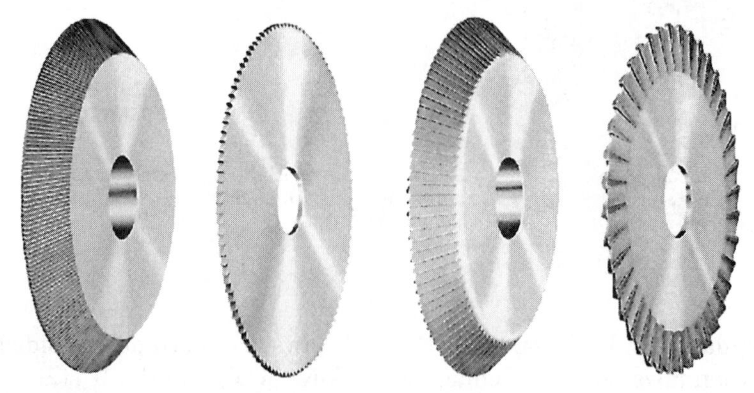

Figure 20.10 Basic replacement key machine cutters, as well as a file, slotting, milling, and a side-milling slotter. (*Taylor Lock Company*)

Figure 20.11 Color key carousel. (*Star Key & Lock Co.*)

In the case of these two items (as with others that you find in various catalogs), the maker isn't just hyping a product in the hope that you will buy it; the manufacturer genuinely wants to provide you, a locksmith, with items that will assist you in the locksmithing trade. By helping you to do your job a little better, easier, faster, and more professionally, and by making available the supplementary business items required, the manufacturer is helping everyone: you, your customer, and itself. When you have problems, questions, or need assistance, the manufacturer or its representative is there to help you. Likewise, the manufacturer is looking ahead to your varying needs and has a variety of items to help you to help yourself.

Another example of a manufacturer's concern for assisting the locksmithing business is shown in Fig. 20.12. Lock displays like this unit are available for setting up and demonstrating specific products within your shop. The customer is interested in a specific type of product—in this case, a lock. With a variety of these displays available, the customer can try out each one and, as the choice of locks narrows, you provide specifics concerning the remaining locks. In this particular case, you might point out the very salient and positive features of the Ultra 800, the Medeco cylinder, the strength and positive locking action of the mechanism, and other specifics.

For narrow stile door installation, Adams Rite has created a specific installation kit. The advantages of this kit should be obvious: extra parts and detailed information for each stile installed. For the locksmith who plans continued installation, repair, and/or replacement of lock units in narrow stile doors, this kit is a must. It allows for a smoother, more rapid installation of numerous units, saving a tremendous amount of time (Fig. 20.13).

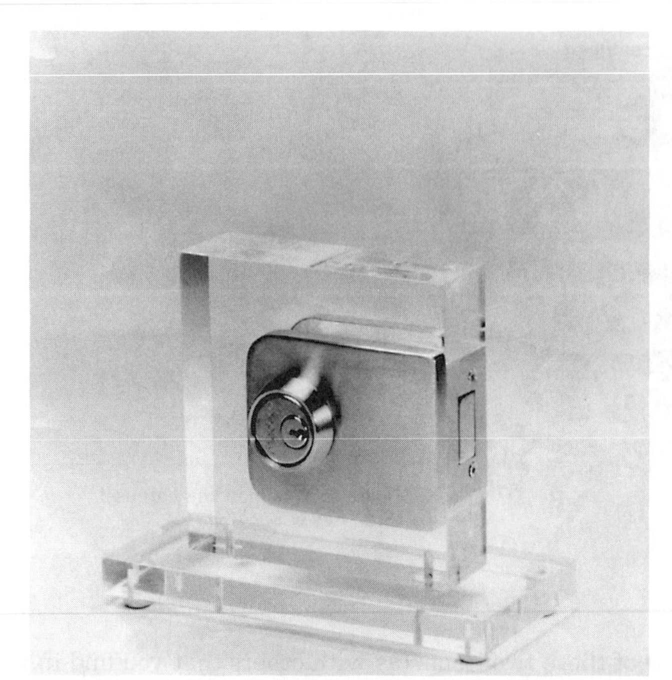

Figure 20.12 Manufacturer-prepared lock display for customer consideration. Every shop should have a variety of such displays for customers to view and touch. (*M.A.G. Engineering and Mfg., Inc.*)

Figure 20.13 Installation kit for the Adams Rite narrow stile door units. (*Adams Rite Mfg. Co.*)

The Locksmith and the Law

As a locksmith, you must have a better understanding of some laws than most people. When setting up your business, it is prudent to consult your attorney regarding all laws that concern your profession. In many jurisdictions there are laws covering licensing, control of locksmithing tools, and the registration of code books. Some local laws regulate the conduct of certain locksmithing business practices, such as duplicating master keys, making bank deposit box keys, opening automobiles, carrying locksmithing tools, etc.

When first entering the business, aside from consulting with your lawyer and possibly the police, you should also contact your area locksmithing association, if one exists. You can also obtain information from national locksmithing organizations through their newsletters and publications. These are excellent sources of information about your legal responsibilities.

Your legal responsibilities demand that you be very careful about the jobs you accept. You must be certain that you are not breaking the law by complying with a customer's wishes. Authorization statements from supervisors, such as from a bank or post office, for duplicating a key should always be double-checked with the main offices. Verification of such written statements is a must; they should also be filed with the job order.

When jobs of this type arise, the following information should be included in your files:

- Name of the person who brought the job in
- Identification (Social Security card or driver's license)
- Address of the person who brought in the job
- Business telephone (call to double-check)
- Name of the firm
- Business address
- Type of service performed
- Type of payment (if a check, it should be a business check, not a personal check)
- Automobile make, model, serial number, license number, state in which it is registered

You may also have forms printed up for your customer to fill in. You can use a single form for all of the situations you run into. Keep the form with the work order.

In some states and cities, laws require that you take an examination administered by your locksmithing peers to show that you are a competent locksmith who meets professional standards. Besides this, the police may check your reputation, qualifications, background, and previous employers to ascertain that you are of good moral character.

Various laws have been enacted both to protect the public from unscrupulous individuals posing as locksmiths and to protect the locksmithing trade from such individuals. The following is extracted from the Los Angeles Code, Ordinances No. 83,128, as an example of a local law regulating the locksmithing profession:

Sec 27.11 Locksmith—Regulating Applicable To

A. Definitions
"Locksmith" shall mean any person whose trade or occupation, in whole or in part, is the making or fashioning of keys for locks, or similar devices, or who constructs, reconstructs or repairs or adjusts locks, or who opens or closes locks for others by mechanical means other than with the regular keys furnished for the purpose by the manufacturers of the locks.

B. Trade of locksmith—permit required
No person shall engage in the business of locksmith, or practice or follow the trade or occupation of locksmith, without a permit therefore from the Board of Police Commissioners.

C. Permit—Application
Such permit shall be issued only on the verified application of the individual seeking the permit. The application shall be on a form prescribed by the Board and shall set forth the proposed location of the applicant's place of business, the names and addresses of five character references, and such other things as the Board may require to determine the character, honesty, and trustworthiness of the applicant. Specimen fingerprints of the applicant shall be furnished with the application.

D. Permits—fees—expiration
Each application for a permit shall be accompanied by a fee of $10 and each application for the annual renewal thereof, by a fee of $5. Each permit, unless sooner revoked, shall expire on December 31st, following the date of issuance. Each permit shall bear a serial number.

E. Permits—issuance and denial
The Board shall cause an investigation to be made on each application, and if the Board finds that the applicant's reputation for honesty is good, that he has not used his skill or knowledge as a locksmith to commit or aid in the commission of burglaries, larcenies, thefts, or other crimes, that he intends honestly and fairly to practice the trade of locksmith in a lawful manner, and that he has not been convicted of a felony, then the permit shall issue. Otherwise it shall be denied.

F. Permittee—must keep record
Each permittee must keep a book, which shall be open to inspection by any police officer at all times, in which the following must be entered:
1. The name and address of every person for whom a key is made by code or number.
2. The name and address of every person for whom a locked automobile, building, structure, house, or store, whether vacant or occupied, is opened, or a key fitted thereto.

G. Keys to be stamped
It shall be unlawful for any locksmith to fail to stamp the serial number of his permit on any key made, repaired, sold, or given away by him.

H. Signs to be displayed

Every locksmith shall display in a conspicuous manner in the place where he is carrying on such business, trade, or occupation, a sign of a style, size, and color to be prescribed by said Board, reading, "Licensed Locksmith," together with the official permit number.

I. Permits—revocation

The Board, on proceedings had as prescribed in Section 22.02, may revoke or suspend any permit issued hereunder on any of the following grounds:

1. Misrepresentation in obtaining such permit.
2. Violation of any provision of this section.
3. That the permittee has committed or aided in the commission of or in the preparation for the commission of any crime by the use of his skill or knowledge as a locksmith or by using or letting the use of his tools, equipment, facilities, or supplies.

Note: To learn more about locksmithing laws, ethics, and business matters, refer to my book, *Locksmithing*, McGraw-Hill (Craftmaster Series).

Chapter

21

Key Duplicating Machines

One of the most important investments you'll make in locksmithing equipment is a key duplicating machine. A wide variety of them are available. Prices range from a few hundred dollars to several thousand dollars. Figure 21.1 shows a few key duplicating machines.

A wide difference in price doesn't necessarily indicate a substantial difference in functions and quality between key duplicating machines. This chapter will help you better understand how the machines work and how to find the best one you can afford.

A key duplicating machine consists of four basic parts:

- Two vises move in unison. One vise holds the key being duplicated; the other holds the key blank.
- A key guide traces the profile of the key.
- A cutter wheel cuts the key blank in accordance with the profile traced by the key guide.

Most inexpensive key duplicating machines are designed to duplicate only cylinder keys, but a few inexpensive models can also duplicate flat keys. More sophisticated models duplicate other types of keys. The most expensive ones can duplicate bit keys, tubular lock keys, angle keys (such as Medeco), and dimple keys. You can save money by purchasing a machine that duplicates only the types of keys you're most interested in duplicating.

Critical Design Factors

When evaluating a key duplicating machine, look carefully at the pivot mechanism. The key should meet the cutter wheel squarely on a dead parallel with

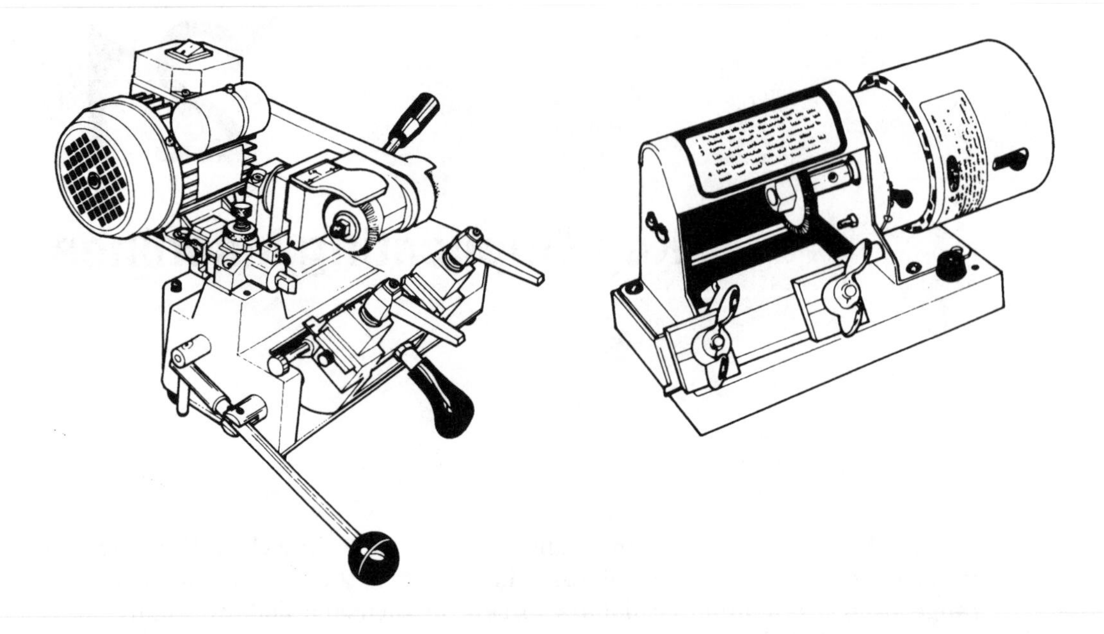

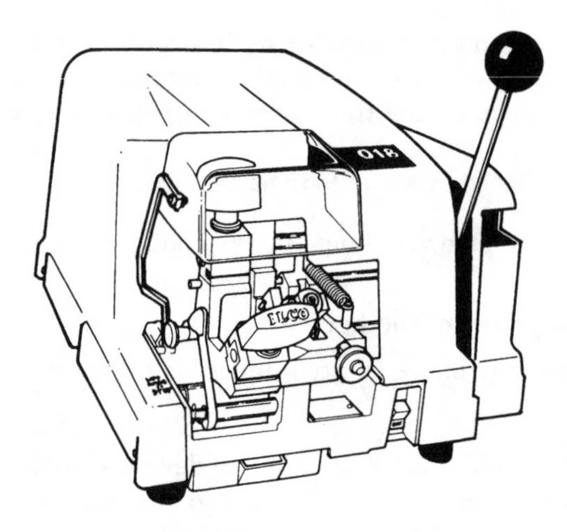

Figure 21.1 Key cutting machines. *(Ilco Unican Corp.)*

the axis of the wheel. A slight angle is enough to upset the dimension of the duplicated key (Fig. 21.2).

Some machines don't have pivots and arrange matters so the vises move laterally into the cutter. As long as the bearings are true, this ensures that the duplicated key is a faithful copy of the original.

The jaws of the vises must be carefully engineered to ensure that both the key and key blank do not shift during the duplication process and that the key and key blank are held squarely against the cutter.

Cutters

There are three basic types of cutters. *Filing* and *milling* cutters are both used to duplicate cylinder keys. A *slotter* cutter is used for square-ended keys (such as bit keys and flat keys). Figure 21.3 shows some examples.

The diameter of the cutter is important. Large-diameter cutters leave a very slight concave on the newly cut key. Smaller cutters make a deeper concave. The most useful key duplicating machines allow you to use various sizes and types of cutters.

Key guide

The key guide should be checked regularly. In the absence of manufacturers' instructions for doing so, check the guide by mounting two identical key blanks in the vises; then, lift the vises up to the key guide and cutter wheel. While keeping the vises in place, slowly rotate the cutter wheel. The cutter wheel should barely scrape one of the key blanks. If the cutter wheel doesn't

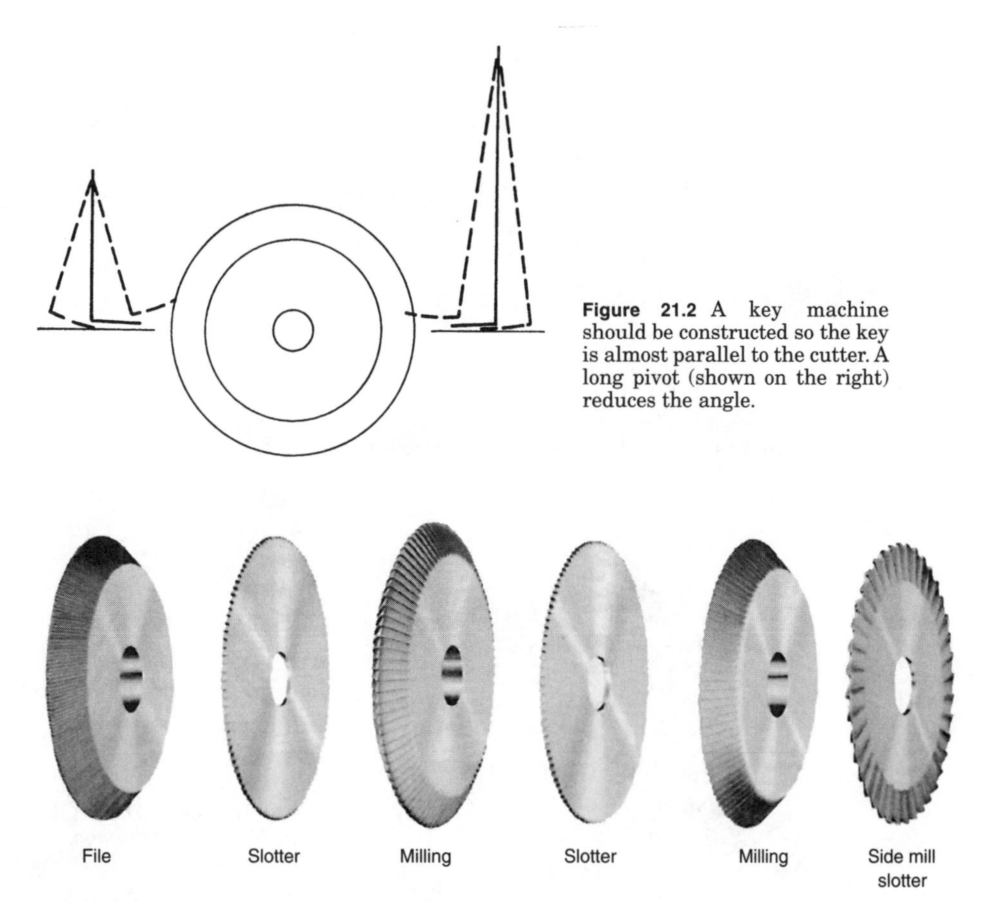

Figure 21.2 A key machine should be constructed so the key is almost parallel to the cutter. A long pivot (shown on the right) reduces the angle.

Figure 21.3 Various cutter wheels.

touch a key blank, or if it digs deeply into a key blank, the machine is out of alignment and needs to be adjusted. Figure 21.4 illustrates how to check a key guide.

Framon's DBM-1 Flat Key Machine

Framon Manufacturing Company's DBM-1 is a high-quality, flat-key duplicating machine. It's designed to duplicate a wide variety of flat and corrugated keys (Fig. 21.5). By understanding how it works, you'll be able to operate many similar machines.

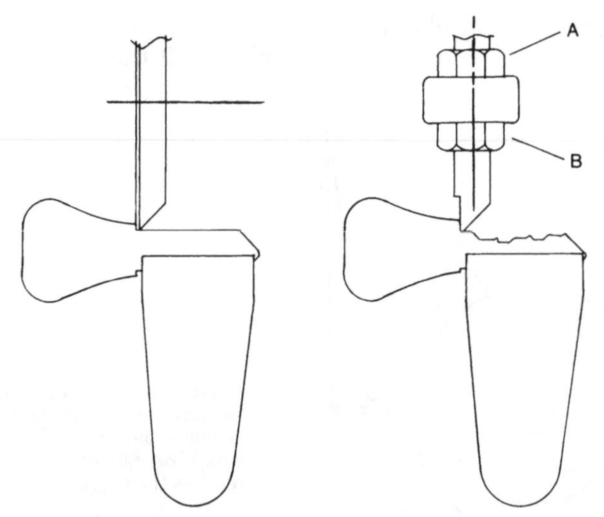

Figure 21.4 The cutter wheel should barely touch the key blank (left). (*Keynote Engineering*)

Figure 21.5 Framon's DBM-1 Flat Key Machine. (*Framon Manufacturing Corp.*)

Cutting procedure

All keys should be set from the tip for spacing. Insert the pattern key in the right-hand vise with the tip of the blank protruding slightly beyond the left side of the vise (Figure 21.6). This position allows cutting of the tip guide on the blank if the blank tip is slightly different from the pattern key. In this position, the tip can be cut without the cutting vise.

Push the guide shaft to the rear and lock it into this position by tightening the locking knob (Fig. 21.7). This relieves spring pressure so the tip setting is easier. Lift the yoke and set the tip of the pattern key against the right side of the guide. While holding this position, insert the blank and set the tip of the blank against the right side of the cutter (Figure 21.8). This procedure assures proper spacing.

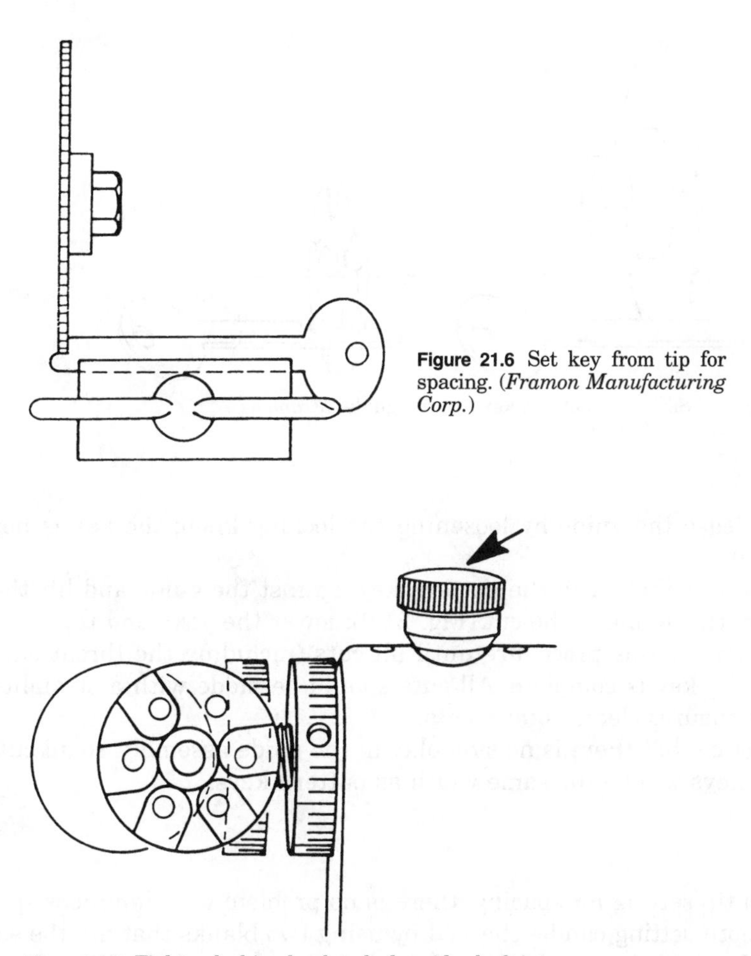

Figure 21.6 Set key from tip for spacing. (*Framon Manufacturing Corp.*)

Figure 21.7 Tighten locking knob to lock guide shaft into position. (*Framon Manufacturing Corp.*)

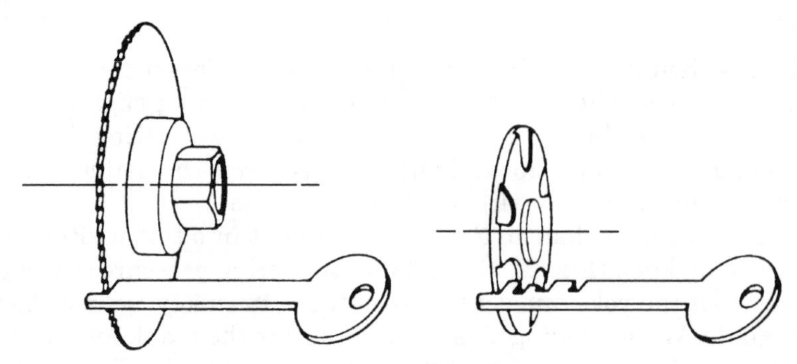

Figure 21.8 Set tip of blank against right-hand side of cutter. (*Framon Manufacturing Corp.*)

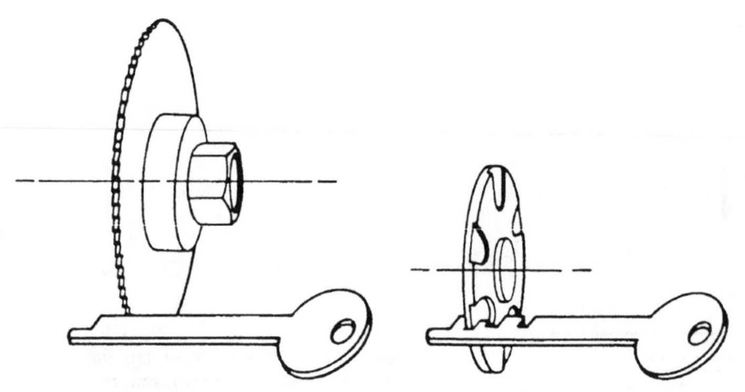

Figure 21.9 Set cut in pattern key against guide. (*Framon Manufacturing Corp.*)

Release the guide by loosening the locking knob; the key is now ready to be cut.

Look at the cut in the pattern key against the guide and lift the yoke into the cutter to make the cut (Fig. 21.9); lower the yoke and repeat for the next cut. Follow this procedure until all cuts (including the throat cut) are made and the key is complete. All cuts should be made with a straight-in motion. This ensures clean, square cuts.

Notice that there is no side play in the guide assembly, so all cuts on duplicate keys will be the same width as pattern keys.

Adjustments

With tip setting for spacing, there is no problem with improper spacing.

Depth setting can be checked by using two blanks that are the same. Check the depth by drawing the blank against the cutter guide. If depth adjustments are needed, simply loosen the set screw on the depth ring and adjust the ring until the cutter barely touches the blank. To make cuts deeper, rotate the depth ring counterclockwise. To make cuts shallower, rotate the depth ring clockwise.

Tighten the set screw after the adjustment is made, but do not tighten it too far.

Another way to check depth is to make one cut on the duplicate key; check the depth cut on both the pattern key and duplicate. As an example, if the cut on the duplicate key is 0.003 deeper than the cut on the pattern key, loosen the set screw on the depth ring, rotate the ring 0.003 clockwise, and tighten the set screw. *Note:* Calibrations on the depth ring are in increments of 0.001.

Check the guide setting; the guide must be set to the same width of the cutter used. The DBM-1 is supplied with one 0.045-width cutter. This is the best width for general work. Cutter widths of 0.035, 0.055, 0.066, and 0.088 are available if needed. Cutter width of 0.100 (LeFebure) can be obtained by using a 0.045 and a 0.055 at the same time. All of these cutters are solid carbide.

To set the guide, simply loosen the cap screw (DBM-42). Rotate the guide to cutter width and tighten the cap screw. The Detent screw in the guide shaft will align the guide. No adjustment is required when changing guide settings.

Maintenance

Lubricate the yoke rod (DBM-09), guide shaft (DBM-09), and vise studs (DBM-18) sparingly using very fine oil. (*Do not use motor oil.*)

Wipe off all excess oil. To lubricate the guide shaft, unscrew the locking knob (DBM-21) and put one or two drops of oil in the opening. Replace the knob.

Other than these parts, cleanliness is the best maintenance.

Ilco's KD50A

Ilco Unican Corporations' KD50A is a high-quality machine for duplicating cylinder keys. By understanding how it works, you'll know the basics of operating most other cylinder key duplicating machines. The following information about the KD50A came from an instruction manual published by the manufacturer.

General operating sequence

The KD50A has a constant power switch that must be turned on. However, the machine motor will not operate until activated by the carriage assembly (Fig. 21.10).

After both key and blank are properly clamped and aligned, pull down on the carriage handle. Use your thumb to depress the carriage release knob; the key setting gauge will automatically spring away. Spring tension will raise the carriage and the motor will automatically start.

Move the lever handle sideways so that the original key touches the key guide in an area between the shoulder and the first cut. Do not let the shoulders touch either the key guide or cutter wheel. Using the lever handle, slide

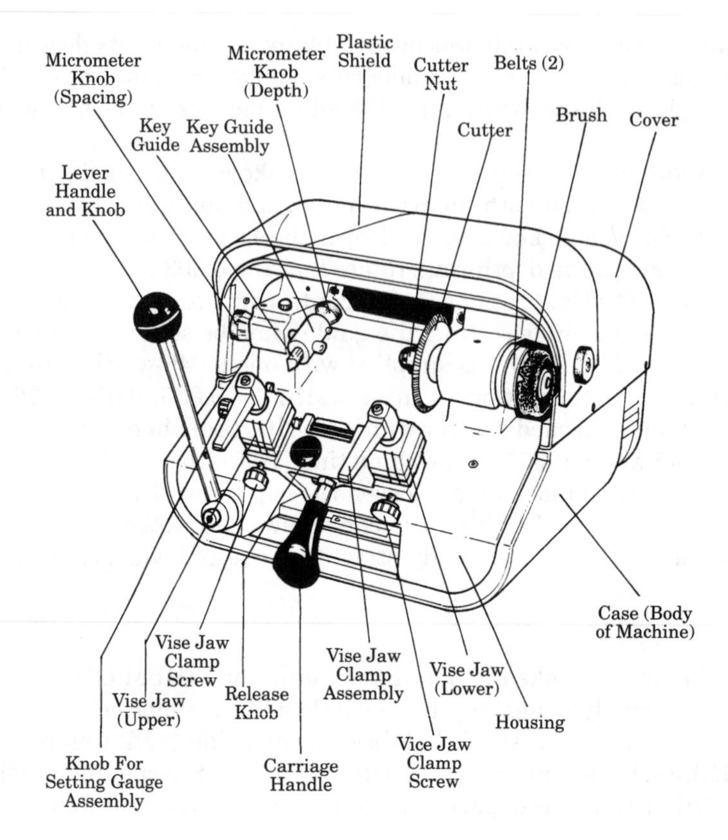

Figure 21.10 Ilco Unican's KD50A Key Duplicating Machine. (*Ilco Unican Corp.*)

the carriage left and then right to complete the cutting operation. Lower the carriage until it locks into the original position, which will automatically stop the motor and the cutter. Remove the new key and deburr with the brush; do not overbrush or run the key into belts.

Adjusting for proper depth of cut

For safety, remove the wire plug from its electrical socket. Clamp the two service bars into the vise jaws as shown in Fig. 21.11, making certain that both bars rest flat against the bottom of the vise and that they are butting against the edge of each vise jaw. Lift the carriage toward the key guide and cutter until a flat portion of the service bar rests against the key guide. (To lift the KD50A carriage, pull down and press the carriage release button between the vise jaws.)

Turn the cutter by hand. The machine is correctly adjusted if the cutter barely grazes the top of the right service bar. If the cutter is stopped from turning or

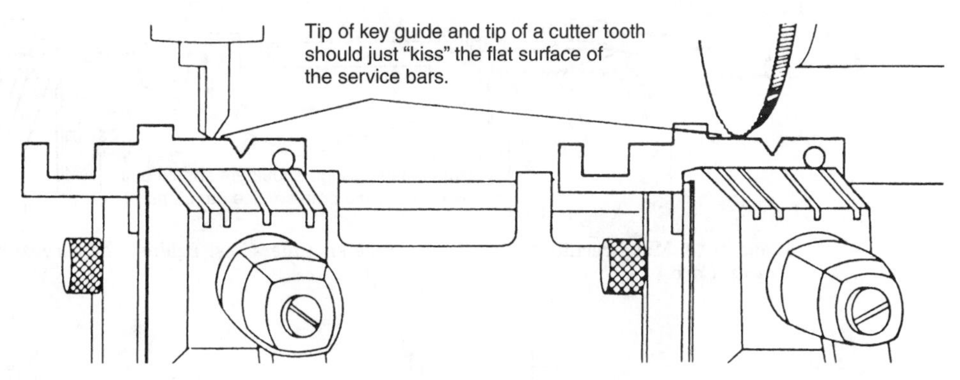

Tip of key guide and tip of a cutter tooth
should just "kiss" the flat surface of
the service bars.

Figure 21.11 Clamp the two service bars into the vise jaws. (*Ilco Unican Corp.*)

turns freely without contacting the service bar, the cutting depth must be adjusted, as follows:

Loosen the Allen screw that holds the key guide.

Turn the cutting depth micrometer adjusting knob behind the guide, either left or right. This moves the key guide in or out. Do this until the cutter just grazes the top of the right service bar when the left service bar is resting against the key guide. Turn the cutter by hand; adjust to the high spot of the cutter.

Tighten the key guide Allen screw.

Note: This adjustment must be made if the cutter is replaced or whenever a test fails to work, indicating that the cutter may have worn down somewhat and resulting in cuts that are too shallow.

Adjusting for proper lateral distance (spacing)

Key cutting accuracy also depends on spacing the key and blank key the same as the distance between the key guide and cutter. To assure that the lateral distance adjustment is correct, refer to Figs. 21.11 to 21.14 and proceed as follows.

Insert the service bars into the vise jaws, making sure that each service bar is butting against the edge of each vise jaw. *This is critical.*

Rotate the key setting gauge up, and make certain that both setting gauge shoulders rest *exactly* against the service bar stops as shown in Fig. 21.12. If there is a discrepancy, loosen the right setting gauge Allen screw and adjust so that both gauge shoulders rest exactly against both service bar stops.

Lift the carriage to the key guide and cutter. Insert the key guide and cutter into the V-shaped grooves in the service bars as shown in Fig. 21.13. Both the key guide and the tip of a cutter wheel tooth must fit exactly into their V grooves or the setting will not be accurate (make certain that you do not seat the space between two cutter teeth into the V groove).

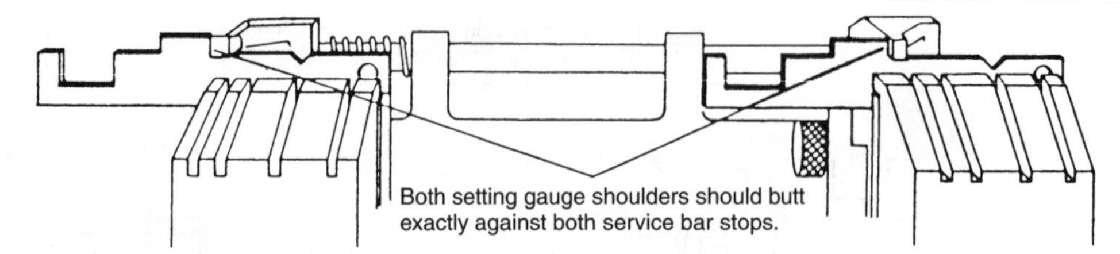

Figure 21.12 Make certain both setting gauge shoulders rest against the service bar stops. (*Ilco Unican Corp.*)

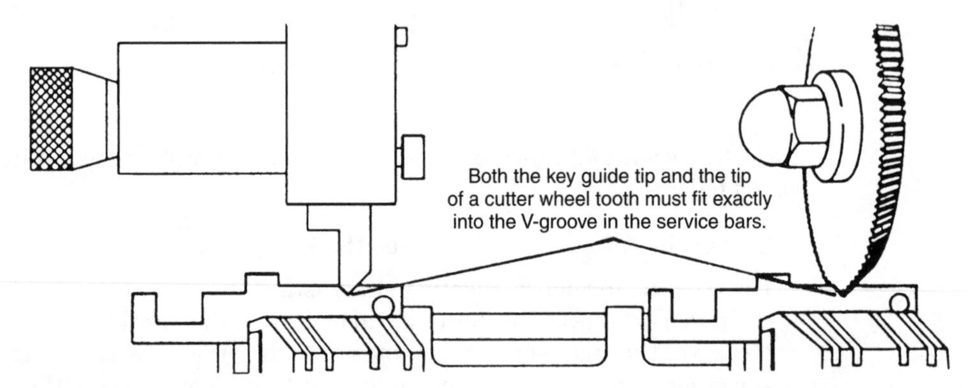

Figure 21.13 Insert key guide and cutter into the V-shaped grooves in the service bars. (*Ilco Unican Corp.*)

If the guide and cutter do not seat exactly into the V grooves, the distance between the cutter and guide must be altered. Loosen the Allen screw in the key guide assembly, and turn the micrometer adjusting knob fore or aft. This action shifts the position of the key guide assembly to the left or right. Continue until the key guide and the cutter both drop into the V notches of the service bars.

Insert the pattern key, left to right, into the left vise. Rotate the key setting gauge up and set its left shoulder against the shoulder of the pattern key. Be sure the key is lying flat along the bottom of the vise. Secure the key by turning the clamp assembly clockwise.

Insert the key blank in the same manner into the right vise and secure. Make sure that the key setting gauge is exactly against both key shoulders. The key and key blank now are spaced the correct distance apart, and they are ready for cutting (Fig. 21.14).

Aligning keys without shoulder (Ford and Best)

On keys without shoulders, the key setting gauge cannot be used. It is necessary to use the service bar to correctly position the key and the blank.

The vise jaws have a series of slots that can be used for the service bar. Also, note the key head rest (KD50A-60), which prevents the key from tilting as the vise jaw tightens. The key head rest can be moved to properly support the key (Fig. 21.15).

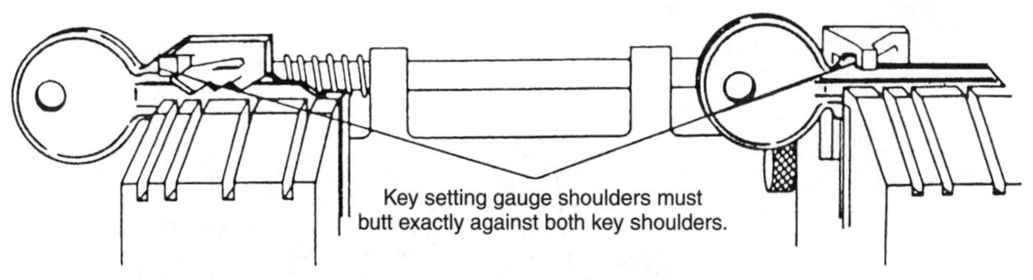
Key setting gauge shoulders must
butt exactly against both key shoulders.

Figure 21.14 The key and key blank are now spaced the correct distance apart for cutting. (*Ilco Unican Corp.*)

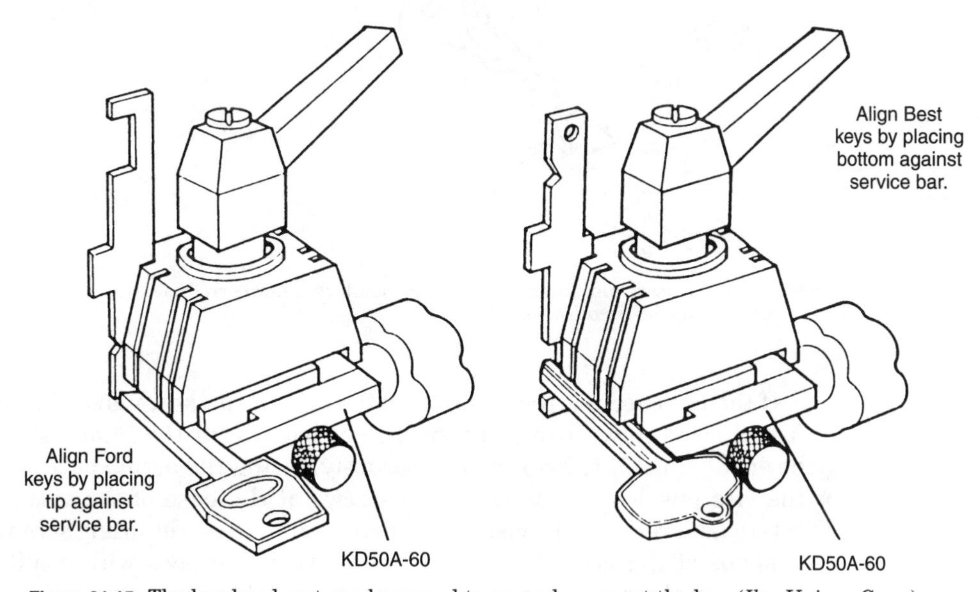
Align Best
keys by placing
bottom against
service bar.
Align Ford
keys by placing
tip against
service bar.
KD50A-60
KD50A-60

Figure 21.15 The key head rest can be moved to properly support the key. (*Ilco Unican Corp.*)

Aligning narrow blade cylinder keys

Some keys have a very narrow blade; therefore they sit deep in the vise jaws with only part of the cuts showing above the vise. This makes it necessary to use the service pins to raise the key for proper cutting.

Insert an equal-size pin under each key and blank on the bottom of the vise jaws. This raises both the key and blank to make the correct depth of cut (Fig. 21.16). Do not cut into the vise jaw.

Aligning double-sided cylinder keys

Before cutting this style of key, examine the key to see if there is a milled groove on either side. If so, reverse the vise jaw and clamp the key using the V jaws. The key will be held securely when only the top or bottom V jaw fits into a milled groove. When there is no V groove on either side of the key, use the flat vise jaw.

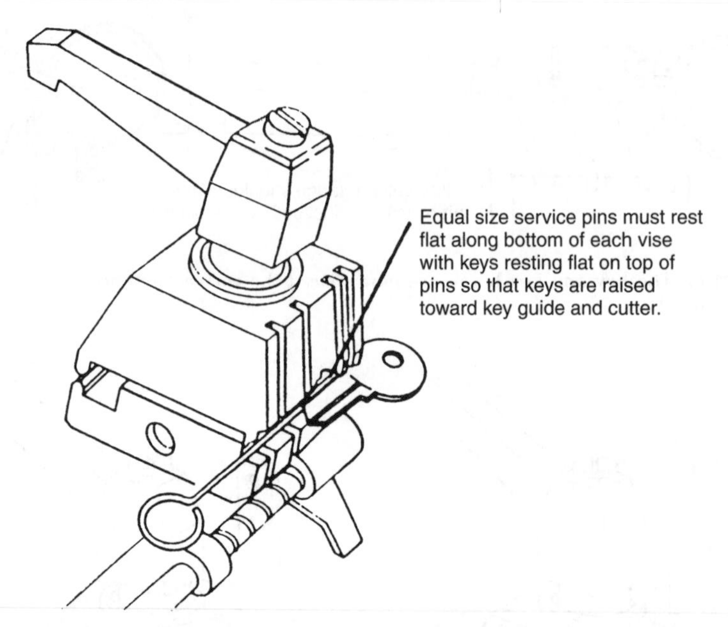

Equal size service pins must rest
flat along bottom of each vise
with keys resting flat on top of
pins so that keys are raised
toward key guide and cutter.

Figure 21.16 Insert pin under key and key blank to allow correct depth of cut to be made. (*Ilco Unican Corp.*)

If the cuts are not the same on both sides of the key, make the shallow cuts first. In this way, when you turn the key over to cut the second side, there will be enough metal to grip the key securely during the actual cutting. To reverse the vise jaw, loosen the retaining screws at the base of the vise jaws. Raise, rotate, and reseat both vise jaws; then tighten their retaining screws. Note the V shape of the jaws. Insert the key between the jaws with a milling groove resting in the point of the V. This holds the blank securely. Align for spacing and proceed out.

Aligning carriage to prevent vise jaw damage

This machine is equipped with a carriage stop that prevents the carriage from moving all the way up to the cutter. When properly adjusted, it stops the cutter from grinding into the vise jaw. Such a condition could occur when reaching the tip of the cut key and the carriage lever continues to move the carriage.

The carriage stop (Part No. KD50A-144) is a U-shaped channel secured to the housing by set screws. It is possible to span the travel of the carriage during the cutting cycle; normally, this position does not change. In addition, there is a carriage stop adjusting screw that is installed in the carriage; this screw controls the distance between the cutter and vise jaw (Fig. 21.17).

The carriage stop adjusting screw is set at the factory to create a clearance of –0.005 inch between the cutter and the vise jaw (Fig. 21.18). This distance is not critical and can be set without measuring instruments. Just loosen the

Chapter

22

Test Your Knowledge

This test is based on the International Association of Home Safety and Security Professionals' Registered Professional Locksmith registration program. If you earn a passing score, you should be able to pass other locksmithing certification and licensing examinations. To receive a Registered Professional Locksmith certificate, see the information after the test.

1. An otoscope can be helpful for reading disc tumbler locks by providing light and magnification.
   a. True
   b. False
2. Kwikset locks come with a KW1 keyway.
   a. True
   b. False
3. Many Schlage locks come with an SCL1 keyway.
   a. True
   b. False
4. The purpose of direct (uncoded) codes on locks is to obscure the lock's bitting numbers.
   a. True
   b. False
5. A skeleton key can be used to open warded bit-key locks.
   a. True
   b. False
6. Typically, the lock on a car's driver side will be harder to pick open than other less often-used locks on the car.
   a. True
   b. False

7. A standard electromagnetic lock includes a rectangular electromagnet and a rectangular wood and glass strike plate.
   a. True
   b. False
8. A blank is a key that fits two or more locks.
   a. True
   b. False
9. One difference between a bit key and barrel key is that the barrel key has a hollow shank.
   a. True
   b. False
10. Parts of a flat key include the bow, blade, and throat cut.
    a. True
    b. False
11. The Egyptians are credited with inventing the first lock to be based on the locking principle of today's pin tumbler lock.
    a. True
    b. False
12. Before impressioning a pin tumbler cylinder, it's usually helpful to lubricate the pin chambers thoroughly.
    a. True
    b. False
13. When you are picking a pin tumbler cylinder, spraying a little lubrication into the keyway may helpful.
    a. True
    b. False
14. If a customer refuses to pay me after I finish the job at his or her house, I have the legal right to remain inside the house until the person pays me.
    a. True
    b. False
15. A long reach tool and wedge are commonly used to open locked automobiles.
    a. True
    b. False
16. It's legal for locksmiths to duplicate a U.S. post office box key at the request of the box renter—if the box renter shows a current passport or driver's license.
    a. True
    b. False
17. The Romans are credited with inventing the warded lock.
    a. True
    b. False

18. Five common keyway groove shapes are left angle, right angle, square, V, and round.
    a. True
    b. False
19. To pick open a pin tumbler cylinder, you usually need a pick and a torque wrench.
    a. True
    b. False
20. Fire-rated exit devices usually have dogging.
    a. True
    b. False
21. Common door lock backsets include
    a. 2⅜ and 2¾.
    b. 2½ and 2¾.
    c. 1¾ and 2½.
    d. 2⅜ and 3¾ inches.
22. How many sets of pin tumblers are in a typical pin tumbler house door lock?
    a. 3 or 4
    b. 5 or 6
    c. 11 or 12
    d. 7 or 8
23. Which lock is unpickable?
    a. A Medeco biaxial deadbolt
    b. A Grade 2 Titan
    c. The Club steering wheel lock
    d. None of the above
24. Which are basic parts of a standard key cutting machine?
    a. A pair of vises, a key stop, and a grinding stylus
    b. Two cutter wheels, a pair of vises, and a key shaper
    c. A pair of vises, a key stylus, and a cutter wheel
    d. A pair of styluses, a cutter wheel, and a key shaper
25. What are two critical dimensions for code cutting cylinder keys?
    a. Spacing and depth
    b. Bow size and blade thickness
    c. Blade width and keyhole radius
    d. Shoulder width and bow size
26. Which manufacturer is best known for its low-cost residential key-in-knob locks?
    a. Kwikset Corporation
    b. Medeco Security Locks
    c. The Key-in-Knob Corporation
    d. ASSA

27. The most popular mechanical lock brands in the United States include
    a. Yale, Master, Corby, and Gardall.
    b. Yale, Kwikset, Master, and TuffLock.
    c. Master, Weiser, Kwikset, and Schlage.
    d. Master, Corby, Gardall, and Tufflock.
28. A mechanical lock that is operated mainly by a pin tumbler cylinder is commonly called a:
    a. Disc tumbler pinned lock.
    b. Cylinder pin lock.
    c. Mechanical cylinder pin lock.
    d. Pin tumbler cylinder lock.
29. A key-in-knob lock whose default position is that both knobs are locked and require that a key be used for unlocking is:
    a. A classroom lock.
    b. A function lock.
    c. An institution lock.
    d. A school lock.
30. Four basic types of keys are:
    a. Barrel, flat, bow, and tumbler.
    b. Cylinder, flat, warded, and V-cut.
    c. Dimple, angularly bitted, corrugated, and blade.
    d. Cylinder, flat, tubular, and barrel.
31. The two most common key stops are:
    a. Blade and V-cut.
    b. Shoulder and tip.
    c. Bow and blade.
    d. Keyway grooves and bittings.
32. Bit keys most commonly are made of:
    a. Brass, copper, and silver.
    b. Aluminum, iron, and silver.
    c. Iron, brass, and aluminum.
    d. Copper, silver, and aluminum.
33. Which of the following key combinations provides the most security?
    a. 55555
    b. 33333
    c. 243535
    d. 35353
34. Which of the following key combinations provides the least security?
    a. 243535
    b. 1111
    c. 321231
    d. 22222

35. A blank is basically just:
    a. A change key with cuts on one side only.
    b. An uncut or uncombinated key.
    c. Any key with no words or numbers on the bow.
    d. A master key with no words or numbers on the bow.

36. You often can determine the number of pin stacks or tumblers in a cylinder by:
    a. Its key blade length.
    b. Its key blade thickness.
    c. The key blank manufacturer's name on the bow.
    d. The material of the key.

37. Spool and mushroom pins:
    a. Make keys easier to duplicate.
    b. Can hinder normal picking attempts.
    c. Make a lock easier to pick.
    d. Make keys harder to duplicate.

38. As a general rule, General Motors' 10-cut wafer sidebar locks have:
    a. A sum total of cut depths that must equal an even number.
    b. Up to four of the same depth cut in the 7, 8, 9, and 10 spaces.
    c. A maximum of five number 1 depths in a code combination.
    d. At least one 4-1 or 1-4 adjacent cuts.

39. When drilling open a standard pin tumbler cylinder, position the drill bit:
    a. At the first letter of the cylinder.
    b. At the shear line in alignment with the top and bottom pins.
    c. Directly below the bottom pins.
    d. Directly above the top pins.

40. When viewed from the exterior side, a door that opens inward and has hinges on the right side is a:
    a. Left-hand door.
    b. Right-hand door.
    c. Left-hand reverse bevel door.
    d. Right-hand reverse bevel door.

41. A utility patent:
    a. Relates to a product's appearance, is granted for 14 years, and is renewable.
    b. Relates to a product's function, is granted for 17 years, and is nonrenewable.
    c. Relates to a product's appearance, is granted for 17 years, and is renewable.
    d. Relates to a product's function, is granted for 35 years, and is nonrenewable.

42. To earn a UL-437 rating, a sample lock must:
    a. Pass a performance test.
    b. Use a patented key.
    c. Use hardened-steel mounting screws and mushroom and spool pins.
    d. Pass an attack test using common hand and electric tools such as drills, saw blades, puller mechanisms, and picking tools.

43. Tumblers are:
    a. Small metal objects that protrude from a lock's cam to operate the bolt.
    b. Fixed projections on a lock's case.
    c. Small pins, usually made of metal, that move within a lock's case to prevent unauthorized keys from entering the keyhole.
    d. Small objects, usually made of metal, that move within a lock cylinder in ways that obstruct a lock's operation until an authorized key or combination moves them into alignment

44. Electric switch locks
    a. Are mechanical locks that have been modified to operate with battery power.
    b. Complete and break an electric current when an authorized key is inserted and turned.
    c. Are installed in metal doors to give electric shocks to intruders.
    d. Are mechanical locks that have been modified to operate with alternating-current (ac) electricity instead of with a key.

45. A popular type of lock used on GM cars is:
    a. A Medeco pin tumbler.
    b. An automotive bit key.
    c. A sidebar wafer.
    d. An automotive tubular key.

46. When cutting a lever tumbler key by hand, the first cut should be the:
    a. Lever cut.
    b. Stop cut.
    c. Throat cut.
    d. Tip cut.

47. How many possible key changes does a typical disk tumbler lock have?
    a. 1500
    b. 125
    c. A trillion
    d. 25

48. Which manufacturer is best known for its interchangeable core locks?
    a. Best Lock
    b. Kwikset Corporation
    c. Ilco Interchangeable Core Corporation
    d. Interchangeable Core Corporation

49. James Sargent is famous for:
    a. Inventing the Sargent key-in-knob lock.
    b. Inventing the time lock for banks.
    c. Inventing the double-acting lever tumbler lock.
    d. Being the first person to pick open a Medeco biaxial cylinder.

50. Which are common parts of a combination padlock?
    a. Shackle, case, bolt
    b. Spacer washer, top pins, cylinder housing

c. Back cover plate, case, bottom pins
d. Wheel pack base plate, wheel pack spring, top and bottom pins.

51. General Motors' ignition lock codes generally can be found:
   a. On the ignition lock.
   b. On the passenger-side door.
   c. Below the Vehicle Identification Number (VIN) on the vehicle's engine.
   d. Under the vehicle's brake pedal.

52. Which code series is commonly used on Chrysler door and ignition locks?
   a. EP 1-3000
   b. CHR 1-5000
   c. CRY 1-4000
   d. GM 001-6000

53. How many styles of lock pawls does General Motors use in its various car lines?
   a. One
   b. Five
   c. Over 20
   d. Three

54. The double-sided (or 10-cut) Ford key:
   a. Has five cuts on each side; one side operates the trunk and door, whereas the other side operates the ignition.
   b. Has five cuts on each side; either side can operate all locks of a car.
   c. Has 10 cuts on each side; one side operates the trunk and door, whereas the other side operates only the ignition.
   d. Has 10 cuts on each side.

55. Usually the simplest way to change the combination of a double-bitted cam lock is to:
   a. Rearrange the positions of two or more tumblers.
   b. Remove two tumblers and replace them with new tumblers.
   c. Remove the tumbler assembly and replace it with a new one.
   d. Connect a new tumbler assembly to the existing one.

56. When shimming a cylinder open,
   a. Use the key to insert the shim into the keyway.
   b. Insert the shim into the keyway without the key.
   c. Insert the shim along the left side of the cylinder housing.
   d. Insert the shim between the plug and cylinder housing between the top and bottom pins.

57. A lock is any:
   a. Barrier or closure that restricts entry.
   b. Fastening device that allows a person to open and close a door, window, cabinet, drawer, or gate.
   c. Device that incorporates a bolt, cam, shackle, or switch to secure an object—such as a door, drawer, or machine—to a closed, locked, on, or

off position and that provides a restricted means—such as a key or combination—of releasing the object from that position.

d. Device or object that restricts entry to a given premise.

58. Which wheel in a safe lock is closest to the dial?
    a. Wheel 1
    b. Wheel 2
    c. Wheel 3
    d. Wheel 0
59. Which is not a type of safe combination wheel?
    a. Hole change
    b. Dial change
    c. Key change
    d. Screw change
60. Which type of cylinder is typically found on an interlocking deadbolt (or "jimmy-proof deadlock")?
    a. Mortise cylinder
    b. Key-in-knob cylinder
    c. Rim cylinder
    d. Tubular deadbolt cylinder

Registered Professional Locksmith Answer Sheet

Make a photocopy of this answer sheet to mark your answers on.

1. _____
2. _____
3. _____
4. _____
5. _____
6. _____
7. _____
8. _____
9. _____
10. _____
11. _____
12. _____
13. _____
14. _____
15. _____
16. _____
17. _____
18. _____
19. _____
20. _____
21. _____
22. _____
23. _____
24. _____
25. _____
26. _____
27. _____
28. _____
29. _____
30. _____
31. _____
32. _____
33. _____
34. _____
35. _____
36. _____
37. _____
38. _____
39. _____
40. _____
41. _____
42. _____
43. _____
44. _____
45. _____
46. _____
47. _____
48. _____
49. _____
50. _____
51. _____
52. _____
53. _____
54. _____
55. _____
56. _____
57. _____
58. _____
59. _____
60. _____

To receive your Registered Professional Locksmith certificate, just submit your answers to this test (a passing score is 70 percent) and enclose a check or money order for $50 (nonrefundable) payable to IAHSSP. And enclose copies of any two of the following items (don't send original documents because they won't be returned):

- City or state locksmith license
- Driver's license or state-issued identification
- Locksmith suppliers invoice
- Certificate from locksmithing or security school or program
- Yellow Pages listing
- Business card or letterhead from your company

- Association membership card or certificate
- Locksmithing bond card or certificate
- Letter from your employer or supervisor on company letterhead stating that you work as a locksmith or security professional
- Letter of recommendation from a Registered Professional Locksmith
- Copy of an article you've had published in a locksmith trade journal
- ISBN number and title of locksmith-related book you wrote.
  Send everything to: IAHSSP, Box 2044, Erie, PA 16512-2044. Please allow 6 to 8 weeks for processing.

Name: ____

Title: ____

Company name: ______________________________________

Address: ____

City, State, Zip Code: ________________________________

Telephone number: ________________________

E-mail address: ___________________________

Chapter

23

Making Locks by Hand

At one time all locks were made by hand. Today most are manufactured using automation. Handmade locks are typically made by hobbyists and blacksmiths for fun and special order. Whether or not you plan to make locks by hand, knowing how to do so will help you better understand and service them. This chapter gives detailed instructions and patterns you can use to make them.

You can make simple lightweight locks in your home or locksmithing shop, using common hand and power tools, such as a caliper, chisels, hammers, prick punches, and files. For complex and large models, you'll need workspace that has a fire-safe floor. You'll also need an anvil, a vise with 6-inch jaws with square faces and square edges, and a forge made of brick 3 feet high by 3 feet square. (Sometimes you can use a torch in lieu of a forge.) The anvil and forge should be located within a couple of feet from each other.

Wear safety glasses, boots, and nonflammable clothing, and keep your work area free of combustibles.

A Warded Bit-Key Lock with Deadbolt*

Anyone who has ever dismantled a warded lock, the type found in homes built before 1950, will find many similarities between the warded lock's deadbolt and the door lock. Though the parts of the warded lock differ greatly from the door lock's parts, the operating principle remains the same. A locking device is placed on the bolt to restrict the movement within the lock. The key lifts the locking device and moves the locking device to once again engage the bolt. Security for these locks is provided by wards or complicated keyhole shapes that prevent the wrong key from entering the lock or turning within the lock.

---

*Excerpt from the *Spruce Forge Manual of Locksmithing: A Blacksmith's Guide* to *Simple Lock Mechanisms* by Bill Morrison and Denis Frechette.

The [following] door lock design (Figs. 23.1 and 23.2) offers a good exercise in fitting several pieces into a working mechanical device.

1. The bolt is made from an 8-inch piece of ¼- × 1-inch square bar. Forge the tail of the bolt first. The finished tail should be ¼ × ½ inch and at least 3 ¼ inches long (Fig. 23.3).
2. Transfer the location of the drill holes used to rough out the bolt and the location of the notches along the top edge of the bolt. Mark the drill locations with a sharp center punch. Use only well-sharpened drill bits with cutting edges that are uniform in length and ground to the same angle. This will minimize any holes from drifting out of position. The accuracy of these holes will determine the success of the lock. Start with a small diameter drill as a pilot hole, then step up to the final size, e.g., ⅛, 3⁄16, ¼, ⅜ inch. After drilling the holes, use a hacksaw to remove the remaining metal. The notches along top edge of the bolt are cut using a hacksaw and chisel (Fig. 23.4).
3. Use a file to clean up any rough edges. A small radius is filed to the bolt teeth. The shape of the teeth is not critical at this point. The final shape of the teeth will later be determined when the bolt is fitted to the key.
4. Cut out both bolt guides from 16-gauge sheet metal and bend to shape. Use the bolt as a mandrel to size the guides. They should slide freely along the entire length of the bolt without being too sloppy.
5. Place the bolt guide over the bolt and clamp in the vise. Use a hacksaw to cut out the tenons. The saw cuts should stop well clear of the bolt. Use a file to refine the shape of the tenons. The shoulders of the tenons should be 1⁄32 inch above the bolt (Fig. 23.5).
6. Cut out the backplate. Transfer all reference lines, bolt guide locations, and the keyhole outline. Rough in the keyhole by drilling out the key

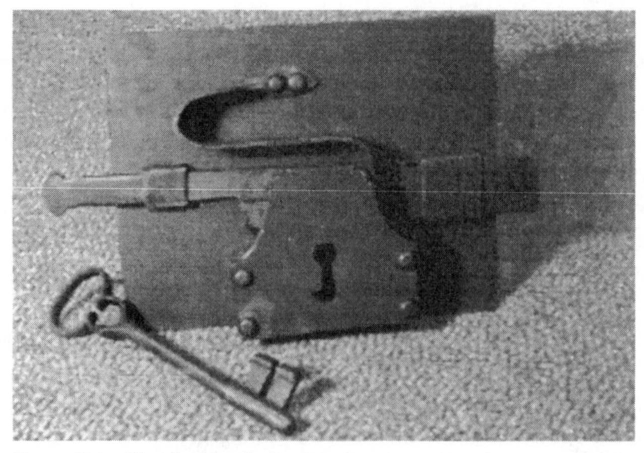

Figure 23.1 Simple door lock.

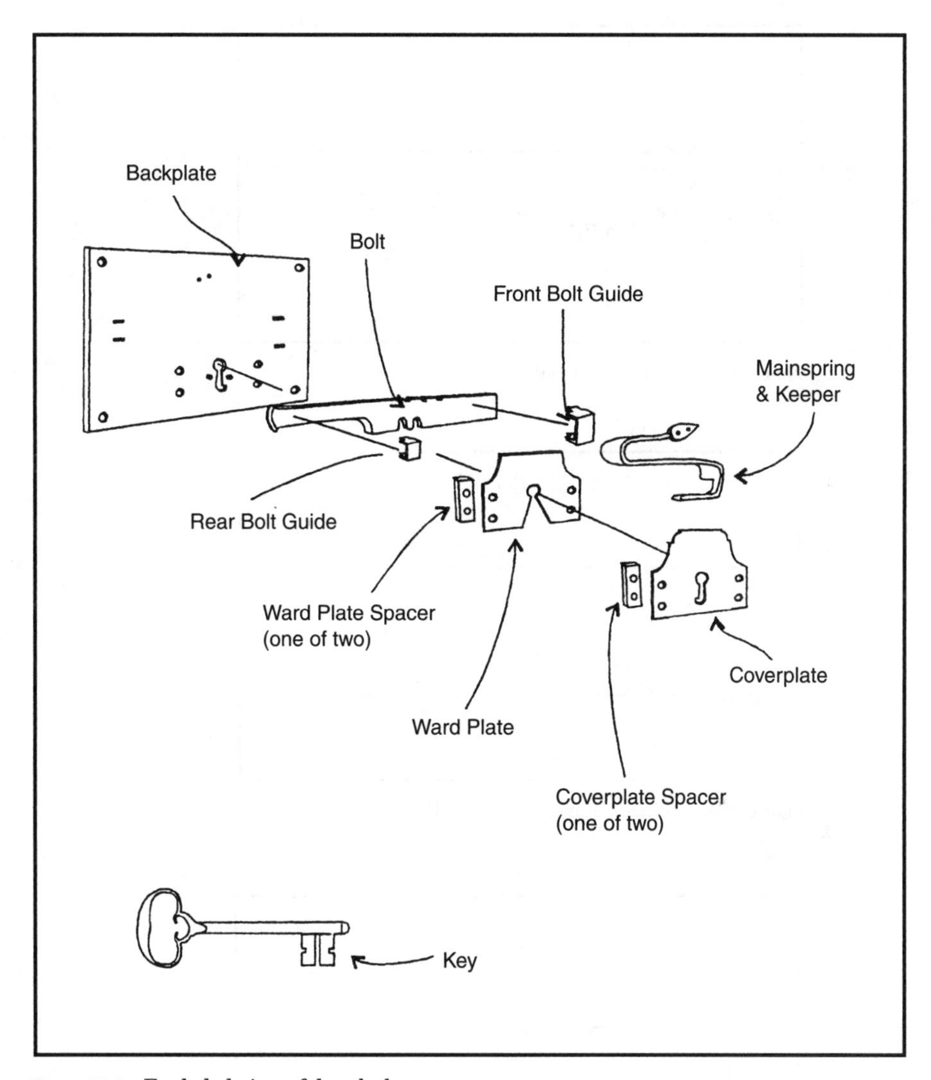

Figure 23.2 Exploded view of door lock.

pivot hole as well as drilling a ¼-inch hole near the base of the key bit. An engraving chisel with a 1⁄16-inch-wide cutting edge is used to remove material from one side of the keyhole. A regular cold chisel can then score the line on the opposite side of the keyhole (Fig. 23.6). The waste can then be broken away from the center of the keyhole. Clean up any rough edges with a file.

7. Rivet the front bolt guide in place. The rear bolt guide is fitted to the backplate but *not* riveted at this time. The bolt will need to be removed later to fit the mainspring and key.

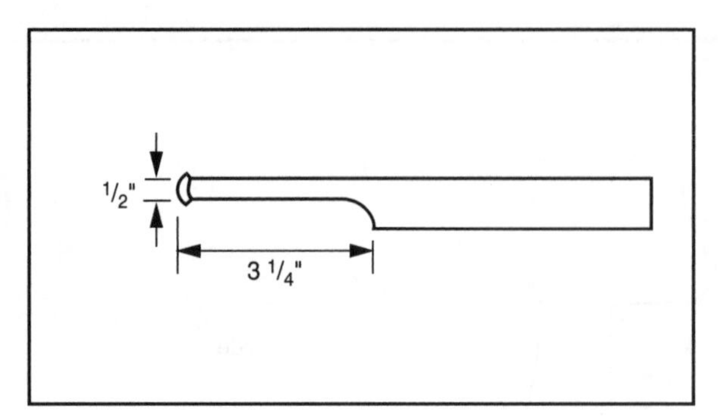

Figure 23.3 The finished tail.

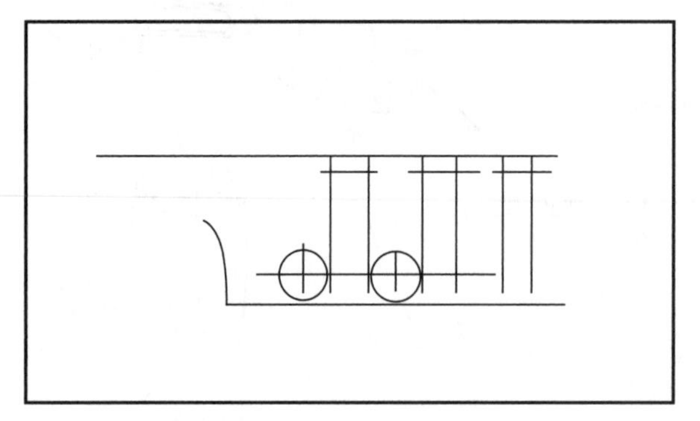

Figure 23.4 Transfer marks for the drill holes.

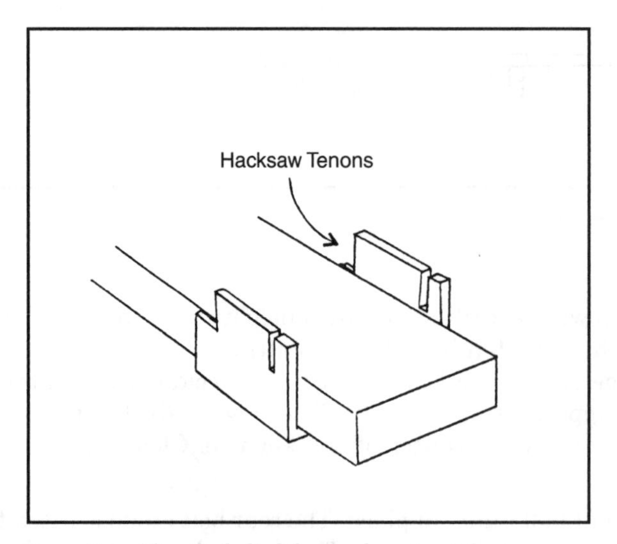

Figure 23.5 After the bolt slides freely, cut out the tenons.

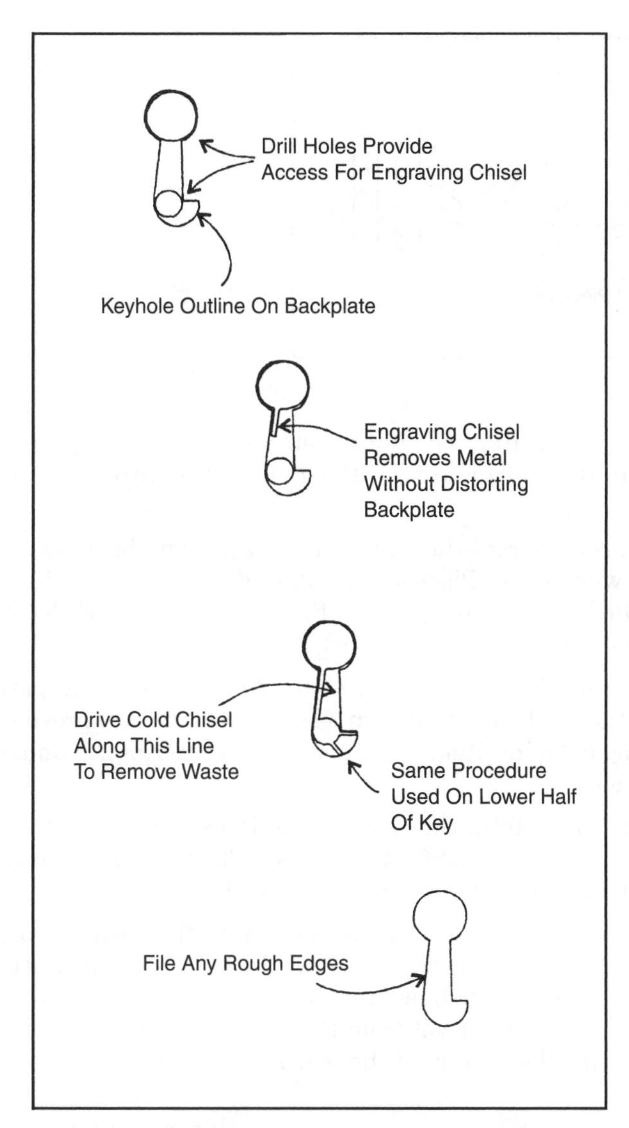
Drill Holes Provide
Access For Engraving Chisel
Keyhole Outline On Backplate
Engraving Chisel
Removes Metal
Without Distorting
Backplate
Drive Cold Chisel
Along This Line
To Remove Waste
Same Procedure
Used On Lower Half
Of Key
File Any Rough Edges

Figure 23.6 Make the keyhole by drilling and chiseling.

8. Insert several layers of paper or a thin cardboard between the bolt guide and the bolt before riveting the bolt guide to the backplate. This is the simplest way to make sure that there will be some play between the bolt guide and the bolt once the bolt guide is riveted in place. Burning away the paper once the riveting is done will release the bolt.

9. Cut out the coverplate and the ward plate from 16-gauge sheet metal. The coverplate and ward plate spacers are 1¾-inch lengths of ⅜-inch square bar. Grind the top edge of each spacer to conform to the top edge of the coverplate. Mark all pieces as shown in Fig. 23.7.

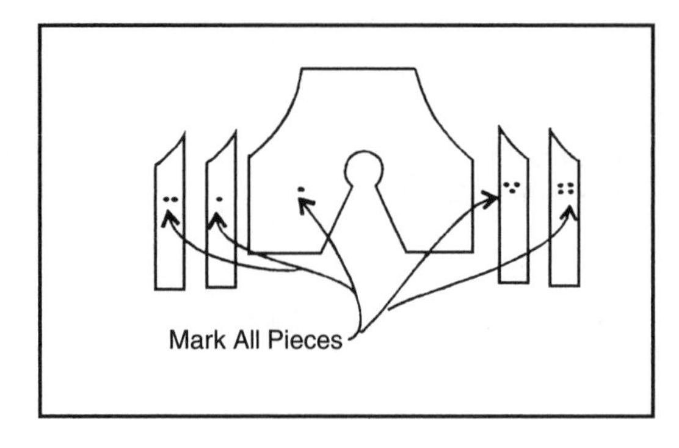

Figure 23.7 Cut out and mark the coverplate.

10. Center-punch the first rivet location on the top of spacer #1. Clamp spacer #1 to the ward plate. Drill a 3/16-inch hole through the spacer and ward plate.
11. Remove spacer #1 from the ward plate. Clean any burrs from the underside of the ward plate. Place spacer #2 under the ward plate. Use the rivet hole drilled in the ward plate to find the center punch location. Clamp spacer #2 to the ward plate and drill.
12. Remove spacer #2 from the ward plate. Place spacer #1 in position on top of the ward plate. Insert a rivet in the first drill hole to prevent the spacer from drifting out of position. Clamp in place, center-punch and drill the second rivet hole.
13. Remove spacer #1 from the ward plate. Place spacer #2 under the ward plate. Insert a rivet in the first drill hole. Clamp, center-punch, and drill. Repeat this procedure for spacers #3 and #4.
14. Assemble the ward plate and spacers by drilling a temporary rivet in each hole. Drill a 1/8-inch rivet hole in the center of each set of spacers. Counterbore the 1/8-inch hole from both sides and rivet the spacers in place. Make the 1/8-inch rivet from the shank of a common nail. Grind the rivet flush with the bottom of the ward plate and clean up all edges with a file.
15. The ward plate and spacers are used as a guide to drill the backplate and coverplate. A 3/8-inch bar is used to align the ward plate with the backplate or coverplate. Align the ward plate so that the 3/8-inch bar is perpendicular to backplate/coverplate. Clamp the ward plate in place and drill. As each hole is drilled, place a temporary rivet in the hole to keep the plate from drifting out of position. Do not rivet the ward plate to the coverplate or backplate at this time.
16. The mainspring is forged from a 6- to 8-inch piece of 5/16-inch square keystock. Forge one end of the bar into a thin, leaf shape as shown in Fig. 23.8.

17. Continue refining the transition between the leaf shape and the spring on a sharp corner of the anvil. The flat section of the spring should be 5 inches long, ¾ inch wide, and ¹⁄₁₆ inch thick. Use the lockspring fuller to draw the spring down to ¹⁄₁₆ inch.
18. The tail of the mainspring should be forged approximately 3 inches long, ³⁄₁₆ inch thick, and ¼ inch wide. Shape the tail of the spring as shown in Fig. 23.9. The bottom edge of the tail should be slightly above the bottom edge of the bolt.
19. Bend the remainder of the spring as shown. Place the mainspring in the middle notch of the bolt and align the center tooth with the centerline of the keyhole. Mark the location of the spring. Drill the rivet holes but do not rivet the spring to the backplate. The spring will need to be removed when adjusting the tension for the lock.

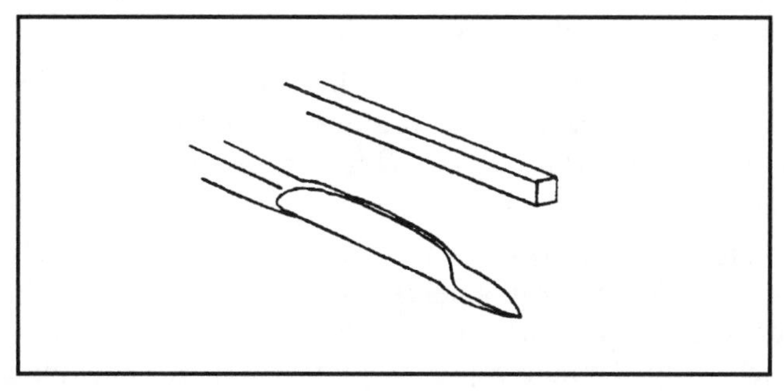

Figure 23.8 The mainspring is made by forging the keystock into a thin, leaf shape.

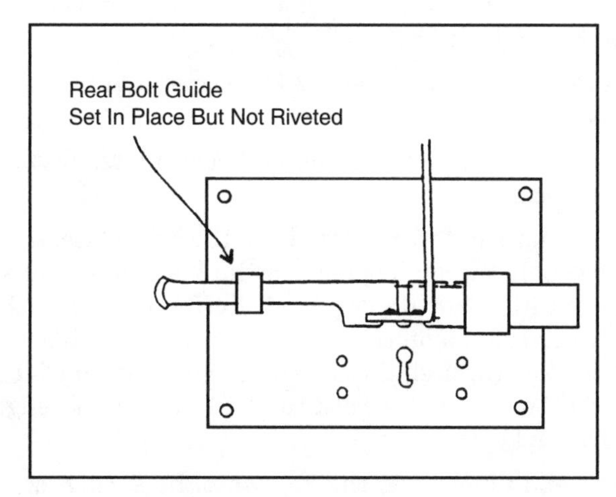

Figure 23.9 Set the tail of the spring.

20. The key is forged from a 4-inch length of ⅜- × 1-inch flat bar. Begin by punching and drifting the two holes for the bow. Punch and drift the 1-inch hole, then the ½-inch hole. Rough-forge the remainder of the key shape, leaving at least 1¼ inches for the key bit. Use a hacksaw to cut out the key bit and pivot pin. Begin refining the overall shape with a file.
21. A rat-tail file is used to carve out the inside bow. The final shape of the bow can be refined with ½-round files.
22. Heat the bow and shape using scroll wrenches or rat-tail tongs. The first step in shaping the key bit is to twist the shaft of the key slightly so that the centerline of the key bit is in line with the centerline of the key. This is an aesthetic adjustment and does not affect the operation of the lock.
23. A hacksaw is used to rough out the profile of the key bit. A narrow cold chisel is used to dig out the waste between the saw cuts.
24. Final shaping is done with a safe-edged file. The use of cardboard or sheet metal templates simplifies the shaping process.
25. Bolt the ward plate, coverplate, and backplate together. Test the key by sliding it into the lock from both sides. Scribe the location of the ward plate on the key bit. Saw it out. Try the key again from both sides. It should turn freely in the lock.
26. The wards for the door lock are located on the inside face of the backplate and coverplate. A pin ward is placed on each side of the keyhole on both of these plates. Begin by cutting a notch in the key bit that is $^{3}/_{16}$ inch wide and $^{3}/_{16}$ inch deep. Unbolt the coverplate from the backplate. Scribe the location of this notch on the backplate and coverplate.
27. Turn the key around and use the scribed lines on the backplate or coverplate to locate the notch on the other side of the key bit. Cut a notch in the key bit that is $^{3}/_{16} \times ^{3}/_{16}$ inch.
28. The pin wards are made from 16-gauge sheet metal. A triangular tenon is filed on the end of the ward.
29. Reassemble the backplate, ward plate, and coverplate. Test the key from both sides of the lock.
30. Remove the coverplate and ward plate from the backplate. Install the bolt and mainspring. Bolt the mainspring to the backplate and set the rear bolt guide into position but do not rivet. Scribe onto the backplate the location of the center notch on the top edge of the bolt. Remove the mainspring. Insert the key through the backplate and turn until it touches the bolt. Remove the bolt from the lock, and file away the edges that are interfering with the key.
31. Align one of the end notches on the key with the spring location marks. Test the key. Remove any unwanted metal. Repeat this process with the

last notch of the key. The key should now be able to move the bolt in and out of the lock without binding. As the key is turned, check that each notch is lined up with the spring location. Do a final test with the ward plate and coverplate in place.

32. Bolt the mainspring in position and test the key. The mainspring should be sitting firmly in the notches of the bolt and it should take little effort to lift the mainspring with the key. If the mainspring has too much tension, the key will tend to snap forward in the hand as it is turned. This will cause the bolt to be thrown out of position and the key will no longer engage the bolt.
33. Adjust the spring tension by opening or closing the main loop in the main spring. Grinding a small amount from the flat face of the mainspring will also reduce the tension as well as increase its range of motion.
34. Now the lock is ready for final assembly. Rivet the rear bolt guide, mainspring, ward plate, and coverplate in place.

The patterns for the lock are shown in Figs. 23.10 to 23.12.

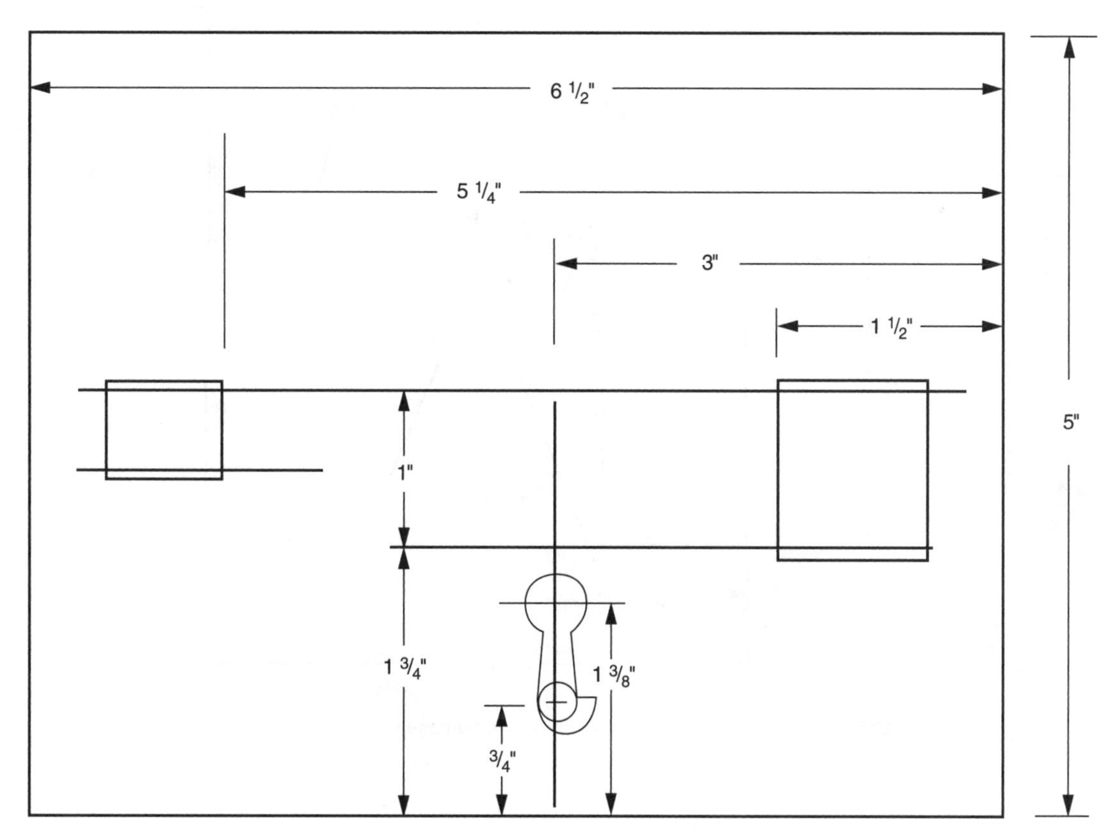

Figure 23.10 Pattern for door lock backplate.

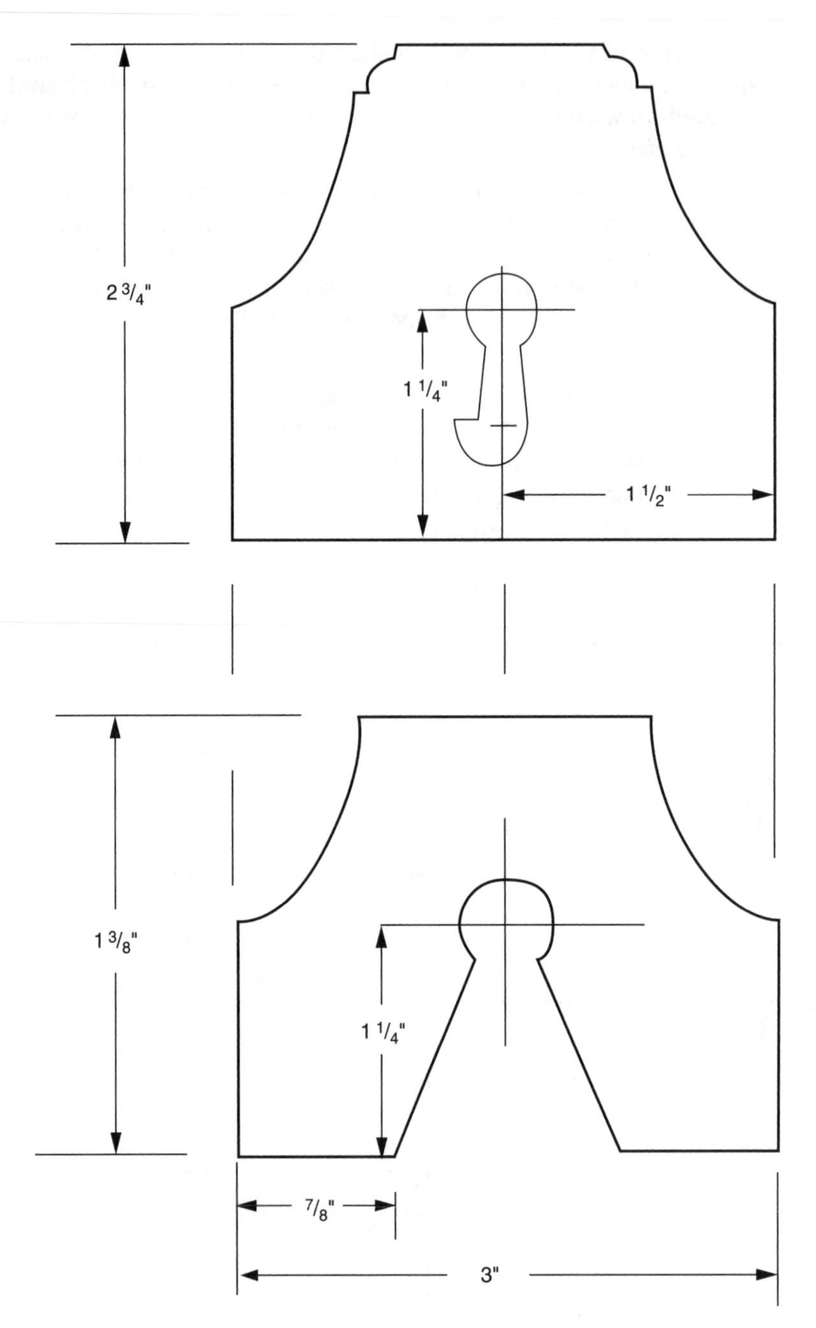

Figure 23.11 Pattern for door lock coverplate and ward plate.

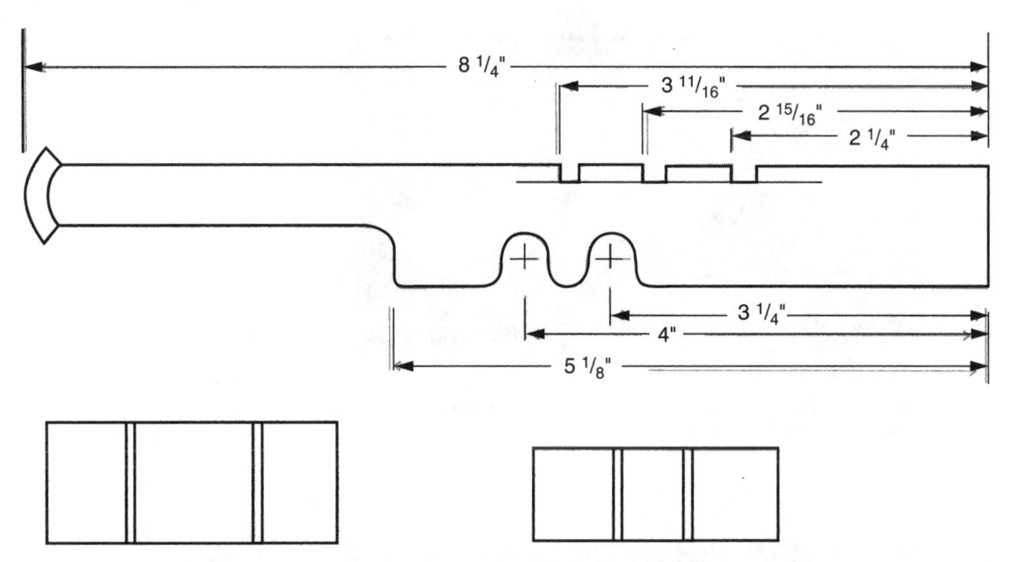

Figure 23.12 Pattern for door lock bolt and bolt guides.

English Iron Rim Lock*

A lock, even in basic form, is a minimachine, with accurately fitted parts that move when activated by a key. It must be so constructed that only those elements that are intended to move can do so, thus stabilizing the mechanical movements in their intended paths.

The key is the heart of the lock, and its beginning; from its measurements all other dimensions stem. This may seem strange, but it is logical. By first making the key, the locksmith has a gauge with which to try the movements as he fits the lock together. The security of the lock depends upon the impediments that the wards can put in the way of a false key, but through which the true key can pass. Although no traditional warded locks are pickproof by modern standards, during their time they were the best protection the craft could offer, with endless varieties of shapes, wards, and devices. Some medieval keys and their boxes of wards could be compared to passing one fine-toothed comb through another.

In locks for passage doors, which must be able to be locked from either side, some means must be taken to assure that the key will rest in its proper position each time it is thrust into the lock either way. With keys with bits of symmetrical profile in section, this is done by providing a collar or shoulder behind the bit, which bears against the lock case or backplate, centering the key in its correct place (Figs. 23.13 and 23.14).

---

*Excerpt from *Professional Smithing: Traditional Techniques for Decorative Ironwork, Whitesmithing, Hardware, Toolmaking & Locksmithing* by Donald Streeter.

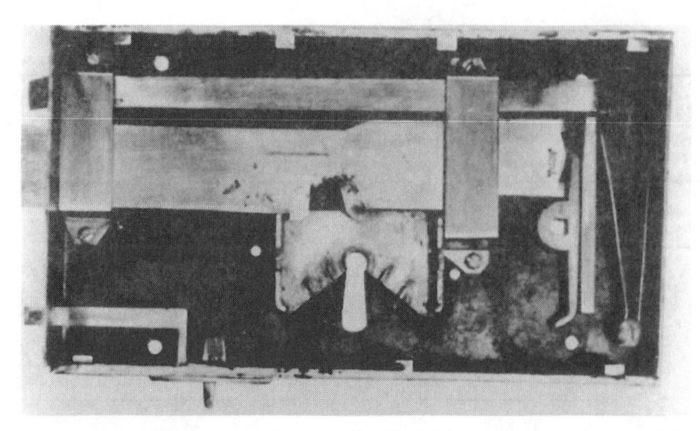

Figure 23.13 Interior of 12-inch English-type rim lock with backplate removed. (*Courtesy of Astragal Press, Mendham, NJ*)

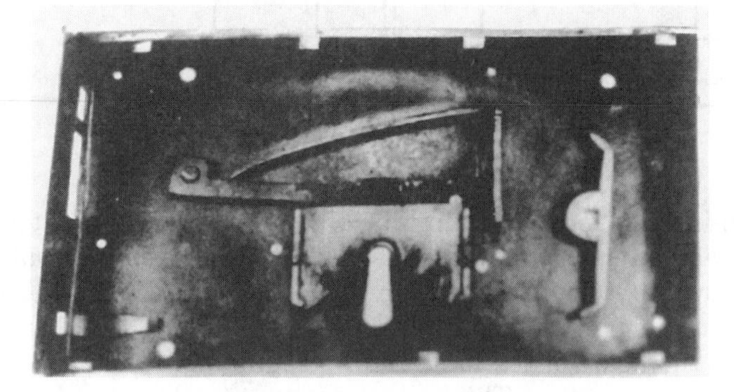

Figure 23.14 Interior of case, showing stud construction. (*Courtesy of Astragal Press, Mendham, NJ*)

To make the key, first forge a blank of proper size, leaving a round section at one end to receive the bit and a flat part at the other end for the bow. Split the flat end, bend the parts, and weld into a ring. Bend the rough bow aside and lathe-turn the stem and shaft to proper size. The bit should be made and fitted to the lock before spending time filing the bow, since if the bit fails, all filing time would be wasted.

Because of the shoulder at the base of the stem, it would be difficult to file the sides of the bit after they are fitted, without damaging the collar. So the bit is filed to shape before being attached to the stem. Two identical pieces must be made, with equally spaced clefts to pass the intended wards of the lock. They can take almost any form, provided they do not weaken the bit or allow passage of a bent wire pick strong enough to move the bolt and raise the tumbler.

In this case, the key is cut so as to pass a main center ward and collar, two minor wards, and one continuous ward. Ideally, all clefts should be cut as arcs concentric with the center of the stem, to avoid weakening the bit and to keep the tolerances as close as possible. They can be sawed straight and curved with chisels, or cut in the lathe. In the absence of a lathe, these were done in a jig at the drill press with hole saws of graduated sizes. The two pieces are attached to the stem with tenons, here ⅛-inch round steel pins, fitted into holes in the stem and bit, and the whole brazed together after clamping, with a spacer sheet between them for the main ward passage. Final finishing was done with thin, sharp cold chisels.

The key made, the lock itself is laid out on a steel sheet that is to be the lock case—here 16-gauge thicknesses—using prick punch, dividers, and scriber. The box of wards also is laid out in identical form, along with the backplate (Fig. 23.15). The main center-ward plate is cut to match them. Tenons are centered at its sides to enter slots cut into the sides of the box, and its center collar and minor wards are silver-soldered in place. Slots are punched in the main and backplates for tenons on the sides of the box, the ward system assembled with trial bolts, and the key filed as need be to make a matching fit while moving easily from either side of the lock. This done, the rest of the parts can be forged and fitted and the lock completed.

The lock illustrated is a three-bolt lock, with latch bolt, dead bolt, and night bolt. It is what is known as a once dead lock, since one revolution of the key throws the bolt through its entire travel. Some locks were made with two turns of the key required for full bolt movement, enabling the use of smaller keys, easier to carry about; they were called twice dead locks.

In this lock, the dead bolt and latch bolt were faggot-welded of flat stock in the traditional method, to avoid having to forge down heavier stock for the thin sections. Two saw cuts were made in the lower edge of the main bolt, in

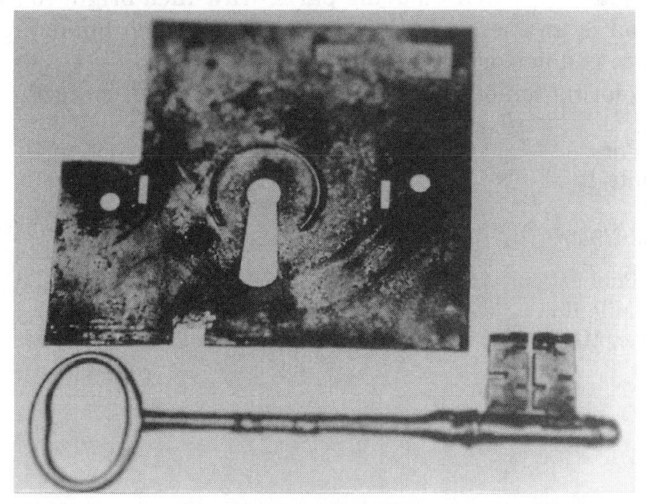

Figure 23.15 Backplate and key. (*Courtesy of Astragal Press, Mendham, NJ*)

position for the key to strike, and the center part bent down at a right angle to form a talon. Thus a talon engages behind a similar talon on the tumbler when in the unlocked position, and ahead of it when locked. The tumbler is raised out of the way as the key turns, by half of the bit, just as the other half strikes one ear of the slot in the bolt, and it drops into place again as soon as the bolt is thrown. A tapered iron spring riveted to the lock case keeps the tumbler in tension, and should be strong enough to function even if, as sometimes happens, the lock must be mounted upside down to accommodate the hand of the door.

The cam that activates the latch bolt, and through which the knob shaft passes, is forged with a round stud riveted through a round hole in the face. The hole is tapered to allow filing smooth with the face and yet allow free movement of the cam. A scissors spring returns the latch bolt to locked position at each operation. Tension springs bear on the face side of the night bolt, to prevent chatter and unwanted movement should the door slam in the wind.

The case is made of three pieces: the face and ends in one sheet, bent at right angles, and top and bottom of tapered flat stock with studs riveted to them and the face plate. The backplate carries the one continuous ward of thin steel sheet, attached with tenons through the plate, and a tension spring to bear on the dead bolt. The continuous ward affords a complete track for the key through its travel, giving support against the pressure of the tumbler, and preventing wear at the keyhole, which would eventually cause the key to wobble and damage other parts.

The keeper for this type of lock is made in similar fashion to the case, with back end of tapered forged stock, and stud construction. Fine locks were fitted with a quarter-round brass striking strip at the lock edge, for the beveled end of the latch bolt, and to match the brass knobs. Cases usually were painted black inside and out, with all moving parts draw-filed bright to avoid friction. In all locks, oil is an enemy, for it attracts dust, which builds up and causes wear. Graphite is a better lubricant.

For more information on the books mentioned above, contact:

The Astragal Press
5 Cold Rd., Suite 12
P. O. Box 239
Mendham, NJ 07945

Spruce Forge Publications
2563 East 5¼ Mile Rd.
Sault Ste Marie, MI 49788
906-635-1316

Chapter

24

Answers to Frequently Asked Questions

The following are my answers to some locksmithing questions I've been asked since the last edition of this book. The questions came from students, apprentices, and professional locksmiths.

Q. How did you learn locksmithing?

ANSWER: My first exposure to locksmithing was through books in high school. Then I took a couple of correspondence courses. I completed a nine-month program at the National School of Locks and Alarms in New York City. Then I apprenticed under two locksmiths—including a Certified Master Locksmith.

Q. How can I get a job as a locksmith?

ANSWER: Locksmithing jobs are plentiful. Every month you'll find dozens of help-wanted advertisements in the national locksmithing trade journals and a lot more on Internet sites. If you want to find a local job, you can send a résumé to nearby locksmith shops and ask for a meeting. It's better to contact them directly rather than waiting for them to place a newspaper advertisement; you'll have less competition. For more information on planning a job search, see Chap. 20.

Q. How can I get started writing about locksmithing?

ANSWER: I started by writing for trade journals. The locksmithing trade journals are always looking for fresh articles and are pretty easy to write for. Subscribe to one or more of them to get a feel for what kinds of material they want. Then write to the managing editor and propose an article for you to write. If you're already working on a job that you want to write about, be sure to take lots of photographs. With good photographs, it will be easier to convince an editor to publish the article.

Q. Do I need to be licensed to work as a locksmith?

ANSWER: Some places, such as New York City, Texas, California, Tennessee, and North Carolina, have licensing laws for locksmiths.

Q. Is it legal for me to carry my lockpicks when I travel to other states?

ANSWER: That depends on the laws of the states you're traveling in and your purpose for having them. Some places have statutes on who may carry lockpicks. However, I've never heard of a locksmith being imprisoned for carrying lockpicks to a locksmith's convention. To avoid being hassled at the airport, you might want to carry them in your baggage. When you carry lockpicks, keep them in their case, and don't flash them around. Don't give people a reason to start asking you about them.

Q. How can I find useful locksmithing information on the Internet?

ANSWER: There are several popular locksmithing sites, and new ones are popping up all the time. If you enter "locksmithing" in Google or another search engine, you'll mostly get Web sites for local locksmithing shops and for lock manufacturers and distributors. Basically, you'll find advertising and little technical information. For more useful information on locksmithing, go to the USENET group *alt.locksmithing*. That's a popular site with lots of locksmithing discussions, but because it's unrestricted—it's open to anybody—locksmiths give little technical information on it. No one is likely to tell you such things as safe drill points or key codes—or even how to open a certain car. But they'll give you their opinion on the best locks or locksmithing schools. Another problem with unrestricted sites is that users are anonymous, so lots of people on them falsely claim to be "master locksmiths."

Some sites have a public area and a restricted area. In the restricted area, locksmiths are much freer in sharing technical information. To get to the restricted area, you need to give evidence that you're a locksmith or safe technician. *www.TheNationalLocksmith.com*, for instance, will show random lock codes and ask you for the bitting numbers. That system is based on the assumption that anyone with code books is probably a locksmith because code books are expensive and for most people are not easy to get. If you answer the code questions, there is no charge to access the site. But the site offers an alternative for locksmiths who don't have the code books, which involves a small fee. You can supply documentation that you're a locksmith.

For documentation, most restricted sites require a copy of your driver's license (or state-issued identification card) and copies of three of the following: your Yellow Pages ad, a membership certificate from a locksmithing or safe technician association (or member number), business letterhead, a letter from a locksmithing school identifying you as a student, a diploma or certificate of completion from a locksmithing school, a supplier's invoice, a letter from your employer identifying you as a locksmith, a locksmith license, or your business license. *www.clearstar.com*, another site with a restricted area, requires similar documentation and charges a fee. It's one of the best locksmith/safe technician sites on the Internet because some of the most knowledgeable locksmiths access it. *www.Locksmithing.com* requires similar documentation, charges a fee, and requires you to pass the Registered Professional Locksmith test (one is included in Chap. 22). The site provides lots of technical information. And I have an "Ask the Lock Professor" message board on it.

You'll have a better chance of getting factual information from a restricted site than from a public site, but not all restricted sites are very good. A lot of them, including those that charge a fee, don't have any useful information. And they seem like they've been accessed by three or four people. Before paying a fee, ask the site to let you access the restricted areas for a little while. Legitimate sites won't mind giving you a few days to look around—but you'll have to send in the documentation first.

Q. Is there a master key that can open any lock?

ANSWER: There can be master keys for certain buildings and for certain building complexes that use locks with the same keyway. But there is no key that can open all locks because the keyway sizes and shapes are too varied and because there are too many different internal configurations among locks and lock cylinders.

Q. Are deadbolt locks hard to pick?

ANSWER: The term *deadbolt* refers to a lock that relies on a rigid bolt for security but says nothing about how easy or hard it is to pick open. That depends on the type of cylinder the deadbolt has. Most low-cost deadbolts (under about $20) come with standard five-pin cylinders and are easy to pick open. Cylinders with six or more sets of pins are harder to pick open. High-security cylinders can be virtually impossible to pick open without intimate knowledge of the lock and without specially designed equipment. By adding a couple of mushroom or serrated pins, you can make almost any pin tumbler cylinder a little harder to pick. (You can pick them by reverse picking—carefully lowering the bottom pins across the shear line.)

Q. Can every lock be picked?

ANSWER: In theory, any lock that's operated with a mechanical key can be picked open because "picking" refers to using tools to simulate the action of a key. (But many locks that don't use a mechanical key can be bypassed easily in ways other than picking.) There are locks that have never been picked open because doing so would take a long time and would require specially designed tools and intimate knowledge of the lock's construction. Most locks that use patented or UL-listed cylinders are impractical to pick open.

Q. Where can I get lockpicks?

ANSWER: You can buy lockpicks and other locksmithing tools from a locksmith supply house. To find a local supplier, look in your Yellow Pages under "Locksmiths' Equipment & Supplies." In some cases, you may be required to supply a copy of your business license, driver's license, letterhead, or newspaper or telephone book advertisement to show that you work as a locksmith (usually you'll have the option of choosing among several things to send). Another option is to buy lockpicks that are advertised in personal safety, detective, and survival magazines. They don't require identification but offer few choices and charge much more for the picks than do locksmith supply houses.

Q. Are pick guns worth the money

ANSWER: Contrary to movies and television shows, pick guns don't quickly open almost any lock. They work by slapping the bottom pins and forcing the top pins into the upper pin chamber. By using a pick gun with a torque wrench, you can often catch a few or all top pins above the shear line. Sometimes you can open a lock just by using the pick gun and torque wrench. Most times you also need to use a standard pick to pick the pins that the pick gun didn't trap.

Q. Can Kryptonite bicycle locks be picked?

ANSWER: Yes. The Kryptonite U-shaped lock can be picked open with a BIC pen. By using the shell of the pen and twisting it (like a tubular lock pick), the lock quickly opens.

Q. Can The Club steering wheel lock be picked open?

ANSWER: The Club can be picked open easily. Picking is a good way for a locksmith to open it quickly. Car thieves rarely pick open a steering wheel lock, however. They're much more likely to saw the steering wheel and slip the lock off.

Q. Is it legal for anyone to carry lockpicks?

ANSWER: In many places, whether or not carrying lockpicks is a crime depends on the intent of the person carrying them. A locksmithing student taking them to school might be fine, for instance, but someone carrying them while committing a burglary may be charged with possessing "burglary tools." In some places, the only people who may legally carry lockpicks are those mentioned by statute (which typically include bona fide locksmiths, repossessors, and law enforcement officers). In such cases, unmentioned persons who have a legitimate need to carry lockpicks (such as building maintenance persons) may have to get a locksmith license. Places that have (or have had) fairly strict laws about carrying lockpicks include Canada, California, Illinois, Maryland, New Jersey, New York State, and Washington, D.C. Talk with an attorney to find out which laws apply to your situation.

Q. What are the best ways to continue learning about locksmithing?

ANSWER: Books, videotapes, correspondence courses, seminars, residential training programs, and talking with other locksmiths are good ways to continue learning about locksmithing. Some restricted Internet locksmithing sites also can be helpful. The best way to continue your learning is by doing. Get a job as an apprentice. And practice installing, servicing, disassembling, and reassembling all types of locks and locking devices on your own.

Q. What's the best locksmithing school?

ANSWER: There are too many locksmithing programs to pick one as being best for all people. The best would depend on where you live, how much money you want to spend, and how much time you can devote to the program. When choosing a residential program, contact several schools, speak to their instructors, and ask for names of former students or of employers who have hired former students. The better schools will provide you with all that information. Before enrolling, visit the school to evaluate the facilities. Choosing a correspondence program is mostly a matter of comparing the course offerings, supplies, and prices. You also may want to post messages on the Internet asking people to share their feelings about certain schools you're considering. Two good places to start are *www.locksmithing.com* and the *alt.locksmithing* newsgroup. Appendix E has a list of locksmithing schools and programs.

Q. Which is the best locksmithing trade journal?

ANSWER: The four primary locksmithing journals in the United States are *Keynotes, Locksmith Ledger International, The National Locksmith,* and *Safe and Vault Technology.* I've written articles for all of them. *Keynotes* includes some good technical information but is only for members of the Associated Locksmiths of America. I've never joined. *Safe and Vault Technology* is another members-only magazine from the Safe and Vault Technicians Association (SAVTA). I never joined them either.

The National Locksmith is the oldest locksmithing trade journal, serving locksmiths since 1929. I think it goes into more depth and has more how-to articles than does *Locksmith Ledger*. If you're interested in articles on safe and vault work, get *The National Locksmith* or join SAVTA. Addresses for all the publications I mentioned are in Appendix C.

Q. What are the best locksmithing organizations to join?

ANSWER: That depends on if you want to specialize in any area. For automobile lock servicing/car entry, join the National Locksmith Automobile Association. If you do safe and vault work, join the National Safeman's Organization or the Safe and Vault Technicians Association (SAVTA). If you do much residential work, join the International Association of Home Safety and Security Professionals. The largest locksmithing trade association in the United States is the Associated Locksmiths of America (ALOA). Addresses of these and other trade associations are in Appendix C.

Q. What's the best drill for locksmiths?

ANSWER: A ½-inch electric drill is most useful for locksmithing. When looking for a drill, don't compare them based on price alone because better drills usually cost several times more than the low-cost models sold in department stores. Look for a ½-inch electric drill with the following features: at least a 5-amp motor, variable-speed reversing (VSR), 600 rpm (or faster), three-stage reduction gearing, and all ball and needle bearings. Popular brands among locksmiths include Bosch, Makita, Milwaukee, Porter-Cable, and Ryobi.

Q. Which lubrication is best for servicing locks?

ANSWER: WD-40 and Tri-Flow are popular lubricants among locksmiths.

Q. What can I do to prevent other locksmiths and key cutters from duplicating a key?

ANSWER: A knowledgeable locksmith can copy any key for which he or she has a blank, but there are things you can do to make it less likely that anyone will duplicate a key. Use a neuter bow blank—one with a bow that gives no information about the key it copied. That alone will stump most key cutters because they don't know how to identify blanks by keyway grooves. Then stamp "Do Not Duplicate" on the bow. Many people won't duplicate a key with such a phrase stamped on it.

Another thing you can do is, after duplicating the key, use your key machine cutter to lightly shave the length of the new key along its bottom. This will make the new key seem to be its normal width, but it will sit too low in a key machine vise so that any new key made from it will be miscut.

Q. Is it illegal to duplicate a key marked "Do Not Duplicate"?

ANSWER: In some places it is illegal to duplicate a key marked "Do Not Duplicate" without getting positive identification and keeping a written record of the work on file. In most places, however, such a phrase stamped on a key is merely a request and doesn't create any legal restrictions or obligations on locksmiths or key cutters. Anyone can have any key marked "Do Not Duplicate." I've had many customers ask to have that phrase stamped on their keys because it at least causes eyebrows to be raised when someone tries to get copies. As a legal matter, however, the type of key is usually more significant than a stamping with respect to which keys may be copied. Whether required or not, for your protection it's best to have a policy of not duplicating any key marked "Do Not Duplicate" unless you know you can legally copy it, and the key holder shows positive identification. You also should maintain a log of all such keys you duplicate.

Q. Which keys are illegal for locksmiths to copy?

ANSWER: It's almost always illegal (or foolish) for a locksmith to duplicate a key that he or she knows or reasonably suspects is being copied without the consent of the owner or to be used for criminal activity. This isn't to say, however, that every time someone wants a key made that the locksmith must track down the owner of the key and figure out the keyholder's intentions. In unusual situations, however, you could be found negligent for not looking further into a matter before duplicating the key.

Here's a situation that happened to me. A woman came into my shop with a clay mold of a Medeco key and wanted me to make that key. This was when the original Medeco was still under patent. She explained that she made the mold because her husband wouldn't let her have a key to their house. The woman was acting nervous and wouldn't show me identification. Nor would she tell me where she lived. I declined to make the key. I think if I had made the key and she had used it to break into a home or business, I could have faced criminal and civil charges because I could have been negligent under that circumstance.

It's also illegal to duplicate post office box keys without permission of the postmaster or post office superintendent.

Q. May locksmiths fit keys to U.S. Navy and U.S. Army locks?

ANSWER: Because thousands of war surplus locks marked "U.S. Navy" and "U.S. Army" have been sold to the general public, such a marking isn't in itself a reason for locksmiths not to fit keys to a lock.

Q. Is it legal to duplicate safe deposit box keys?

ANSWER: It's common for locksmiths to duplicate safe deposit box keys. Such keys aren't restricted because safe deposit box security isn't based primarily on a box holder's key. (It's also dependent on a second key, personnel, and building security.) Unless there's something suspicious surrounding a request to copy a safe deposit box key, there's generally no problem with making copies.

Q. If I rekeyed a lock on a home and later learned that the person who ordered the work wasn't authorized to do so, would I be liable?

ANSWER: You might be liable for doing unauthorized work, especially if you messed up a landlord's master key system or if your actions prevented someone from gaining lawful entry during an emergency. Anytime you do work on location, you need to be especially careful about being properly authorized to do the work. Before rekeying a lock on a home, ask to see a driver's license or other positive identification. Then ask the person why he or she wants the lock rekeyed, and ask who is the owner of the home. If the person is going through a divorce or separation and wants to lock out a spouse or partner, it's safer to decline the job (even though the lockmith may be legally allowed to do the work without the consent of the other spouse or partner). If the person doesn't own the home, get permission to call the landlord and ask if it's all right for you to rekey the lock. Generally, a locksmith isn't obligated to call a landlord. But that phone call could help you avoid a lawsuit.

Q. If a tenant wants me to install a new lock or other hardware on a door, do I need to get permission from the landlord?

ANSWER: Many locksmiths simply have the tenant sign an authorization form and have had no problems with that practice. In most cases, unless the locksmith has to drill holes or otherwise permanently alter the door, there's no requirement for the locksmith to get a landlord's permission to install a lock. However, it's a good idea to get a landlord's permission whether you drill or not. If you don't seek the landlord's permission, at least leave all the locks and hardware you remove with the tenant (but if you still get sued, don't blame me).

Q. If I mistakenly opened an automobile or house door for a thief, would I be held liable?

ANSWER: When unlocking doors, you have to act in good faith and exercise due care. Generally, that means checking a driver's license or car registration. It also might mean asking neighbors to verify the identity of the person who wants you to unlock the door. But acting in good faith and exercising due care may not help you to avoid civil liability (losing a lawsuit), especially if someone is harmed or his or her property is stolen as a result of your actions. It isn't enough just to come up with excuses for why you were tricked; you just have to not open doors for unauthorized persons.

Establish and consistently follow a policy that minimizes your risk. For instance, when someone asks you to open a building or car door, immediately fill out a work order that includes all relevant information, such as the customer's name, address, telephone number (get that even if the caller isn't at home), driver's license number, license plate number, and the make, model, and year of the car. If the situation seems suspicious, decline to do the job. Be sure you're making consistent decisions based on a clear written policy that's reasonable; otherwise, you may end up facing a lawsuit (or bad press) for discriminatory practices.

Q. If I install a lock and the customer refuses to pay, may I take my locks back?

ANSWER: Once a lock is installed on a door, it becomes part of that door and is the property of the door owner. The proper remedy would be to file a lawsuit. If the lock were on an apartment and you had gotten the landlord's consent to install it, you might want to include the landlord in your lawsuit.

Q. If I finish a job in someone's home and the customer refuses to pay me, would I be within my rights to refuse to leave until I'm paid?

ANSWER: Not being paid doesn't give a locksmith the right to be in someone's home. Staying after being asked to leave could quickly become criminal trespassing. The proper remedy to not being paid is to file a lawsuit promptly (usually with small claims court or with a district justice).

Q. If I go to someone's home or car to unlock a door and find the door already open when I get there, can I still charge for my time?

ANSWER: It isn't uncommon that after a customer calls a locksmith to open a lock on a home or car, the customer gains entry before the locksmith arrives. Sometimes to get faster service, people call several locksmiths for the same job. To charge for your time even if the door has already been opened, you need to clearly make such an agreement before you leave for the job. When the customer is requesting the work, you have to tell the customer of your minimum service charge and that it is payable on your arrival. Then you'll be able to collect that fee (either before you leave the job or later in court).

Q. How can I make sure of getting paid after opening a car?

ANSWER: It isn't unusual for customers to renege after the locksmith performs the service. This is why you need to be vigilant during that critical period between opening the door and closing your hands around the payment. When I open a car door, I stand close to the door. If the person gives me any reason to believe that I'm not getting paid, I quickly close the door. Then I tell the person that he or she has to pay an additional fee in advance for me to open the door again.

If a customer doesn't have cash, I ask him or her to postdate a check. Or I'll ask to hold something as collateral. I also have the person sign an IOU. I consider these steps necessary because the customer knew of my fee before I arrived and knew that it was to be paid in full on completion of the work. Because the person has already reneged on our agreement, I need to deal with him or her differently if I'm to be paid.

Q. Are you concerned that criminals might use your books to commit crimes?

ANSWER: Most information in my books has little practical use for committing crimes. There are much faster ways to break into places than to pick or impression locks, for instance. Locksmithing is an honorable profession, and people should be allowed to learn and read about it. Honest people shouldn't be denied technical information because someone might use it to commit crimes.

Q. Is there anything you wouldn't share in your books?

ANSWER: I like to offer readers more information than other general locksmithing books offer. I don't include information that would likely aid criminals but that has no practical value to locksmiths. And I don't include proprietary information that I accepted in confidence.

Appendix

Manufacturers

A&B Safe Corporation
114 S. Delsea Drive, Suite 3
Glassboro, NJ 08028-6237
(800) 253-1267; (856) 863-1186
www.a-bsafecorp.com

A-l Security Manufacturing Corporation
3001 W. Moore Street
Richmond, VA 23230
(804) 359-9003
www.demanda1.com

Makes more than 100 locksmithing tools. Brand names: Pak-A-Punch, TB3, and The Block.

Abloy Door Security
9360 Trans Canada
Montreal, Quebec, Canada H4S 1V9
(800) 465-5761; (514) 335-9500
www.abloy.ca

Abus Lock Company
23910 N. 19th Avenue, #56
Phoenix, AZ 85027
(800) 352-2287; (623) 516-9933
Woburn, MA 01888
www.abuslock.com

Adams Rite Manufacturing Co.
260 Santa Fe Street
Pomona, CA 91767
(800) 872-3267; fax: (909) 632-2369
www.adamsrite.com

Adesco Safe Mfg. Co.
16720 S. Garfield Avenue
Paramount, CA 90723
(800) 821-6803; (562) 630-1503
www.adesco.com

Almont Lock Co., Inc.
P.O. Box 568
Almont, MI 48003-0568
(810) 798-8950

Makes rekeyable brass padlocks and cylinders.

American Lock Company
137 W. Forest Hill Avenue
Oak Creek, WI 53154
(800) 323-4568
www.americanlock.net

Founded in 1912 by John Junkunc as the Junkunc Safe and Lock Company.

American Security Products Company (or AMSEC)
11925 Pacific Avenue
Fontana, CA 92337-6963
(800) 421-6142; (909) 685-9680

Makes a wide variety of safes under the brand names AMSEC, Star, and Major. Founded in 1946 as Star Safe Company (or Starco). In 1964, the company changed its name to American Security Products Company. In 1987, the company purchased one of its biggest competitors, Major Safe Company.

Arrow Lock Mfg. Co.
325 Duffy Avenue
Hicksville, NY 11236
(800) 221-6529; (516) 704-2700
www.arrowlock.com

An ASSA ABLOY Group Company.

ASSA, Inc.
110 Sargent Drive
New Haven, CT 06511
(800) 235-7482; fax: (800) 892-3256
www.assalock.com

Baton Lock & Hardware Company, Inc.
14275 Commerce Drive
Garden Grove, CA 92643
(800) 395-8880; (714) 265-3636; fax: (714) 265-3630
www.batonlockusa.com

Makes padlocks, cam locks, and assorted door hardware.

Best Lock Corporation
P.O. Box 50444
6165 E. 75th Street
Indianapolis, IN 46250
(317) 849-2250; fax: (317) 595-7620

BiLock North America, Inc.
2420 Carson, Suite 125
Torrance, CA 90501
(800) 328-7030; (310) 328-7030
www.bilock.coni

Offers a line of patented high-security lock cylinders.

Canadian Safe Manufacturing
170 Chatham Street
Hamilton, Ontario, Canada L8P 2B6
(800) 267-7635; (905) 528-7233; fax: (905) 528-7728

Cannon Safe, Inc.
216 S. 2nd Avenue, Bld. 932
(800) 242-1055; (909) 382-0303
www.cannonsafe.com

CCL Security Products
301 W. Hintz Road
Wheeling, IL 60090
(800) 733-8588; (847) 537-1800; fax: (847) 537-1881
www.cclsecurity.com

Makes combination padlocks, cam locks, and other types of locks and security products. Brand names: Sesamee and Prestolock. Corbin Cabinet Lock (CCL) was for many years a division of Emhart Industries. In 1987, the Eastern Company bought the division and reorganized it under the name CCL Security Products.

Chicago Lock Company
4311 W. Belmont
Chicago, IL 60641
(708) 747-1235

Corbin Russwin Architectural Hardware
225 Episcopal Road
Berlin, CT 06037
(800) 543-3658; fax: (800) 447-6714
www.corbin-russwin.com

Makes a complete line of door hardware products, including locks, exit devices, door closers, and key systems. In May 1993, Corbin and Russwin brands merged.

Corby Industries
1501 East Pennsylvania Street
Allentown, PA 18103
(800) 652-6729; (610) 433-1412; fax: (610) 435-1963
www.corby.com

Makes tubular-key locks and assorted locks and keys. Trade names: Ace and Acell.

CorKey Control Systems, Inc.
846 Mahler Road
Burlingame, CA 94010
(800) 622-2239; (650) 692-9495; fax: (650) 692-9410
www.corkey.com

Detex Corp.
302 Detex Drive
New Braunfels, TX 78130
(800) 729-3839; (830) 629-2900; fax: (830) 620-6711

Diebold, Inc.
P.O. Box 8230
Canton, OH 44711-8230
(800) 999-3600; (330) 489-4000; fax: (330) 588-3794
www.diebold.com

Makes safes. Trade name: Banklock.

Dynalock
P.O. Box 2728
Bristol, CT 06011
(877) DYNALOCK; (860) 582-4761; fax: (860) 585-0338

Falcon Lock
111 Congressional Boulevard, Suite 200
Carmel, IN 46032
(800) 266-4456; (317) 613-8150; fax: (800) 840-7735
www.falconlock.com

Federal Lock Company
RR3, 3466 Fords Road
Honesdale, PA 18431
(888) 562-5562 (570) 253-5424; fax: (570) 253-4292
www.federallock.com

Makes padlocks and high-security hasps. Founded in 1992 by Ken Erickson and some of his associates.

Fichet-Brauner USA
5855 Oakbrook Parkway
Norcross, GA 30093
(404) 448-5593

Folger Adam Electric Door Controls
9100 West Belmont Avenue
Franklin Park, IL 60131
(800) 260-9001; (704) 283-2101; fax: (800) 338-0965
www.folgeradamedc.com

A leading manufacturer of electric strikes, electric locks, detention hardware, and assorted security products.

Fort Knox Security Products
933 North Industrial Park Road
Orem, UT 84057
(800) 821-5216; (801) 224-7233; fax: (801) 226-5493
www.ftknox.com

Fort Lock Corporation
3000 N. River Road River
Grove, IL 60171
(708) 456-1100

Framon Manufacturing Company, Inc.
909 Washington Avenue
Alpena, MI 49707
(989) 354-5623; fax: (989) 354-4238
www.framon.com

Makes key-cutting and code machines.

Gardall Safe Corporation
P.O. Box 240
Syracuse, NY 13206
(800) 722-7233; (315) 432-9115; fax: (315) 434-9422
www.gardall.com

Makes a wide range of Underwriters Laboratories (UL)–listed safes for commercial and residential use.

Hayman Safe Company, Inc.
1291 North S.R. 426
Oviedo, FL 32765
(800) 444-5434; (407) 365-5434; fax: (407) 365-8958
www.haymansafe.com

Hudson Lock, Inc.
81 Apsley Street
Hudson, MA 01749
(800) 434-8960; (978) 562-3481; fax: (978) 562-9859

Ilco Unican Corporation
P.O. Box 2627
Rocky Mount, NC 27802
(252) 446-3321

A leading manufacturer of a wide range of locks and security products.

InstaKey Lock Corporation
1498 S. Lipan Street
Denver, CO 80223
(303) 761-9999

Makes high-security locks.

International Locking Devices, Ltd.
104-4 Branford Road
North Branford, CT 06471
(800) 863-9600; (203) 481-6738; fax: (203) 481-8490
www.gatelock.com

Makes assorted gate locks and latches.

Jet Hardware Manufacturing Corporation
800 Hinsdale Street
Brooklyn, NY 11207
(718) 257-9600; fax: (718) 257-9073
www.jetkeys.com

Kaba Access Control
2941 Indiana Avenue
Winston Salem, NC 27105
(800) 334-1381; (252) 446-3321; fax: (252) 446-3321
www.kaba-ilco.com

Keedex Manufacturing
12931 Shackelford Lane
Garden Grove, CA 92641-5108
(714) 636-5657; fax: (714) 636-5680
www.keedex.com

Makes locksmithing tools and equipment.

Kustom Key, Inc.
1010 Aviation Drive
Lake Havasu City, AZ 86403
(800) 537-5397; (520) 453-8338; fax: (520) 453-8733
www.kustomkey.com

Makes neuter bow key blanks incised or embossed with company names or custom messages.

Kwikset Corporation
19701 DaVinci
Lake Forest, CA 92610-2622
(800) 327-LOCK; (949) 672-4000; fax: (949) 672-4001

A leading manufacturer of deadbolt locks, key-in-knob locks, lever locks, handle sets, and other security hardware. Major brand and trade names: Kwikset and Titan. Over 2000 employees. Operating facilities in Anaheim, CA, Bristow, OK, Denison, TX, and Irvine, CA. Founded in the late 1940s by Adolf Schoepe and Karl Rhinehart as a joint partnership and named Gateway Manufacturing Company. In 1947, Gateway Manufacturing changed its name to Kwikset. In 1952, Kwikset became publicly owned; about 50 percent of its stock was placed on the market. In July 1957, the American Hardware Corporation acquired controlling interest in Kwikset. By 1958, Kwikset became an operating division of American Hardware. In 1964, the American Hardware Corporation merged with Emhart Manufacturing Company and established the Emhart Corporation. In 1989, the Black & Decker Company purchased the Emhart Corporation, including the Kwikset division.

Liberty Safe & Security
10700 Jersey Boulevard, #660
Rancho Cucamonga, CA 91730
(909) 948-3514

Locknetics Security Eng.
575 Birch Street
Forestville, CT 06010
(866) 322-1237; (860) 584-9158; fax: (860) 584-2136
www.irsecurityandsafety.com

Locksoft
P.O. Box 129
Hastings, NE 68902-0129
(402) 461-4149; fax: (402) 461-4359
www.locksoft.com

Makes computer software for locksmiths.

M.A.G. Eng. & Mfg., Inc.
15381 Assembly Lane
Huntington Beach, CA 92649
(800) 624-9942; (714) 891-5100; fax: (714) 892-6845

A leading manufacturer of high-security strike boxes, door reinforcers, and other door and window hardware.

Major Manufacturing, Inc.
1825 Via Buron
Anaheim, CA 92806
(714) 772-5202; fax: (714) 772-2302

Makes locksmithing tools and equipment.

Master Lock Company
P.O. Box 927
137 W. Forest Hill Avenue
Oak Creek, WI 53154
(414) 571-5625; fax: (414) 766-6333
www.masterlock.com

A leading manufacturer of padlocks, key-in-knob locks, deadbolts, and other types of locks and related products. Major brand and trade names: Durashine finish and The Pro Series line of weather-resistant padlocks. Over 1800 employees at its 800,000-square-foot manufacturing plant (warehouse) headquarters in Milwaukee, WI, and at its production facility in Auburn, AL. Founded in 1921 by Harry E. Soref, who introduced the Master Laminated Keyed Padlock. In 1970, Master Lock became a subsidiary of American Brands, Inc. In 1985, Master Lock acquired Dexter Lock Company. In 1988, American Brands formed Master-Brand Industries, Inc., a subsidiary for its hardware and home improvement products firm—including Master Lock. In 1990, Master Lock International was formed. Master Lock started its Door Hardware Division and introduced a line of easy-to-install door hardware in 1991. In the same year, Master Lock International opened an office in Paris and a warehouse in Rotterdam.

MAXIS Security Locks
103-950 Powell Street
Vancouver, British Columbia, Canada V6A-1H9
(604) 255-2005

Medeco Security Locks (Canada)
141 Dearborn Place
Waterloo, Ontario, Canada N2J 4N5
(888) 633-3264; (519) 888-7000; fax: (519) 888-6134

Medeco Security Locks, Inc.
P.O. Box 3075
Salem, VA 24153
(800) 839-3157; (540) 380-5000; fax: (540) 380-5010
www.medeco.com

A leading manufacturer of high-security locks and cylinders.

Meilink Safe Company
101 Security Parkway
New Albany, IN 47150
(800) 634-5465; (812) 941-0024; fax: (812) 948-0205
www.fireking.com

Miwa Lock Canada
Box 3708, Station D
Edmonton, Alberta, Canada T5L 4J7
(403) 453-5226

Miwa Lock USA
71 North Pecos Road, #106
Las Vegas, NV 89101
(800) 397-4631

Monarch Tool & Mfg. Co., Inc.
P.O. Box 427
Covington, KY 41011
(800) 462-9460; (859) 261-4421; fax: (859) 261-7403
www.monarchcoin.com

Makes coin boxes and assorted vending machine products. Established in 1903.

Mosler
8509 Berk Boulevard
Hamilton, OH 45012
(513) 870-1900; fax: (513) 867-4062

A leading manufacturer of UL-listed burglary safes, vaults, vault doors, safe deposit boxes, and related products. Established in 1867.

Mul-T-Lock USA, Inc.
300-1 Route 17S, Suite A
Lodi, NJ 07644
(800) 562-3511; (973) 778-3222; fax: (973) 778-4007

Makes high-security locks.

Multi Lock Corp.
P.O. Box 70277
Ft. Lauderdale, FL 33307-0277
(800) 879-5645; (954) 563-2148; fax: (954) 563-5644
www.multilock.com

Makes ornamental iron-gate locks used on homes and swimming pools and makes iron-mesh locks used in jails and prisons.

New England Lock & Hardware Company, Inc.
46 Chestnut Street
South Norwalk, CT 06856
(203) 866-9283; fax: (203) 838-4837

Makes assorted vertical deadbolts, rim locks, and other locks and cylinders. Brand name: Segal.

Norden Lock Corp.
2043 Wellwood Avenue
East Farmingdale, NY 11735-1283
www.nordenlock.com

Ohio Travel Bag Mfg. Co.
811 Prospect Avenue
Cleveland, OH 44115
(216) 498-1955

Makes luggage and trunk keys, locks, and hardware.

Precision Hardware Inc.
38100 Jaykay Dr.
P.O. Box 74040
Romulus, MI 48174-0040
(734) 326-7500
www.precisionhardware.com

Preso-Matic Keyless Locks
100 A Commerce Way
Sanford, FL 32771
(800) 269-4234; (407) 324-9933; fax: (407) 328-9977

Makes mechanical pushbutton door locks for commercial and residential installations.

Qualtec Data Products, Inc.
47767 Warm Springs Boulevard
Fremont, CA 94539
(800) 628-4413; (510) 490-8911; fax: (510) 490-8471

Makes assorted locks and antitheft devices for personal computers. Established in 1984.

Rofu International Corporation
7107 28th Street, E
Tacoma, WA 98424
(800) 255-ROFU; (253) 922-1828; fax: (253) 922-1728
www.rofu.com

A leading manufacturer of electric strikes, electromagnetic locks, and other security products.

Sargent & Greenleaf, Inc.
P.O. Box 930
Nicholasville, KY 40356
(800) 826-7652; (859) 885-9411; fax: (859) 887-2057
www.sargentandgreenleaf.com

A leading manufacturer of locks for safes, vaults, and safe-deposit boxes. Occupies 10,000 square feet of manufacturing space and 22,000 square feet of offices. The company also serves European and other international customers from a facility in Switzerland. In 1857, James Sargent, then an employee of Yale & Greenleaf Lock Manufacturers, devised a magnetic bank lock that quickly gained acceptance from safe manufacturers and the U.S. Treasury Department. That lock brought financial stability to the company and paved the way for James Sargent to continue devising new security products. In 1865, one of Mr. Sargent's former employees, Halbert Greenleaf, became his equal partner in the firm of Sargent & Greenleaf.

Schlage Lock Co.
111 Congressional Boulevard, Suite 200
Carmel, IN 46032
(800) 847-1864; (317) 613-8150; fax: (800) 452-0663
www.schlage.com

Makes assorted door locks and security products.

Securitech Group, Inc.
54.45 44th Street
Maspeth, NY 11378
(800) 622-LOCK; (718) 392-9000; fax: (718) 392-8944

Makes high-security door locks and related products. Brand names include MAGLATCH, MP Police Lock, Double-throw, and Iderbolt.

Securitron Magnalock Corp.
550 Vista Boulevard
Sparks, NV 89434
(800) 624-5625; (800) MAGLOCK
www.securitron.com

Security Control Systems
20202 S. Holly
Frankfort, IL 60423
(815) 469-0532; fax: (815) 469-6425

Makes GSA-approved key control software.

Security 777, USA
P.O. Box 398
Chandler, AZ 85244-0398
(602) 899-7770

Select Hardware Corporation
675 Emmett Street
Bristol, CT 06010
(203) 589-7303; fax: (203) 589-6245

Makes deadbolts that can be fitted with many popular key-in-knob or interchangeable core cylinders.

Sentry Group
900 Linden Avenue
Rochester, NY 14625
(800) 828-1438; (716) 381-4900; fax: (716) 381-8559
www.sentrysafe.com

A leading manufacturer of UL-listed fire safes and containers for the home and office.

Silca Keys USA, Inc.
9049 Dutton Drive
Twinsburg, OH 44087
(216) 487-5454; fax: (216) 487-5459

Makes keys and key machines.

Simplex Access Controls
2941 Indiana Avenue
P.O. Box 4114
Winston-Salem, NC 27115-4114
(910) 725-1331; fax: (800) 346-9640

Southern Steel
4634 S. Presa Street
San Antonio, TX 78223
(210) 533-1231; fax: (210) 533-2211

A leading manufacturer of detention locks, hinges, and other door hardware.

Trine Access Technology
1430 Ferris Place
Bronx, NY 10461
(718) 829-2332; fax: (718) 829-6405

Makes electric strikes and other products.

Von Duprin, Inc.
2720 Tobey Drive
Indianapolis, IN 46219
(800) 999-0408; (317) 613-8944; fax: (800) 999-0328

A leading manufacturer of door exit devices, electric strikes, and related security products.

Weiser Lock Company
6660 S. Broadmoor Road
Tucson, AZ 85746
(520) 741-6200; fax: (520) 741-6241

Makes assorted residential locks. Established in 1904.

Weslock
8301 E. 81st Street, Suite A
Tulsa, OK 74133
(800) 575-2658; (918) 294-3888; fax: (918) 294-3869
www.weslock.com

Makes assorted residential locks.

Wilson Bohannan Co.
P.O. Drawer 504
Marion, OH 43301
(800) 382-3639; (740) 382-3639; fax: (740) 383-1653
www.padlocks.com

Makes solid-brass pin tumbler padlocks. Employs nearly 70 people in a 50,000-square-foot facility. Major brand and trade names: WB and Top Brass. Founded in 1860 by Wilson Bohannan as a machine shop; he ran the business out of the back of his home in Brooklyn, New York. In the 1880s, the business grew to the point where it had to move to its own four-story 10,000-square-foot building. In 1927, Wilson Tway, who was related to the Bohannan family through marriage, moved the company to Marion, Ohio. He remained president of the company until he died in 1982. His son, Dick Tway, was president until his untimely death in 1989. Dorthea Tway Norris and other family members maintain an active interest in the company.

Appendix

B

Locksmith Suppliers' Profiles and Addresses

A-1 Security Lock
2630 S. Virginia Street
Reno, NV 89502
(702) 828-5625; fax: (702) 828-5623

Accredited Lock Supply Co.
P.O. Box 1442
Secaucus, NJ 07096-1442
(201) 865-5015

Complete line of locks and supplies; some access control and other electronic security supplies; 25 employees; 35,000-square-foot warehouse. Founded in 1974 as a locksmith shop by Rudy Weaver and his son Ron.

Acme Wholesale Distributors, Inc.
P.O. Drawer 13748
New Orleans, LA 70185
(800) 788-2263; fax: (504) 837-7321

Full-line distributor of most major lock lines, supplies, safes, and access control devices. Has branch distribution centers in New Orleans, Houston, San Antonio, and Ft. Worth. Founded in the early 1970s. Joined the LSDA Group in 1985.

Aero Lock
3675 New Getwell Road, Suite 9
Memphis, TN 38118
(800) 627-9433; (901) 368-2147
www.aerolock.com

American Lock & Supply, Inc.
4411E. LaPalma Avenue
Anaheim, CA 92807
(800) 844-8545; (714) 996-8882; fax: (714) 579-3508

Full-line distributor of most major lock lines. More than $20 million in inventory at 11 locations throughout the United States.

Aristo Sales Company, Inc.
27-24 Jackson Avenue
Long Island City, NY 11101
(800) 221-1322; (718) 361-1040; fax: (718) 937-5794

Supplier of automotive locks, alarms, safes, and supplies.

Armstrong's Lock & Supply, Inc.
1440 Dutch Valley Place, NE
Atlanta, GA 30324
(800) 726-3332; (404) 875-0136; fax: (404) 888-0834

Full-line distributor of locks, safes, supplies, and electronic security devices. Carries more than 13,000 different products from more than 250 manufacturers. Branches in Miami, Tampa, Jacksonville, and Norfolk.

Canada Lock Products, Ltd.
70 Floral Parkway
Toronto, Ontario, Canada M6L 2C1
(416) 248-5625

Capitol Lock Co.
9815 Rhode Island Avenue
College Park, MD 20740
(202) 882-9502

Central Lock & Hardware Supply
95 NW 166 Street
Miami, FL 33169
(305) 947-4853

Commonwealth Lock Co.
1853 Massachusetts Ave.
Cambridge, MA 02140
(800) 442-7009; (617) 876-3301
www.commonwealthlockco.bizonthe.net

D. Silver Hardware Co., Inc.
591 Ferry Street
Newark, NJ 07105
(201) 344-3963

Doyle Lock Supply
2211 W. River Road N
Minneapolis, MN 55411-2228
(612) 521-6226
www.doylesecurity.com

Dugmore & Duncan, Inc.
30 Pond Park Road
Hingham, MA 02043
(800) 225-1595

Foley-Belsaw
6301 Equitable Rd.
P.O. Box 419593
Kansas City, MO 64141-6593
(800) 821-3452

Fried Brothers, Inc.
467 N. 7th Street
Philadelphia, PA 19123
(800) 523-2924; (21 5) 627-3205; fax: (215) 523-1255

Products from more than 120 different manufacturers. Established in 1922.

Kenco Locksmith Supply Co.
P.O. Box 4160
Omaha, NE 68104
(402) 397-8291

Mayflower Sales Co., Inc.
614 Bergen Street
Brooklyn, NY 11238
(718) 622-8785
www.mfsales.com

Maziuk & Company, Inc.
1125 W. Genesee Street
Syracuse, NY 13204
(800) 777-5945; (315) 474-3959; fax: (315) 472-3111

Stocks wide range of locks and supplies from more than 135 manufacturers. Has a 5300-square-foot branch office and warehouse in the Buffalo area. Established in 1943.

McDonald Dash Locksmith Supply, Inc.
5767 East Shelby Drive
Memphis, TN 38141-6804
(901) 797-8000; fax: (901) 366-0005

Distributor of more than 15,000 locks, tools, supplies, and related security products from 125 manufacturers. Established in 1945.

McManus Locksmith Supply, Inc.
P.O. Box 9231
Central Avenue
Charlotte, NC 28299
(704) 333-9112; fax: (704) 332-8664

M. Taylor, Inc.
5635-45 Tulip Street
Philadelphia, PA 19124
(800) 233-3355; fax: (215) 288-5588

More than 11,000 items from more than 130 manufacturers. Established in 1912.

MD Kramer Locksmith Supply
7 Aldgate Drive W
Manhasset, NY 11030-3940
(718) 647-7800

Security Lock Distributors
59 Wexford Street
P.O. Box 815
Needham Heights, MA 02194
(800) 847-5625; (617) 444-1155; fax: (800) 878-6400

Sells electronic and mechanical locking devices.

Wilco Supply
5960 Telegraph Avenue
Oakland, CA 94609
800-745-5450; (510) 652-8522; fax: (510) 653-5397

More than 32,000 different locks, tools, supplies, and related products and 52 employees. Founded in 1951.

William Silver & Company, Inc.
42 Ludlow Street
New York, NY 10002
(212) 982-8720

Wm. B. Allen Supply Co. Inc.
Allen Square 300 Block N
New Orleans, LA 70112
(504) 525-8222

Sells burglar/fire alarm systems, equipment, and supplies. Stocks more than 72,000 different products from more than 300 manufacturers. Established in 1940.

ZipfLockCo.
830 Harmon Avenue
Columbus, OH 43223
(800) 848-1577; (614) 228-3507; fax: (800) 228-6320

Stocks more than 16,000 different locks, lock parts, tools, and supplies. Established in 1908.

Appendix

C

Miscellaneous Important Addresses, Phone Numbers, and Web Sites

American Lock Collectors Association
c/o Robert Dix
8576 Barbara Drive
Mentor, OH 44060

American Society for Industrial Security
1655 North Fort Myer Drive, Suite 1200
Arlington, VA 22209
(703) 522-5800

Associated Locksmiths of America, Inc.
3003 Live Oak Street
Dallas, TX 75204

The Association of Ontario Locksmiths
2220 Midland Avenue, Unit 106
Scarborough, Ontario, Canada M1P 3E6

Builders Hardware Manufacturers Association
355 Lexington Avenue, 17th floor
New York, NY 10017
(212) 297-2122

The Canadian Locksmith Magazine
Arnold Sintnicolaas Publisher
137 Vaughan Road
Toronto, Ontario, Canada M6C 2L9
(905) 294-0660

Canadian Security Association (CANASA)
610 Alden Road, Suite. 100
Markham, ON L3R9Z1
(905) 513-0622

ClearStar Security Network
www.clearstar.com

Has a restricted area that is one of the best information sites for locksmiths and safe technicians.

Door and Hardware Institute
14170 Newbrook Drive
Chantilly, VA 22021-0750
(703) 222-2010

International Association of Home Safety and Security Professionals
Box 2044
Erie, PA 16512-2044

International Association of Investigative Locksmiths
P.O. Box 144
Mt. Airy, MD 21771

Lock Museum of America, Inc.
230 Main Street
Terryville, CT 06786-0104
(203) 589-6359

The Locksmith Ledger
850 Busse Highway
Park Ridge, IL 60068
(847) 692-5940

Locksmithing World
www.locksmithing.com

One of the best Web sites for locksmithing information. Has a restricted area for security professionals only.

The National Locksmith
1533 Burgundy Parkway
Streamwood, IL 60107
(630) 837-2044
www.TheNationalLocksmith.com

One of the best free sites for locksmithing information.

National Locksmith Automobile Association
P.O. Box 77-97592
Chicago, IL 60678-75002

National Locksmith Suppliers Association
1900 Arch Street
Philadelphia, PA 19103-1498
(215) 564-3484

National Safeman's Organization
P.O. Box 77-97592
Chicago, IL 60678-7592

Safe and Vault Technicians Association (SAVTA)
3003 Live Oak Street
Dallas, TX 75204-6186
(214) 821-7233

Security Industry Association
1801 K Street Northwest, Suite L
Washington, DC 20006
(202) 466-7420

Underwriters Laboratories, Inc (UL)
333 Pfingsten Road
Northbrook IL 60062
(847) 272-8800

Appendix

D

Locksmithing Schools and Training Programs

(In the United States, Canada, and Australia)

Abram Friedman Occupational Center
1646 S. Olive Street
Los Angeles, CA 90015
(213) 742-7657

Academy of Locksmithing
2220 Midland Avenue, Unit 106
Scarborough, Ontario, Canada MlP 3E6
(416) 321-2219
E-mail: *Office@TAOL.net*

Acme School, Locksmithing Division
11350 S. Harlem
Worth, IL 60482
(708) 361-3750

American Locksmith Institute of Nevada
875 S. Boulder Highway
Henderson, NV 89015
(702) 565-8811

Bay Area School of Locksmithing
942 Industrial Avenue
Palo Alto, CA 94303
(650) 493-3022
E-mail: *locksmithschool@earthlink.net.*

California Institute of Locksmithing
14721 Oxnard Street
Van Nuys, CA 91411
(818) 994-7426

Charles Stuart School
1420 Kings Hwy, 2d Floor
Brooklyn, NY 11229
(718) 339-2640

Colorado Locksmith College
4991 W. 80th Avenue
Westminster, CO 80030
(303) 427-7773

Commercial Technical Institute
116 Fairfield Road
Fairfield, NJ 07004
(201) 575-5225

Cothron's School of Lock Technology
509 Rio Grande
Austin, TX 78701
(800) 294-6273; (512) 472-6273

Foley-Beisaw Institute
A Division of Foley-Beisaw Co.
6301 Equitable Road
Kansas City, MO 64120
800-821-3452

Golden Gate School of Locksmithing
3722 San Pablo Avenue
Oakland, CA 94608
(510) 654-2677

Granton Institute of Technology
263 Adelaide Street
West Toronto, Ontario, Canada M5H lY3
(416) 977-3929

Greater Regional Technical College
6830 Burlington Avenue
Burnaby, British Columbia, Canada V5J 4H1

HPC Learning Center
3999 N. 25th Avenue
Schiller Park, IL 60176
(708) 671-6280

IAHSSP
Box 2044
Erie, PA 16512-2044

Lock and Safe Institute of Technology, Inc.
1650 N. Federal Highway
Pompano Beach, FL 33062
(305) 785-0444

Lockmasters Security Institute
1014 S. Main Street
Nicholasville, KY 40356
(800) 654-0637

Locksmith Business Management School
P.O. Box 8525
Emeryville, CA 94662
(510) 654-2677

Locksmith School
51 Beverly Hills Drive
Toronto, Ontario, Canada, M3L 1 A2
(416) 960-9999

Locksmith School, Inc.
3901 S. Meridian Street
Indianapolis, IN 46217
(317) 632-3979

LTC Training Center
P.O. Box 3583
Davenport, IA 52808-3583
(800) 358-9393

Master Locksmith Training Courses
Units 4–5 The Business Park
Woodford Halse, Daventry
Northants NN22 6PZ, England
(44) 01327 262255

Messick Vo-Technical Center
703 South Greer
Memphis, TN 38111
(901) 325-4840

North Bennet Street School
39 N. Bennet Street
Boston, MA 02113
(617) 227-0155

Northern Metropolitan College of TAFE, Security Technology Section
Waterdale Road and Bell Street
Heidelberg, Victoria 3081
Australia
(62) 3 9269 8687

NRI School of Locksmithing
4401 Connecticut Avenue N.W.
Washington, DC 20008
(202) 244-1600

Prince Georges Community College
301 Largo Road, Room K-205
Largo, MD 20772-2199
(301) 322-0871

Quintilian Institute Services (QIS)
5001-A Lee Highway, Suite 101
Arlington, VA 22207
(703) 525-7525

San Francisco Lock School
4002 Irving Street
San Francisco, CA 94122
(415) 566-5545; (415) 347-2222

School of Lock Technology
1049 Island Avenue
San Diego, CA 92101
(619) 234-1036

School of Security Technology
302 W. Katelia
Orange, CA 92867
(714) 633-1366

Security Education Plus
400-B Etter Drive
P.O. Box 497,
Nicholasville, KY 40356
(606) 887-6027

Security Systems Management School
116 Fairfield Road
Fairfield, NJ 07004
(201) 575-5225

Southern Locksmith Training Institute
1387 Airline Drive
Bossier City, LA 71112
(318) 227-9458

South Metropolitan College of TAFE
Security Technology Section
15 Grosvenor Street
Beaconsfield, Western Australia 6162
(61) 09-239-8386
E-mail: *Tissa@newton.dialix.com.au*

Stotts Correspondence College
Australian College of Locksmithing
140 Flinders Street
Melbourne, Victoria 3000
Australia

Sydney Institute of Technology Ultimo
Locksmithing Section
Building M, Mary Ann St.
Ultimo, NSW 2007
Australia
(61) 02-217-3449

Universal School of Master Locksmithing
3201 Fulton Avenue
Sacramento, CA 95821
(916) 482-4216

Valley Technical Institute
5408 N. Blackstone
Fresno, CA 93710
(209) 436-8501

Appendix

E

Depth and Space Charts

Arrow

Shoulder to first cut: 0.265 inch
Center to center: 0.155 inch
MACs: 7

	Root depth	Bottom pins	Master pins
#0	0.312	0.178	
#1	0.298	0.192	
#2	0.284	0.206	0.028
#3	0.270	0.220	
#4	0.256	0.234	0.056
#5	0.242	0.248	
#6	0.228	0.262	0.084
#7	0.214	0.276	
#8	0.200	0.290	0.112
#9	0.186	0.304	

Corbin

	Z Bow	X Bow
Shoulder to first cut:	0.250 inch	0.197 inch
Center to center:	0.156 inch	
MACs: 8		

	Root depth		Bottom pins		Master pins
	Z Bow	X Bow	Z Bow	X Bow	
#1	0.343	0.333	0.160	0.171	
#2	0.329	0.319	0.174	0.185	0.028
#3	0.315	0.305	0.189	0.198	0.042
#4	0.301	0.291	0.203	0.212	0.056
#5	0.287	0.277	0.218	0.226	0.069
#6	0.273	0.263	0.231	0.241	0.084
#7	0.259	0.249	0.246	0.256	0.099
#8	0.245	0.235	0.259	0.269	0.112
#9	0.231	0.221	0.273	0.284	0.127
#10	0.217	0.207	0.287	0.297	

Dexter

Shoulder to first cut: 0.216 inch

Center to center: 0.155 inch

MACs: 7

	Root depth	Bottom pins	Master pins
#0	0.320	0.165	
#1	0.305	0.180	
#2	0.290	0.195	0.030
#3	0.275	0.210	
#4	0.260	0.225	0.060
#5	0.245	0.240	
#6	0.230	0.255	0.090
#7	0.215	0.270	
#8	0.200	0.285	0.120
#9	0.185	0.300	

Emhart

Shoulder to first cut: 0.250 inch

Center to center: 0.156 inch

MACs: Dependent on adjacent angles

	Root depth	Bottom pins	Master pins	Drivers
#1			0.193	
#2	0.305	0.242	0.097	0.158
#3	0.277	0.270	0.125	
#4	0.249	0.298	0.153	
#5	0.221	0.326		
#6	0.193	0.354		

Falcon

Shoulder to first cut: 0.237 inch

Center to center: 0.156 inch

MACs: 7

	Root depth	Bottom pins	Master pins
#0	0.315	0.168	
#1	0.297	0.186	
#2	0.279	0.204	0.036
#3	0.261	0.222	
#4	0.243	0.240	0.072
#5	0.225	0.258	
#6	0.207	0.276	0.108
#7	0.189	0.294	
#8	0.171	0.312	0.144
#9	0.153	0.330	

Ilco

Shoulder to first cut: 0.277 inch

Center to center: 0.156 inch

MACs: 7

	Root depth	Bottom pins	Master pins
#0	0.320	0.180	
#1	0.302	0.198	
#2	0.284	0.216	0.036
#3	0.266	0.234	
#4	0.248	0.252	0.072
#5	0.230	0.270	
#6	0.212	0.288	0.108
#7	0.194	0.306	
#8	0.176	0.324	0.144
#9	0.158	0.342	

Kwikset

Shoulder to first cut: 0.247 inch

Center to center: 0.150 inch

Titan

Shoulder to first cut: 0.097 inch

Center to center: 0.150 inch

MACs: 4

	Root depth	Bottom pins	Master pins
#1	0.329	0.172	0.023
#2	0.306	0.195	0.046
#3	0.283	0.218	0.069
#4	0.260	0.241	0.092
#5	0.237	0.264	0.115
#6	0.214	0.287	*
#7	0.191	0.310	

* Kwikset does not make a No. 6 master wafer.

Lockwood

Shoulder to first cut: 0.277 inch

Center to center: 0.156 inch

MACs: 7

	Root depth	Bottom pins	Master pins
#0	0.320	0.150	
#1	0.305	0.165	
#2	0.290	0.180	0.030
#3	0.275	0.195	0.045
#4	0.260	0.210	0.060
#5	0.245	0.225	0.075
#6	0.230	0.240	0.090
#7	0.215	0.355	0.105
#8	0.200	0.370	0.120
#9	0.185	0.385	0.135

Medeco

Shoulder to first cut: 0.244 inch

Center to center: 0.170 inch

MACs: 4

	Root depth	Bottom pins	Master pins	Drivers
#1	0.266	0.236	0.030	0.270
#2	0.236	0.266	0.060	0.240
#3	0.206	0.296	0.090	0.210
#4	0.176	0.326	0.120	0.180
#5	0.146	0.356	0.150	0.150
#6	0.116	0.386		0.120

Note: Medco does not allow change keys with a No. 6 cut next to the shoulder.

Medeco Biaxial

Shoulder to first cut:

Fore: 0.213 inch

Aft: 0.275 inch

Center to center:

Aft–fore: 0.108 inch

Fore–fore: 0.170 inch

Aft–aft: 0.170 inch

Fore–aft: 0.232 inch

MACs: Aft–fore: 2

Fore–fore: 3

Aft–aft: 3

Fore–aft: 4

	Root depth	Bottom pins	Master pins	Drivers
#1	0.264	0.239	0.025	0.270
#2	0.239	0.264	0.050	0.240
#3	0.214	0.289	0.075	0.210
#4	0.189	0.314	0.100	0.180
#5	0.164	0.339	0.125	0.150
#6	0.139	0.364		

Medeco Keymark

Shoulder to first cut: 0.195 inch

Tip to first cut: 0.090 inch

Cut to cut: 0.150 inch

MACs: 9

	Root depth	Bottom pins	Master pins
#0	0.1385	0.110	—
#1	0.1260	0.122	—
#2	0.1135	0.135	0.025
#3	0.1010	0.147	0.037
#4	0.0885	0.160	0.050
#5	0.0760	0.172	0.062
#6	0.0635	0.185	0.075
#7	0.0510	0.197*	0.087
#8	0.0385	0.210*	0.100
#9	0.0260	0.222*	0.112

* Spool Pins.

Russwin

Shoulder to first cut: 0.250 inch

Center to center: 0.156 inch

MACs: 7

	Root depth	Bottom pins	Master pins
#0	0.341	0.160	
#1	0.326	0.175	
#2	0.311	0.189	0.030
#3	0.296	0.203	0.045
#4	0.281	0.220	0.060
#5	0.266	0.234	0.075
#6	0.251	0.248	0.090
#7	0.236	0.263	0.105
#8	0.221	0.279	0.120
#9	0.206	0.294	0.135

Sargent

Shoulder to first cut: 0.216 inch

Center to center: 0.156 inch

MACs: 7

	Root depth	Bottom pins	Master pins
#1	0.328	0.170	
#2	0.308	0.190	0.040
#3	0.288	0.210	0.060
#4	0.268	0.230	0.080
#5	0.248	0.250	0.100
#6	0.228	0.270	0.120
#7	0.208	0.290	0.140
#8	0.188	0.310	0.160
#9	0.168	0.330	0.180
#10	0.148	0.350	

Schlage

Shoulder to first cut: 0.231 inch

Center to center: 0.156 inch

MACs: 7

	Root depth	Bottom pins	Master pins
#0	0.335	0.165	
#1	0.320	0.180	
#2	0.305	0.195	0.030
#3	0.290	0.210	
#4	0.275	0.225	0.060
#5	0.260	0.240	
#6	0.245	0.255	0.090
#7	0.230	0.270	
#8	0.215	0.285	0.120
#9	0.200	0.300	

Segal

Shoulder to first cut: 0.262 inch

Center to center: 0.156 inch

MACs: 5

	Root depth	Bottom pins	Master pins
#0	0.315	0.166	
#1	0.295	0.186	0.020
#2	0.275	0.206	0.040
#3	0.255	0.226	0.060
#4	0.235	0.246	0.080
#5	0.215	0.266	0.100
#6	0.195	0.286	*

* No. 6 master wafer.

System 70

Shoulder to first cut: 0.250 inch

Center to center: 0.156 inch

MACs: 4

	Root depth	Bottom pins	Master pins
#1	0.339	0.160	0.028
#2	0.311	0.189	0.056
#3	0.283	0.217	0.084
#4	0.255	0.245	0.112
#5	0.227	0.273	0.140
#6	0.199	0.301	

Weiser

Shoulder to first cut: 0.237 inch

Center to center: 0.156 inch

MACs: 7

	Root depth	Bottom pins	Master pins
#0	0.315	0.168	
#1	0.297	0.186	
#2	0.279	0.204	0.036
#3	0.261	0.222	
#4	0.243	0.240	0.072
#5	0.225	0.258	
#6	0.207	0.276	0.108
#7	0.189	0.294	
#8	0.171	0.312	0.144
#9	0.153	0.330	

Weslock

Shoulder to first cut: 0.220 inch

Center to center: 0.156 inch

MACs: 7

	Root depth	Bottom pins	Master pins
#0	0.330	0.156	
#1	0.314	0.172	
#2	0.299	0.187	0.030
#3	0.283	0.202	
#4	0.268	0.219	0.060
#5	0.252	0.234	
#6	0.236	0.250	0.090
#7	0.221	0.265	
#8	0.205	0.281	0.120
#9	0.190	0.297	

Yale

Shoulder to first cut: 0.206 inch

Center to center: 0.165 inch

MACs: 7

	Root depth	Bottom pins	Master pins
#0	0.320	0.182	
#1	0.301	0.201	
#2	0.282	0.220	0.038
#3	0.263	0.239	
#4	0.244	0.258	0.076
#5	0.225	0.277	
#6	0.206	0.296	0.114
#7	0.187	0.315	
#8	0.168	0.334	0.152
#9	0.149	0.353	

Index

Note: **Boldface** numbers indicate illustrations; *t* indicates a table.

ABOUT THE AUTHOR

Bill Phillips is the leading author of locksmithing books, including this best-selling classic. He also wrote McGraw-Hill's *Locksmithing*, a part of the "Craftmaster" series, and the *Complete Book of Electronic Security* (McGraw-Hill). Mr. Phillips is the author of the "Lock" article in the 1998–2001 editions of the *World Book Encyclopedia*, and has written hundreds of security-related articles for professional and general-circulation periodicals, including *Home Mechanix*, *Los Angeles Times*, *Consumers Digest*, and others. He is president of the International Association of Home Safety and Security Professionals.